Comprehensive Treatise of Electrochemistry

Volume 5
Thermodynamic and Transport Properties
of Aqueous and Molten Electrolytes

COMPREHENSIVE TREATISE OF ELECTROCHEMISTRY

Comprehensive Treatise of Electrochemistry

Volume 5
Thermodynamic and Transport Properties
of Aqueous and Molten Electrolytes

Edited by

Brian E. Conway
University of Ottawa
Ottawa, Ontario, Canada

J. O'M. Bockris
Texas A&M University
College Station, Texas

Ernest Yeager
Case Western Reserve University
Cleveland, Ohio

PLENUM PRESS · NEW YORK AND LONDON

Library of Congress Cataloging in Publication Data
Main entry under title:

Thermodynamic and transport properties of aqueous and molten electrolytes.

   (Comprehensive treatise of electrochemistry; v. 5)
   Includes bibliographical references and index.
   1. Electrolytes — thermal properties. I. Conway, B. E. II. Bockris, J. O'M.
(John O'M.), 1923–   . III. Yeager, Ernest B., 1924–   . IV. Series.
   QD552.C64 vol. 5 [QD565]   541.3′7s   82-22372
   ISBN 0-306-40866-X   [541.3′72]

© 1983 Plenum Press, New York
A Division of Plenum Publishing Corporation
233 Spring Street, New York, N.Y. 10013

Printed in the United States of America

Contributors

B. E. Conway, Chemistry Department, University of Ottawa, Ottawa, Ontario, Canada K1N 9B4

Jacques E. Desnoyers, Department of Chemistry, Université de Sherbrooke, Sherbrooke, Quebec, Canada J1K 2R1

Carmel Jolicoeur, Department of Chemistry, Université de Sherbrooke, Sherbrooke, Quebec, Canada J1K 2R1

Jean-Claude Justice, Laboratoire d'Electrochimie, Université Pierre et Marie Curie, Place Jussieu, Paris, France

S. Lengyel, Hungarian Academy of Sciences, Budapest, Hungary

G. N. Papatheodorou, Chemical Engineering Division, Argonne National Laboratory, 9700 South Cass Avenue, Argonne, Illinois 60439

Preface to Comprehensive Treatise of Electrochemistry

Electrochemistry is one of the oldest defined areas in physical science, and there was a time, less than 50 years ago, when one saw "Institute of Electrochemistry and Physical Chemistry" in the chemistry buildings of European universities. But, after early brilliant developments in electrode processes at the beginning of the twentieth century and in solution chemistry during the 1930s, electrochemistry fell into a period of decline which lasted for several decades. Electrochemical systems were too complex for the theoretical concepts of the quantum theory. They were too little understood at a phenomenological level to allow the ubiquity in application in so many fields to be comprehended.

However, a new growth began faintly in the late 1940s, and clearly in the 1950s. The growth was exemplified by the formation in 1949 of what is now called The International Society for Electrochemistry. The usefulness of electrochemistry as a basis for understanding conservation was the focal point in the founding of this Society. Another very important event was the choice by NASA in 1958 of fuel cells to provide the auxiliary power for space vehicles.

With the new era of diminishing usefulness of the fossil fuels upon us, the role of electrochemical technology is widened (energy storage, conversion, enhanced attention to conservation, direct use of electricity from nuclear–solar plants, finding materials which interface well with hydrogen). This strong new interest is not only in the technological applications of electrochemistry. Quantum chemists have taken an interest in redox processes. Organic chemists are interested in situations where the energy of electrons is as easily controlled as it is at electrodes. Some biological processes are now seen in electrodic terms, with electron transfer to and from materials which would earlier have been considered to be insulators.

It is now time for a comprehensive treatise to look at the whole field of electrochemistry.

The present treatise was conceived in 1974, and the earliest invitations to authors for contributions were made in 1975. The completion of the early volumes has been delayed by various factors.

There has been no attempt to make each article emphasize the most recent situation at the expense of an overall statement of the modern view. This treatise is not a collection of articles from *Recent Advances in Electrochemistry* or *Modern Aspects of Electrochemistry*. It is an attempt at making a mature statement about the present position in the vast area of what is best looked at as a new interdisciplinary field.

Texas A & M University	J. O'M. Bockris
University of Ottawa	B. E. Conway
Case Western Reserve University	Ernest Yeager
Texas A & M University	Ralph E. White

Preface

Traditionally, electrochemistry has encompassed the fields of electrolyte solutions and molten salt electrolytes, as well as electrode processes and phenomena at charged interfaces. Indeed, the first topic occupied much of the literature of physical chemistry in the 1930s, somewhat at the expense of the kinetic and mechanistic aspects of the other main area of electrochemistry, electrode processes. Whatever the historical background, the behavior of electrolytes, both in solution and in the molten state in the absence of solvent, has also held a central position in practical aspects of electrochemistry.

The present volume on electrolytes first covers the subject of ionic hydration in Chapter 1, by Desnoyers and Jolicoeur. This is a basic topic in many aspects of electrochemistry, since the solvation state of ions determines the dynamic behavior of ions in electron-transfer reactions and in conductance, as well as thermodynamically in adsorption, double-layer phenomena, and ionic properties in solution.

Chapter 2, by Conway, gives an account of the long- and short-range interactions between ions in solution as manifested in activity behavior and ion pairing. Emphasis is placed on relating modern treatments of activity behavior to the classical Debye–Hückel framework for dilute solutions. The role of hydration in determining ionic activity behavior in strong aqueous solutions is treated in some detail, as well as are hydration effects in ion pairing.

One of the basic properties of electrolytes is their conductivity. Justice, in Chapter 3, gives a comprehensive account of modern treatments of conductance of electrolyte solutions, emphasizing the hierarchy of equations now developed for representing the concentration dependence of conductance; the limitations of various levels of approximation involved in various published equations are critically examined.

The behavior of the solvated proton has played a central part in much of electrochemistry, e.g., in Brønsted acid–base equilibria and proton-transfer kinetics, as well as in the mechanisms of cathodic hydrogen evolution. Its unusual conductance behavior has also attracted much attention for many years. Lengyel and Conway, in Chapter 4, deal with the state of the solvated proton in solution and the mechanisms of its transfer in conductance and diffusion. Relations between homogeneous proton transfer and heterogeneous transfer in electrode processes are examined.

Finally, various aspects of the important field of molten salt electrochemistry are covered in Chapter 5 by Papatheodorou, where various models and treatments of electrolytes in the molten state are discussed. This chapter is complementary to the one by Inman and Lovering in the volume *Kinetics and Mechanisms of Electrode Processes*, where molten salt electrode processes are treated.

This present volume therefore selectively covers several main topics in the field of physical chemistry of electrolytes that are of interest in the various broader areas of electrochemistry treated in other volumes of this treatise. Nonaqueous solutions are not covered in this volume.

University of Ottawa   B. E. Conway
Texas A & M University   J. O'M. Bockris
Case Western Reserve University   E. Yeager

May 1981

Contents

1. Ionic Solvation

Jacques E. Desnoyers and Carmel Jolicoeur

2. Ionic Interactions and Activity Behavior of Electrolyte Solutions

B. E. Conway

3. Conductance of Electrolyte Solutions

Jean-Claude Justice

4. Proton Solvation and Proton Transfer in Chemical and Electrochemical Processes

S. Lengyel and B. E. Conway

5. Structure and Thermodynamics of Molten Salts

G. N. Papatheodorou

Notation

$a_\pm$	mean activity; a_i, a_j activities of species i, j	E_{CB}	energy of conduction band
c	concentration (molar); velocity of light (cm s^{-1})	E_F	Fermi level
		E_H	measured potential on the hydrogen scale in the *same* solution
C_1, C_2, etc.	differential capacities of regions 1, 2, etc.	E_{NHE}	measured potential on the scale of the normal hydrogen electrode
cn	coordination number		
d	thickness, e.g., of a film, or of a dielectric	E_{SS}	energy of surface states
D	diffusion coefficient		
D_{x_ψ}	dissociation energy for molecule x_ψ	E_{VB}	energy of valence band
$\mathbf{D}$	dielectric displacement	$\mathscr{E}$	electrostatic field
e	electron charge	$f_\pm$	rational activity coefficient (mean)
E	potential (cf. electrode, on metal–solution potential difference, in kinetics)	F	Faraday constant
		g	interaction parameter, in non-Langmuir isotherms
E_{cal}	measured potential on the scale of the normal calomel electrode	$g_{ij}(r_{ij})$	radial distribution function (of distance r_{ij}); pair correlation function

G, H, S	free energy, enthalpy, and entropy (per mole)	$n_p{}^0$	concentration of holes in bulk
h	Planck's constant	N_A	concentration of charge acceptors
i	current density	N_D	concentration of charge donors
I_0	intensity of light		
I	current moment of inertia	N_{SS}	concentration of surface states
J	flux; quantum number for rotation	P	pressure (Pa), e.g., P_{O_2}, presence of a gas, O_2; momentum
k	with subscript, rate constants		
k_s	salting out (Setschenow) coefficient	$P(E)$	probability (for state of energy E)
k	Boltzmann constant	q, Q	partition function
K	thermodynamic equilibrium constant	Q_i	charge for some species, i, e.g., on a surface
K_1, K_2, etc.	integral capacities of regions 1, 2, etc.	r_i	radius of an ion
m	concentration (molal); mass of particle	r_{ij}	distance between particles i, j
M	molarity; N no longer used; number of particles	R	molar gas constant; resistance
		t	time
n	solvation number; quantum number for vibration	T	absolute temperature (K); with subscript, nmr relaxation times (T_1, T_2)
n_{CB}	density of electronic states in the conduction band	U	internal energy
		v	velocity (usually of a reaction); mobility of ion under $1\ V\,cm^{-1}$ charge
n_e	concentration of electrons		
$n_e{}^s$	concentration of electrons at the surface	V	volume; partial molar volume
$n_e{}^0$	concentration of electrons in bulk	x, y, z	coordinate system; distances
n_p	concentration of holes	$y_{\pm}$	stoichiometric activity coefficient (mean, molar)
$n_p{}^s$	concentration of holes at the surface	$\neq$	activated state (used as superscript)

Greek Symbols

α light absorption coefficient; transfer coefficient; specific expansibility; degree of dissociation; Brønsted factor

β charge-transfer symmetry factor; specific compressibility

γ surface tension

$\gamma_{\pm}$ stoichiometric activity coefficient (mean) molal

δ diffusion-layer thickness; barrier thickness

$\Delta_i^{i,b}\varphi$ potential inside a metal $(i = m)$, semiconductor $(i = sc)$, or insulator $(i = ins)$

$\Delta_1^i\varphi$ potential drop at the inner Helmholtz plane φ $(i = M, sc, ins, etc.)$

$\Delta_b^2\varphi$ potential in the diffuse (Gouy) double layer

$\Delta_2^i\varphi$ potential in the Helmholtz layer $(i = M, sc, or ins)$

Γ_i surface excess of species i

ε permittivity; quantum efficiency

ζ zeta potential

η overpotential; viscosity

θ fractional surface coverage; relative permittivity;

dielectric constant

κ conductivity; Debye–Hückel parameter

$\Lambda_{\pm,c}$ molar ionic conductivity at concentration c

Λ_c molar conductivity at concentration c

Λ_∞ molar conductivity at infinite dilution

$\Lambda_{\pm,\infty}$ molar ionic conductivity at infinite dilution

μ electric dipole moment or chemical potential

μ_e mobility of electrons

μ_p mobility of holes

μ^0 standard chemical potential

$\tilde{\mu}$ electrochemical potential

ν stoichiometric number; frequency of vibration (s^{-1})

$\tilde{\nu}$ wave number (cm^{-1})

ρ density of space change; resistivity

$\rho(E)$ volume charge density

$\rho_i(E)$ density of states $(i = M, sc, or ins)$

σ surface charge density in distribution; charge in double-layer region (subscripted) divided by area

σ_e capture cross section of electrons

σ_m charge on metal surface, divided by area

σ_p capture cross section of holes

τ relaxation time

ϕ double-layer potential (subscripted for indication of region)

ϕ_x apparent molar function of x; with subscript $\bar{x}$, partial molar function of x

φ inner potential

$\Delta\varphi$ Galvani potential

χ surface potential

$\Delta\chi$ surface potential difference

ψ outer potential; potential, in an electrolyte solution

$\Delta\psi$ Volta potential

ω angular frequency

1

Ionic Solvation

JACQUES E. DESNOYERS and CARMEL JOLICOEUR

1. Introduction

Electrochemistry deals mostly with charged interfaces, one of which is often an electrolyte solution. The equilibrium distribution of ions near the interface and the motion of these species in the field or concentration gradients depend on the forces acting on them. Consequently, a thorough understanding of electrolyte solution properties remains essential to the electrochemical sciences.

As a general definition, electrolytes may be taken as a class of compounds that, upon dissolution into a polar solvent, dissociate at least partially into ionic species. In discussing this process we usually distinguish between solvation behavior and solute–solute interactions. The present chapter deals exclusively with the former phenomenon, and concentration dependences will be examined only in cases where experimental measurements cannot be performed at concentrations sufficiently low to neglect ion–ion interactions.

To set a perspective for the present discussion it is useful to note that the limiting excess thermodynamic (e.g., activity coefficients, enthalpies of dilution) and transport properties (e.g., conductance, diffusion) are predicted quantitatively with the now famous Debye–Hückel theory. Modern statistical and nonequilibrium thermodynamics have extended the range of applicability of theoretical predictions, but the situation is still in a turmoil at intermediate

JACQUES E. DESNOYERS and CARMEL JOLICOEUR • Department of Chemistry, Université de Sherbrooke, Sherbrooke, Quebec, Canada J1K 2R1.

and high concentrations, where the distributions and properties of ions become much more specific to each system. The ions can no longer be considered free but exist as pairs or complex ions, and the mathematical treatment of thermodynamic and transport models becomes extremely complex. Interactions between a charged electrode and the neighboring ionic atmosphere are similar to ion–ion effects. Research in these areas is still very active and many years will pass before we get a full understanding of concentrated electrolyte solutions.

The strong field of the ions not only affects the ion–ion interactions but will also polarize the solvent near the ions. Even at infinite dilution, where there are no ion–ion interactions, manifestation of these interactions is obvious. The standard partial molal thermodynamic properties, such as volumes, entropies, and heat capacities, are significantly different from the expected molar values of free ions in solution. Similarly, infinite dilution transport properties (equivalent conductance, mobility, diffusion coefficients) show strong deviations from classical hydrodynamic theories, e.g., Stokes law and Zwanzig's equation. There is also more direct information on the solvation phenomenon coming from scattering studies, which give access to radial distribution functions from which coordination numbers and intermolecular distances can be derived. Spectroscopic data (IR, Raman, NMR) are very informative of the structure of stable solvation complexes and of changes in the structure of the solvent. Relaxation studies will determine the dynamics of the interactions.

The ion–solvent interactions will in turn influence the ion–ion and ion–electrode interactions and affect the concentration dependence of transport properties at finite concentrations. Strong solvation will reduce the effective quantity of free solvent. More subtle effects arise with highly structured solvents like water. Perturbation of the hydrogen-bond distribution near an ion may be reduced or enhanced as other ions get closer, and this is reflected in the excess properties. A good example of these effects is the hydrophobic interactions or bonding that are often evoked in aqueous biological systems.

Progress towards a quantitative understanding of the solvation phenomenon has been slower than that of excess functions. We are dealing here with shorter-range forces, and the molecular nature of the interactions does not lend itself as readily to electrostatic or statistical thermodynamic treatment. Until recently the more quantitative studies were mostly concerned with aqueous systems, and authors such as Bernal and Fowler,[1] Frank and Evans,[2] Gurney,[3] Frank and Wen,[4] and Samoilov[5] have set up the modern conceptual basis of hydration and of the structure of water. In the last 20 years there has been much improvement in the experimental and theoretical tools for the study of solvation. In particular, studies of solvation in the gas phase and the introduction of molecular dynamics and computer simulations have all contributed to a better understanding of solvation. The importance of water in biological systems and environmental problems is also responsible for the vast number of studies that have been recently made on the structure of aqueous solutions. Most of the recent work has been reviewed in the series on water edited by Franks[6] and in other monographs[7-11] or reviews.[12,13]

Studies on the solvation of ions in nonaqueous systems have not received the same attention as the solvation of ions in water. However, in view of the importance of these systems in inorganic chemistry and electrochemistry, this gap is rapidly being filled and there are now several excellent recent monographs on this topic.[14-16]

Because of the enormous literature concerning ionic solvation, a highly selective approach is compulsory. Most existing reviews deal with specific topics (types of data, properties, or methods), so we felt that it would be more appropriate here to place more emphasis on the complementarity of various experimental approaches. However, this requires that we compare the various properties and experimental data for the same chemical systems under comparable conditions. For lack of both experimental data and space in this chapter we are forced to limit ourselves to a few families of model inorganic and organic electrolytes in a few typical solvents or solvent mixtures and to try to reconcile the evidence coming from thermodynamic, transport, spectral, relaxation, and diffraction properties.

To set up a framework for our studies a brief overview of the various solvent systems for electrolyte systems will be given in the next section. Here special emphasis will be given on water and heavy water in view of the important role played by their three-dimensional structure on ion–solvent interactions. In the following section thermodynamic and transport properties will be examined in detail. In Section 4 an overview of our state of knowledge of ion–solvent interactions from spectroscopic and diffraction studies will be given, and finally in the last section some of the main theoretical approaches and new trends in theories will be briefly reviewed. Despite its limitations, we do hope that the present chapter will give the reader a reasonable idea of what occurs when an ion is dissolved in a solvent and what various properties can tell us about this process.

2. Solvents

The macroscopic and molecular properties of solvents play an important role in ion–solvent interactions. Although water is by far the most widely used solvent for several obvious reasons, it is not always appropriate for electrochemical systems. For example, high-energy batteries based on alkali metals cannot use protic solvents. Similarly, metallic deposition is often carried out in nonaqueous media. Even corrosion is not limited to aqueous systems. A large variety of solvents are therefore of interest in electrochemistry. The properties of some of the most common ones will be given here.

2.1. Classification of Solvents

To be useful in electrochemistry solvents must be able to dissolve inorganic electrolytes in reasonable quantities, i.e., at least $10^{-3}\,\mathrm{mol\,l^{-1}}$. This automatically excludes apolar solvents such as benzene and hexane. The classification of polar solvents is not easy and depends to some extent on the properties investigated. Brønsted distinguishes eight types of solvents, depending on the dielectric constant and acidic and basic characters.[17] Others prefer to speak of hard and soft acids and bases.[18] Also, a number of solvent polarity scales have been based on the influence upon spectroscopic transitions[19] (see also Section 4.4).

The properties of strong acidic and basic solvents have been reviewed in recent monographs.[14–16] Unfortunately, relatively few studies have been made on these systems with the aim of elucidating ion–solvent interactions. Weaker protic solvents such as the alcohols have been extensively investigated and will be considered here. A summary of some of their characteristic properties is given in Table 1.

Aprotic solvents are probably of greatest interest for electrochemistry, especially those with a high dielectric constant; some of the properties of the most common ones are compared in Table 1. In this list water is in a class of its own, since it is a double proton donor and proton acceptor.

Some indication of the extent of molecular interaction in protic solvents can be obtained from the properties in Table 1. With most solvents there is a close relationship between the dipole moment and the dielectric constant, but solvents that can form intermolecular hydrogen bonds have higher dielectric constants than expected from the theories of Debye or Onsager, e.g., H_2O, MeOH, and FA. Similarly, these solvents have higher freezing and boiling points and viscosities than do aprotic solvents of similar molecular weight. Because of these strong intermolecular forces, a certain degree of order is present in these solvents, which in turn should have some influence on the properties of dissolved electrolytes. Water is unique in this respect and will be treated in detail below.

The solubility of alkali halides in some of these solvents is given in Table 2. In aprotic solvents the solubility is in general larger the higher the dielectric constant. However, in hydrogen-bonded solvents the solubility is larger than in aprotic solvents of similar dielectric constant. For example, water, with a dielectric constant of 79, can dissolve many more inorganic electrolytes than can NMA, which has a dielectric constant of 111. The change in solubility from one electrolyte to another depends primarily on the lattice energies of the solid electrolytes. As a rough guide, these energies are high for ions of the same size and low for electrolytes made up of small and large ions. Therefore LiF and CsI have low solubilities in most solvents, but LiI and CsF are very soluble. Of course ion–solvent interactions will also have some effect on solubility, as will be discussed in Section 3.

Table 1
Physical Properties of Some Common Polar Solvents at 298 K[a]

Solvent	Abbreviation	T_f (K)	T_b (K)	Density (g cm^{-3})	Viscosity (cP)	Dipole moment (D)	Dielectric constant
Water	H$_2$O	273.15	373.15	0.9971	0.8903	1.84	78.54
Heavy water	D$_2$O	276.81	374.42	1.1044	1.094	1.84	78.25
Methanol	MeOH	175.32	337.75	0.787	0.542	1.67	32.6
n-butanol	n-BuOH	182.80	390.7	0.806	2.4	1.68	16.1
Formic acid	FA	281.3	373.7	1.214	1.64	1.19	57.0
Formamide	F	275.55	483.7	1.129	3.30	3.68	111.3
Methylformamide	NMF	267.6	453.0	0.999	1.65	3.86	182.4
Dimethylformamide	DMF	212.0	431.0	0.944	0.796	3.86	36.7
Methylacetamide[b]	NMA	302.5	479.0	0.950	3.89	4.39	178.9
Dimethylacetamide	DMA	293.0	438.0	0.937	0.92	3.79	37.8
Acetonitrile	AN	227.3	354.6	0.777	0.339	3.97	36.0
Dimethylsulfoxide	DMSO	291.6	462.0	1.096	1.096	3.90	46.6
Nitromethane	NMT	244.4	374.2	1.131	0.627	3.46	35.9
Ethylene carbonate[c]	EC	313.0	521	1.322	0.185	4.87	89.6
Propylene carbonate	PC	223.8	514.7	1.206	2.53	4.94	64.4
Tetramethylsulfolane	TMSO	239	—	1.173[b]	5.2[b]	—	42.5
Sulfolane[b]	SO	301.9	556	1.262	9.87	4.81	43.2

[a] Data taken from compilations of Janz and Tomkins[16] and Krishnan and Friedman.[20]
[b] At 303 K.
[c] At 313 K.

Table 2
Solubility of Alkali Halides in Some Common Solvents[a]

Salt	Water	Methanol	Acetonitrile	Formic acid
LiF	0.3	—	—	—
LiCl	45	41.0	0.14	27.5
NaF	4	0.03	3.0×10^{-3}	—
NaCl	36	1.4	2.5×10^{-4}	5.2
NaBr	95	17.4	4.0×10^{-2}	19.4
NaI	200	83.0	24.9	61.8
KCl	35	0.53	2.4×10^{-3}	19.2
RbCl	91	1.34	3.6×10^{-3}	56.9
CsCl	190	3.01	8.4×10^{-3}	130.5
CsBr	124	2.25	0.14	71.7
CsI	50	3.79	0.99	29.5
ε	78	33	36	57

[a] Data given in grams per 100 g of solvent, data in methanol, acetonitrile, and formic acid from Price.[21]

The consideration of solvent properties on a more detailed level is rapidly faced with highly specific behavior originating in the molecular structure and chemical functionality of the solvent molecules. Helpful comparisons can be extended somewhat, as done below for water and several other solvents, but it should be noted already that the details of solvent–solvent interactions in polar liquids are generally unknown. Except for water, the self-association of solvent molecules (leading to solvent structure) is only characterized by structure factors (e.g., Kirkwood g factor) that have little molecular meaning. Therefore, although solvent–solvent interactions may play an important part in solvation processes, their effect is very difficult to establish in nonaqueous solutions. Water is rather privileged in this respect: It has a high degree of molecular order and has been characterized with a high level of experimental and theoretical sophistication. The approach in assessing and interpreting solvent structure can thus be illustrated with some detail.

2.2. Water and Deuterium Oxide

In the past two decades water has been scrutinized with every available method to understand the energetic, geometrical, and dynamic aspects of its molecular interactions. This level of activity has generated a wealth of data that has been periodically reviewed by several workers in the field.[5,7,22–25] In the present section we will emphasize some of the distinguishing features of liquid water and try to illustrate how and to what extent these features can be understood at the molecular level.

2.2.1. Macroscopic Properties

Several bulk physical properties of liquid water are compared with those of other common solvents in Tables 1 and 3. The solvents in Table 3 are chosen to cover the main categories of solvents: nonpolar, polar aprotic, and protic. Examination of the macroscopic properties points out several distinctive characteristics of liquid water, from which some aspects of the molecular interactions can be inferred.

The first obvious, but often overlooked, feature of water is its very small molar volume, the smallest of all common solvents; it consequently has the highest particle density (molecules cm^{-3}). (The significance of this last feature, together with free volume, becomes clear in some of the theoretical calculations discussed in Section 5.) The latent heat of vaporization of water is also several times greater than its latent heat of fusion, contrary to the case of nonpolar solvents, for which these quantities are usually of similar magnitudes. This observation already suggests intermolecular interactions in liquid water that are stronger than those in other solvents. The same is observed from a comparison of the surface tension of the liquids, which is again much greater for water. Also, considering the relative molecular weights of the solvents, the viscosity of liquid water appears anomalously high, further indicating strong attractive forces between water molecules.

Table 3

Selected Properties of Water and Some Typical Nonaqueous Solvents

	Benzene[a]	AN	MeOH	H_2O[b]	D_2O[b]
Molecular weight	78	41	32	18	20
Latent heat of fusion ($kJ\ mol^{-1}$, T_{melt})	34.7	—	—	6.02	6.36
Latent heat of vaporization ($kJ\ mol^{-1}$, 298 K)	25.1	7.14[c]	37.2[d]	44.0	45.4
Surface tension (dyn cm^{-1})	28.2	29[e]	28[e]	71.97	71.93
Expansibility ($10^3\ K^{-1}$, 298 K)	1	1.3[c]	1.2[f]	0.26	0.22
Compressibility ($10^5\ bar^{-1}$, 298 K)	8.7	—	12[g]	4.45	4.59
Heat capacity ($J\ mol^{-1}\ K^{-1}$, 298 K)	35.6	90.4[h]	81.2[d]	75.3	84.3

[a] F. Franks.[26]
[b] Arnett and McKelvey.[27]
[c] *Chemical Rubber Company Handbook.*
[d] Wilhoit and Zwolinski.[31]
[e] Timmermans.[28]
[f] Philip and Jolicoeur.[29]
[g] *Lange's Handbook of Chemistry.*[30]
[h] Zana et al.[32]

More striking perhaps is the large molar heat capacity of liquid water; on a molar basis water and methanol have comparable heat capacities, though methanol has three more atoms than the water molecule and thus six more internal modes of vibration. To explain this difference multistate equilibria must be invoked that can, in liquid water, provide energy-storing devices to increase the heat capacity.[33] From basic concepts of statistical thermodynamics the heat capacity is related to energy fluctuations in the molecular ensemble; in the same way density fluctuations are related to the compressibility. The magnitudes of such fluctuations have been evaluated and their significance discussed recently by Frank.[34,35]

The bulk properties of liquid water thus imply a high degree of molecular association, which makes its thermodynamic properties (e.g., ΔH^0_{vap}, ΔS^0_{vap}) often closer to those of ice than to those of other common liquids. A more detailed examination of these bulk properties reveals still other peculiar features, such as the well-known density maximum at 277 K and a shallow minimum in the temperature dependence of the heat capacity at constant pressure near 308 K. An explanation of these various behaviors lies most certainly in a hydrogen-bonding equilibrium, which may be represented in the simplest terms as

$$\begin{cases} -O-H\cdots\overset{|}{O}- & \rightleftharpoons & -O-H+ & \overset{\diagdown}{O}\diagup \\ \quad (OH)_b & & (OH)_f & + \ (LP) \end{cases}$$

where $(OH)_b$ and $(OH)_f$ stand for the bonded and free OH groups, respectively, and (LP) represents a lone electron pair on the oxygen. Because of the double donor–acceptor character of water molecules, nine possible states must be envisaged, as depicted in Figure 1. Although the actual existence of each, or any, of these states can only be confirmed through spectroscopic techniques,

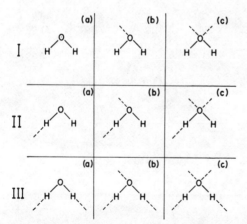

Figure 1. Possible hydrogen-bonding states of the water molecule in liquid water (reproduced with permission from Reference 36).

some particular states may be postulated and incorporated into models to describe the physical and thermodynamic properties of liquid water. A variety of such models have been proposed to account for the properties of liquid water, and their merit has been compared in several recent reviews.[34,35,37]

The molecular representations suggested for liquid water can be divided into two broad categories. In a first group, referred to as "uniformist" or "continuum" models, all molecules participate in hydrogen bonding and there is no significant fraction of broken hydrogen bonds. The latter may be bent with a harmonic restoring force constant, and thus there will be a continuous range of hydrogen-bonding energies.[38] In the second group, called "mixture" models, it is assumed that a limited number of distinguishable species can be identified. These species may be hydrogen-bonding energy states (e.g., molecules bonded to zero, one, two, three, or four neighbors) or they may involve specific molecular structures chosen by analogy to other states of water, namely, vapor, various forms of ice, and hydrates. The simplest model of this type was proposed by Frank and Wen[4] and pictures water as a mixture of icelike labile clusters ("flickering clusters") and monomeric water molecules. A number of more elaborate mixture models use well-defined molecular arrangements. For example, the model proposed by Pauling and Marsh[39] consists of regular dodecahedron structures, each capable of hosting an intersticial water molecule; other models such as those proposed by Bernal and Fowler[1] and John *et al.*[40] involve an equilibrium between two known structural varieties of ice.

It appears so far that, at least qualitatively, most experimental data on liquid water and aqueous solutions can be explained equally well from either the continuum or mixture models. A shift in a continuous distribution of hydrogen-bonded states can yield the same overall effect as does a shift in the equilibria involving only several sharply defined states or molecular species. As pointed out by various authors, each of the proposed models have been devised for the purpose of explaining a specific property of water; so it would be unrealistic to expect any single model presently available to be truly representative of the real behavior of water molecules in liquid water. At this stage it is clear that further refinement of the molecular description will require microscopic information as can only be obtained through scattering and spectroscopic studies.

2.2.2. Scattering Methods and Short-Range Order

As is apparent from the discussion above, the controversy over molecular models of water first originated in a simple but fundamental question regarding the existence of preferred hydrogen-bonding states for the OH groups in liquid water. If such states are present and observable, with limitingly free and bonded OH groups [$(OH)_f$ and $(OH)_b$], their ratio will provide a basis for a thermodynamic treatment. If, in addition, some information can be obtained on the actual molecular structures in liquid water, then a more specific molecular representation can be suggested. The local order that results from hydrogen-bonding association in liquid water can be characterized most directly through X-ray and other scattering techniques. The oxygen–oxygen pair correlation function $g_{O-O}(r)$ derived from such studies exhibits many important features.[41] A first maximum is observed near 0.3 nm (at 0.284 nm at 277 K, moving to 0.294 nm at 473 K), the area under which can be accounted for by 4.4 nearest-neighbor oxygen atoms. At low temperatures two other maxima occur in $g_{O-O}(r)$ near 0.45 and 0.7 nm, corresponding respectively to second and third neighbors. As the temperature is raised, every peak in $g_{O-O}(r)$ is broadened and the maxima at 0.45 and 0.7 nm progressively disappear below 373 K. These observations provide the strongest evidence in support of short-range structure in liquid water, where, on the average, the tetrahedral coordination of ice I is preserved to a significant extent; at room temperature and below, the range of positional correlation extends to ~0.8 nm from a central reference molecule.

The temperature-dependent structure evidenced from $g_{O-O}(r)$ has also been confirmed independently by electron diffraction studies.[42] Hence there can be little doubt that some degree of resemblance to ice is maintained in liquid water, at least over distances of several water molecules.

Neutron diffraction studies[43] on D_2O yield complementary information regarding this short-range structure. From the neutron data and $g_{O-O}(r)$ given by X-ray scattering the other atom pair correlation functions $g_{O-D}(r)$ and $g_{D-D}(r)$ can be obtained for distances ranging between 0 and ~0.4 nm. The peaks in these correlation functions are rather broad (rms variations of 0.03–0.04 nm), indicating that local or instantaneous departures from tetrahedral coordination do occur as required for an average coordination number of 4.4 as quoted above. Although these results do not specify the nature of the species present in liquid water, they provide a highly critical test for evaluating molecular models. Models that will accurately match the experimental g_{O-O}, g_{O-D}, and g_{D-D} will most probably also reproduce many properties of the real liquid.

2.2.3. Vibrational Spectroscopy and Hydrogen Bonding

Infrared and Raman spectroscopy can be used to further specify the hydrogen-bonding equilibria in water; these techniques measure the intra- and intermolecular vibrational modes of water molecules and the force constant for those modes that depend markedly on the energy of the hydrogen bonds.

The isolated H_2O molecules have C_{2v} symmetry and thus three internal modes of vibration: ν_1, symmetric stretch, ν_3, antisymmetric stretch; and ν_2, bending. The frequencies of these modes are given in Table 4 for H_2O, HOD, and D_2O in the dilute gas and liquid water. The large shift observed in

Table 4

Infrared Vibrational Frequencies of Isotopic Species of Water in the Gas and Liquid Phases[a]

	ν_1		ν_2		ν_3	
	Vapor	Liquid	Vapor	Liquid	Vapor	Liquid
H_2O	3656.65	$(3280)^b$	1594.59	1645	3755.79	$(3490)^c$
D_2O	2671.46	$(2450)^b$	1178.33	1215	2788.05	$(2540)^c$
HOD	2726.73	(2500)	1402.20	1447	3707.47	3400

[a] Data from Eisenberg and Kauzmann.[23]
[b] The infrared absorption is broad and exhibits two apparent maxima.
[c] The data for liquid is at or near 298 K.

stretching frequencies upon condensation is comparable to that found with alcohols and can most certainly be attributed to hydrogen bonding. However, the quantitative interpretation of these effects for H_2O or D_2O is complicated owing to the proximity of ν_1 and ν_3 (and also $2\nu_2$). Thus we will focus mostly on the spectra of HOD diluted in H_2O (or D_2O), where intramolecular coupling is absent.

In Figure 2 the Raman OD spectra of HOD in H_2O as reported by Lidner[44] are shown from room temperature to 673 K, well above the critical temperature of 647 K. It is instructive to note that even above the critical point, where hydrogen bonding can be excluded, ν_{OD} is still much lower (~100 cm^{-1}) than in the dilute vapor state. Thus the OH groups of randomly oriented water molecules at constant density near 1 g cm^{-3} are still subject

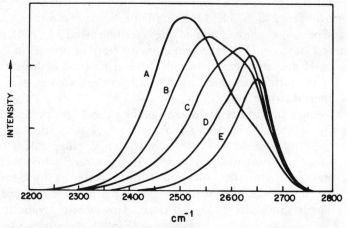

Figure 2. High-temperature and high-pressure Raman spectrum of HOD in H_2O obtained by Lidner: (A) 298 K, 100 bar; (B) 373 K, 1000 bar; (C) 473 K, 2800 bar; (D) 573 K, 4700 bar; (E) 673 K, 3900 bar. Spectra A–D at densities of 1.0 g cm^{-3}; spectrum E at 0.9 g cm^{-3} (reproduced with permission from Reference 44).

to strong perturbations from their neighbors. The absence of distinct absorption bands and the smooth spectral changes with temperature indicate that at room temperature the hydrogen-bonding states of the OH groups cannot be very sharply defined. However, the shape of the Raman band profile is clearly suggestive of two or several overlapping components, and indeed the spectra could be accurately decomposed into three Gaussian components.[45] From earlier studies carried out over a more restricted temperature range (303–366 K) the band deconvolution yielded only two Gaussian components, which were assigned to distinct types of OD groups, bonded and unbonded. From the intensities of these components the van't Hoff enthalpy of hydrogen-bond breakage was obtained as $10 \ kJ \ mol^{-1}$.[46]

Similar two-state equilibria can also be inferred from the infrared absorption spectrum,[47] although the asymmetry of the band envelope is much less pronounced than in the Raman spectra. The hydrogen-bonding enthalpy derived from these studies is also on the order of $10 \ kJ \ mol^{-1}$.

Further evidence of two-state behavior is also available from extensive investigation of overtones and combination bands in the near infrared. Although this spectral region lacks a quantitative understanding of absorption intensities, the conclusions are supported by comparative studies on several overtones and combination bands of H_2O and HOD[48–50] and also on several alcohols as neat liquids or diluted in nonpolar solvents.[51] The temperature dependence of the near-infrared absorption bands is easily analyzed when the spectra are recorded differentially,[52] as shown in Figure 3. Over the temperature range investigated the differential absorption consists of two overlapping Gaussians that, again, were assigned to different classes of OH groups. Analysis of these and other similar spectra also yields hydrogen-bonding enthalpies of $10–12 \ kJ \ mol^{-1}$. From spectra recorded over broader temperature ranges (273–673 K) Luck and Ditter[49] have shown that, as with the Raman spectra, deconvolution of the overtone band of HOD or H_2O require three Gaussians. The three components were attributed to free OH groups and OH groups engaged in either linear or bent hydrogen bonds. Evidence for the latter was based on low-temperature studies of H_2O and CH_3OH in inert matrices.[53]

The existence of distinct hydrogen-bonding states or species in liquid water has often been inferred from the observation of isosbestic points in the spectra, i.e., frequencies at which the absorption is independent of temperature, as seen, for example, in Figure 3. The general conditions required for observing such isosbestic points have been reviewed by Senior and Verrall[47] and the significance of the isosbestic point in the infrared O—D absorption has been discussed in great detail. However, the validity of this criterion as proof of the existence of distinguishable hydrogen-bonded states has been questioned[54,55]; indeed, over broad temperature ranges only isosbestic regions are observed. Thus the isosbestic criterion is not sufficient in itself, but it must be acknowledged that it was shown valid in other hydrogen-

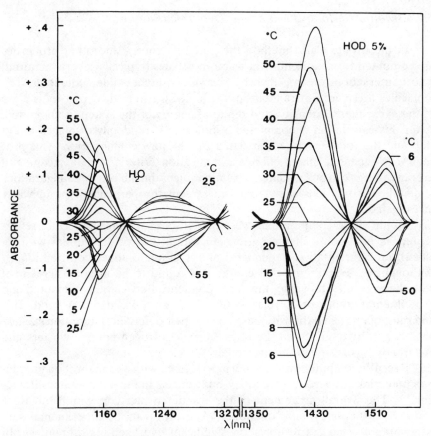

Figure 3. Near-infrared differential spectra of water as functions of temperature. The reference cell (10.00 cm) is maintained at 273 K and the length of the working cell is adjusted to account for density variations (reproduced with permission from Reference 50).

bonding systems (e.g., alcohols[56,57]) where the equilibrium species are more readily identified.

Finally, we note that the two-state interpretation of infrared and Raman spectra of liquid water also provides the equilibrium concentration of the species involved. Typically the derived fraction of the "free" OH groups (those having the weakest perturbation) is on the order of 10% at the melting point, increasing quite linearly to ~30% at 373 K. At constant density near 1 g cm^{-3} this fraction increases smoothly with temperature up to unity at the critical point.

For polyatomic fluids, and especially for water, the calculation of atom–atom pair correlation functions through statistical mechanics is extremely difficult; however, computer simulation methods provide an alternate route through Monte Carlo and molecular dynamics calculations.

2.2.4. Monte Carlo and Molecular Dynamics Methods

In the Monte Carlo method the potential energy and other properties are computed for an ensemble of water molecules (typically several hundred) whose interactions are specified by pairwise or higher-order potentials. The molecules are placed at random in a volume such that their density is close to the experimental density, and the total energy of the system is computed. After this calculation is completed a molecule is randomly displaced a unit step and the total energy recalculated for the new configuration. This procedure is repeated under energy minimization criteria until a meaningful average energy can be evaluated. Examination of the molecular configurations under these conditions shows the structure of the particular type of water investigated.

Monte Carlo computations were carried out for water molecules represented as four-point charges, using an interaction pair potential with an electrostatic (orientation-dependent) and a Lennard-Jones contribution.[58] Although the average internal energy was found to be in good agreement with the experimental value, the correspondence between the computed and experimental atom pair correlation functions $[g(r)]$ was not very good. This and subsequent calculations[59] based on the pair potential assumption exclude any cooperativity effects in the formation of hydrogen-bonded clusters and apparently favor the continuum models.

Recently computations have been performed with a configuration interaction potential that accounts for many-body effects in the water–water interaction.[60] The overall agreement of the computed pair correlation functions and several thermodynamic quantities (A, C_v, K) with the experimental data is remarkable, and the results show significant hydrogen-bond breakage. In these model simulations the many-body effects appear to play a minor role in determining the structure factors, but they are believed to lower the internal energy by approximately $4\,kJ\,mol^{-1}$. Since this is a key issue in the model debate, confirmation of this result is of greatest importance.

The molecular dynamics method has been devised to study both the interaction energy and the dynamic behavior of ensembles of water molecules. In these "computer experiments" the position and velocity of the molecules are initially defined and computed at fixed time intervals, using a specified potential and classical laws of motion. As with the Monte Carlo methods, the molecular dynamics computations yield the atom pair correlation functions and, in addition, the molecular velocity and orientation correlation functions.

The atom pair distribution function, obtained from computations based on a four-point charge model and a combined electrostatic–Lennard-Jones pair potential, are in qualitative agreement with those obtained from scattering data.[61] The equilibrium molecular configurations show no evidence of distinct types of water molecules or structures, but free OH groups are found to be present for times longer than the period of the OH stretching vibration. The suggested hydrogen-bond breaking process is characterized by an excitation energy of about $10 \, kJ \, mol^{-1}$.

As with the Monte Carlo calculation, it can be anticipated that with interaction potentials accounting for triplet and higher interactions the molecular dynamics simulation will yield highly realistic pictures of the molecular motions and interactions in liquid water.

2.2.5. Deuterium Oxide

The macroscopic properties of liquid D_2O have been included in Table 3 for comparison with those of liquid H_2O. An extensive discussion of the relative properties of these two solvents has been given by Swain and Bader,[62] Arnett and McKelvey,[27] and Owicki et al.,[63] so we summarize here only several points pertinent to the solvation effects discussed in later sections.

Following the discussion of the relative properties of water and other solvents, it seems quite straightforward that at a given temperature the molecular interactions in D_2O are greater than in H_2O. This is generally attributed to a zero-point energy effect on the hydrogen bonds that results in a more stable hydrogen bond among D_2O molecules as compared to among H_2O molecules; this effect has been estimated from librational frequencies at about $1 \, kJ \, mol^{-1}$.[34] Since the $O—D \cdots O$ bond is more stable than the $O—H \cdots O$ bond, it must be expected that more such bonds will be present in liquid D_2O. Hence the structural organization in liquid D_2O, defined here as the product of the number of hydrogen bonds and the hydrogen-bonding enthalpy, will be greater than in H_2O.

It must be emphasized, however, that structural differences between water and heavy water must be quite subtle, since the oxygen–oxygen correlation function in D_2O is not significantly different from that of H_2O at 277 K.[41] The similarities of the molar volumes and dielectric constants of these two liquids also caution against interpreting their properties, or the properties of their solutions, in terms of simple structural differences.

2.3. Concluding Remarks

The results discussed above represent a minute fraction of all the data available from investigations on liquid water. Many other experimental studies using NMR spectroscopy, dielectric and ultrasonic absorption, viscosity, diffusion, and other techniques yield information on molecular motions or processes and the activation energies for these processes. So far, however, these methods have been of limited assistance in improving the structural model, though, as will be illustrated in later sections, they have proven very powerful in characterizing ionic solvation. It must also be remarked that even among the selected results quoted earlier there remain active debates concerning, for example, spectral band deconvolution, the existence of identifiable intermediate hydrogen-bonding states (e.g., bent hydrogen bonds), and the extent of cooperativity in hydrogen-bonded aggregates involving more than two water molecules. Summarizing what appear to be the conclusions from the most reliable results, the following picture of liquid water has emerged.

At room temperature positional correlation, or short-range order, among water molecules extends several diameters away from a given reference molecule. Although tetrahedral coordination is, on the average, predominant, significant local (or instantaneous) departure from the icelike structure must occur to account for the width of the peaks in the atom pair correlation functions $g_{O-O}(r)$, $g_{O-H}(r)$, and $g_{H-H}(r)$, as well as for a coordination number of 4.4. Such fluctuations are provided by a hydrogen-bond breaking process that involves at least two distinct categories of OH groups evidenced by vibrational spectroscopy. Within each category there exists a distribution of energy states, and on the average the two types of OH groups differ by a van't Hoff enthalpy of approximately 10 kJ mol^{-1}. However, the spatial distribution of these OH groups in structural frameworks cannot yet be specified. In a limiting situation the free OH groups would occur at the interface between clusters of hydrogen-bonded molecules; in another extreme case the free OH may belong to water molecules instantaneously trapped in cavities of hydrogen-bonded networks. In still another situation different local domains of structured and unstructured water could coexist in the liquid. For the time being this aspect remains unsettled, but recent Monte Carlo and molecular dynamics calculations promise valuable clues as to what the real situation might be like.

The difficulties encountered in trying to specify the molecular aspects of water structure illustrate rather well the complexity of molecular interactions occurring in this liquid. In other solvents, such as those in the dipolar aprotic category, the structural aspect may be less important, but other factors (e.g., dispersion forces, molecular geometry) can play a more significant role in determining the properties of these solvents. Hence a similar level of difficulty can be anticipated in any quantitative treatment of solvent properties or of the perturbations of solvent–solvent interaction by molecules, ions, and neutral or charged surfaces. It should therefore be realized that, except in cases where the solvation effects are similar to chemical reactions (e.g., in the first coordination sphere of transition metal ions), a model description of solvation effects must often remain qualitative.

3. Thermodynamic and Transport Properties

Thermodynamic properties such as stability constants, electromotive forces, and solubilities are functions of the chemical potential of the species in solution, and their variations with temperature and pressure are related to the corresponding entropies, enthalpies, heat capacities, volumes, etc. All these functions are related or influenced by the solvation of ions.

Classical thermodynamics deals with macroscopic properties and as such can only give information on averaged interactions in solution. It is therefore necessary to go through models and theories to obtain information on the molecular nature of the interactions. Still these properties are very useful, since they can be defined in such a way that solute–solvent and solute–solute interactions can be separated and they lend themselves readily to quantitative tests of various theoretical models.

Different thermodynamic properties give complementary information on solvation. Free energies reflect the forces acting between molecules, enthalpies, the energetics of the interactions, entropies, the statistical configuration of the solvent molecules near the ions, volumes, the compactness of the spatial arrangement of the solvent molecules, heat capacities, energy fluctuations, etc. Consequently, a particular type of interaction will not influence all properties equally, so some discrimination can be achieved. Free energies will reflect mostly the strong interactions (e.g., electrostatic) between molecules, whereas heat capacities will be more sensitive to interactions that are very temperature dependent (e.g., hydrogen-bond distribution). On the other hand, information on strong charge–dipole interactions that are not too temperature dependent are now lost with this property. Therefore, to get a complete picture of the interactions, all thermodynamic properties should be investigated whenever possible. Only if a model or theory accounts for all thermodynamic properties do we have hopes of being close to reality.

The interpretation of thermodynamic data of solvation in the gas phase is straightforward, since solvation can be represented as a chemical equilibrium. The situation is somewhat more complicated in solution, since the interactions are not so well defined. It is therefore convenient to introduce various solvation functions and models to express the thermodynamic data. Then it will be possible to examine trends among the functions of alkali and tetraalkylammonium halides for which reliable data exist. We will therefore systematically examine the data in a typical protic solvent, methanol (MeOH), a typical aprotic solvent, acetonitrile (AN), whenever the data are available, and water. As we will show later, transfer functions between solvents and from a pure solvent to a mixed solvent are often very useful in studies of solvation. Some of the best-characterized transfer functions are from H_2O to D_2O and from H_2O to urea solution, and these will be presented whenever available.

Transport properties are also very useful for the study of ionic solvation. Like thermodynamic properties, they are macroscopic in nature but, through models, can give information on the effective size of a moving particle in solution. These properties are sensitive to strong ion–solvent interactions, which increase the effective size of the ions, and also to any modification in the structure of the solvents. In general, properties such as conductance can be measured very precisely at low concentrations and reliable infinite dilution values are readily obtained. Also, one of the main advantages of transport over thermodynamic data is that true ionic values are measurable, as, for example, in conductance and self-diffusion. Whenever possible, the same trends in transport properties will be examined as in the case of thermodynamic functions.

It is hoped that a systematic coverage of many properties for a family of typical ions and solvents or mixed solvents will give an overall view of the main aspects of solvation. It would have been useful to also examine the transition metal ions, which are very more relevant to electrochemical systems, but reliable solvation data are often lacking, primarily in view of the difficulty in measuring standard thermodynamic functions as a result of problems arising from hydrolysis and ion association. Still it is hoped that the present overview will give a framework from which the solvation of these more interesting electrolytes can be investigated.

3.1. Standard Thermodynamic Functions

In the gas phase solvation can be represented as a chemical equilibrium between the ions and solvent molecules:

$$M^+ + nA \rightleftarrows MA_n^+ \tag{1}$$

This equilibrium can be defined by the equilibrium constant K_p and the corresponding thermodynamic functions $\Delta G^0 = RT \ln K_p$, $\Delta H^0 = \partial(\partial G^0/T)/\partial(1/T)$, etc.

In solution the situation is more complicated, since molar thermodynamic functions are not additive and partial or apparent molal quantities must be used. A partial molal quantity $\bar{Y}_i$ of the component i is defined by

$$\bar{Y}_i = \left(\frac{\partial Y}{\partial n_i}\right)_{T,P,n_1,n_2,\ldots} \tag{2}$$

The number of moles n_i can usually be replaced by the molality m_i, the number of moles per kilogram of solvent. These partial molal quantities or changes in these quantities can sometimes be determined directly, e.g., from chemical potentials from the emf, but it is usually only possible to measure integral quantities. For this purpose it is often convenient to introduce apparent molal quantities ϕ_Y that are derived from experimental measurements: For a binary system of a solute (2) in a solvent (1)

$$\phi_Y = (Y - n_1 Y_1^0)/n_2 \tag{3}$$

where Y is the total property of the solution, n_1 is the number of moles of solvent having the molar property Y_1^0, and n_2 is the number of moles of solute. The partial molal quantities can readily be derived from ϕ_Y with the relation

$$\bar{Y}_2 = \phi_Y + n_2 \left(\frac{\partial \phi_Y}{\partial n_2}\right)_{T,P,n} \tag{4}$$

Here again n_2 can be replaced by m. From Eq. (4) it is obvious that at infinite dilution $\bar{Y}_2^\theta = \phi_Y^\theta$.

For the investigation of ion–solvent interactions it is sufficient to consider standard thermodynamic functions, since they are by definition independent of ion–ion interactions. For all thermodynamic properties except for free energies and entropies the standard state is infinite dilution, therefore the experimental ϕ_Y have to be extrapolated to infinite dilution graphically or through a suitable equation. Such standard partial molal quantities are known in many solvents for volumes $\bar{V}_2^\theta$, heat capacities $\bar{C}_{P2}^\theta$, expansibilities $\bar{E}_2^\theta$, and compressibilities $\bar{K}_2^\theta$. Most literature data are for isentropic compressibilities. Although these can be converted to isothermal ones with $\bar{C}_{P2}^\theta$ and $\bar{E}_2^\theta$ data,[64] usually the difference is small and can often be ignored.

The standard state for the chemical potential $\mu_2 = (\partial G/\partial n_2)_{T,P,n_1}$ is the hypothetical ideal $1\,m$ solution, since

$$\mu_2 = \mu_2^\theta + RT \ln \gamma_2 m_2 \tag{5}$$

The standard state for the entropy is defined by

$$\bar{S}_2^\theta = (\bar{H}_2^\theta - \mu_2^\theta)/T \tag{6}$$

Since the standard states for $\bar{H}_2^\theta$ and μ_2^θ are defined differently, it is difficult to give a physical significance to the standard state for $\bar{S}_2^\theta$.

Absolute values of μ_i^θ and $\bar{H}_i^\theta$ are not measurable because of the existence of zero point energies, and only changes in these properties are accessible experimentally. For this reason it is convenient to use the gas phase as a reference state, since properties in the gas phase are additive and usually measurable. The *standard functions of solvation* are therefore defined for a solute by

$$\Delta Y_s^\theta = \bar{Y}_2^\theta - Y_2^0 \text{(gas)} \tag{7}$$

where Y_2^0(gas) is the standard function for an ion in the gas phase. The standard enthalpies of solvation ΔH_s^θ can be calculated from the standard enthalpies of solution and lattice energies, the standard entropies of solvation ΔS_s^θ from emf or solubility measurements combined with third law or statistical thermodynamic entropies in the gas phase, and free energies ΔG_s^θ from ΔH_s^θ and ΔS_s^{θ}.[65] In water these functions are called hydration functions.

It is not always convenient or possible to use the gaseous phase as a reference state, especially with organic ions. In such cases standard transfer functions between solvents α and β are used:

$$\Delta Y_2^\theta(\alpha \rightarrow \beta) = \bar{Y}_2^\theta(\beta) - \bar{Y}_2^\theta(\alpha) \tag{8}$$

These functions are derived directly from the corresponding properties in both solvents. For example, $\Delta H_2^\theta(\alpha \to \beta)$ is calculated from the difference between the standard enthalpies of solution in solvents α and β. These functions are especially useful in comparing various solvents, since the large contribution to the functions related to the change in standard states (gas to solution) is avoided. In the case of two nonaqueous solvents Krishnan and Friedman[20] suggested propylene carbonate as a reference solvent, since it appeared to be a fairly typical aprotic solvent, although, as will be shown later, this choice might not always be the best one.

The transfer functions from H_2O to D_2O are of particular interest.[66,67] The similarity of the dipole moments and dielectric constants of these two liquids will tend to cancel short-range charge-dipole interactions in the transfer functions. On the other hand, all interactions related to the structural state of water before the addition of ions should be reflected in this transfer function.

Reasonably accurate thermodynamic data are available for simple electrolytes in common solvents. For example, Friedman and Krishnan[68] have compiled hydration functions, and Millero[69] $\bar{V}_2^\theta$ and $\bar{E}_2^\theta$ data. Data for D_2O have been reviewed by Arnett and McKelvey[27] and Krishnan and Friedman.[70] Thermodynamic data for electrolytes in nonaqueous solvents are not as complete as for those in water, and most of the studies have been limited to enthalpies.[20] However, data on volumes and heat capacities are now being measured in many laboratories.[32,71-73]

Transfer functions can also be used to discuss mixed solvents. In such cases the pure solvent is taken as the reference. The standard transfer function of solute 2 from solvent 1 to the mixed solvent $1 + 3$ is given by

$$\Delta Y_2^\theta(1 \to 1 + 3) = \bar{Y}_2^\theta(1 + 3) - \bar{Y}_2^\theta(1) \tag{9}$$

There is an enormous literature on these transfer functions, especially for aqueous-organic mixed solvents, and adequate coverage is beyond the scope of this chapter. However, some typical cases will be discussed in Section 3.7.

3.2. Basic Solvation Concepts

Ionic solvation can lead to rather stable complexes, as in the case of some transition metal ions, or to more ill-defined structural effects, as with some aqueous electrolytes. Although the theoretical models will be examined primarily in Section 5, it is necessary to define at this point the various types of solvation and the basic concept to establish a framework from which thermodynamic and transport data can be discussed.

3.2.1. Coulombic or Primary Solvation

The high radial electric field of ions will strongly polarize the solvent molecules in their vicinity. This polarization will affect all the thermodynamic properties, and the magnitude of its effect can be estimated from simple electrostatic theories, as discussed below. If the field is high enough (small ions, multivalent ions), some of the solvent molecules will lose part of their translational and rotational degrees of freedom. This will again be reflected in the thermodynamic, transport, and spectral properties. For example, an electrolyte solution will be denser and less compressible than the pure solvent, the equivalent conductance will decrease, and the IR and Raman solvent bands will be shifted. In cases where the solvation complex is stable enough (e.g., transition metal ions), new bands in the IR and Raman spectra and characteristic NMR relaxation times may be observed. This type of solvation exists in all solvents to various degrees and will be referred to henceforth as coulombic or primary solvation.

There are two classical theoretical approaches to the quantitative treatment of coulombic solvation. If a solvent is considered a dielectric continuum polarized by the field of an ion, then the free energy of solvation can be calculated from the difference in the work of charging of the ion in the solvent and in vacuum.[74] This work in a solvent of dielectric constant ε is given by

$$W = \int_0^q \psi \, dq = \int \frac{q}{\varepsilon r} \, dq = \frac{q^2}{2\varepsilon r} \tag{10}$$

The final charge q is given by ze. In vacuum $\varepsilon = 1$, and this leads to a standard free energy of solvation given by

$$\Delta G_s^\theta = \frac{(ze)^2 N}{2r}\left(\frac{1}{\varepsilon} - 1\right) \tag{11}$$

where N is Avogadro's number. This relation is known as the Born equation.

The corresponding ΔH_s^θ and ΔS_s^θ are obtained from the temperature derivative of ΔG_s^θ:

$$\Delta S_s^\theta = -\left(\frac{\partial \Delta G^\theta}{\partial T}\right)_P = \frac{(ze)^2 N}{2r}\left[\frac{1}{\varepsilon^2}\left(\frac{\partial \varepsilon}{\partial T}\right)_P\right] \tag{12}$$

and

$$\Delta H_s^\theta = -T^2\left(\frac{\partial \Delta G_s^\theta/T}{\partial T}\right)_P = \frac{(ze)^2 N}{2r}\left[\frac{1}{\varepsilon} - 1 + \frac{T}{\varepsilon^2}\left(\frac{\partial \varepsilon}{\partial T}\right)_P\right] \tag{13}$$

Similarly, the change in heat capacity is given by

$$\Delta C_{Ps}^\theta = \left(\frac{\partial \Delta H_s^\theta}{\partial T}\right)_P = -\frac{(ze)^2 NT}{2r\varepsilon^3}\left[2\left(\frac{\partial \varepsilon}{\partial T}\right)_P^2 - \varepsilon\left(\frac{\partial^2 \varepsilon}{\partial T^2}\right)_P\right] \tag{14}$$

ΔC_{Ps}^θ can be calculated from $\bar{C}_{P2}^\theta$ if the intrinsic heat capacity of solute 2 is known:

$$\Delta C_{Ps}^\theta = \bar{C}_{P2}^\theta - C_P(\text{in}) \tag{15}$$

The intrinsic heat capacity should normally be the molar heat capacity of the gaseous ion. When it is not available, it can be approximated by the value of the corresponding pure liquid or solid, since the change in heat capacity of a pure substance is usually not very large when going from the gas phase to a condensed state.

The volume change associated with solvation is derived from the pressure derivative of the free energy:

$$\Delta V_s^{\theta} = \left(\frac{\partial \Delta G_s^{\theta}}{\partial P}\right)_T = -\frac{(ze)^2 N}{2\varepsilon}\left[\frac{1}{\varepsilon^2}\left(\frac{\partial \varepsilon}{\partial P}\right)_T\right] \tag{16}$$

and

$$\Delta V_s^{\theta} = \bar{V}_2^{\theta} - V(\text{in}) \tag{17}$$

In the case of volumes $V(\text{in})$ cannot be defined unambiguously in the gas phase, and it is therefore assumed to be the volume of the hypothetical pure liquid solute. This volume is that of the bare ion plus a contribution for the free space around the ion arising from the finite size of the solvent molecules. Various equations have been proposed to estimate $V(\text{in})$ from the ionic radii.[69] The solvation contribution to expansibilities and compressibilities are again obtained from the temperature or pressure derivatives of Eq. (16). For monatomic ions $K(\text{in})$ and $E(\text{in})$ are often assumed to have a zero value, since the volume of the ions should be independent of T and P. However, the void space near the ions in the solvation shell changes with T or P, causing a positive contribution to $E(\text{in})$ and $K(\text{in})$. With organic ions a small residual intrinsic value might also be present owing to internal rearrangement of the atoms.

In this simple theoretical approach all thermodynamic functions of solvation are related to the charge and radius of the ion and to the dielectric constant and its derivatives with respect to temperature and pressure. The limitations of this theoretical approach will be discussed in Section 5. Still, this simple theory can be used to predict the signs and trends of the thermodynamic changes due to coulombic solvation.

The assumption that the solvent is a continuum is hardly valid when discussing strong ion–solvent interactions, since the sizes of solvent molecules are comparable to those of the solute molecules. A more realistic approach is to consider the solvent molecules as spheres characterized by their polarizability and dipole and quadrupole moments. Again, these theories will be described in Section 5. For the purpose of discussing coulombic solvation in terms of discrete solvation complexes it is sufficient to introduce here the concept of solvation numbers, which are defined as the effective numbers of solvent molecules that are bound to ions or that have suffered certain changes in property. In this approach all the solvent molecules around an ion are assumed to be unaffected, except for the ones in the solvation shell. Following

Padova,[75] the solvation effect can then be represented for the property Y by

$$\bar{Y}_2^\theta - Y_2(\text{in}) = n(y_1^s - y_1) \tag{18}$$

where n is the solvation number, by definition always positive, y_1 is the molar value of the pure solvent, and y_1^s is the molar value of the solvent in the solvation shell. The value of n can be evaluated if $Y_2(\text{in})$ and y_1^s are known or estimated from models, and it can be used to interpret trends in solvation effects.

3.2.2. The Structure-Breaking Effect

In highly structured solvents such as water and heavy water, modifications in the structure of the liquid by ions may significantly influence the properties of the solutions. For example, aqueous solutions of KCl are less viscous than pure water, the partial molal heat capacity of the electrolyte is very negative, and the NMR chemical shifts and modifications in the IR spectra are in the same direction as an increase in temperature. These phenomena are generally attributed to structure-breaking or negative hydration, which has been discussed in detail by Frank and Evans,[2] Gurney,[3] Samoilov,[5] and Frank and Wen.[4] In bulk water, through hydrogen bonding, most water molecules are coordinated to four other molecules, two dipoles being oriented toward the central molecule and two away from it. On the other hand, near an ion the field tends to orient all the water molecules radially.

It is therefore reasonable to imagine that at some distance from the ion there will be competition between the two orienting influences and the water molecules will be more labile than in pure water, i.e., they will behave as if they were at a higher temperature. While this effect should be observable in all highly structured solvents, it has been evidenced mostly in aqueous systems.

A schematic representation of the Frank and Wen model[4] is shown in Figure 4. In the inner solvation shell of the ion there are n_h water molecules tightly held by the field of the ion (coulombic hydration) and having the molar property $\bar{y}_h$. Next to it there is an intermediate structure-broken region

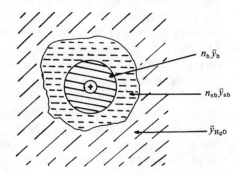

Figure 4. Frank and Wen model for the hydration of hydrophilic ions (reproduced with permission from Reference 67).

where n_{sb} water molecules have the properties of non-hydrogen-bonded water, $\bar{y}_{sb}$. Outside this shell water has its normal structure, $\bar{y}_{H_2O}$.

The process of hydration is visualized as follows.[67] An ion interacts with n water molecules from the bulk to form a structure-broken region. The ion then further interacts with n_h of the n water molecules to form a hydration complex, leaving n_{sb} water molecules in the structure-broken region. The overall hydration function is therefore

$$\bar{Y}_2^\theta - \bar{Y}(\text{in}) = (n_h + n_{sb})(\bar{y}_{sb} - \bar{y}_{H_2O}) + n_h(\bar{y}_h - \bar{y}_{sb}) \tag{19}$$

For ΔG_h^θ, ΔH_h^θ, and ΔS_h^θ the left-hand side of Eq. (19) would be modified to allow for changes in the reference states. The first term on the right-hand side of Eq. (19) represents the structure-breaking effect and occurs only with hydrogen-bonded solvents. The second term is the contribution from coulombic interactions, which exists for all polar solvents [see Eq. (18)]. This equation can be used to predict or interpret trends in the hydration function if the properties of water in the two hydration shells can be estimated. For example, the molar heat capacity of pure water is about twice as large as that which would be expected for unbonded water molecules, and a negative contribution is predicted for the structure-breaking effect.

It should be noted that Eq. (19) is applicable to $\bar{C}_{P2}^\theta$, $\bar{E}_2^\theta$, and $\bar{K}_2^\theta$ only if the number of water molecules in the two hydration cospheres is assumed to be independent of temperature or pressure. Also, this equation predicts that there is no structural contribution to ΔG, since in an equilibrium situation the chemical potential of the water molecules is equal in all cospheres.[12] Obviously the function that should be examined is the standard ΔG_s^θ, which is related to the equilibrium constant of the hydration process. Unfortunately, such an equilibrium constant is not easily derivable from the simplified model above. This representation of the Frank and Wen model is readily extended to transfer functions and particularly to the transfer functions from H_2O to D_2O.[67] If it is assumed that $\bar{Y}(\text{in})$, $\bar{y}_{sb}$, n_h, and n_{sb} are the same in H_2O and D_2O, then the leading terms of the transfer functions are

$$\Delta Y_2^\theta (H_2O \rightarrow D_2O) = n_h(\bar{y}_h^{H_2O} - \bar{y}_h^{D_2O}) + (n_h + n_{sb})(\bar{y}_{H_2O} - \bar{y}_{D_2O}) \tag{20}$$

The first term on the right-hand side represents the solvent isotope effect on the coulombic hydration, and the second term the effect on the structure-breaking. This second term is expected to dominate with most functions. Therefore, if the assumption that n_h and n_{sb} are the same in both solvents is correct, the sign and trends of the structure-breaking influence on the transfer functions can be predicted simply from the molar properties of pure H_2O and D_2O. For example, $\bar{C}_{pD_2O}$ is larger than $\bar{C}_{pH_2O}$ and $\Delta C_{P2}^\theta (H_2O \rightarrow D_2O)$ of the structure breakers should be negative.

This model can also be extended to mixed solvents[76]; if it is assumed that the only effect of a cosolvent is to modify the structure of water in the outer region represented in Figure 4, then the leading terms of the transfer functions are given by

$$\Delta Y_2^\theta (H_2O \rightarrow H_2O + 3) = (n_h + n_{sb})(\bar{y}_{H_2O} - \bar{y}_{H_2O}^{(3)}) \tag{21}$$

where $\bar{y}_{H_2O}^{(3)}$ is the partial molal function of water in the mixed solvent. This function can readily be calculated from the concentration dependence of the apparent molal function of the cosolvent in water[76]:

$$\bar{y}_{H_2O} - \bar{y}_{H_2O}^{(3)} = \frac{m^2}{55.51} \frac{\partial \phi_{y,3}}{\partial m} \tag{22}$$

where $\phi_{y,3}$ is the apparent molal property of solute 3 in water.

3.2.3. Hydrophobic Hydration

Rare gases, alkanes, or large organic ions generally show anomalous properties in water. Compared with their properties in nonaqueous solvents, their partial molal heat capacities are much larger, their viscosities higher, and their equivalent conductance lower; also, changes in the IR spectra are often in the same direction as a decrease in temperature. This type of hydration is therefore explained on the basis of an increase in the structure of water, as we would expect, for example, by analogy with clathrate hydrates.[77,78] Since this hydration is related to the hydrophobic character of the molecule or side chains, it is designated as such.[77] Hydrophobic hydration should not be confused with hydrophobic interactions or bonding, which refer to the solute–solute interactions between two hydrophobic solutes.[7,78,79]

3.3. Solvation in the Gas Phase

Gas-phase studies offer an elegant way of investigating small cluster interactions without complicating factors arising from the structure of the solvent.

The pioneering work in this area was mostly done by Kebarle,[80,81] using mass spectrometry to study the equilibrium constants K for reactions of the type

$$M^+ + A \rightleftarrows M^+A$$
$$\underline{M^+A + \quad \rightleftarrows M^+A_2}$$
$$M^+A_n + A \rightleftarrows M^+A_{n+1}$$

for n equal to 1 to about 6. From these K's the corresponding standard free-energy changes are calculated, $\Delta G^0 = -RT \ln K$, and from van't Hoff plots the corresponding ΔH^0 and ΔS^0 values can be derived.

A detailed description of various experimental approaches has been given by Kebarle.[80,81] Essentially, ions produced by a pulsed electron beam or by thermionic emission from a filament coated with a suitable electrolyte are injected in a constant-temperature gas chamber containing molecules at pressures in the range 4–10 torr, and the various association species are analyzed by mass spectrometry. One of the major advantages of this technique is that the solvations of cations and anions can be investigated separately.

The thermodynamic functions of ionic solvation in the gas phase can never be equal to the ones in solution, since the initial states for the solvent molecules are different. Still Kebarle has shown[80] that the difference between the solvation functions of cations and anions in the gas phase approaches that between the solvation functions of the same ions in solution as the number of molecules bound to the ions increases. For example, in Figure 5 the difference in ΔH^0 of cations and anions solvated by H_2O has the same sign and approaches the same value as the ΔH_h^θ of these ions in liquid water if the ionic scale of Randles is used (see Reference 13). Therefore, from a study of the interactions of ions with the first few solvent molecules very useful information can be obtained on the strength and specificity of the interactions leading to the electrostatic contribution to thermodynamic solvation functions in solution. Such studies indicate that in general cations are more solvated by aprotic solvents than are anions. On the other hand, the difference in the interactions between cations and anions of the same size with water is much smaller when the number of water molecules in the cluster increases, as seen in Figure 6.

3.4. Ionic Thermodynamic Functions in Solution

The solvations of cations and anions are necessarily different unless the solvent is a pure dipole, and the gas-phase studies generally confirm this difference. The manifestation of the specificity in ionic solvation will depend on the property being investigated, since different properties emphasize different types of interaction, as discussed earlier. It would therefore be advantageous to obtain the ionic contributions to the thermodynamic functions. Some methods that seem quite convincing have been developed to measure or evaluate ionic thermodynamic values. This is especially true with volumes[82] and free energies.[13] The various approaches to ionic values have been reviewed by Conway,[83] and reasonably good values now exist for free energies, enthalpies, entropies, and volumes in water, but the situation is quite different with nonaqueous solvents; for example, simple self-consistency tests of the ionic thermodynamic functions (e.g., $\Delta G_s^\theta = \Delta H_s^\theta - T\Delta H_s^\theta$, $(\partial \Delta G_s^\theta / \partial T)_P = -\Delta S_s^\theta$, and $(\partial \Delta H_s / \partial T)_P = \Delta C_{Ps}^\theta$) fail even if ionic data from the same author are used.

In the few cases where reliable ionic values exist they will be used to discuss ionic solvation. In most cases, however, the effect of cations and anions will be examined through a study of MBr or NaX, where M represents an alkali metal or tetraalkylammonium ions and X represents halide ions. This approach gives the correct relative dependence on the size of the cations or

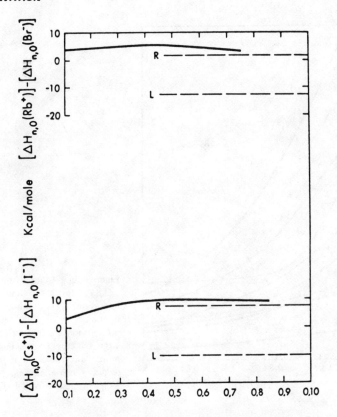

Figure 5. Comparison of the gas-phase enthalpy changes $\Delta H_{n,0}$ corresponding to the process $M(H_2O)_n = M + nH_2O$ with single-ion hydration enthalpies due to Latimer (L) and Randles (R). $\Delta H_{n,0}[Cs^+] - \Delta H_{n,0}[I^-]$ extrapolates for high n to the difference $\Delta H_h[Cs^+] - \Delta H_h[I^-]$ corresponding to Randles single ion hydration energies. A similar result is obtained for the Rb^+Br^- pair (reproduced with permission from Reference 80).

anions as well as the influence of the solvent. Ionic values would in addition give the difference between cations and anions, but a wrong choice of scale would lead to wrong conclusions, not only on the relative solvation of anions and cations but also on the relative effects of various solvents.

3.5. Free Energies and Thermochemical Properties

The attraction between an ion and a solvent molecule should be related to the dipole moment of the solvent molecule and consequently to the dielectric constant ε of the solvent. The dependence of ΔG_s^θ of two electrolytes on $1/\varepsilon$ is shown in Figure 7. As expected from the Born equation, Eq. (11), ΔG_s^θ is more negative for smaller ions and solvents of higher dielectric constant; however, the points are quite scattered and obviously other contributions

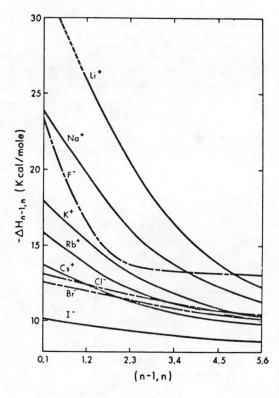

Figure 6. Comparison between $\Delta H_{n-1,n}$ for alkali and halide ion hydration. The initial hydration interactions of the isoelectronic positive ion are higher; however, for large n a crossover occurs, indicating more favorable interactions for negative ions at large n (reproduced with permission from Reference 80).

such as solvent multipoles, size, and hydrogen bonding must play an important role.

The ionic dependence of the three basic hydration functions ΔG_h^θ, ΔH_h^θ, and $T\Delta S_h^\theta$ is shown in Figure 8. For this purpose the Waddington[13,84] radii were used, but any other ionic scale yields a similar picture.

The signs and trends of these functions are consistent with the Born equation. The high affinity of the ions for the solvent causes a negative ΔG_h^θ, the potential energy is decreased when water solvates the ions, and the molecular order in the solvation complex increases. The entropy term $T\Delta S_h^\theta$ is small compared with ΔH_h^θ, indicating that the solvation, as seen by these functions, is primarily energy driven. As expected, the hydration functions are more negative for smaller ions. Anions appear to be more solvated than cations of the same size (assuming that the Waddington scale is correct), but this specificity with respect to the sign of the charge vanishes as the size of the ions increases. It is also noticeable that the large R_4N^+, whenever reasonably good ΔH_h^θ data exist, do not show any anomaly on this type of graph.

The main difficulty with these solvation functions is that the leading effect is the large contribution of the work of charging in solution and in the gas phase, and this obscures the finer details of solvation functions. In this respect

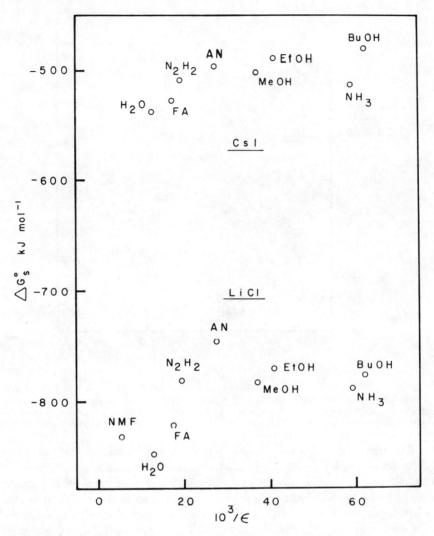

Figure 7. Standard free energies of solvation of LiCl and CsI in various solvents of dielectric constant ε (data from Padova[75]).

transfer functions are more sensitive. Typical examples of enthalpies of transfer are shown in Figure 9. Here propylene carbonate (PC) was used as a reference, as suggested by Krishnan and Friedman.[20] As typical aprotic and protic solvents dimethylformamide (DMF) and methanol (MeOH) were used. The ionic values are based on the assumption that Ph_4As^+ and Ph_4B^- contribute equally to the transfer functions.[20] These ionic values can be seriously in error and not too much weight should be placed on the relative magnitudes in different solvents.

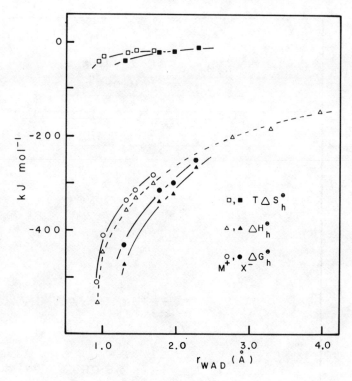

Figure 8. Thermodynamic hydration functions of ions at 298 K: $\bigcirc$, $\square$, $\triangle$, alkali metal and tetraalkylammonium cations; $\bullet$, $\blacksquare$, $\blacktriangle$, halide anions (data from Desnoyers and Jolicoeur[13] and Ahrland, in Reference 14, Vol. 5A, Chap. 1).

The dependence on the cationic size of the enthalpies of transfer from PC to another solvent are all qualitatively similar, which suggests that PC may not be as ideal a solvent as often suggested and may show specific solvation effects. Still, some useful information can be obtained from the trends. Using the enthalpy criteria, inorganic ions appear to be more solvated in nonaqueous solvents than in water, and all large R_4N^+ show relatively small solvation differences in nonaqueous solvents. Anions seem more hydrated than cations. There is no evidence for solvent structural effects for alkali halides in water, since the data in MeOH are similar to those in H_2O. On the other hand, $\Delta H^{\theta}_{R_4N^+}$ (PC $\rightarrow$ H_2O) are anomalously negative. This would be consistent with the statements that these hydrophobic ions promote the structure of water, an increase in structure is equivalent to the freezing of some water, and this change is exothermic.

The evidence for structural interactions is better seen with ΔH^{θ}_2 (H_2O $\rightarrow$ D_2O). Based on the Frank and Wen model expressed by Eq. (20), all alkali halides appear as structure breakers. Physically speaking, this solvent isotope effect can be visualized as follows: more hydrogen bonds must be broken in

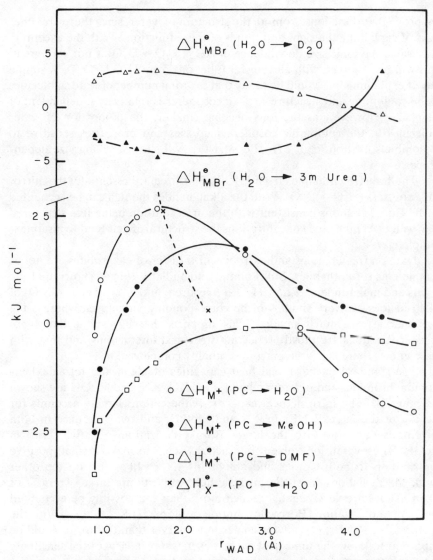

Figure 9. Enthalpies of transfer of 1 : 1 electrolytes at 298 K (data for the transfer from propylene carbonate to various solvents from C. V. Krishnan and H. L. Friedman,[20] with ionic values derived from the assumption that the functions are equal for Ph_4As^+ and Ph_4B^-; data for H_2O to D_2O from H. L. Friedman and C. V. Krishnan[68]; data for H_2O to 3 m urea from Y. Pointud and J. Juillard,[85] M. Y. Schrier, P. J. Turner, and E. E. Schrier,[86] and R. B. Cassel and W. Y. Wen[87]).

D_2O than in H_2O, since D_2O is more structured than H_2O at 25°C. With the hydrophobic salts R_4NBr the enthalpies of transfer become negative as the size of the cations increases. This is consistent with hydrophobic hydration concepts. However, the sign of $\Delta H_2^\theta (H_2O \to D_2O)$ of hydrophobic ions does not prove that these ions promote the structure of water, since there are other ways of explaining the signs and trends of these functions. With the exception of Li salts, the ionic size dependence of $\Delta H_2^\theta (H_2O \rightarrow D_2O)$ is not very great. This will be observed with all transfer functions from H_2O to D_2O. A simple way of explaining this would be to say that the total number of water molecules in the coulombic and structure-broken cospheres is relatively independent of ionic size. However, as the ions become smaller, the proportion of water molecules in the coulombic cosphere increases and properties sensitive to coulombic solvation, e.g., $\Delta H_2^\theta (PC \rightarrow H_2O)$, will show a strong size dependence.

The transfer enthalpies from H_2O to $3\,m$ urea are essentially the mirror image of $\Delta H_2^\theta (H_2O \rightarrow D_2O)$. With the alkali halides the sign can be accounted for by Eq. (21) and is consistent with the urea solution being less structured than water. At high urea concentrations the structural effects tend to disappear completely.[87]

Data on free energies and entropies of transfer of electrolytes are not as complete as for enthalpies. Still Pointud and Juillard[85] have shown that the trends and magnitude of $T\Delta S_2^\theta (H_2O \rightarrow 3\,m$ urea) and $T\Delta S_2^\theta (H_2O \rightarrow D_2O)$ of alkali halides are very similar to the corresponding enthalpies, whereas the free energies are much smaller. This supports the suggestions that these transfer functions are mostly reflecting structural interactions and that with these effects there is a large entropy–enthalpy compensation.

The standard partial molal heat capacities of alkali and tetraalkylammonium bromides and potassium halides in H_2O, AN, and MeOH are shown in Figure 10. The intrinsic heat capacity of the cations largely accounts for the $\bar{C}_P^\theta$ of the large R_4NBr in AN and MeOH, and the difference in both solvents is due mostly to the larger Born polarization (negative term) in MeOH. The overall ion–solvent interactions seem to make a small positive contribution to both cations and anions in AN.[32] In water, on the other hand, the deviations from the intrinsic heat capacity are large. The $\bar{C}_P^\theta$ of alkali halides are very negative, much more than can possibly be accounted for by the polarization [Born contribution $= -(55/r)/J\,K^{-1}\,mol^{-1}$]. With the hydrophobic R_4NBr, $\bar{C}_P^\theta$ now becomes much larger than $C_P(in)$. It would be important to know the ionic heat capacities in water in view of the sensitivity of this function, and some progress in this direction is being made.[89]

As discussed earlier, coulombic solvation is not very sensitive to temperature and therefore has little effect on $\bar{C}_{P2}^\theta$. On the other hand, effects that involve perturbation in the equilibrium between bonded and free water molecules (relaxation contribution) will have a large effect on $\bar{C}_{P2}^\theta$. The transfer functions $\Delta C_{PMBr}^\theta (H_2O \rightarrow D_2O)$ have the same sign, and $\Delta C_{PMBr}^\theta (H_2O \rightarrow 3\,m$ urea) the opposite sign, as $\bar{C}_{P2}^\theta - C_{P2}(in)$, and this adds supporting evidence to the importance of structural interactions on $\bar{C}_{P2}^\theta$.

As in the case of the enthalpies of transfer $\Delta H_2^\theta (H_2O \rightarrow D_2O)$ and $\Delta H_2^\theta (H_2O \rightarrow 3\,m$ urea), the heat capacities of transfer $\Delta C_P^\theta (H_2O \rightarrow D_2O)$ and

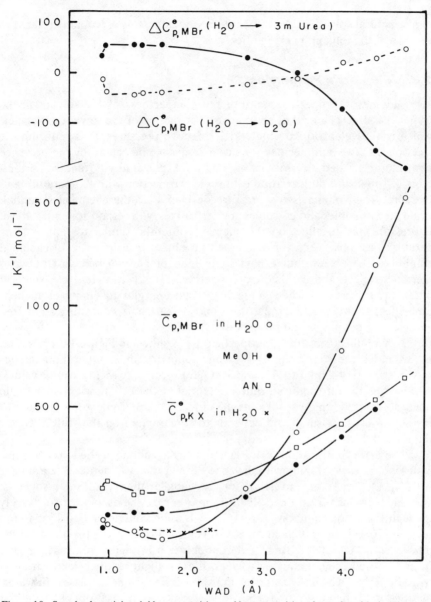

Figure 10. Standard partial molal heat capacities and heat capacities of transfer of 1:1 electrolytes at 298 K (data for H_2O, D_2O, and 3 m urea from Perron et al.[88] and references cited therein; data for AN from Zana et al[32]; and data for MeOH from Shin and Criss[72]).

$\bar{C}_{P2}^{\theta} - C_P(\text{in})$ in H_2O do not show a great size dependence with alkali halides. Again, this is consistent with Eqs. (20) and (21), if it is assumed that the number of water molecules in the total hydration cosphere is constant. As with enthalpies and entropies, the Li salts are out of line and appear to show less structural solvation. As will be discussed later, this is probably related to the size of the ion and packing effects.

3.6. Volumetric Properties

Volumetric properties measure the compactness of the solvent in the vicinity of ions. The field of an ion exerts attraction of the solvent molecules, which is equivalent to extremely high effective pressures, thus reducing the free space between molecules. The change in volume caused by the presence of a charge is called electrostriction, and the expansibility $\bar{E}_2^\theta$ and compressibility $\bar{K}_2^\theta$ measure the variation of this electrostriction with temperature and pressure. The volume is the pressure derivative of the chemical potential, while the enthalpies and entropies are derivatives with respect to temperature. If pressure has an effect on the chemical potential similar to that of temperature, as observed, for example, in phase transitions and chemical equilibria, then we would expect some kind of parallel behavior between thermochemical and volumetric properties. The sensitivity to structural interactions increases when going from free energies to entropies to heat capacities; so we would expect the same to be true when going from free energies to volumes to compressibilities.

As with thermochemical properties, $\bar{V}_2^\theta$ is shown in Figure 11 for a series of bromides for some typical solvents. Data for most other nonaqueous solvents are very similar to AN and MeOH. Reasonably accurate ionic values, obtained by the ultrasonic vibration potential method,[82] now exist for $\bar{V}_2^\theta$ in many solvents. With most solvents the difference between cations and anions of the same size is small and depends to some extent on the scale used for the ionic radii.

The main contribution to $\bar{V}_2^\theta$ is the intrinsic volume of the ions and since $V(\text{in})$ increases essentially with the cube of the radii, $\bar{V}_2^\theta$ increases in a similar way. The $\bar{V}_2^\theta$ of alkali and halide ions are smaller than $V(\text{in})$ in all solvents, and the difference from one solvent to another can be accounted for largely through the Born equation [Eq. (16)]. The ionic volume of Bu_4N^+ is found to be $270 \pm 2 \text{ cm}^3 \text{ mol}^{-1}$ in most solvents, including water,[32] and the difference in $\bar{V}_2^\theta$ of R_4NBr in different solvents in Figure 11 comes mostly from the anions. In nonaqueous solvents there is slight residual electrostriction with the R_4N^+, which decreases as the size of the cations increases. In water, on the other hand, there seems to be some compensation between the electrostriction coming from the charge (negative contribution), an increase in the structure of water around the hydrophobic groups (positive contribution), and economy of space resulting from the fitting of the solute in the natural cavities of liquid water (negative contribution). This latter effect seems to be characteristic of hydrophobic hydration;[77] nearly all hydrophobic nonelectrolytes exhibit $\bar{V}_2$ values in water that are smaller than the value of the molar volume of the pure liquid solute.

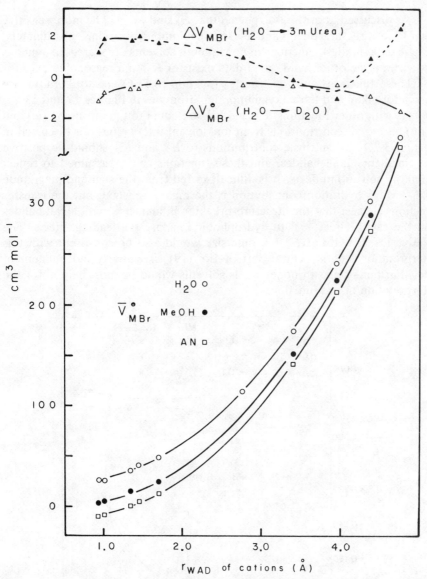

Figure 11. Standard partial molal volumes and volumes of transfer of alkali bromides at 298 K (data for H_2O, D_2O, and 3 m urea from Perron et al.[88] and references cited therein; data for MeOH from Shin and Criss[72]; and data for AN from Zana et al.[32]).

The structural influence on $\bar{V}_2^{\theta}$ is better seen from transfer functions, as shown in Figure 11. The observed functions are small, but their signs are consistent with the structure-breaking effect of alkali halides (negative contribution coming from the melting of icelike structures) and the hydrophobic effect with R_4N^+ (negative contribution coming from the filling of cavities).

As discussed above, it is expected that $\bar{E}_2^\theta$ and $\bar{K}_2^\theta$ will be more sensitive than $\bar{V}_2^\theta$ to structural effects and less so to coulombic solvation. Unfortunately, precise data on these functions in nonaqueous solvents are scarce. In aqueous systems, on the other hand, good data exist for $\bar{K}_2$, and reasonable ones for $\bar{E}_2^\theta$. These functions and the transfer function to D_2O and to $3\,m$ urea are shown for alkali and tetraalkylammonium bromides in Figures 12 and 13.

As with other functions, $\bar{E}_2^\theta$ and $\bar{K}_2^\theta$ are equal to the intrinsic $E(\text{in})$ and $K(\text{in})$ and to all contributions from ion-solvent interactions. As discussed in Section 3.2.1, the intrinsic contributions to $\bar{E}_2^\theta$ and $\bar{K}_2^\theta$ should be positive but small for alkali halides, and these functions can be assumed to reflect primarily solvation effects. It is difficult to tell from the sign and magnitude of $\bar{E}_2^\theta$ which hydration contribution is the most important, but the transfer functions suggest that the structure-breaking influence[76] with alkali halides and the cavity effect[88] with hydrophobic ions are the leading effects. The relative insensitivity of $\bar{E}_2^\theta$ on ionic size would also be consistent with this interpretation [the structural part of Eq. (19) is relatively independent of size]. Further studies in nonaqueous solvents would be most helpful for the interpretation of this function.

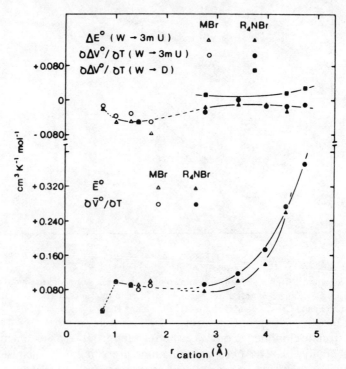

Figure 12. Standard partial molal expansibilities of $1:1$ electrolytes in water at 298 K and transfer functions from H_2O to D_2O and to $3\,m$ urea (reproduced with permission from Reference 88).

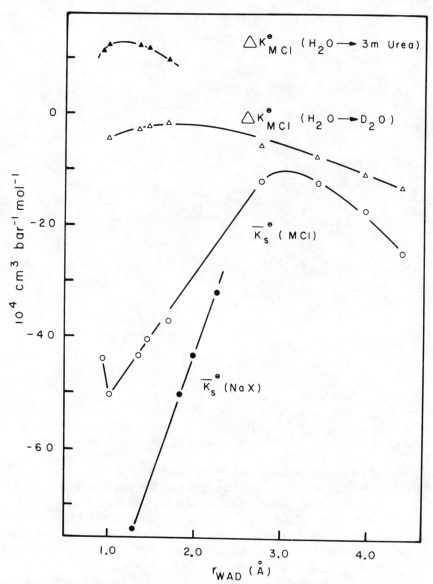

Figure 13. Standard partial molal isentropic compressibilities of 1:1 electrolytes in water at 298 K and transfer functions from H_2O to D_2O and to $3\,m$ urea solution (data for H_2O and D_2O from Mathieson and Conway[90] and data for $3\,m$ urea from Desrosiers *et al.*[76]).

The strong size dependence of $\bar{K}_2^{\theta}$ of alkali halides (Figure 13) suggests that electrostriction is important in determining the sign and magnitude of this function, since the water in the hydration cosphere is expected to be less compressible than bulk water. The signs of the transfer functions indicate that the structure-breaking effect also contributes to the negative value of $\bar{K}_2^{\theta}$. The

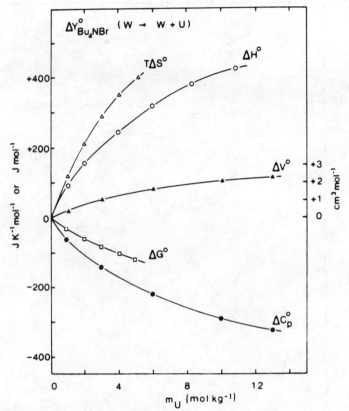

Figure 14. Thermodynamic functions of transfer of Bu_4NBr from H_2O to H_2O + urea mixtures at 298 K (reproduced with permission from Reference 88).

$\bar{K}_2^\theta$ of R_4NBr is also negative and becomes more so the more hydrophobic the cation. Studies under way on the compressibility of electrolytes in $AN^{(91)}$ indicate that the $\bar{K}_2^\theta$ of alkali halides are more negative in AN than in H_2O, but those of the larger R_4NBr become large and positive. The negative values of $\bar{K}_2^\theta$ of R_4NBr in H_2O are therefore related to hydrophobic hydration. Conway and Verrall[92] have interpreted the negative sign of $\bar{K}_2^\theta$ of these large cations through the lack of free space in the hydrophobic cosphere. Thus the compressibility of the hydration cosphere, e.g., of a clathrate type, is less than that of bulk water. Cabani *et al.*[93] have also discussed the negative contribution of hydrophobic hydration to $\bar{K}_2^\theta$ through a two-state model for water.

Neglecting the data for LiX, the size dependence of $\bar{K}_2^\theta$ of alkali halides is slightly larger for anions than for cations.[76] This would again be consistent with anions being slightly more hydrated than cations of the same size. With $\bar{E}_2^\theta$ the difference in trends between cations and anions is small if the lithium salts are ignored. This is also consistent with the suggestion that the magnitude of $\bar{E}_2^\theta$ is primarily due to the structure-breaking effect.

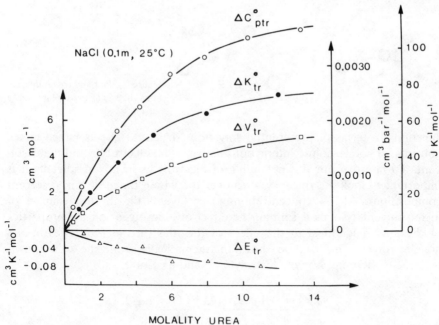

Figure 15. Thermodynamic functions of transfer of NaCl from H_2O to H_2O–urea mixtures at 298 K (reproduced with permission from Reference 76).

3.7. Thermodynamic Properties in Mixed Aqueous Solvents

A large number of studies have been made on the thermodynamic behavior of electrolytes in mixed solvents.[94,95] Unfortunately, it is difficult to compare these studies, since they have often been limited to a few mixed solvent compositions and again the aqueous-organic mixtures were the most studied systems. The transfer functions from water to these mixtures sometimes vary in a very regular fashion with the concentration of the cosolvent. Urea is a typical example of this kind of behavior. As seen in Figures 14 and 15, as the concentration of urea increases, the transfer functions vary in the direction expected were the main role of urea to decrease the structural hydration. However, with most systems the transfer functions exhibit maxima and minima; this is especially the case with protic cosolvents having hydrophobic alkyl groups, e.g., alcohols.

The interpretations of the transfer functions vary enormously from one author to another. Many consider the mixed solvent as a uniform medium characterized by its dielectric constant. This approach may be valid in discussing ion–ion interactions when the ions are far apart, as in the case of low-concentration conductance data, but it is hardly applicable to ion–solvent interactions, where short-range forces are involved. Others will interpret these

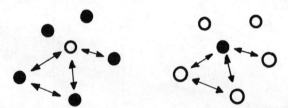

Figure 16. Pair and triplet interactions between two solutes in a solvent.

functions in terms of preferential solvation, structural effects in the mixed solvents, or solute–solute interactions between the electrolyte and the cosolvent. To improve our understanding of electrolytes in mixed solvents it is important to look at typical systems over the whole range of mixed solvent compositions. At low concentrations in cosolvents the thermodynamics of mixed solvent systems is simple; the effect of a cosolvent on the properties of the ions is identical to the effect of the ions on the properties of the nonelectrolyte. This can be shown mathematically from the well-known reciprocity theorem. For an electrolyte (2) and a cosolvent (3)

$$\frac{\partial \Delta Y_2^\theta (1 \to 1 + 3)}{\partial m_3} = \frac{\partial \Delta Y_3^\theta (1 \to 1 + 2)}{\partial m_2} \qquad (23)$$

The physical meaning of this is illustrated in Figure 16. If interactions are pairwise, the interactions between solutes 2 and 3 are identical if either solute 2 or solute 3 is in excess. At higher concentrations triplet interactions become important and this reciprocity does not hold any more. A good example of this is the system NaCl–*tert*-butanol–H_2O. As seen in Figure 17, the limiting slopes of the transfer function of NaCl from H_2O to H_2O + *tert*-butanol (TBA) is equal to that of TBA from H_2O to H_2O + NaCl. At higher concentrations deviations are observed. These limiting slopes are related to the salting-in and salting-out constants of nonelectrolytes by electrolytes, and the system electrolyte–alcohol–water has recently been thoroughly studied in this respect.[97]

At higher concentrations the triplet and higher interactions will reflect the influence of solute 2 on the interactions in the binary system $1 + 3$. For example, in Figure 17 NaCl enhances the TBA–TBA interactions in its solvation cosphere, especially beyond $3\ m$ in TBA. At still higher concentrations solute 2 becomes preferentially solvated by cosolvent 3, and 1 cannot be considered as the main solvent anymore.

3.8. Transport Properties

Transport properties are all related to the effective size of the moving particles: electric conductance to the movement in an electric field, diffusion

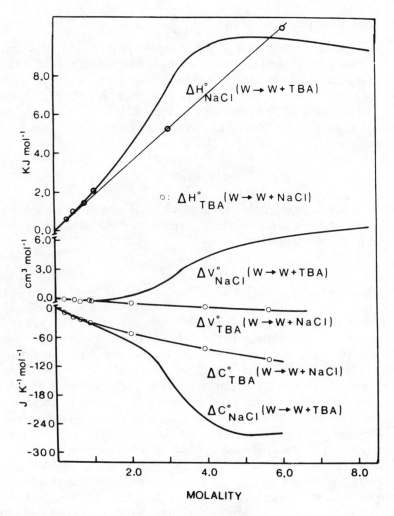

Figure 17. The standard thermodynamic transfer functions of the system NaCl–*tert*-butanol–H$_2$O (NaCl–TBA–W) at 298 K (reproduced with permission from Reference 96).

in a concentration gradient, thermal conductance in a temperature gradient, and viscosity in a liquid shear flow. In many cases ionic values are known or estimated and the transport functions can often be defined in such a way that solute–solvent and solute–solute interactions can be distinguished. The use of transport data for the determination of ion–solvent interactions has been well reviewed in the literature[10,98–100] and only a summary of the main observations and conclusions will be given here. This section will be restricted to ionic conductances and viscosities.

3.8.1. Equivalent Conductance

The equivalent conductance Λ can be measured very precisely[10] in dilute solutions. Information on ion–solvent interactions can be obtained both from Λ_0, the infinite dilution value, and from the association constant K_A derived from the concentration dependence of Λ. This latter procedure has been well described by Kay et al.[99] Essentially, K_A is related to the distance of closest approach of the ion pair through a suitable theory. For example, the Bjerrum equation gives

$$K_A = \int_a^q 4\pi N r^2 \exp\left(\frac{|z^+z^-|e^2}{r\varepsilon kT}\right) dr \tag{24}$$

where a is the distance of closest approach, N Avogadro's number, z^+ and z^- the ionic changes, r the interionic distance, k the Boltzmann's constant, q a critical distance defined by

$$q = \frac{|z^+z^-|e^2}{2\varepsilon kT} \tag{25}$$

and e the electronic charge. If this theory is valid, information on the size of the solvated ion can be obtained from a. Unfortunately, as Kay et al.[99] have shown, $\ln K_A$ is not a simple function of $1/\varepsilon$ as predicted by Bjerrum and other theories. If four types of solvents are distinguished (basic, strongly acidic, hydrogen-bonded, and neutral solvents) better relationships are observed; but still, no simple correlation exists for the dependence of $\ln K_A$ on ionic size. Depending on the type of solvents, K_A values increase, remain constant, or decrease with ionic size. Also, the a values are not always realistic, since they are sometimes found to be smaller than the sum of the crystallographic radii. The situation is not better with mixed solvents. Even with isodielectric mixtures K_A varies significantly with solvent composition.

The failure of equilibrium constants to reveal unambiguous information on ion–solvent interactions is due to the fact that these parameters are measuring solute–solute interactions, and there are more to these interactions than can be predicted from a simple association model.

There is better hope of obtaining solvation information from Λ_0. Ionic conductivities can be calculated from transport numbers: $\Lambda_0 = \lambda_0^+ + \lambda_0^-$. The simplest way of relating $\lambda_0^\pm$ to the effective ionic size in solution $r^\pm$ is through the Stokes' equation:

$$\lambda_0^\pm = \frac{|z|e\,\mathrm{F}}{A\pi\eta r^\pm} \tag{26}$$

where e is the electronic charge, F the Faraday, η the viscosity of the solvent, and A is equal to 4 or 6, depending on the assumptions made in the derivation of this simple hydrodynamic theory; $A = 6$ for a perfect stick between the sphere and the solvent and $A = 4$ if there is a perfect slip. If the ions are not solvated, $\lambda_0^{\pm}\eta$ should be a constant (Walden product) for various solvents at various temperatures. This is approximately so for Bu_4N^+ in nonaqueous solvents and $\lambda_0^{\pm}\eta$ has a value of 0.21 ± 0.01. In water, on the other hand, $\lambda_0^{\pm}\eta = 0.17$, which is consistent with the hydrophobic hydration, which increases the effective size of this ion. The Walden product also increases with temperature in water, as expected from a decrease in hydrophobic hydration.

The ionic size dependence of $\lambda_0^{\pm}\eta$ is shown in Figure 18. If Stokes' law with $A = 4$ is used, then the R_4N^+ are closer to the theoretical slope (see Figure 20). Qualitatively, it can be said that the structure-breaking ions in water show a positive deviation and the strongly hydrated ions (Li^+ and Na^+) show negative deviations. Also, most ions in acetonitrile appear to be strongly solvated, more so than in water.

However, Stokes' law is really valid only for nonsolvated spheres. If ion–solvent interactions exist, then orientation of the solvent dipoles will result in a dielectric friction force related to the relaxation time of the solvent dipoles. This effect is taken into account in the Zwanzig equation[10]

$$\eta\lambda_0^{\pm} = \frac{|z|e\,F}{A\pi r + Z^2 B/r^3} \tag{27}$$

where F is the Faraday and B has the values $\frac{3}{8}$ or $\frac{3}{4}$ for perfect sticking or perfect slipping. As seen in Figure 19, this equation overcorrects the size

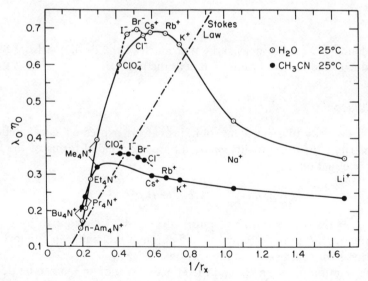

Figure 18. The limiting ionic conductance-solvent viscosity product plotted against the reciprocal ionic radii for several ions in water and acetonitrile at 298 K (reproduced with permission from Reference 100).

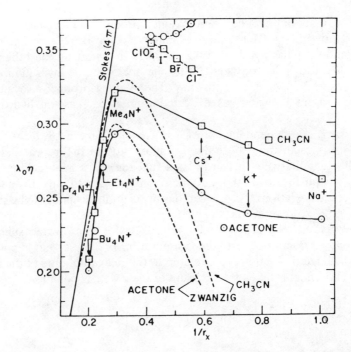

Figure 19. The limiting conductance-viscosity product for the alkali metal, halide, and tetraalkylammonium ions in acetonitrile and acetone as a function of the reciprocal ionic radii. The broken curves show the Zwanzig theory for the slip case (reproduced with permission from Reference 99).

dependence of $\lambda_0^{\pm}\eta$ but does show that much of the deviation from Stokes' law is due to the solvent orientation near the ions.

3.8.2. Viscosity

The viscosity of electrolyte solutions can be expressed as a function of the molarity c or the molality m by different extended forms of the Jones and Dole equation, e.g.,

$$\eta/\eta_0 = 1 + A_\eta c^{1/2} + B_\eta c + \cdots \qquad (28)$$

where η_0 is the viscosity of the pure solvent, A_η is related to long-range coulombic forces and can be calculated through Falkenhagen's theory, and B_η accounts for solute–solvent interactions.[98]

The relation between B_η and the effective size of an ion can be derived from Einstein's relation,

$$\eta/\eta_0 = 1 + 2.5\phi \qquad (29)$$

where ϕ, the volume fraction of solute molecules, can be expressed in terms of the hydrodynamic volume (in $cm^3\,mol^{-1}$) of the solute by

$$\phi = Vc/1000 \tag{30}$$

which leads to

$$\eta/\eta_0 = 1 + 0.0025\,Vc \tag{31}$$

The coefficient B_η is therefore given by $0.0025\,V$. In nonaqueous solvents the B_η of large ions like Bu_4N^+ appears to be relatively constant[102] but is again larger in water, as expected. With alkali halides in nonaqueous solvents V is significantly larger than the intrinsic volume of the ions, as expected from strong ion–solvent interactions.

3.8.3. Structural Effects

The structure-breaking and hydrophobic effects of ions in aqueous solutions have a significant effect on transport properties. As in the case of thermodynamic functions, these are best illustrated through solvent isotope effects and temperature dependences, as shown in Figure 20. The function $B_\eta - 0.0025\,\bar{V}_2^0$ is also shown: Desnoyers and Perron[103] have used this function as a measure of solute–solvent interactions in water, assuming that for an ideal solution obeying Einstein's equation the hydrodynamic volume should be given by the partial molal volume. Ionic values for B_η were taken assuming that $B_\eta(K^+) = B_\eta(Cl^-)$.[98]

The functions in Figure 20 are all plotted in such a way that the sign of the deviation illustrates the same trend in the structural effects. Hydrophobic hydration increases the size of ions and decreases with temperature; there is therefore a positive deviation from Einstein's equation, B_η decreases with temperature, and the Walden product is slightly smaller in D_2O (larger radius) and increases with temperature. All structure-breaking ions appear as negative deviations in Figure 20 and strong coulombic hydration (Li^+, F^-) causes a positive deviation. As with thermodynamic data, the transport data suggest that the coulombic hydration of anions is larger than that of cations of the same size. There is therefore a qualitative agreement between the solvation effects measured through thermodynamic and transport data.

3.9. Concluding Remarks

In this section trends in the thermodynamic properties of alkali and tetraalkylammonium halides were examined in aqueous and typical nonaqueous solvents. With a single property in a particular solvent or a mixed solvent there can be many models, even conflicting ones, that can account for the observed trends. If many properties and solvent systems are

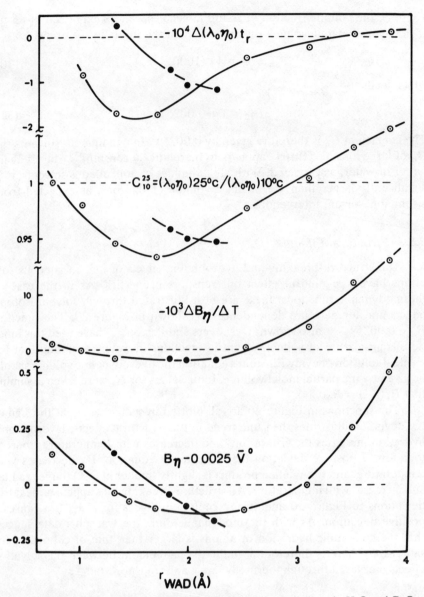

Figure 20. Transport properties of symmetrical monovalent electrolytes in H_2O and D_2O at 298 K and the temperature dependence of some properties in H_2O (reproduced with permission from Reference 12).

investigated, the number of suitable models decreases rapidly. All the present trends were shown to be qualitatively consistent with the well-known Frank and Wen model for the solvation of ions and the following generalities can be advanced.

Coulombic Solvation. This type of solvation, present in all solvents, increases in magnitude with decreasing ion size and increasing ion charge. The sign of its contribution to the solvation functions is given in Table 5 and is generally that predicted by the Born equation with two possible exceptions. First, Born polarization predicts a negative contribution to $\bar{C}^{\theta}_{P2}$, while experimentally there seems to be an additional positive contribution with smaller ions, at least in some nonaqueous solvents. Second, the coulombic contribution to $\bar{E}^{\theta}_{2}$ cannot be evaluated unambiguously owing to a lack of good experimental data. Further work on $\bar{C}^{\theta}_{P2}$, $\bar{E}^{\theta}_{2}$, and $\bar{K}^{\theta}_{2}$ in nonaqueous systems and on higher-valent electrolytes would be very useful for the characterization of this type of solvation.

Structural Solvation. With aqueous systems, and possibly other highly structured solvents, coulombic solvation is not sufficient to account for all trends. With hydrophilic ions, univalent and multivalent, additional interactions in the direction of a decrease in the icelike structure of water must be added. With large hydrophobic ions another type of hydration, now in the direction of increasing icelikeness of the solvent, becomes important. The predicted influence of these interactions on the solvation properties are given in Table 5. Experimentally, the structure-breaking effect is relatively independent of the size of the hydrophilic ions, while hydrophobic hydration generally increases with hydrophobic character. The contribution of structural interactions increases in relative importance with functions that are higher derivatives with respect to temperature and pressure; it decreases rapidly with increasing temperature and the addition of certain cosolvents such as urea; on the other hand, it increases when going from H_2O to D_2O, at least at low temperatures. Of course, in this qualitative analysis of thermodynamic and transport data

Table 5
Signs of Solvation Contributions to Thermodynamic and Transport Properties[a]

	Coulombic	Structure breaking	Hydrophobic hydration
ΔG	−		
ΔH	−	+	
ΔS	−	+	−
ΔC_P	(+)	−	+
ΔV	−	−	(−)
ΔK	−	−	−
ΔE	(+)	+	+
Λ_0	−	+	−
B_η	+	−	+

[a] Experimentally, all the structural interactions are enhanced in D_2O and decreased in urea–water mixtures, at least at low temperatures.

we do not need to define the actual structure of the icelike structure that is broken down by hydrophilic ions or that around hydrophobic solutes.

Ionic Specificity. From thermodynamic data it is difficult to get unambiguous information on the relative solvation of cations and anions. Absolute ionic values are required for this purpose, and in the few cases where reliable ionic data exist, the specificity does not seem very large. Some idea of the relative importance of ionic solvation comes from the size dependence of the functions. In water these data would suggest that the coulombic hydration of anions is slightly more important than for cations of the same size; on the other hand, the structural effects do not seem to show much specificity on the sign of the charge of the ions. With many properties Li^+ seems to be out of line with the other alkali metal ions and often appears less solvated, although the solvent molecules can get closer to the ion. The usual explanation is that on account of geometrical packing requirements the coordination number of Li^+ is smaller than that of the other alkali metal ions. Consequently, fewer solvent molecules are affected even though the interactions with each solvent molecule in the first coordination shell can be quite strong.

Solvation Numbers. The present comparison of all the thermodynamic and transport properties of electrolytes in various solvents illustrates very well that different properties measure different extents of the ion–solvent interactions. The solvation numbers being defined as the effective numbers of solvent molecules that have undergone certain changes in property can be useful in comparing various electrolytes with the same property but have no absolute significance. Tables of such parameters are therefore useless unless they are given with reference to a property and to a definition of solvation for that property.

4. Spectroscopic and Diffraction Methods

Following the examination of the thermodynamic and transport properties of electrolytes, a number of fundamental questions remain concerning the molecular aspects of ionic solvation. These questions pertain, for instance, to the true radii of ions in solution, the coordination number of the ions, the molecular orientation in the solvation cosphere, the range of influence of the electric field of the ions, the lifetime of the ion–solvent interactions, and the nature of the hydrophobic hydration shell. In principle, the combined use of appropriate spectroscopic and diffraction methods would provide answers to all of these unknown aspects and thus clarify solvation concepts at the molecular level.

The requirement for such methods to fulfill our expectation may be understood through the following simple approach. Many physical quantities measured with electrolyte solutions (e.g., thermodynamic data) must be viewed as a result of integration over three variables: time, sample volume, and species (solutes, solvent). Consequently, one cannot obtain direct information on the molecular nature of the solvated complex. The advantage of spectroscopic and diffraction methods is that often the technique allows the observation of a single species over a period that is short compared to the lifetime of the ion–solvent complex. This reduces the multiplicity of the integrals from three to two and in some favorable cases from two to one.

It must be recognized, of course, that as long as the measured property averages over more than one variable, the extraction of molecular information from the data will still require a chemical model, and the end result may be strongly dependent on the choice of the model for the interpretation. This is, however, not too serious a limitation because different spectroscopic methods average over different variables, so a proper combination of techniques can still yield much pertinent information. A more serious limitation to the applicability of these methods is the high concentrations of electrolyte that quite often have to be used. Thus low-solubility, solute–solute interactions, and ion pairing may complicate studies of solvation in the solvents of greatest practical interest.

In this section we attempt to illustrate each type of data and information that can be obtained from a variety of methods falling in the category of spectroscopic techniques. In view of the complexity of interpretation of each of these highly specialized methods, as well as of the limited space available here, we have chosen to emphasize the two most widely applicable methods: vibrational spectroscopy and NMR spectroscopy. Typical results obtained from these techniques will be exposed in some detail, whereas for other methods only qualitative descriptions and key references will be provided.

The strategies followed in NMR and vibrational spectroscopy have much in common, although these techniques probe vastly different microscopic properties of ion–solvent complexes. In both methods the specific interactions between an ion and a solvent molecule can be studied in ternary systems where the electrolyte and solvent of interest are diluted in a weakly interacting solvent such as carbon tetrachloride or methylene chloride. This type of study provides fundamental data that should be somewhat related to gas-phase solvation and that are essential for the understanding of overall solvation effects. The solvation properties of individual ions are further investigated by choosing electrolytes for which the solvation of one of the ions may be considered negligible; for example, with R_4NX salts the solvation of tetraalkylammonium ions in nonaqueous solvents will be much weaker than for the halide ions, so, to some extent, solvation of the latter may be examined independently. Finally, the overall solvation effects of a particular ion or electrolyte may be compared to the effect of temperature or pressure upon the spectrum of the pure solvent. This approach is particularly useful in studies in hydrogen-bonded solvents, where ionic solvation requires the breaking or making of a significant number of hydrogen bonds in the solvent. At least qualitatively, the latter type of investigation can serve to characterize the dominant structural influence of different categories of ions.

With thermodynamic and transport properties systematic trends were examined for standard functions of alkali and tetraalkylammonium halides in typical solvents. Such an approach is more difficult with spectroscopic data, since in many cases systematic studies have not been carried out owing to experimental problems or to the intrinsic nature of the measurements. Consequently, with many types of spectroscopic data quantitative comparisons are limited to a discussion of the relative behaviors observed in rather few systems. Moreover, as found in several cases below, some of the results obtained with different techniques, or from different observations by means of the same technique, differ or conflict with other information derived from thermodynamic or transport data. Hence some details of the methods and interpretation must also be included in the discussion.

4.1. Infrared and Raman Techniques

Infrared and Raman methods are among the most powerful tools available to study the molecular aspects of solvation phenomena. The measurement of vibrational transition energies for solvent molecules or ions provides direct information on the interactions in which the observed species are involved. Hence examination of the many possible normal mode frequencies of the solvent, ions (if polyatomic), and ion–solvent complexes allows a highly selective approach to characterize ionic solvation. This selectivity may be further enhanced through isotopic substitution, as, for example, H/D, $^6Li/^7Li$.

The infrared spectroscopic region of interest for the study of electrolyte solutions extends between approximately 100 and 14,000 cm^{-1}, corresponding to wavelengths of 100 and ~0.7 μm; the frequencies corresponding to these extrema are such that the periods of the vibrational modes range approximately between 3×10^{-14} and 3×10^{-12} sec. Consequently, solute–solvent interactions that persist for times longer than the period of the mode investigated should lead to distinct bands. In relation to the above discussion, it is clear that the time scale of the measurement is short compared to the expected residence time of a solvent molecule in a solvation complex. The method is also specific for the species, but in many cases it will only integrate over a sample volume, i.e., it gives a statistical average over allowed configurations.

The Raman technique measures transition energies that correspond to the far and fundamental infrared through the observation of emission lines shifted from the Rayleigh wavelength. For the present purpose we will thus examine IR and Raman results concurrently, though it should be recalled that these methods are complementary rather than interchangeable owing to their respective selection rules (IR, $d\mu/dr \neq 0$, Raman, $d\alpha/dr \neq 0$, μ and α being respectively the bond dipole moment and polarizability). Since in electrolyte solutions one is usually dealing with highly polar molecules, Raman intensities will tend to be low, but, except for specific symmetry restrictions, both IR and Raman spectra can generally be observed.

In the foregoing sections we will examine a number of selected results of solvation studies grouped according to the types of normal modes investigated; the latter may be subdivided as shown schematically in Figure 21. Although the measurable spectral parameters comprise maximum absorption frequencies, linewidths, and molar absorptivity, we will refer mostly to frequency and intensity data that are the most complete. Topic reviews dealing with some of the subjects discussed here can be found in the literature for aqueous electrolytes[13,104-108] and for other solvent systems.[109,110] In particular, the review by Irish[110] gives a useful compilation of references to band assignment for a variety of common solvents.

4.1.1. Solvent Vibrational Modes

4.1.1.1. Solutions in Water and Other Protic Solvents

The changes in the vibrational spectrum of liquid water upon addition of an electrolyte have been studied both by IR and Raman techniques. For reasons inherent to the water spectrum discussed earlier, i.e., proximity of ν_1 and ν_3 and large bandwidths, the study of H_2O or D_2O is not particularly rewarding. Trends in solvation effects can be deduced from the IR spectra of H_2O in, e.g., alkali halide solutions, but the situation is greatly simplified if the O—H or O—D stretching vibrations of HOD (in either D_2O or H_2O, as appropriate) are examined instead.

In Figure 22 we reproduce the infrared O—H and O—D absorption bands in water and 9.1 mol% NaCl solutions[111] recorded at 283 and 353 K. Except for specific cases discussed below, the spectral changes are usually weak and without distinctive new absorption caused by ion–water interactions. From spectra similar to those in Figure 22, Hartman[112] reported the maximum absorption frequency ν_{O-H} (HDO) as a function of concentration

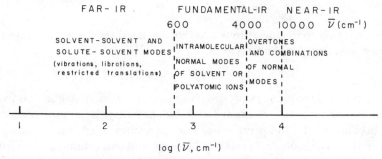

Figure 21. Schematic representation of IR spectral regions with vibrational modes of interest for studies of electrolyte solutions.

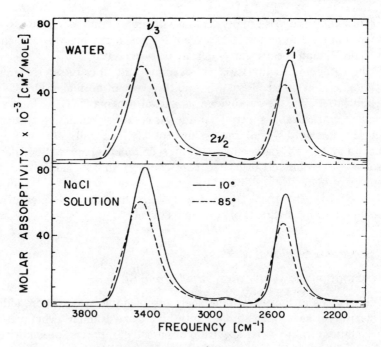

Figure 22. IR spectra of HOD (2000–4000 cm^{-1}) at 283 and 385 K in water and aqueous 9.1 mol% NaCl solutions (reproduced with permission from Reference 111).

for various electrolytes; his results are summarized in Figure 23. Judging from the frequency shifts, it would appear that the average hydrogen-bonding interactions in aqueous solutions of many 1 : 1 electrolytes are weaker than in pure water.

From these and similar other studies, it is generally found that cations have little influence on the O—H stretching frequency (usually weak high-frequency shifts), while in the presence of anions the O—H stretching frequency is shifted markedly and follows the sequence[113]:

$$F^- < H_2O < NO_3^- < Cl^- < Br^- < I^- < ClO_4^-$$

As a basis for comparison, we may add that a related sequence of ν_{O-H} values has been observed for electrolytes in methanol[114]:

$$H_2O < F^- < Cl^- < SO_4^{2-} < NO_3^- < ClO_4^-$$

The positive frequency shift observed in water with all halide ions except F$^-$ should not be taken as immediate evidence that the anion solvation is weak because of the following other considerations. First, the coordination of the anion must involve several O—H groups, and thus a number of hydrogen bonds in water (or methanol) must be ruptured to accomodate the primary solvation requirements. Second, if the Frank and Wen model of ionic solvation

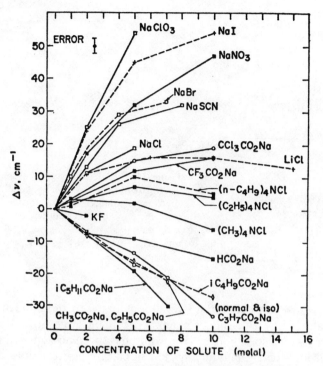

Figure 23. IR frequency shift of ν_{O-H} (HDO) in D_2O salt solutions versus the molal concentration of various electrolytes (reproduced with permission from Reference 112).

is correct (Figure 4), then the structure-breaking effect will result in more broken hydrogen bonds contributing to a high-frequency shift.

To visualize various changes that may occur in the O—H (or N—H) stretching spectrum owing to ionic solvation, Figure 24 illustrates a schematic spectrum of a hydrogen-bonded solvent such as an alcohol and the spectrum of an electrolyte solution in the same solvent. The hydrogen-bonding groups are assumed to be either "free" or "bonded," so the solvent is a two-state system as described earlier for water and hypothetically distinct bands are resolved for each species (spectrum A). Upon addition of a sufficient quantity of a simple 1:1 electrolyte MX, various spectral changes may be expected, as shown in spectrum B. The direct anion–solvent interaction (O—H$\cdots$X$^-$) should lead to a new band at frequencies lower than the absorption of the bulk solvent. The coulombic solvation of the cation should increase the O—H stretching frequencies of the coordinating solvent molecules by virtue of the electrostatic polarization of the O—H bond. Following the Frank and Wen solvation model (Figure 4), we further expect that the intensity of the free O—H absorption will increase owing to the disruption of solvent–solvent hydrogen bonds. As a consequence of solvation and structure-breaking effects, the intensity of the absorption due to bonded O—H groups should decrease,

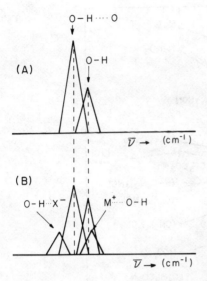

Figure 24. Schematic IR hydration spectrum of an electrolyte MX: spectrum A, hypothetical two-state spectrum of water (O—H_{bonded} and O—H_{free}); spectrum B, shows hypothetical new bands for anion- and cation-polarized O—H groups, reduction of the intensity of the O—H_{bonded} band and increase in intensity of the O—H_{free} due to the structure-breaking effect.

subject, of course, to equilibrium conditions between the bonded and free states. Finally, if the electrolytes examined contain large organic ions, e.g., R_4N^+, the expected solvation spectra would have weaker contributions from electrostatic ion–solvent interactions, but significant perturbations of the solvent hydrogen-bonding equilibrium might still be anticipated.

In view of the various contributions to solvation spectra that in most experimental cases are not resolved as distinct bands, the interpretation of solvation spectra is in no way straightforward. Consequently, we should first examine cases where one (or the fewest possible) solvation effect dominates the spectra. In aqueous solutions of several salts of very strong acids the IR and Raman O—H stretching bands exhibit a new high-frequency component. This effect is illustrated in Figure 25 for ν_{O-D} in $NaBF_4$ solutions of increasing salt concentrations.[115] The same type of band splitting has been reported with ClO_4^-, CCl_3COO^-, PF_6^-, and SbF_6^-,[116–118] though it was not observed in the presence of many other polyatomic ions.[116] These spectral changes have generally been interpreted as due to a marked structure-breaking influence of these large anions in water. The band-splitting phenomenon, however, is not exclusive to aqueous solutions; investigations of salts of ClO_4^- and BF_4^- anions in methanol have shown a similar ν_{O-H} band[115,118] behavior. The various spectroscopic data quoted above indicate that the solvation of the large monovalent anions in hydrogen-bonded solvents will generally occur with an overall reduction of hydrogen-bonding energy. We draw attention to the results in Figure 23 for another important category of electrolytes, the tetraalkylammonium halides. Their influence on ν_{O-H} is weak and the shift is to higher frequencies, as was seen with simple electrolytes such as NaCl. As discussed before with respect to other properties, the large R_4N^+ ions are

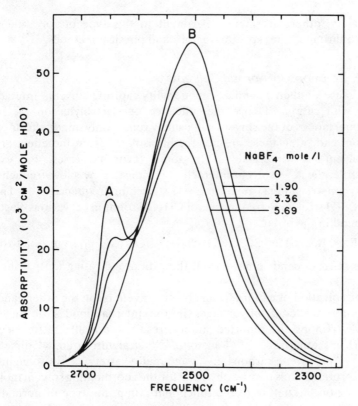

Figure 25. IR O —D stretching vibration of HDO in aqueous NaBF$_4$ solutions at 301 K (reproduced with permission from Reference 115).

expected to have a low surface charge density and, to some extent, exhibit the hydration properties of hydrocarbons (e.g., the structure-promoting influence). The net positive $\Delta\nu_{O-H}$ gives no evidence of such effects, except perhaps through comparisons of alkali and tetraalkylammonium halides. To a first approximation, a comparison of, e.g., R$_4$NCl and NaCl eliminates the anion hydration spectra and indicates that, relative to Na$^+$, the R$_4$N$^+$ ion shifts ν_{O-H} in the direction corresponding to increased hydrogen bonding.

Further information on the hydration of this important class of compounds has been obtained from Raman investigations.[116] The profile of the Raman O—H stretching band in Bu$_4$NCl solutions is quite different from that observed in other 1:1 electrolyte solutions, having a few component near the O—H stretching frequency of HOD in D$_2$O ice. The appearance of this new component has been attributed to an important water structure-enhancing influence of the Bu$_4$N$^+$ ion.

The few results quoted here illustrate the elusive character of the solvent structural changes associated with the hydration of apolar residues.

Fortunately, other information pertinent to this type of solvation can be obtained through other spectral regions and physical methods.

4.1.1.2. Solutions in Aprotic Solvents

Because of their chemical functionality, aprotic solvents interact most readily with cations. Hence the influence of electrolytes on the internal vibrational modes of the solvent will usually reflect only minor effects of anion solvation, and often these are not readily isolated from the effect of triplet $M^+ \cdots$solvent$\cdots X^-$ interactions. In some favorable cases, for example, acetonitrile, the $X^- \cdots$solvent interaction is indicated by a low-frequency shift in the symmetric and asymmetric C—H stretching frequencies.[119] The symmetric C—H stretching frequencies of CH_3CN solutions of various electrolytes were found in the order

$$Cl^- < F^- < Br^- < I^- < SCN^- \sim BPh_4^- < ClO_4^- \ (\Delta\nu_{C-H} \text{ from } -25 \text{ to } 0 \text{ cm}^{-1})$$

and the shifts depend somewhat on the cation, indicating ion pairing of the type $X^- \cdots H_3C—C\equiv N \cdots M^+$.

The solvation of cations by an aprotic solvent is, on the other hand, quite readily investigated from changes in the internal modes of the solvent molecule. Among the reported studies, results are available for electrolyte solutions in acetonitrile,[119,120] acetone,[121] tetrahydrofuran,[122] dimethylsulfoxide,[123] propylene carbonate,[124] and methyl esters.[125] In some instances new absorption bands may be observed for the coordinating solvent molecules, depending on the nature of the functional group, the type of normal mode, and the energy of the $M^+ \cdots$solvent interaction.

IR studies of cation solvation in acetonitrile have provided evidence of strong $M^+ \cdots N\equiv C—CH_3$ interactions from changes of both the $C\equiv N$ and C—C stretching frequencies.[119] For alkali metal ions these *increase* according to the sequence

$$Cs^+ < Rb^+ < K^+ < Na^+ < Li^+ \ (\text{e.g., } \Delta\nu_{C-N} \text{ from } 9 \text{ to } 29 \text{ cm}^{-1})$$

The observed shifts are not significantly dependent on the anion. Larger effects are found with Mg^{2+} and Co^{2+} (35–40 cm^{-1}), but Bu_4N^+ induces no noticeable change in ν_{C-N}.[126]

The relative magnitude of frequency shifts observed in studies of series of salts having a common anion appears well correlated with other types of measurements, namely, solvation enthalpies. However, the interpretation of the frequency shifts in terms of ion–molecule interaction energies is in no way straightforward. Even in the simple case of the stretching mode of a functional group that is shifted primarily as a result of direct ion–dipole interaction, the shift also depends on the redistribution of electron density within the solvent molecules (polarizability). This latter effect may be quite important, as, for example, in acetonitrile, where the $C\equiv N$ stretching frequency increases upon interaction with cations.

4.1.1.3. Hydrogen Bonding between Anions and Proton Donors

To elucidate the type of interactions involved in the coulombic solvation of anions in protic solvents spectroscopic investigations have been carried out with ternary systems consisting of an electrolyte and a proton donor, both diluted in an "inert" solvent. Typically, the stretching absorption frequency of a proton donor (e.g., R—O—H) diluted in an inert solvent appears as a single narrow band. Upon addition of a proton acceptor (e.g., pyridine, Br^-) a distinct new band or shoulder is observed at lower frequencies; the latter is usually broad and very intense and is readily assigned to the donor species involved in hydrogen bonding.

Detailed investigations of direct anion–alcohol interactions have been carried out.[109,127,128] With MeOH in dichloromethane, the ν_{O-H} absorption in the presence of R_4NX salts is shifted $\sim$200–300 cm^{-1}, increasing in the halide series in the order $I^- < Br^- < Cl^- \sim F^-$.[128] Complementary data for perchlorate salts and the dimerization of methanol lead to the following overall sequence of frequency shifts:[109]

$$ClO_4^- < MeOH < I^- < Br^- < Cl^- < F^-$$

These results further confirm the weak $O—H\cdots ClO_4^-$ interactions as inferred from earlier studies discussed above.

The same approach may also be followed to investigate anion–water interactions. This requires a more polar diluting solvent because of the low water solubility in CCl_4 or $C_2H_2Cl_2$, but comparisons with MeOH results can still be made if the latter is studied under the same conditions. Such data have been reported by Kuntz and Cheng[129] for HDO and MeOH in acetonitrile, propylene carbonate, tetramethylurea (TMU) and *N,N*-dimethyl-formamide containing various added electrolytes. The ν_{O-H} spectra of ternary systems may exhibit up to four different bands, which, in order of decreasing frequency, have been assigned respectively to $O—H\cdots S$, $M^+\cdots O—H\cdots S$, $O—H\cdots X^-$, and $M^+\cdots O—H\cdots X^-$, where S is a solvent molecule. Although the anion-sensitive band sometimes depends on the cation and on the salt and water concentrations, the observed frequencies yield the familiar sequence $ClO_4^- < I^- < Br^- < Cl^-$. Furthermore, compared to ν_{O-H} in liquid water (or liquid MeOH), the $\nu_{O-H\cdots X^-}$ band occurs at *lower* frequencies for the Cl^- and Br^- solvates, whereas the opposite is found for I^- and ClO_4^-. This again suggests that the Cl^- and Br^- ions have hydrogen-bonding interactions that are stronger than the hydrogen-bonding interactions occurring in pure water (or methanol). Clearly, these results have an important bearing on the interpretation of anion solvation in water and other protic solvents.

4.1.1.4. Overtones and Combination Bands

For protic solvents (e.g., water, alcohols, and amines) overtones of the O—H or N—H stretching vibrations occur in the near-infrared region, which offers many experimental advantages summarized recently by Luck.[107]

Moreover, in hydrogen-bonded systems the overtone absorption intensity of the functional groups (OH, NH) is greater for the weakly interacting species (unbonded), compared to the hydrogen-bonded component. In the case of water, for example, this enhanced sensitivity towards the free O—H groups results in a more pronounced temperature dependence of the near-infrared absorption spectrum, compared to changes observed in the fundamental region. It can thus be expected that the near-IR spectral features will be more sensitive towards hydration effects, and indeed, much lower electrolyte concentrations are required to induce measurable changes.

On the other hand, the lack of quantitative understanding of additivity rules for the intensities of overtones and combinations suggests greater caution in relating the spectral changes to the ion–solvent interaction energies or other solvation parameters. Consequently, most of the investigations rely on a comparison of the influence of temperature and solutes on the near-IR bands. The bulk of the results presently available in this spectral region concerns aqueous electrolyte solutions, though, from the work of Luck and Ditter on various alcohols,[130,131] the technique would be applicable to solutions in these solvents as well.

Reports of early investigations described the frequency shifts of the overtones and combination bands upon addition of electrolytes. For example, the data of Williams and Millet[132] on the 1.45- $(\nu_1 + \nu_3)$ and 1.96-μm $(\nu_{1,3} + \nu_2)$ bands of H_2O showed that various alkali and alkaline earth halides shifted the absorption maximum in the same direction as an increase in temperature. In a more recent investigation Bunzl[133] applied this strategy to study the hydration of the tetraalkylammonium bromides, using the 0.97-μm $(\nu_1 + 2\nu_3)$ band of H_2O. In all solutions of homologous R_4NBr (R=Me→Bu) the salt-induced shift was of the same sign as that found when decreasing the temperature of water, and the influence of the various salts was characterized by a "structural temperature" increment: $\Delta T_{str} = T_{str} - T_{solution}$. As remarked by Bunzl,[133] however, the observed shift originates from two distinct contributions: a charge-induced shift due to direct ion–water interactions (particularly the Br^- ion) and a structural shift due to changes in the hydrogen-bonding equilibrium in water. The variation of ΔT_{str} as a function of temperature was believed to depend mostly on the structural shifts. Since for Me_4NBr ΔT_{str} increased with temperature, whereas the opposite was found with Pr_4NBr and Bu_4NBr, it was thus concluded that only the Pr_4N^+ and Bu_4N^+ ions could act to stabilize hydrogen-bonding structures in water.

While the interpretation of frequency-shift measurements may be qualitatively correct, it must be recalled (e.g., from examination of the water spectrum) that these shifts occur with important changes in the shape of the absorption bands. Hence in recent studies more emphasis has been given to intensity measurements and detailed analysis of band shapes, either by deconvolution of underlying components or by the use of differential methods.

Measurement of the absorption intensities of the overtones and combination bands as a function of temperature and added electrolytes also yields solution structural temperatures useful in characterizing the relative influence of various ions. Using the 0.97- and 1.15-μm H_2O bands, Luck[134,135] measured absorbances at the wavelength corresponding to the free O—H groups in pure water at 473 K for some 15 1:1, 1:2, 2:2, and 2:1 electrolytes. His results indicate that all electrolytes examined increase the structural temperature of the solution and in an order identical to the Hofmeister series.

In related work Worley and Klotz[136] examined the intensity of the $2\nu_{O-H}$ (1.4-μm) band of HDO in D_2O solutions as a function of temperature and various added electrolytes. (In connection with the above remarks, it is worth noting that no significant frequency shifts could be detected in this case. The ratios of absorbances at the wavelengths attributed to free and bonded O—H groups indicate that concentrations near 2 m of most inorganic 1-1 or 2-1 electrolytes increase the concentration of free O—H groups from the level present in pure water. Quaternary ammonium salts larger than Me_4NBr and sodium carboxylates larger than acetates decrease the concentration of free O—H groups in a way roughly proportional to the size of the alkyl residue of the ions.

Other evidence of the structure-promoting influence of ions bearing large aliphatic groups has been presented by Jolicoeur and coworkers from studies of the 0.97-μm band of H_2O and on the 1.4-μm band of HOD.[137–139] In these investigations the experiments were designed to record directly the differential hydration spectrum, using variable-length cells to maintain the same number of moles of water in each compartment of the spectrometer. The method was systematically applied to study the hydration of nonelectrolytes, common 1:1, 2:1, and 1:2 inorganic electrolytes, alkylcarboxylates, and various quaternary ammonium salts. Some of the relevant results are shown in Figures 26a and 26b.

The difference spectra for hydration obtained with various sodium salts (Figure 26a) exhibit marked spectral changes qualitatively similar to an increase in temperature. The same behavior was observed for the halides, except NaF, which shows spectral changes typical of a temperature decrease.[138] The magnitude of these effects was shown to be well correlated with the "structural entropies" defined by Frank and Evans[2] and reexamined by Friedman and Krishnan.[68] This and other correlations[50,138] indicated quite conclusively that the differential near-IR spectra monitor the structure-breaking influence of the halides (except F^-) and other anions.

The difference spectra obtained with alkali chlorides[137] or alkaline earth chlorides (Figure 26b) shows only a minor dependence on cation size, as noted earlier in fundamental IR studies. Furthermore, evidence of cation charge-induced effects is apparent from the shape of the differential absorption compared to a temperature difference spectrum. These charge effects will invalidate quantitative comparisons between the influence of temperature and

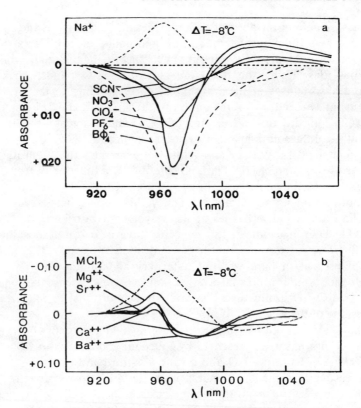

Figure 26. Differential near-IR spectra of various aqueous $1:1$ and $2:1$ 0.5 m electrolytes versus water at 298 K: the dotted curve shows a temperature difference spectrum obtained with water at 290 K versus water at 298 K (adapted from Reference 50 with permission).

electrolytes, but the spectra strongly suggest that the extent of structure-breaking by cations and anions is different and more important for the latter.

The structure-promoting influence of the large tetraalkylammonium ions is evident from the spectra shown in Figure 27. After accounting for solute volume and absorption the difference hydration spectrum of NaBr is subtracted from that of Bu_4NBr. The resulting spectrum thus illustrates the structural influence of the Bu_4N^+ ion *relative to Na^+*, and this is seen to be similar both in shape and position to the effect of decreasing the temperature of water. Qualitatively, this behavior was found common to all the R_4N^+ (R=Me→Bu) ions, and was not dependent on the counterion.[50] Moreover, the magnitude of the spectral difference is very sensitive to temperature for the larger homologues, but much less so for Me_4N^+ and Et_4N^+. The temperature dependence of these differential spectra could further be used to calculate a large structural (relaxational) contribution to the partial molal heat capacity, e.g., Bu_4N^+ in water.[140] The general results of near-IR investigations on tetraalkyl-

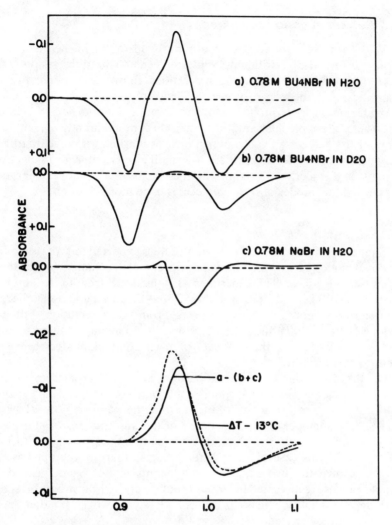

Figure 27. Differential near-IR spectra of various aqueous solutions recorded against H_2O or D_2O at 298 K: (a) 0.78 M Bu_4NBr versus H_2O; (b) 0.78 M Bu_4NBr in D_2O versus D_2O; (c) 0.78 M NaBr in H_2O versus H_2O. The spectra are recorded using a variable path length adjusted to have identical water content in the reference and sample beams. The spectrum labeled $a - (b + c)$ is compared with a temperature difference spectrum (260 K) obtained with pure water; the compound spectrum $a - (b + c)$ shows the influence of the Bu_4N^+ ion on the water spectrum relative to Na^+ (reproduced with permission from Reference 137).

ammonium and alkali halides were also found to exhibit features similar to those observed with other functions sensitive to solvent structural perturbations. This is illustrated in Figure 28, where the ion size dependence of the near-IR, ΔT^* ($\equiv$ molal ΔT), and deuterium isotope effects on the standard enthalpies of solution and on the ionic Walden products are compared.

4.1.2. Solvent Intermolecular Modes

With many solvents far-IR absorption bands occur in the $50-150 \text{ cm}^{-1}$ region, which can be attributed to intermolecular vibrations resulting from dipole–dipole or dipole–clustering interactions. In hydrogen-bonded solvents these modes may be assigned more specifically to hydrogen-bond bending and stretching (restricted translation) and librational motions (restricted rotation). These intermolecular modes should reflect more directly the nature and energy of the solvent–solvent interactions responsible for solvent structure, so it would seem highly appropriate to examine their behavior in electrolyte solutions. In this section we summarize the results obtained for the librational and translational modes of water molecules in aqueous solution.

4.1.2.1. Librational Modes

The tetrahedrally bonded water molecule is expected to give rise to three librational modes, all of which should be observable in Raman spectra though only two are IR active. The experimental Raman spectrum in the librational region $(400-1000 \text{ cm}^{-1})$ does not show three distinct bands, but the band profile shows evidence for subcomponents. From a deconvolution of the band envelope Walrafen identified the librational components centered at ~450, ~550, and $\sim720 \text{ cm}^{-1}$, the intensity of each component decreasing with increasing temperature.[141]

On the other hand, the IR librational absorption spectrum of water consists of a single broad absorption band centered at $\sim685 \text{ cm}^{-1}$. This is illustrated in Figure 29, as reproduced from the paper of Armishaw and James.[142] The temperature dependence of the absorption, in water as well as in electrolyte solutions, exhibits a shift of the band maximum to lower frequencies. In this case of pure water it seems that no attempt has been made to resolve the two expected components in this band; instead, the spectral changes are discussed in terms of completely hydrogen-bonded species absorbing at high frequencies and incompletely hydrogen-bonded species contributing to lower-frequency absorption.

Summarizing the results of the most extensive IR[113,142,143] and Raman[141,144] investigations, the influence of $1:1$ electrolytes on the librational bands may be stated as follows:

All electrolytes except NH_4F decrease the IR librational frequencies in a way roughly proportional to concentration up to the highest concentrations investigated (e.g., $\sim8 \, M$).

The influence of electrolytes on both IR and Raman band shapes are often markedly different according to the changes observed upon varying the temperature; in the case of chloride and bromide salts, for example, the Raman spectra suggest a new species contributing to the librational intensity at $\sim600 \text{ cm}^{-1}$.

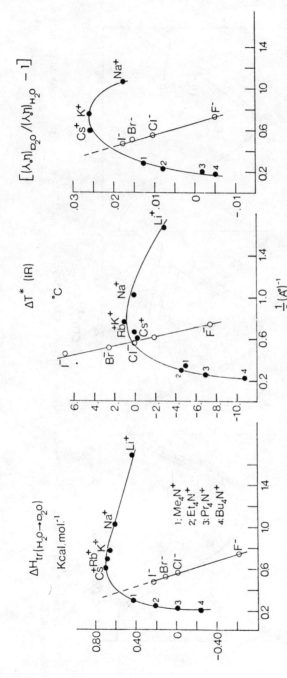

Figure 28. Comparison of the size dependence at ΔT^* (IR) and other properties of alkali and tetraalkylammonium chloride and sodium halides: $\Delta H_{tr}(H_2O \to D_2O)$; standard enthalpies of transfer of the electrolytes from H_2O to D_2O; $(\lambda_0\eta)_{D_2O}/(\lambda_0\eta)_{H_2O}{}^{-1})$, solvent isotope effect on ionic Walden product (reproduced with permission from Reference 50).

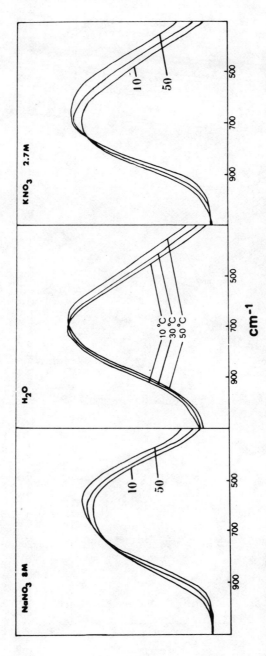

Figure 29. Temperature dependence of the librational IR band of H_2O in 8 M NaNO$_3$, H_2O, and 2.7 M KNO$_3$ (reproduced with permission from Reference 142).

Both the cations and anions affect the librational spectra (IR or Raman), the changes being greater for small cations and large halide anions; the influence of the ions do not appear to be additive.

The general trend of ionic effects on the IR librational frequencies has been given as[113]:

Cations: $H_2O > Li^+ > NH_4^+ > Na^+ > K^+$

Anions: $F^- > H_2O > NO_3^- > Cl^- > Br^- > ClO_4^- > I^-$

Considering the high electrolyte concentrations involved, the spectral changes should be interpreted from the following considerations: a disruption of $H_2O\cdots H_2O$, the appearance of modes involving $X^-\cdots H_2O$ or $X^-\cdots H_2O\cdots M^+$ interactions, and, possibly, $M^+\cdots(H_2O)_n$ "cage vibrations" (see Section 4.1.3.). Since the ionic sequences above follow roughly those observed in the fundamental and near-IR regions, it may be suggested that the structure-disrupting influence plays the major role, with the exception of F^-. However, the fact that the cation and anion effects are not additive indicates that $X^-\cdots H_2O\cdots M^+$ interactions must be taken into account. This has been recognized by James and Armishaw,[113] who analyzed the librational spectrum of electrolyte solutions (e.g., Figure 28), assuming two types of water molecules: those involved in the first hydration layer and the remainder, which are taken as spectroscopically identical to pure liquid water.

In a recent study involving nonelectrolytes and tetraalkylammonium nitrates James *et al.*[145] found that alkyl ureas, for example, increase the IR vibrational frequencies, whereas the total intensity decreases. The magnitude of these changes increases with the number or size of the alkyl substituents. The frequency and intensity changes are assigned respectively to an increase in the concentration of the completely bonded water molecule and to a decrease in the overall extent of hydrogen bonding.

With the homologous R_4N NO_3 (R=Me→Bu), the frequency and intensity variations are opposite to those found with alkyl ureas: The total intensity increases in a way related to the size of the cations. These observations are taken as evidence that the R_4N^+ ions reduce the fraction of tetrahedrally bonded water molecules, so James *et al.*[145] concluded that "in terms of the structure of water, there is no doubt that the R_4NX salts are acting as structure breakers." This is clearly in conflict with the conclusions of near-IR investigations discussed earlier, but it should be recalled that at the concentrations used in the far-IR studies, major spectral changes were noted in near-IR bands.[139] In view of this, and because of the importance attributed to anion hydration effects, a far-IR investigation of other tetraalkylammonium salts would seem most appropriate.

4.1.2.2. Translational Modes

The central water molecule of a tetrahedral hydrogen-bonded structure undergoes six restricted translational modes: four of these correspond to hydrogen-bond stretching motions, the other two involve hydrogen-bond bending. Each group of modes leads to single broad bands located respectively near 170 and 60 cm^{-1}. This region is not easily accessible, but the stretching modes have been investigated both by Raman[141] and IR techniques.[143] Although no quantitative analysis of the spectra has yet been possible, the qualitative behavior of the bands shows interesting features.

In D_2O solutions the presence of various inorganic electrolytes changes both the frequency (ν_T) and intensity of the IR absorption due to translational modes. The shift and line shape changes depend markedly on the nature of the anion and, to some extent, on the cations. With potassium salts, the maximum absorption frequencies follow the order

$$F^- > D_2O > Cl^- > Br^- > I^-$$

indicating similar trends in the average hydrogen-bonding energies, in agreement with results of studies in other spectral regions. Various other sodium salts appear to cause either an increase or decrease of ν_T in the following order:

$$NO_3^- > OD^- > I^- > ClO_4^- > Cl^- > D_2O > ClO_3^- > Br^-$$

although the frequency shifts are small and accompanied by important changes in line shape which might account for the unexpected anionic sequence.

The influence of electrolytes (salts of Li^+, Na^+, K^+, and NH_4^+) on the translational Raman band was generally observed as a decrease in intensity. Qualitatively, there appeared to be no influence of the cations, whereas the anions decrease in the sequence $Cl^- > ClO_4^- > OH^- > SO_4^{2-} > F^-$.

Although there is sufficient variation among the far-IR results (e.g., anionic sequences) to preclude any attempt at detailed rationalization, some broad statements can be advanced concerning the hydration of anions. Generally, and particularly from the intensity behavior, the dominant influence of the Br^- and I^- ions appears as a disruption of hydrogen-bonding interactions. With F^- and Cl^- the coulombic hydration also requires a disruption of water–water interactions, but most results agree that these are replaced by $X^- \cdots HOH$ interactions of comparable energies. With F^- the latter consistently appear to be stronger than the water–water interactions, while for Cl^- the net effect depends on the environment. This observation might have been expected from the results of anion hydration in aprotic solvents as described earlier.

4.1.3. Other Vibrational Modes

In concluding the examination of IR methods we should draw attention to two other types of normal modes that may be investigated to characterize solvation effects: the ion–solvent vibrational modes and the internal vibrational modes of polyatomic ions.

4.1.3.1. Ion–Solvent Modes

Ion–solvent vibrational modes are readily observed in aqueous solutions of various di- or trivalent metal salts and can be assigned to $M^{+n}\cdots OH_2$ interactions by comparison with the spectra of the crystalline hydrates.[146–148] For many of the ions investigated (Be^{2+}, Mg^{2+}, Sn^{2+}, Cu^{2+}, Ni^{2+}, Fe^{2+}, Mn^{2+}, Cd^{2+}, Hg^{2+}, Al^{3+}, Cr^{3+}, In^{3+}) the vibrational modes of the solvation complexes are observed in the $200–500$ cm^{-1} region, and the results are readily amenable to normal coordinate analysis. The situation is quite different when the ion–solvent complex is short-lived, as is usually the case with monovalent ions.

Interestingly, however, with several monovalent cations new absorption bands have been observed in many aprotic solvents in the spectral region where the solvent itself exhibits only low absorption. For various alkali metal salts in, for example, dimethyl sulfoxide, pyridine, and tetrahydrofuran[104] the new bands occur as follows: $Li^+ \sim 400$ cm^{-1}, $Na^+ \sim 200$ cm^{-1}, $K^+ \sim 150$ cm^{-1}, and $Rb^+ \approx Cs^+ \sim 110$ cm^{-1}. In solvents of high dielectric constant the absorption frequencies appear to be independent of the anion and, also, relatively independent of the solvent. While the origin of these bands does not seem to be fully understood, it has been suggested that they arise from vibrations of the alkali metal ions within a solvation shell or "cage." It was further suggested[149] from Raman investigations that the anion can participate in the solvation cage to form either contact- or solvent-separated ion pairs.

At this point it should be asked whether alkali metal ions undergo similar "cage" vibrations in protic solvents. This possibility is usually overlooked in discussions of the far-IR spectra of electrolyte solutions, since no distinct band can be observed. However, recent Raman studies on aqueous solutions of lithium halides in the librational region provide evidence of a tetrahedral $Li(OH_2)_4^+$ complex,[150] so it seems likely that cation "cage" vibrations can also contribute to the librational spectra of other alkali halides in water; possibly such effects could account for line shape variations and irregular ionic sequences found in this spectral region.

4.1.3.2. Internal Modes of Polyatomic Ions

Except for a few particular cases, the internal vibrational modes of polyatomic ions are not very sensitive to changes in solvent composition. For instance, in the aprotic solvents dichloroethane, dimethylsulfoxide, and acetonitrile the IR frequencies of the ν_3 mode of NO_3^- were found to be identical within 5 cm^{-1}.[151] In protic solvents, e.g., water and methanol, somewhat larger shifts are found and with the NO_3^- ion a splitting of the ν_3 band may also be observed. This latter phenomenon has stimulated great interest toward the IR and Raman spectra of NO_3^- and other oxyanions, namely, SO_4^{2-}, ClO_3^-, and ClO_4^-.[108,152–155] Although splittings of the ν_3 mode of several oxyanions can be observed in solid-state studies of matrix-isolated ion pairs,[155,156] the NO_3^- ion appears unique in exhibiting such splitting in dilute solutions at room temperature. This behavior has been thoroughly investigated and interpreted as due to a lowering in symmetry, from D_{3h} for the unperturbed nitrate ion to C_{2v} for the solvated ion, removing the degeneracy of the ν_3 mode. Based on the experimental observations that splitting is independent of concentrations below $1 M$, independent of the cation ($\phi_4 As^+$ or Na^+), and absent in aprotic solvents, it was concluded that the lowering in symmetry of the ion is due to asymmetric solvation of the nitrate ion.[151,157]

4.2. Nuclear Magnetic Resonance

Among other experimental techniques applicable to the investigation of ionic solvation effects, nuclear magnetic resonance is probably the most widely used method in present times. Much as the IR and Raman methods, NMR techniques offer a diversity of probing approaches via the magnetic properties of the solvent and solute nuclei. With current instrumental sensitivity many common nuclei of broad interest (e.g., ^1H, ^2H, ^{13}C, ^{19}F, ^{23}Na, etc.) can be observed in their natural abundance, and isotopic enrichment allows the studies of other important species such as oxygen and nitrogen (^{17}O and ^{15}N). Therefore highly specific aspects of ionic solvation can be examined, which should prove rewarding for solvation studies of atomic or molecular ions.

In addition to its versatility in terms of observable nuclear species, the NMR method offers several advantages over many other physical techniques. NMR measurements can usually be performed in relatively dilute solutions ($<1 M$) so that "infinite dilution" parameters can be reliably obtained. Furthermore, as will be shown below, the measurable NMR parameters can provide information on both the energy and kinetics of the ion–solvent interactions. Hence molecular rate parameters can often be extracted from the data, such as the lifetime of the ion–solvent interaction or the rotational diffusion rates of the ions and solvent molecules. The possibility of reducing the experimental data to single bulk or molecular parameters enables a direct comparison of the relative solvation characteristics of various ions. Consequently, this rapidly expanding field has been periodically reviewed in a consistent manner, and during the past decade excellent topic reviews have appeared at an average rate of one per year.[158-167] In particular, the reviews of Covington and Newman,[158] Hertz and Zeidler,[166] and von Goldammer[159] provide thorough discussions of NMR methods and many recent investigations. It is therefore sufficient for the present purpose to present a phenomenological introduction to quantities of interest in solvation studies, typical data for categories of electrolyte solutions, and, finally, correlations of NMR results with other properties of electrolytes. Details of the theoretical developments concerning various topics outlined here can be found in several textbooks.[168-170]

4.2.1. Experimental Quantities of Interest

NMR experiments are usually aimed at measuring two types of observables: "chemical shifts' and "relaxation rates." The chemical shift is related to the magnetic screening of the nuclei by the electrons, while the relaxation rates describe the kinetics of equilibration of the spin system.

4.2.1.1. Chemical Shifts

The electronic shielding effects are operationally defined as a shielding constant σ from the expression:

$$\omega = \gamma H_0(1 - \sigma) \tag{32}$$

where ω is the measured resonance frequency of the nuclei having a gyromagnetic ratio γ when in the external field H_0. The dimensionless chemical shift parameter δ will reflect changes in nuclear shielding through changes in ω:

$$\delta = \left(\frac{\omega - \omega^r}{\omega^r}\right) \times 10^6 \tag{33}$$

where ω^r is the resonance frequency of a reference nucleus.

The shielding constant σ has positive and negative components that are termed diamagnetic (>0) and paramagnetic (<0); σ_{dia} depends on the wave functions of the electronic ground state, and σ_{para} is related to the excited electronic states. While the former may be calculated quite accurately in many systems, the excited-state wave functions for such ensembles as ion–solvent complexes cannot be reliably obtained. Hence the prediction or quantitative account of chemical shift data is extremely difficult for the systems of interest here.

In condensed phases the following contributions to the shielding constant have been identified:

$$\sigma = \sigma_g + \sigma_b + \sigma_a + \sigma_W + \sigma_E + \sigma_c \tag{34}$$

where each term corresponds respectively to the following: σ_g, the gas-phase value; σ_b, the bulk susceptibility effect; σ_a, residuals from incompletely averaged magnetic anisotropy of the shielding tensor; σ_W, the van der Waals interaction term; σ_E, a reaction-field term; and σ_c, specific chemical effects (chemical environment or molecular interactions). Clearly, before the chemical shift can be assigned to a particular molecular phenomenon (e.g., solvent structural effect) great care must be exercised to account for all other contributions. Appropriate choices of internal standards can account for changes in σ_b and to some extent in σ_W, but the other factors (σ_a, σ_E) may be difficult to eliminate in order to obtain the chemically relevant term σ_c.

4.2.1.2. Chemical Exchange Rates

Solvation studies are concerned with the description of the molecular environment of the ions, and, ideally, the chemical shift of the nuclei could be monitored for each of the local environments in the solution. The observation of distinct resonance signals for each of these microstates requires, however, that their lifetime be very long compared to that of average molecular interactions in liquids at room temperature. Considering a simple two-state (A, B) system with resonant frequencies ω_A and ω_B, the following general conditions apply:

If the exchange rate k_{AB} of the nucleus (or molecule) between states A and B is small compared to $|\Delta\omega|$ (i.e., $\omega_A - \omega_B$), two distinct lines will be observed (slow exchange limit).

If $k_{AB} \gg |\Delta\omega|$, the system exhibits a single absorption centered at frequency $\omega = p_A\omega_A + p_B\omega_B$, where p is the fractional population (or probability) of each state (exchange-narrowing limit).

In the intermediate situation the absorption line will be broadened, and the exchange broadening can be used to determine the average lifetime of the states.

As an example of chemical shift interpretation, we may briefly consider the case of water that has been investigated through ^1H and ^{17}O resonances over a broad temperature range; with ^1H this range extends between the supercooled and supercritical states. In the gas phase the proton chemical shift is independent of temperature and pressure. Condensation at 373 K produces a large negative shift ($\sim$5 ppm) due to deshielding effects of the water–water interaction. Upon further cooling the chemical shift decreases with temperature in a nearly linear fashion at a rate $d\delta/dT \sim 0.95 \times 10^{-2}$ ppm K^{-1}. Since the lifetime of the $H_2O \cdots H_2O$ interactions is on the order of picoseconds, the observed chemical shift is averaged over the various states of the water molecules; then, assuming a two-state hydrogen-bonding equilibrium, δ may be represented as

$$\delta = X_b\delta_b + (1 - X_b)\delta_u \qquad (35)$$

where the subscripts b and u refer to hydrogen-bonded and unbonded species, and X to the mole fraction of the species.

In a thorough investigation of this problem Hindman[171] assigned a chemical shift to the unbonded state (δ_u) by reference to data for water in organic solvents and theoretical calculations of the contributions from the dispersion forces and the reaction field; the chemical shift of the bonded state δ_b was defined relative to ice. After showing that the main contribution to the difference $\delta_b - \delta_u$ was due to polarization of the O—H bond (deshielding the proton) upon hydrogen bonding, the chemical shift data was used to calculate the temperature dependence of the concentrations of the b and u species in liquid water.

It has been remarked, however, that the two-state model of water is not required to account for the temperature dependence of the observed chemical shift.[171] The bending and stretching of the hydrogen bonds (librations and restricted translations) all involve low-frequency modes that are easily excited, changing the OH$\cdots$O distances. On this bases alone an important variation of the water chemical shift to a higher field with increasing temperature should be expected.

The ^{17}O resonance in liquid water also exhibits a positive frequency shift with increasing temperature, with a slope $d\delta/dT \sim 5 \times 10^{-2}$ ppm K^{-1}.[172-173] The deshielding of the oxygen nucleus at low temperatures has been explained in terms of the tetrahedral charge distribution resulting when a water molecule is hydrogen-bonded to four neighbors. It was recognized, however, that the ^{17}O and ^{1}H shifts upon hydrogen bonding arise from different mechanisms and that the fractions of bonded and unbonded species in liquid water derived from $d\delta/dT$ for ^{1}H and ^{17}O will be different.[158]

4.2.1.3. NMR Relaxation Rates

Following the absorption of energy (pulse or continuous wave) the spin system will "relax" to its initial equilibrium state, i.e., a Boltzmann distribution of the spin populations among the different energy levels. This relaxation process is assisted by fluctuations in the local magnetic fields at the observed nuclei. In most cases of interest here the fluctuating fields are due to atomic and molecular motions that modulate the magnetic interactions of the nuclear spin with, for instance, the external magnetic field or other nuclear spins. Generally, the efficiency of the field-modulation mechanisms for inducing nuclear relaxation depends on their frequencies, with an optimum effect near the precession frequency of the nuclear spin.

The rates of magnetic relaxation processes are reported in terms of two characteristic times T_1 and T_2. The rate of exponential decay of the sample magnetization in the direction of the magnetic field is characterized by $1/T_1$; T_1 is thus referred to as the longitudinal relaxation time and is usually measured from the recovery rate of the spin system following a short energy pulse.

On the other hand, $1/T_2$ specifies the loss of sample magnetization in the plane perpendicular to the external field; hence T_2 is called the transverse relaxation time. The magnitude of T_2 may be obtained from pulse experiments or from the width of the resonance absorption line. From the definitions of T_1 and T_2 we must have $T_1 \geq T_2$ and, except in a few cases (e.g., highly viscous solutions, presence of paramagnetic ions), T_1 and T_2 are equal in liquids. Consequently, the remaining discussion will refer mostly to T_1 data.

4.2.1.4. Relaxation Rates and Molecular Motions

Since the NMR relaxation rates generally depend on molecular motion, direct examination of T_1 data for solvent or ion nuclei can yield qualitative information on solvation phenomena. However, in many cases quantitative relationships have been derived between the relaxation rates and specific intra- or intermolecular motions, so the description of solvation "microdynamics" can be greatly refined.

As discussed above, the relaxation rate of a given nucleus depends on local field fluctuations induced by fast random motions of the nucleus examined and of the neighboring nuclei (or electrons). Following von Goldammer,[159] the general situation may be represented by

$$1/T_1 = \Delta^2 f(\omega_0, \tau_c) \tag{36}$$

where Δ^2 is an energy parameter representing the coupling of the nuclear spin to the local fluctuating fields, τ_c is a correlation time describing the terminal motion responsible for relaxation, and ω_0 is the resonance frequency of the nuclei. If the "frequency" of thermal motions $(1/\tau_c)$ is much greater than ω_0 (i.e., $\omega_0^2 \tau_c^2 \ll 1$), the results are independent of frequency and $1/T_1$ may be directly proportional to τ_c. To proceed further in the description of relaxation phenomena the relaxation mechanism must be identified and the correlation time τ_c related quantitatively to the molecular motion. The principles involved may be illustrated for the case of proton relaxation in water.

For protons and other nuclei of spin $\frac{1}{2}$ the dominant relaxation mechanism is usually the result of interactions between the nuclear magnetic dipoles. Since in water such interactions can occur between two protons either on the same molecule or on neighboring molecules, the intra- and intermolecular motions must be considered separately. The rate contributions are additive, so we may write

$$1/T_1 = (1/T_1)_{\text{intra}} + (1/T_1)_{\text{inter}} \tag{37}$$

The intramolecular contribution in this case is given by

$$(1/T_1)_{\text{intra}} = (\tfrac{3}{2})(\gamma^4 \hbar^2 / b^6) \tau_c \tag{38}$$

where γ is the gyromagnetic ratio of the proton, b is the intramolecular proton–proton distance, and τ_c is the correlation time for the rotational diffusion of a vector joining the two protons. τ_c may thus be different from other measurements of molecular reorientation times, such as the dielectric correlation time τ_d; generally $\tau_d/3 \leq \tau_c < \tau_d$. In liquid water at 298 K, τ_c is found to be 2.5 psec, while τ_d is ~8 psec.[23,174]

The intermolecular part of the relaxation rate is obtained as

$$(1/T_1)_{\text{inter}} = \pi \gamma^4 \hbar^2 \frac{C\tau_D}{a^3}\left(1 + \frac{2a^2}{5D\tau_D}\right) \tag{39}$$

where C is the concentration of protons per unit volume, a is the distance of proton–proton closest approach, D is the self-diffusion coefficient of water, and τ_D is the correlation time for translational diffusion. The latter may be taken as $\tau_D = \langle l^2 \rangle / 6D$, where $\langle l^2 \rangle$ is the mean-square length of a diffusion step. Taking $(\langle l^2 \rangle)^{1/2}$ as a molecular diameter (0.275 nm), τ_D in water at 298 K is found to be 5.5 psec.

The separation of the intra- and intermolecular components, although more difficult in water than in aprotic solvents, has been achieved using a combination of data for ^1H, ^2H, and ^{17}O resonances. However, most recent NMR relaxation studies of ionic hydration have been concerned with rotational diffusion effects, so intramolecular relaxation is of greatest relevance here.

In addition to the dipole–dipole interaction discussed above, mention should be made of other intramolecular relaxation mechanisms, namely, spin-rotation and quadrupolar interactions.

Spin-rotation interaction arises from the coupling of the nuclear dipole with the magnetic moment resulting from the rotation of the entire dipole. This provides a relaxation mechanism that may be dominant at high temperatures or in low-viscosity solvents. The quadrupolar interaction, on the other hand, provides the main relaxation mechanism for nuclei having an electric quadrupole moment, i.e., nuclei with spin $> \frac{1}{2}$ (e.g., ^2H, ^{14}N, ^{17}O, ^{23}Na, etc.). The nuclear quadrupole can interact with electric field gradients in the molecule, and modulation of this interaction by molecular motion will induce nuclear relaxation. Since the electric field gradients are fixed with respect to the intramolecular coordinates, the correlation times derived from quadrupolar (e.g., ^2H) or dipolar relaxation (e.g., ^1H) data may be different. In fact, τ_c values will be identical only for the isotropic rotational diffusion of rigid molecules.

4.2.2. Chemical Shifts

4.2.2.1. Solvent Chemical Shifts

With the general background given above, we may examine several results of NMR studies of electrolyte solutions of diamagnetic ions. For the greater part we will restrict our discussion to proton resonance, since the latter has been investigated most systematically; reference to interesting studies involving the ^{17}O, ^{13}C, and ^{14}N nuclei may be found in the reviews of von Goldammer[159] and Covington and Newman.[158]

As discussed previously, depending on the lifetime of the solvation complex, NMR spectra of the solvent may exhibit either new lines characteristic of the solvated species or a shift of the bulk solvent resonance lines. The latter situation is much more frequent and only with cations having a large charge–radius ratio do we observe distinct NMR signals due to bulk and ion-bound solvent molecules. In the latter case, however, the chemical shift and coordination number in the coulombic solvation sphere may be accurately determined, and many such studies (e.g., for Mg^{2+}, Al^{3+}) have been reported in protic, aprotic, and mixed solvents, generally at low temperatures.[158,165] These are proving to be highly valuable in understanding the origin of solvent chemical shifts and preferential solvation effects; such data are also useful in the assignment of molal ionic chemical shifts, as discussed below.

In the case of $1:1$ electrolytes at room temperature only a shift of the solvent resonance line is observed, to higher or lower field, depending on the solvent and on the electrolyte. In the aprotic solvents acetonitrile, sulfolane, and dimethylsulfoxide all $1:1$ electrolytes investigated produce a low-field shift, with the exception of NaBPh$_4$, for which the chemical shifts are large and to higher fields.[119,175] In such solvents, however, the protons observed are usually not on the interaction site of the solvent molecule, so the chemical shift arises from polarization effects through several chemical bonds. In view of the difficulties involved in the interpretation of such effects and the scarcity of available data, the remaining discussion will be concerned only with protic solvents.

In water or methanol at room temperature the majority of $1:1$ electrolytes induce an upfield shift of the —OH proton resonance, in the same direction as that caused by an increase of temperature. Generally, it is found that salts of the largest ions yield the most positive shifts, while salts containing small ions (Li$^+$, F$^-$) produce weak positive or negative chemical shifts. These observations were first interpreted by Shoorley and Alder[176] as resulting from the combined effects of the electrostatic polarization of the solvent molecule (low-field shift) and the breaking of hydrogen bonds in the solvent (high-field shifts). This interpretation was thoroughly elaborated by Hindman,[177] who considered the following contributions to proton chemical shifts in aqueous solutions.

Low-Field Shifts

δ_p: Proton deshielding due to electrostatic polarization of the water molecule.
δ_{ne}: Effect of nonelectrostatic interactions (e.g., partially covalent M$^{+n}\cdots$OH$_2$ interactions).

High-Field Shifts

δ_{bb}: Arising from the breaking of a minimum number of hydrogen bonds to orient water molecules radially in the first coordination sphere of the ion.
δ_{sb}: Due to additional breakage of hydrogen bonds as in the structure-broken region.

More recent investigations have further demonstrated the importance of electrostatic polarization effects for cations and anions.

The influence of anions has been obtained from studies of ternary systems of the type R$_4$NX/ROH/aprotic solvent[178] in a way similar to that described earlier in IR studies. The MeOH$\cdots$X$^-$ complex in CCl$_4$ or acetonitrile exhibits a large downfield shift of the O—H proton relative to the dilute methanol solutions in the aprotic solvents, the order being I$^- <$ Br$^- <$ Cl$^-$. However, near room temperature only the MeOH$\cdots$Cl$^-$ solvate exhibits a proton chemical shift to fields lower than the O—H resonance in pure MeOH.

The solvent polarization shift due to cations has been investigated as a function of ionic charge in binary and ternary systems[179,180] and in solutions of various multiply charged ions. Generally, the electrostatic shift of the O—H proton is a low-field shift and proportional to the charged density of the ion, i.e., z/r. In water or methanol cations having d^{10} electronic structure (e.g., Ag^+, Zn^{2+}, Cd^{2+}) exhibit larger shifts than alkali or alkaline earth ions owing to the additional effect of a partial covalency of the $M^{+n} \cdots OHR$ interaction.[181–183]

From the above observations a hypothetical proton NMR hydration spectrum of a $1:1$ electrolyte may be sketched as illustrated in Figure 30. If we choose an anion and cation with similar radii, (e.g., K^+, Cl^-), the electrostatic polarization in the coulombic hydration cosphere will shift the resonance of the O—H protons to lower field and the effect should be more important for the anion because of direct interactions with the proton. At high field a line due to protons engaged in hydrogen bonds that are weaker than those in the bulk solvent (structure-breaking effects) is shown. The averaging of these various situations due to the fast exchange of water molecules may thus lead to very small net chemical shifts. Therefore ionic molal values of chemical shifts are required in order to attempt any interpretation of chemical shift data in terms of polarization and structure-breaking effects.

4.2.2.2. Ionic Molal Chemical Shifts

For electrolyte solutions in water and methanol two different methods have been suggested by Davies and co-workers for obtaining molal ionic values of the proton chemical shift.[183] The first method is based on the observation of a distinct signal for cation bound solvent (e.g., low temperature, high concentration) and involves some assumptions concerning the temperature and concentration dependence of the resonance of "free" and "bound" solvent. The second method utilizes an empirical relationship between the observed shift and the cation charge density, as illustrated for aqueous solutions in Figure 31. Since the data for divalent and most

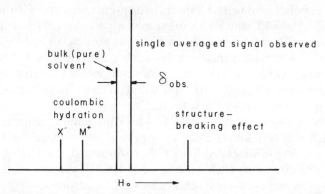

Figure 30. Hypothetically resolved proton NMR spectrum of an aqueous electrolyte solution. The lines at low field from the pure solvent (labeled X^- and M^+) result from deshielding of the proton owing to ion–water polarization effects; the structure-breaking effect leads to proton shielding and resonance at higher field; the unique signal observed will average out these various effects.

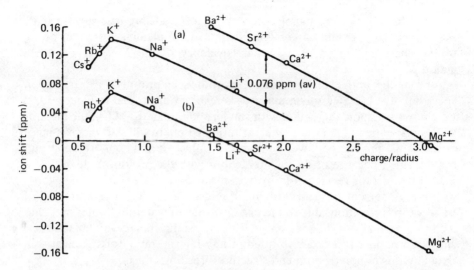

Figure 31. Ionic proton molal chemical shifts of cations in water at 273 K: (a) assuming $\delta_{Cl^-} = 0$; (b) taking $\delta_{Cl^-} = 0.076$ ppm (reproduced with permission from Reference 183).

monovalent cations follow parallel lines when plotted against z/r, the difference between the behavior of $1:2$ and $1:1$ salts may be taken as due to the effect of one anion. Values derived from both methods appear in good agreement, leading to $\delta(Cl^-) = 0.076$ and -0.01 ppm in water and methanol, respectively.

Using these ionic scales, the molal chemical shift values for a series of monovalent ions in water and methanol can be compared as shown in Figure 32, where the data are plotted against $1/r$. In both MeOH and H_2O the electrostatic polarization effect appears to be predominant only in the presence of the smallest ions Li^+ and F^-; all other monatomic ions exhibit weak positive shifts, which would imply a net reduction (or weakening) of hydrogen bonding in the solvent. For cations this contribution is greater in MeOH than in H_2O, while the reverse order observed for anions, suggesting a different mechanism for hydrogen-bond disruption by cations and anions. (These effects have been discussed in some detail by Davies et al.[183]).

The homologous R_4N^+ ions (R=Me→Bu) induce negative chemical shifts, which may be taken as evidence for the ability of these ions to enhance hydrogen bonding in water. Although it is surprising that similar δ values are found for the three ions Et_4N^+, Pr_4N^+, and Bu_4N^+, the sign of the effect is in general agreement with most other results discussed throughout this chapter. The magnitude of the structural effect as evidenced by proton chemical shifts may be compared with that obtained by IR spectroscopy, using the temperature dependence of the proton chemical shift of water. For example, $\delta(Bu_4N^+)$ relative to Na^+ is ~ 0.06 ppm $d\delta/dT = 0.95 \times 10^{-2}$ ppm K^{-1} and

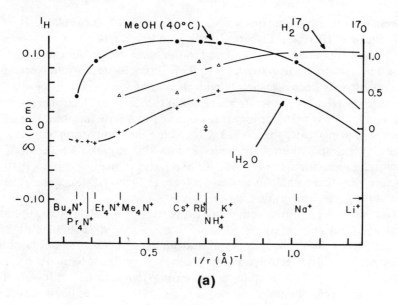

(a)

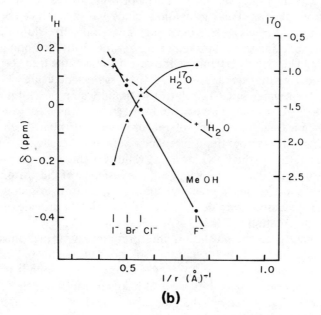

(b)

Figure 32. (a) Molal ^{1}H chemical shift of cations in water and methanol solutions and ^{17}O shifts in water against the inverse radii of the ions. (b) Molal ^{1}H chemical shift of halide ions in water and methanol solutions and ^{17}O shifts in water against the inverse radii of the ions.

the relative structural temperature increment of Bu_4N^+ is calculated as -6 K, while the near-IR results discussed earlier led to $\Delta T_{str} = -13$ K.

Interesting behavior is also found in the temperature dependence of the molal ionic shifts, particularly for the R_4N^+ homologues. While δ values of monovalent ions decrease slightly with increasing temperature, the data for the larger R_4N^+ ions change from negative to positive. This suggests that a reversal of their structure-enhancing influence to a structure-breaking effect at temperatures greater than ~313 K. Pursuing the comparison with near-IR results, an extrapolation of the near-IR data also suggests a reversal of the structural effect, though near 353 K. Similarly, it may be noted that the structure-sensitive enthalpies of transfer from H_2O to D_2O (Section 3) change signs in the same temperature range.

As is frequently done with other properties of the R_4N^+ ions, the ionic molal chemical shift of these ions may be compared to similar data for nonelectrolytes. With the exception of alcohols, which may undergo proton exchange, most nonelectrolytes shift the water proton resonance to higher fields,[162] i.e., in the direction of increasing temperature, although from various other types of measurements, including NMR relaxation, these solutes appear to enhance hydrogen bonding. A suggested explanation[158] for the chemical shift behavior is that some hydrogen bonds are broken in forming a cavity to accommodate the solute particle, while the remaining hydrogen bonds are strengthened. This explanation seems plausible if it can be assumed that the protons of free O—H groups have a large positive shift, dominating the low-field shift due to the strengthening of other hydrogen bonds near the apolar solute. This could certainly be the case if some of the free O—H groups were pointed (on the average) toward the organic molecule. The proton chemical shift of water in CCl_4, relative to liquid water, is $\sim+4$ ppm; hence if 1% of the O—H groups were oriented toward an organic environment, the resulting chemical shift could be of the order of 0.05 ppm. Such an effect would significantly narrow the discrepancy between the chemical shift data and other properties of the R_4N^+ ions. The validity of this explanation requires, however, more quantitative knowledge of the response of the proton chemical shift towards short-range interactions and changes in microscopic environment. Similarly, temperature dependence studies should provide useful complementary information.

The difficulty of chemical shift interpretation is further illustrated in Figure 32 from the ^{17}O data for aqueous solutions of several 1:1 electrolytes (ionic values assuming $\delta_{NH_4^+} = 0$).[172] On electrostatic grounds it might be expected that cations would shift the ^{17}O resonance to lower fields in a way related to the charge density of the ions. Since the data exhibits the opposite behavior, other effects must be important and remain to be elucidated. Recalling that the ^{17}O resonance moves to higher fields with increasing temperature, the influence of anions is also contrary to that expected on the basis of the structure-breaking ability of the large halide ions.

In spite of these problems some interesting correlations have been found between chemical shift data and other ionic properties. For instance, the molal shift of anions on the O—H resonance of MeOH is related quite linearly to the pK_a of the conjugate acids for a broad variety of organic and inorganic ions. A similar relationship was also observed in water, though the correlation is not linear and only a general trend is apparent.

4.2.2.3. Solute Chemical Shifts

NMR investigations have recently been extended to measurement of the chemical shift of ionic nuclei. These include studies on alkali and alkaline earth metal ions, halides, and many others, namely, Tl^+, Cd^{2+}, Al^{3+}, with an emphasis, perhaps, on biologically important species such as ^{23}Na.[158-161,165,167] Polyatomic ions are also being investigated from NMR studies of their constituent atoms. Generally, these investigations have been conducted on one of the following topics:

The relationship between ion chemical shift and solvent properties.
The preferential solvation of ions in mixed solvent systems.
The ion–ion interactions from the concentration dependence of chemical shifts.

The sensitivity of solute chemical shifts to solvation effects is readily noted from the deuterium solvent isotope effect on the resonance of alkali and halide ions in aqueous solutions. The ion chemical shifts are much greater (by approximately an order of magnitude) for the halide ions than for the alkali metal ions.[159] Although these results are not yet fully understood, they will likely provide extremely valuable information on the subtle changes in ionic solvation upon solvent isotope substitution.

For the halide ions further insight into the origin of solute chemical shifts has been obtained from studies on $^{35}Cl^-$, $^{81}Br^-$, and $^{127}I^-$ in various solvents.[184,185] In H_2O, MeOH, AN, DMSO, and DMF the ion chemical shifts were found to be linearly related to the energy of the charge transfer to the solvent (CTTS) as measured by UV absorption. This suggests that an important part of these shifts arises from the excitation energy in the paramagnetic contribution to the shielding constant.

Investigation of alkali ion resonance in various protic and aprotic solvents has revealed some degree of correlation with solvent basicity or polarizability. ^{23}Na chemical shifts were reported to correlate linearly with Guttman's donor number,[165,186,187] i.e., absolute values of the enthalpy of complex formation between the solvent and $SbCl_5$. On the other hand, ^7Li and ^{133}Cs chemical shifts do not exhibit such a behavior; ^7Li shifts appear to be more related to solvent polarizabillities.[188,189] This type of relationship has been taken as evidence for a covalent contribution to the ion–solvent interactions.

The concentration dependences of the alkali and halide ion chemical shifts exhibit downfield shifts that are very sensitive to the counter ions.[158,159] However, the sequence of halide ion effects on cation resonance is identical for all alkali metal ions, and the same is true for the influence of alkali ions on halide resonance. These sequences are found to be $I^- < Br^- < Cl^- < F^- < H_2O$ and $Cs^+ < Rb^+ < Li^+ < K^+ < Na^+$ and have been interpreted as due to overlap repulsive forces in ion–ion interactions. Although solvation effects contribute markedly to the ion chemical shifts, they appear to be of secondary importance in determining solute chemical shifts upon ion–ion interactions, except, possibly, with the anion chemical shifts in the presence of Li^+. Interestingly, it has been pointed out that the ionic sequences given above follow quite closely the order of ^{17}O chemical shifts.[158] Hence, for the latter also, overlap repulsive forces may be dominant, with only secondary contributions due to structure-breaking effects.

Finally, mention should be made of ionic chemical shift studies in solvent mixtures; these provide highly incisive data on the preferential solvation of each ion by one of the solvent components. Such effects are readily detected and analyzed from a curvature in the plot of ion chemical shifts versus the mole fractions of the components in the binary mixtures. In some cases, for example, in a water–acetonitrile mixture, the halide ions exhibit a marked preference for water over acetonitrile, as expected. For several multivalent cations (Mg^{2+}, Al^{3+}, and Ti^{4+}) in aqueous organic mixtures the following sequence of solvating abilities of organic solvents has been established:[190] DMSO > alcohols > amides > tetramethylurea > THF > acetone > acetonitrile > dioxane. In other mixtures, however, the selectivity could hardly be predicted; for instance, in H_2O–H_2O_2 mixtures ions such as Rb^+, Cs^+, and F^- are preferentially solvated by H_2O_2, but Li^+ prefers H_2O.[191] Furthermore, cases have been found, e.g., H_2O–DMSO mixtures, where the preferential solvation changes from one component to the other throughout the mole fraction scale.

4.2.3. NMR Relaxation

4.2.3.1. Relaxation of Solvent Nuclei

Although the measurement of relaxation rates usually involves experimental methods that are more sophisticated than those used in chemical shift studies, the results are generally easier to interpret. As discussed earlier, NMR relaxation mechanisms are well understood, so the relaxation rates can be quantitatively accounted for in terms of kinetic parameters, such as correlation times. Hence studies on electrolyte solutions have yielded a greatly refined description of the "microdynamics" of the ion–solvent complex; they permit the specification of the reorientation times and activation energies (rotation, diffusion) of solvent molecules in the solvation shell and the average exchange rates of solvent molecules between the solvation shell and the bulk. As with chemical shift data, the most extensive and precise investigations have been performed on aqueous solutions, but recent literature indicates growing interest for electrolytes in nonaqueous solvents.

Before discussing electrolyte solution data it might be useful to examine relaxation rates and derive correlation times in several other aqueous mixtures. In dilute solutions of water in various aprotic solvents the rotational correlation time of the H_2O molecule increases in a manner proportional to the change in the IR O—H stretching frequency of water in these solvents: $\Delta\nu_{gas\rightarrow soln}$.[160] This clearly confirms that the rotational correlation time reflects the energy of the interactions that hinder the rotational motion of the water molecule.

On the other hand, upon addition of nonelectrolytes to water the τ_c value of the water molecule increases with the solute concentration and with the size of the aliphatic groups on these nonelectrolytes.[162,192,193] Since results obtained from ^1H, ^2H, and ^{17}O relaxations are in agreement regarding this behavior, there is little doubt that the motion of the water molecules in the water-rich region of hydro-organic mixtures is restricted compared to that of water molecules in pure liquid water. It was further shown from ^1H, ^2H, and ^{13}C relaxation measurements that the rotational correlation times of the nonelectrolyte molecules are also greater than in solvents (or pure liquid) of comparable viscosity.[162,192,194] These findings provide substantial evidence supporting the concept of the stabilization of water structure by apolar groups restricting the motions of both the water and nonelectrolyte molecules.

In aqueous solutions the water proton (or deuteron) relaxation rate is decreased by salts of the larger monovalent ions, except in the presence of organic ions such as alkyl carboxylates and tetraalkylamonium. In the latter case $1/T_1$ may increase greatly, as is the case with ions having high charge densities. The general trends in the relaxation data are thus in qualitative agreement with those obtained through several other properties, namely, solvent immobilization by ions having a large z/r ratio, structure-breaking effects for the larger alkali and halide ions, and solvent structural enhancement by large organic ions. The relaxation data were found to correlate well with the inverse of the self-diffusion coefficient of water in the electrolyte solutions[195] and were also related to viscosity B coefficients.[160] Furthermore, if the concentration dependence of $1/T_1$ is expressed as

$$1/T_1 = \alpha + \beta m + \gamma m^2 + \cdots \tag{40}$$

then the β coefficient exhibits a good linear correlation with the "structural entropies" calculated from the models of Eley and Evans and Frank and Evans.[159] From the intramolecular contribution to the relaxation rates a refined model for the ion solvent complex can be proposed, as pictured in Figure 33. This representation, elaborated by Hertz,[196] describes the electrolyte solution in terms of three microscopic "subliquids" or environments: the solvation shells of the anion and the cation and the bulk solvent. Each microstate is characterized by a rotational correlation time of the solvent

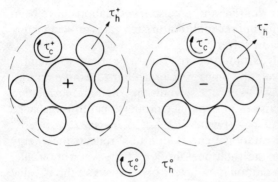

Figure 33. Illustration of several characteristic times for solvent molecular motion in electrolyte solutions: τ_c and τ_h respectively describe the rotational and translational motion of the water molecules in the hydration cosphere; τ_c^0 and τ_h^0 represent the correlation times for these molecular motions in the bulk solvent.

molecule (τ_c^0, τ_c^+, and τ_c^-) and its mean lifetime (τ_h^0, τ_h^+, and τ_h^-). For completeness, τ_c values should be assigned to the ions as well, but these may be omitted here. When $T_1 \gg \tau_h^{0,\pm}$ (fast exchange), the relaxation rates in the various states are additive in the same way as are the solvent chemical shifts discussed earlier. Mean values of the solvent correlation time in the solvation shell of the ions can then be calculated if the stoichiometry of the solvation complex is specified. The assignment of τ_c values to molecules in the solvation shell of the individual ions requires a particular assumption, and, generally, $\tau_c^{K^+} = \tau_c^{Cl^-}$ has been used.

The $\tau_c^\pm$ values obtained for various ions in aqueous solution[193,195] are collected in Figure 34, where the ratio $\tau_c^\pm/\tau_c^0$ is plotted against ion charge

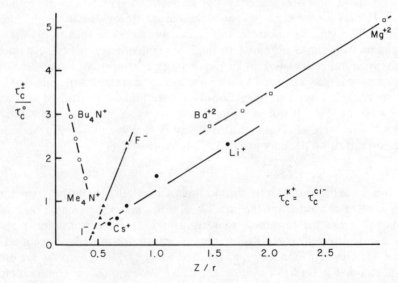

Figure 34. Ratio of $\tau_c^\pm$ for water molecules in the hydration cosphere to τ_c^0 in bulk water plotted against the charge–radius ratio. The data is obtained assuming that $\tau_c^{K^+} = \tau_c^{Cl^-}$.

density as given by z/r. Several interesting trends can be noted. The reorientational correlation time of water molecules in the hydration shell of monatomic cations appears to be linearly related to z/r (except Na$^+$), with very similar dependences for monovalent and divalent cations. The same is true for the anions, but the variation of τ_c^- with z/r appears to be much greater than for τ_c^+. With the large R$_4$N$^+$ ions τ_c^+ increases with the size of the ions which is in good agreement with the structure-enhancing ability of the larger ions, although a net structure-breaking character has usually been attributed to Me$_4$N$^+$. On the other hand, the structure-breaking ability of the larger monovalent ions is clearly indicated by $\tau_c^-/\tau_c^0 < 1$. The latter behavior has also been observed in water, ethylene glycol, and glycerol, but not in other nonaqueous solvents. Thus it may be concluded that it is related to a breaking of solvent hydrogen-bonded structures by the ions. Finally it should be remarked that in all cases examined here the rotational correlation times of water molecules near the ions are not drastically different from those in pure water. Clearly, the solvation complex must be regarded as a highly labile entity, even for those ions called structure makers.

4.2.3.2. Relaxation of Solute Nuclei

The investigation of the relaxation rates of solute nuclei offers other interesting possibilities of characterizing ionic solvation effects. Since most monatomic ions whose nuclei are observable by NMR spectroscopy have a nuclear quadrupole moment, their relaxation rates depend on field gradients caused by the solvent molecules and the motions of the latter in the solvation shell. However, as pointed out by Hertz,[160] it was difficult to extract additional information from these relaxation rates concerning the ion–solvent interactions. Consequently, most studies have focused on the relationship between the relaxation of ionic nuclei and ion–ion interactions. References and discussions of work in this area can be found in the reviews by Hertz,[160] von Goldammer,[159] and Popov.[165]

4.3. Scattering Methods

X-ray and neutron diffraction currently provide the most decisive means of defining the structural features of solids or liquids. With solids the possibilities of these techniques have been amply demonstrated by the resolution of highly complex molecular and crystal structures. With liquids the scattering experiment can, in principle, define each of the various atom–atom radial distribution functions; hence scattering studies of electrolyte solutions should specify the molecular arrangements in the solvation shell of the ions.

Because of the varying abilities of different atomic species to scatter radiation and particles, X-ray and neutron diffraction are optimally applicable to different systems, and the two techniques can be used in a complementary way. This is particularly evident with protons, which are poor scattering centers for X-rays. Fortunately, the proton pair distribution functions can be obtained from neutron diffraction experiments, since the protons (or deuterons) scatter neutrons efficiently. Electron diffraction has also been employed recently[42] in studies of water, but the potential of the method for solvation studies is not yet clear.

As is often the case, however, the great sensitivity of scattering techniques is also the origin of its limitations. In what might be considered a simple system, for example, aqueous sodium chloride solutions, there are 10 different atom–atom pair distances that can contribute to the scattering intensities: $O \cdots O$, $O \cdots H$, $O \cdots Na^+$, $O \cdots Cl^-$, $H \cdots H$, $H \cdots Na^+$, $H \cdots Cl^-$, $Na^+ \cdots Na^+$, $Cl^- \cdots Cl^-$, and $Na^+ \cdots Cl^-$. While some of these are obviously expected to contribute more than others, the difficulty in extracting the distribution functions relevant to the hydration problem is quite formidable. Consequently, these methods are presently best suited for studies in solvents having a simple molecular geometry and at salt concentrations above $1\ m$. For these reasons, and also because of the long-range order indicated from diffraction studies in liquid water, most of the data available on electrolyte solutions pertain to aqueous solutions. The main results of X-ray and neutron diffraction studies to date have been critically reviewed by Enderby and Neilson,† with an emphasis on recent advances.[197] A description of the methods can be found in reviews by Narten and Levy (X-rays)[198] and Page (neutrons),[199] while a discussion of various earlier X-ray data has been given by Kruh.[200]

X-ray diffraction studies on aqueous electrolyte solutions were initiated by Bernal and Fowler,[1,201] who introduced the concept of "structural temperature" much in the sense used earlier in this section, i.e., comparing scattering data of electrolyte solutions to those obtained with water as a function of temperature. This type of study was refined by Brady and Krouse,[202] who discussed the X-ray in terms of order-producing and order-destroying ions, indicating that the hydration cosphere of the ions could exhibit more or less structure than pure water, depending on the intensity of the electric field of these ions. However, it is only rather recently that the structure of the hydrated complex could be elucidated in some detail.

In a thorough study of LiCl solutions (involving both X-ray and neutron scattering) Narten and co-workers[203,204] were able to conclude that the chloride ion is surrounded by six water molecules with predominantly linear $Cl^- \cdots H - O$ interactions, while the lithium ion is coordinated by four oxygen atoms. Other investigations on various salts of di- and trivalent metal ions have shown well-defined first hydration spheres with a typical coordination of six water molecules, e.g., Cr^{3+} and Ca^{2+}.[205] It was further suggested that for these multiply charged ions structural order persists beyond the first

---

† We are indebted to Professor Enderby for making his manuscript available to us prior to publication.

coordination sphere. The exact orientation of water molecules in the solvated complex could not, however, be specified, owing to the limitation of X-ray diffraction for protons.

Neutron diffraction techniques were introduced much later than X-ray methods for obvious instrumental reasons, so there is much less available data on electrolyte solutions; most studies have thus far focused on solid hydrates and clathrates and pure liquids, including water.[199] In addition to the technical complexity of the experiments the extraction of structural information from the scattering data is intrinsically more difficult with neutrons than with X rays. The neutrons interact with atomic nuclei, so their scattering will depend on the mass and spin of these nuclei; since a solvent will usually contain atoms of different atomic masses and nuclear spins, significant scattering (incoherent) will occur that is unrelated to the structural features of the solutions. This, for instance, precludes the use of H_2O in such experiments, so aqueous electrolytes are usually studied in D_2O.

On the other hand, the dependence of scattering efficiency on nuclear mass has been exploited advantageously by Enderby and Neilson[197] through the isotopic substitution of each ion in an aqueous salt solution. Selectively changing to different isotopes of a given ion enables "difference" scattering functions to be obtained, which have been shown to be highly sensitive towards short-range ion–water interactions. From studies of $NiCl_2$ solutions Soper *et al.*[206] have determined that the first coordination number of Ni^{2+} is 5.8 ± 0.2, is independent of salt concentration. In concentrated solutions ($\sim 4.5\ m$) the plane through the D_2O molecule was shown to be significantly tilted away from the $Ni^{2+} \cdots O$ direction, i.e., $\sim 40°$, but the tilt angle gradually decreases as the solution is diluted. These results provide strong indications that in dilute solution the cation–water interaction occurs mainly through the axis of the water–dipole. In these and later experiments Enderby and co-workers also found that the scattering data for the hydration of the chloride ion is consistent with a coordination number between 5 and 6 and a linear $Cl^- \cdots H—O$ arrangement.

4.4. Other Physical Methods

As stated at the beginning of this section, numerous other physical methods have contributed to improve our concepts of ionic solvation phenomena. Although no attempt at a significant coverage or comparison of these methods can be undertaken here, we should point out some of their possibilities, referring to typical papers or to reviews when available.

4.4.1. Electronic Spectroscopy

As indicated by many analytical applications, electronic absorption spectroscopy (UV visible) and fluorescence are widely used to characterize the state of a variety of ionic species in solution. The electronic spectra of a number of metal ions having d-shell electrons (transition metals), of various molecular ions (e.g., NO_3^-, MnO_4^-), and of many organic ions (e.g., aromatic anions or cations) are dependent on the type of ligands or solvent.[207] Clearly this can be used to study the microenvironment of ions in, for example, mixed solvent systems in a manner analogous to NMR chemical shift measurements. In many cases the absorption coefficients are large so that very dilute solutions (e.g., 10^{-2}–$10^{-4}\,M$) can be investigated.

At a more quantitative level the electronic spectra can be used to characterize the extent of charge transfer between the ion and its ligands or solvation shell.[109,207,208] This type of study may be pursued to either establish the nature and energy of the ion–ligand interactions or determine the electron-donating ability of solvents. In the case of aromatic anions or cations, changes in the electronic spectra may be examined in relation to electron redistribution within the molecular ion upon solvation. Similarly, if the solvent exhibits electronic transitions, the latter may be used to evidence interaction with ions.

The dependence of the electronic transition energies on solvent property has been used to define several solvent polarity scales. One such scale (referred to as the Z scale) was derived by Kozower,[209] based on the energy of a charge-transfer absorption band for a contact ion pair of substituted pyridinium iodide. Other scales of solvent polarity have also been proposed by Kamlet and co-workers[19] from the UV spectra of several probe aromatic molecules in various solvents. Still another solvent classification has been suggested from the energy of CTTS bands, which are observed for many molecular anions and the larger halide ions Cl^-, Br^-, and I^-.[210,211] The CTTS bands of the halides have also been widely investigated in electrolyte solutions and mixed solvents in relation to solvation and ion-pairing effects.[207,208]

Finally, the limitations of this method should also be mentioned and they consist mainly in the assignment of spectral changes. Since the electronic distribution within the ion (or within the ion–solvent complex) is different in the ground and excited states, solvation can perturb each state differently, so the origin of spectral changes is often difficult to assess.

4.4.2. ESR Spectroscopy

Electron spin (or paramagnetic) resonance spectroscopy (ESR) monitors the magnetic properties of unpaired electron spins; since the latter interact with other spins (electronic or nuclear), this technique constitutes another highly refined tool for probing the electronic distribution in ion–solvent or ion–ligand complexes. The direct applicability of the method is, of course, limited to paramagnetic species, mostly transition metal ions and organic radical anions or cations. It is of interest, however, that in many cases the sensitivity is such that very dilute solutions (10^{-3}–$10^{-6}\,M$) can be investigated.

The basic principles of ESR spectroscopy are identical to those of NMR spectroscopy, so the experimental quantities of interest in the two methods are related.[169,212] The ESR equivalent to the NMR chemical shift is the isotropic g factor (Landé factor), intra- and intermolecular ESR coupling constants, which arise from interactions of the electronic spin with all nuclei having nonzero spin, may be observed. The ESR relaxation times T_1 and T_2 are defined in the same way as in NMR relaxation; both quantities are experimentally accessible, though ususally with more difficulty in the case of ESR. The ESR relaxation rates also have components that depend on molecular motion, and since these relaxation rates are usually much faster than nuclear relaxation, (i.e., 10^5–$10^8 \sec^{-1}$) higher-rate processes can be studied.[212–215]

The investigations concerned with the coordination of metal ions have been pursued mostly in the solid state and much of the work has been directed towards biological problems.[216] Solution studies have also been reported for, e.g., Mn^{2+} salts in various solvents[217,218] and Fe^{3+}, Mn^{2+}, and Cr^{3+} in aqueous solutions[219] VO^{2+} in basic aqueous solutions,[220] and Cr^{3+} and Mn^{2+} in the presence of various anions.[221,222] On the other hand, the ESR spectra of various transition metal ions (e.g., VO^{2+}, Mn^{2+}) (free ions or complexed) have been found to be extremely useful in studying the rotational diffusion of the ions or respective ion-complexes.[223–227]

ESR spectroscopy has also contributed to the investigation of various solvation effects through the spin probe or spin label approach. A number of stable nitroxide radicals have been used in this way to monitor the microenvironment and dynamics in electrolyte solutions, solvent mixtures, and more complex systems.[208,212,216,228–231] If the probe radicals used has functional groups capable of solvating ions, the method can then yield information pertinent to ion solvation and coordination.[232]

The study of organic radical ions (e.g., aromatic anions) has also been pursued quite actively with respect to ion-pairing and solvation effects.[208,233] The results of such investigations have provided a very detailed picture of the mechanisms and rates of ion pairing in these systems.

Finally, with regard to electrochemical investigations, ESR spectroscopy offers the interesting possibility of observing the behavior of transient ionic species that are electrochemically generated in the spectrometer cell. This approach makes ESR spectroscopy a particularly useful tool and the combined technique is rapidly gaining importance.[234]

4.4.3. Other Relaxation Methods

Ultrasonic and dielectric absorption spectroscopy are two other powerful physical methods of investigating ionic interactions. The frequency dependence of ultrasonic absorption allows the measurement of the rates of processes, with relaxation frequencies typically in the range 10^5–10^9 Hz.[235] These processes must exhibit nonzero volume and enthalpy changes, which is generally the case with ion–solvent, ion–ligand, and ion–ion interactions. Hence the method has found broad applications in solute–solvent and solute–solute interactions.[236] Of particular interest in the present context are studies of the kinetics of ligand or solvent substitution for metal ions. For water the substitution rate in the first coordination shell of various metal ions is typically 10^8–10^{10} sec^{-1} for Li$^+$–Cs$^+$ and 10^2–10^9 sec^{-1} for Be^{2+}–Ba^{2+}. For trivalent metal ions the rate of water substitution for, e.g., Al^{3+} is on the order of 1 sec^{-1} and may be much longer for other ions. Data on these exchange rates are also available for other solvents and many common ligands[236] and, recently, for some macrocyclic ligand complexes.[237]

The measurement of dielectric dispersion, on the other hand, enables the evaluation of the correlation time for the reorientation of molecular dipoles. This technique has also been applied extensively to electrolyte solutions and excellent accounts of the results for aqueous systems can be found in a book by Hasted[238] and in a review by Pottel.[239] Dielectric relaxation studies have yielded much quantitative information on the influence of ions upon solvent dipole reorientation, and the results generally agree with those obtained by NMR relaxation, discussed earlier.

4.5. Concluding Remarks

In view of the very decisive nature of the spectroscopic and diffraction data in so many structural and chemical problems, the examination of any single spectroscopic result on ionic solvation may seem rather deceptive. The reason for this is simple and twofold. First, with most ions of interest solvation is far from a static event, so that, instead, the various solvation parameters (coordination number, nearest neighbor distances, relative orientation, and interaction energies) undergo rapid fluctuations and each parameter must be viewed as having a distribution of values. Second, with most of the physical methods examined the theory that relates the spectral features to the solvation parameters is not sufficiently developed to incorporate proper fluctuations and/or distributions of the ion–solvent interaction parameters. Following the topics discussed in this section, we therefore cannot issue definite statements for many of the questions raised at the beginning. Nonetheless, spectroscopic and scattering methods have supported important advances in the definition of ionic solvation and, in conclusion, the most prominent ones should be recalled for each of the three main aspects: energy, geometry, and kinetics.

With regard to the energetics of ion solvation physical methods have clearly identified ions with firmly bound solvation shells from those for which the ion–solvent interactions are too weak to yield a stable ion solvate, even on the time scale of molecular collisions. Various results generally show that, with the exception of small di- or higher-valent ions, the solvation shell remains a highly labile unit, in spite of large solvation free energies. For ions with high charge densities (z/r) the solvation shell is often distinguishable from the bulk solvent, so the properties of the molecules near the ions can be studied directly. In many cases the energy of the internal modes of the ion–solvent complex can be specified as required for a statistical-mechanical analysis of solvation thermodynamics. With low-charge-density ions (e.g., Cl^-, Br^-, ClO_4^-, etc.) in protic solvents the spectral data show that the ion–solvent interaction energies are comparable (often weaker) to those of the solvent–solvent interactions.

With respect to geometry, physical methods have enabled an easy distinction between the solvent coordination behaviors of anions and cations. Generally, in polar solvents the interaction site of the ion on the solvent molecule can be readily identified; hence in favorable cases the importance of ion–dipole interactions relative to other effects (ion–quadrupole interaction, functional group specificity, geometrical packing) may be qualitatively understood. On the other hand, there appears to be no scattering data on the coordination geometry for electrolytes in nonaqueous solution. In aqueous solution the preferred geometrical arrangement of water molecules in the hydration shell is intrinsically difficult to specify, mainly because of the high degree of solvent structure and the consequences of structural perturbation by the ions. However, in a number of key examples, e.g., divalent metal ions and Cl^-, the geometrical hydration parameters have been accurately determined from scattering experiments.

Concerning the dynamic aspects of ion–solvent interactions, our knowledge has been greatly improved with regard to three main areas: the internal motions within ion–solvent complexes, the solvent exchange kinetics, and the rotational diffusion of both the ion (or solvated ions) and the solvent molecules. Relaxation methods have secured a large quantity of reliable data on the solvation dynamics of many different monatomic and molecular ions in water and several nonaqueous solvents. For high-charge-density ions the kinetic parameters confirm an important solvent immobilization by the ions themselves and provide an accurate description of this effect. With low-charge-density ions (e.g., K^+, Cl^-) the motion of neighboring solvent molecules does not appear to be greatly hindered. Furthermore, in water the rate of some dynamic processes (e.g., rotational motion of water molecules) appears, in fact, faster in the presence of these ions, which confirms the structure-breaking effects.

Finally, spectral methods provide additional information on hydrophobic effects occurring in solutions of large organic molecules and ions and, possibly, at neutral interfaces. Although there is no direct structural evidence to characterize the hydrophobic hydration, there is now much spectral data showing that relative to other simple ions, e.g., Na^+, or compared to temperature effects, the larger R_4N^+ ions increase the extent of hydrogen bonding in water. As with the structure-breaking effects, it is possible that similar "solvophobic solvation" may occur in other highly structured solvents; hitherto solvation studies have given little clues of such effects.

5. Theoretical Aspects of Solvation

5.1. Introduction

Most quantitative treatments of ionic solvation are based on electrostatic theories. This applies to the calculation of the potential energy of an ion in a dielectric continuum as well as to computations of the interaction energy of a collection of electrons and nuclei of the solute–solvent complex. Therefore the level of sophistication of solvation theories will vary greatly, depending on the model used. Simplifying to the extreme, the systems are pictured as charged spheres in a polarized dimensionless dielectric medium or as rigid solvation complexes. While these approaches lend themselves readily to quantitative treatment (continuous solvent and molecular theories), they are necessarily oversimplified.

In the other extreme the ion and a few neighboring solvent molecules are treated as a many-electron many-nucleus system for which the charge distribution and potential energy are computed under various configurations of the ion–solvent complex. Such *ab initio* treatments can provide very detailed information on the ion–solvent interactions, such as ion–solvent equilibrium distances, the most probable orientation of the solvent molecule in the solvation shell, and the importance of solvent–solvent interactions in the solvation complex. However, the feasability of the computations involved presently limits the investigations to systems much smaller than required for a good representation of the ion in a molecular liquid. At the present stage the results of such computations are best related to gas-phase solvation data.

The theoretical treatment of solvation effects that involve perturbation of the solvent structure requires quantitative knowledge on the structure of the solvent. Therefore, until successful theories for the latter are developed, it is difficult to expect too much progress in the quantitative interpretation of the structural effects of solutes.

Finally, a number of properties of electrolyte solutions have been computed using hard-sphere theories such as the "scaled-particle" theory. These usually involve an equation of state, so they cannot basically be considered solvation theories. However, properties calculated from these equations can provide useful information, and they will be discussed briefly below.

In this section a brief outline will be given of the various theoretical approaches, with an emphasis on continuum and one-shell molecular theories, since their advantages and limitations are now reasonably well understood. Other theoretical methods such as quantum-mechanical theories and computer simulations will be dealt with on a more qualitative level, since they are still in an active state of development. In principle, all solvation properties can be interpreted quantitatively.

5.2. Continuous Solvent Theories

The Born equation [Eq. (11)] and all other thermodynamic relations derived from it have been the basis for much of the earlier theories of solvation. This approach relates all solute–solvent interactions to the dielectric properties of the solvent. In their simplest form these equations give the sign and magnitude of many solvation functions, especially in cases where coulombic solvation predominates. Unfortunately, quantitative agreement is rarely achieved. Generally, the predicted free energies of solvation are too negative and various improvements have been suggested over the years. Some of the earlier studies have been reviewed by Desnoyers and Jolicoeur[13] and the main attempts to improve on the Born treatment are as follows:

1. Stokes[240] claims that van der Waals radii should be used for the free ions in the gas phase, and crystal radii in solution.
2. Stokes[240] and Jain[241] assumed the first solvation shell of ions to be dielectrically saturated.
3. Noyes[242] suggested that a small nonelectrostatic term should be added to the Born equation to account for the energy necessary to move the ions through the solvent interface. This term is necessary only if ionic values are considered.
4. Desnoyers and Jolicoeur[13] have suggested using an effective radius for ions in solution to account for the free space near the ions resulting from the finite size of solvent molecules.

If all these corrections are incorporated, the modified Born equation becomes

$$\Delta G_s^{\theta} = \frac{z^2 e^2 N}{2} \left[\frac{1}{\varepsilon_s} \left(\frac{1}{r_e} - \frac{1}{r_s} \right) + \frac{1}{\varepsilon_0 r_s} - \frac{1}{r_v} \right] + \Delta G_s^{ne} \qquad (41)$$

where ε_s and ε_0 are respectively the dielectric constants in the saturated region (approximately equal to the square of the refractive index) and in the bulk, r_e, r_s, and r_v are respectively the effective, solvation, and van der Waals ionic radii, and ΔG_s^{ne} is the nonelectrostatic contribution. The effective radii r_e are estimated from the intrinsic volumes. Desnoyers and Jolicoeur[13] have applied this equation to ions in water and derived the radii r_h of hydrated ions from ΔG_h^{θ}. The values obtained are of the correct magnitude, but r_h appears to be slightly too small for small ions and slightly too large for large monatomic ions.

More rigorous derivations of the free energy of solvation were made by Padova[75,243] and Beveridge and Schmielle.[244] Padova considers the free energy of a dielectric continuum in an electrostatic field to be

$$\Delta G = \frac{1}{4\pi} \int\int \mathbf{E} \, d\mathbf{D} \, dv \qquad (42)$$

where E and D are vectors representing the field strength and dielectric displacement in a volume element dv. For spherical symmetry $dv = 4\pi r^2 dr$ and the differential dielectric constant is defined by $\varepsilon_d = dD/dE$. Equation (42) can therefore be rewritten as

$$\Delta G = \frac{1}{2} \int_r^\infty \int_0^{E_r^2} \varepsilon_d \, d(E^2) \, r^2 \, dr \qquad (43)$$

If ε_d is taken as field independent and E is replaced by the electrostatic field near a point charge, Eq. (43) can be solved in the solvent and in the gas phase ($\varepsilon_d = 1$) and reduces to the Born equation. Some of the molecular changes in the solvation shell may be introduced indirectly by taking ε_d as field dependent:

$$\varepsilon_d = n^2 + (\varepsilon_0 - n^2)/(1 + bE^2) \qquad (44)$$

where n is the refractive index, ε_0 is the bulk static dielectric constant, and b is a parameter independent of E. Padova[243] has solved Eq. (43) numerically for the transfer of electrolytes from water to methanol, using effective ionic radii as discussed in relation with Eq. (41). As seen from Table 6, his predicted free energies of transfer are quite close to the observed ones. If ionic transfer functions are considered, then the true free energy of transfer is given by

$$\Delta G_{tr}^\theta(\text{true}) = \Delta G_{tr}^\theta(\text{elect}) + z \, \Delta F \chi \qquad (45)$$

where F is the Faraday and χ is the surface potential of the vacuum–solvent interface. Padova has shown that this second term is close to -22 kJ mol^{-1} for cations and $+22 \text{ kJ mol}^{-1}$ for anions for the transfer from water to methanol.

The equation of Beveridge and Schmielle[244] is much more elaborate than Padova's and is essentially a two-layer model. The first layer is dielectrically saturated and the total polarization is the sum of charge, dipole, and quadrupole contributions. Abraham and Liszi[245] have used this equation, retaining only the charge polarization term; they use the ionic radii of Goldschmidt and Pauling, a dielectric constant of 2 in the first solvation shell and a shell thickness equivalent to the diameter of the solvent molecule. They have further assumed that most of the nonelectrostatic contribution to ΔG_s^θ (change in standard state, cavity formation, etc.) could be given by the free

Table 6
Standard Free Energies of Transfer of Electrolytes from Water to Methanol at 298 K [244]

Electrolyte	ΔG_h^θ(kJ mol^{-1})	Experimental	Electrolyte	ΔG_h^θ(kJ mol^{-1})	Experimental
HCl	22.6	22.4	KCl	27.6	26.4
LiCl	20.0	18.8	KBr	24.3	23.8
NaCl	23.6	22.2	KI	20.7	
KCl	27.6	26.4			

energy of solvation of rare gases of similar size. The treatment has essentially no adjustable parameter. Calculations have been carried out for many solvents and a summary of their results in methanol and acetonitrile is given in Table 7. The predicted values are often of comparable precision to the measured values.

The continuum theories have also been used successfully to predict the electrostriction of ions. Here the basic equation has been derived by Frank,[246] who developed the thermodynamics of a fluid in the presence of an electrostatic field. The molar change in volume due to the charging of an ion is given by

$$\Delta V = \frac{N}{2} \int_r^\infty \int_0^{E_r^2} \left(\frac{\partial \varepsilon}{\partial P} \right)_{E,T} d(E^2)\, dr \qquad (46)$$

As in the case of Eq. (43), if $(\partial \varepsilon / \partial P)_{E,T}$ is assumed constant and E is replaced by the simple electrostatic field near a point charge, Eq. (46) reduces to Eq. (16) derived from the Born equation. Desnoyers et al.[247] have solved Eq. (46) numerically for aqueous systems, calculating the effective pressure that would, in the absence of the field, produce the same change in volume as would the field; dielectric saturation was also considered. They have shown that the correct order of magnitude can be predicted for the electrostriction. Dunn[248] has extended this treatment to high temperatures and Padova[243] has used a similar approach for nonaqueous solvents.

Much less has been done with other thermodynamic functions. The expression for the entropy change due to the charging of an ion is[75]

$$\Delta S = \frac{N}{2} \int_r^\infty \int_0^{E_r^2} \left(\frac{\partial \varepsilon}{\partial T} \right)_{E,P} d(E^2)\, dr \qquad (47)$$

Unfortunately, it is difficult to predict the dependence of $(\partial \varepsilon / \partial T)_{E,P}$ on E. By comparison with Eq. (46), it would be expected that a simple relation should exist between ΔV_s^θ and ΔS_s^θ, and such a relation is observed experimentally.[249] The predictions of the enthalpies of solvation are faced with

Table 7

Free Energies of Solvation (kJ mol^{-1}) of Ions at 298 K

Ion	Methanol		Acetonitrile	
	Calculated	Observed	Calculated	Observed
Na^+	−430	−410	−403	−392
K^+	−334	−334	−329	−325
Rb^+	−314	−313	−309	−305
Cs^+	−277	−282	−273	−274
Me_4N^+	−170	−177	−170	−174
Cl^-	−258	−271	−254	−254
Br^-	−243	−244	−239	−237
I^-	−220	−211	−218	−213

the same difficulties as ΔS_s^θ. Therefore Eq. (47), which was derived from the Born equation and which implicitly assumes that $(\partial \varepsilon / \partial T)_{E,P}$ is a constant, is not expected to hold in the first coordination shell of the ions. Cobble,[250] on the other hand, has recently shown that the heat capacity expression [Eq. (14)] derived from the Born equation can predict nearly quantitatively $\bar{C}_{P2}^\theta$ of NaCl at high temperatures. Zana et al.[32] have also shown that the differences in $\bar{C}_{P2}^\theta$ of electrolytes in various nonaqueous solvents are due largely to the Born contribution. Little seems to have been done on compressibilities and expansibilities.

The continuum theories have not been very popular in recent years. The models are not realistic, since the molecular nature of the solvent is largely ignored, and there is no provision in the theory for specificity of the solvation of a cation and anion or for structural effects. Therefore our understanding of ion–solvent interactions is not greatly improved through this approach. On the other hand, if we are interested in predicting solvation functions in cases where coulombic solvation predominates (e.g., nonaqueous solvents) and where specific effects are not too important (at high temperatures and pressures), these theories are simple and work remarkably well. Unfortunately, in the cases where the theories are expected to work best good data on $\partial \varepsilon / \partial T$ and $\partial \varepsilon / \partial P$ are often very scarce.

5.3. Molecular or Chemical Models

The short-range nature of ion–solvent interactions can be used to advantage in developing models for solvation. The first coordination shell of the ions may be considered as consisting of rigid spherical dipoles or quadrupoles, while beyond this shell the solvent can be taken to be continuous. Once a structure for the one-shell solvation complex is assumed, the solvation functions can be calculated for the various interactions (charge-dipole, charge-induced dipole, etc.). Bernal and Fowler[1] were the first to use such an approach, and since then various improvements have been proposed, namely, those of Eley and Evans,[251] Buckingham,[252] Muirhead-Gould and Laidler,[253] Goldman and Bates,[254] and Bockris and Saluja.[225]

All of these theories make use of a Born–Haber cycle for the calculation of the standard thermodynamic solvation functions ΔY_s^θ. Following the scheme of Goldman and Bates,[254] the cycle can be broken down into four processes:

$$\text{Ion } r_g(\text{g}) + nS(\text{l}) \xrightarrow{\Delta Y_s^\theta} (\text{Ion } r_c \cdot ns)(\text{sol})$$

$$\text{1 atm} \qquad\qquad\qquad\qquad \text{hyp } 1\, m$$

$$\Delta Y_1^\theta \downarrow \qquad \Delta Y_2^\theta \downarrow \qquad\qquad \uparrow \Delta Y_4^\theta$$

$$\text{Ion } r_c(\text{g}) + nS(\text{g}) \xrightarrow{\Delta Y_3^\theta} (\text{Ion } r_c \cdot ns)(\text{g})$$

$$\text{1 atm} \qquad \text{1 atm} \qquad \text{1 atm}$$

Most of these calculations have been made for ΔH_s^θ and some for ΔG_s^θ and ΔS_s^θ. The extension of this approach to other functions becomes a formidable task, since a knowledge of all the variations of the interaction energies with temperature and pressure is required. The following description of the various terms will be given primarily with reference to enthalpies.

Process 1. Ions in the gas phase have radii r_g that are larger than those in solution, r_c, as discussed in Section 5.2. This contribution can be calculated from the Born equation. Since ΔS_1^θ from this process is zero,[254]

$$\Delta H_1^\theta = \Delta G_1^\theta = \frac{N(ze)^2}{2}\left(\frac{1}{r_c} - \frac{1}{r_g}\right) \tag{48}$$

and this term is always positive and can be appreciable.

Process 2. This process refers to the evaporation of n solvent molecules and can be obtained precisely for most solvents.

Process 3. This is the most important contribution, since it involves all the strong electrostatic interactions between the ion and the solvent molecules in the first coordination shell. Results vary significantly, depending on the model used for the solvation complex; for example, the solvent molecule can be considered a point dipole, a point dipole plus a point quadrupole, an array of fixed point charges, etc. Bockris and Saluja[255] have also introduced the concept of nonsolvated coordinated solvent molecules. We will describe here only the approach of Goldman and Bates,[254] since it appears to be the most complete one at the moment. For the enthalpy

$$\Delta H_3^\theta = E + \tfrac{1}{2}hc\sum_i w_i + RT^2\left(\frac{\partial \ln Q}{\partial T}\right)_P \tag{49}$$

where E is the sum of all potential energies at 0 K, $\tfrac{1}{2}hc\sum_i w_i$ corrects for the zero point vibrational energy, and Q is the ratio of the partition functions Q_{trans}, Q_{rot}, and Q_{vib} for the products and reactants of reaction 3.

The quantity $\mathbf{E}$ consists of the sum of potential-energy terms:

$$\mathbf{E} = \sum \mathbf{E}_i \tag{50}$$

The main terms are the ion-dipole, the ion-induced dipole, and London dispersion energies, lateral energies between solvent molecules, repulsive energies, ion-quadrupole energies, etc. The original papers should be consulted for a detailed description of these terms.

The error in calculating $\mathbf{E}$ can be appreciable, since it is the resultant of a number of large terms of different signs. Goldman and Bates evaluated these terms in water, using the gas-phase solvation energies (Section 3). Since there are now reliable values available for thermodynamic data of ion solvation in the gas phase, especially in nonaqueous solvents, it should be possible to examine more critically the various models for process 3 for the three main functions.

Process 4. This is the least accurately known contribution in such models; it involves the formation of a cavity in the solvent, rearrangement of the solvent molecules around the solvation complex, long-range polarization (Born term), and any surface potential terms when individual ions are considered. This is also where the structure-breaking effect would intervene. To our knowledge Bockris and Saluja[255] were the only ones to attempt to estimate this term for enthalpies and entropies. They have estimated ΔH_{sb} for the passage of a water molecule from the bulk to the structure-broken region, using the free-volume theory of liquids. Unfortunately, this leads to a *negative* value for ΔH_{sb}, whereas obviously we would expect the breakage of hydrogen bonds to lead to a *positive* value. On the other hand, they do obtain a positive value for ΔS_{sb}, as expected.

It is unfortunate that so few calculations have been carried out on nonaqueous solvents. Most of the complications with cavity formation and structural effects do not arise in these cases, and process 4 should be easier to evaluate. Somsen[256,257] was one of the few to calculate the enthalpy of solvation in some nonaqueous solvents, using the approach of Buckingham,[252] and he obtained a reasonably good agreement with experimental data, assuming an octahedral solvation complex. Since much more reliable data are now available in the gas and condensed phases, such calculations could be extended significantly.

At present the one-layer molecular theories still involve many ill-defined contributions, so their level of precision cannot be taken as high. Their predictive use is therefore quite limited; they can be useful in formulating models that account for the structure of the solvation complex and the specificity of ionic solvation. Also, if these models can be extended to include more sensitive functions such as heat capacities, compressibilities, and expansibilities, much progress would then be made in the interpretation of the weaker interactions, which are largely responsible for the peculiar behavior of aqueous systems.

5.4. Scaled-Particle Theory

The continuous and chemical models for ion solvation can be applied to many systems in which coulombic solvation is the leading effect. At present such models cannot account for the structural effects; however, some recent attempts have been made to evaluate these, using various simple theories for the structure of water. On the other hand, the scaled-particle theory has been particularly successful in accounting for the thermodynamic properties of hydrophobic solutes in water and merits some attention. It appears that a significant part of the interactions and hydrophobic hydration can be accounted for by the thermodynamic functions for cavity formation in the solvent. As such, it is similar to Eley's theory,[258] which is based on Maxwell's relation,

$$\left(\frac{\partial S}{\partial V}\right)_T = \left(\frac{\partial P}{\partial T}\right)_V = \frac{\alpha}{\beta} \tag{51}$$

where α and β are the coefficients of thermal expansion and isothermal compressibility. Integrating at constant temperature, the entropy for the formation of a cavity of volume V is given by

$$\Delta S_c = \int dV = \frac{\alpha}{\beta} V \tag{52}$$

Similarly, other thermodynamic functions of cavity formation can be derived. For water α is much smaller than for other liquids, and ΔS_c and ΔH_c are also smaller, in agreement with experiment.

The scaled-particle theory was developed by Reiss *et al.*[259] and extended to aqueous solutions by Pierotti.[260] The standard free energy of solvation is given by

$$\Delta G_s^\theta = \bar{G}_C + \bar{G}_I + RT \ln RT/V_l \tag{53}$$

where $\bar{G}_C$ and $\bar{G}_I$ are the molar free energies of cavity formation and of interaction with the surrounding solvent molecules; the third term of Eq. (53) takes into account the change in standard states when the solute is transferred from the gas phase to the liquid of volume V_l. Other thermodynamic functions are obtained from the various derivatives of Eq. (53) with respect to T and P. The cavity term can be calculated from an approximate form of the radial distribution function and is a function of the diameter of the cavity and of the various physical properties of the solvent (particle density, α, β, $d\alpha/dT$, etc.). The interaction terms can be calculated from the Lennard-Jones potential or other similar potentials.

With this theory Pierotti[260] predicted surprisingly well the enthalpies, entropies, volumes, and heat capacities of nonpolar gases in water. Since then Lucas *et al.*,[261-263] Masterton *et al.*,[264] and Shoor and Gubbins[265] have successfully extended these calculations to gases in electrolyte solutions. Also, Philip and Jolicoeur[29] have calculated the thermodynamic functions of transfer of hydrophobic solutes from H_2O to D_2O, and Desrosiers and Desnoyers[266] the thermodynamic functions of transfer from water to mixed solvents. Even though this theory does not assume any structure or modification of the structure of the solvent by the solute, the properties of the hydrophobic solutes are predicted nearly quantitatively. Of course, the structure of the solvent is taken into account indirectly by the experimental parameters α, $d\alpha/dT$, and β, and part of the solute–solvent interactions may be absorbed in the various solute and solvent diameters. As it appears at this stage, the scaled-particle theory provides an elegant means of correlation between the density of the solvent (and its derivatives) and the thermodynamic properties of the solutes. However, it does not give us much insight into the molecular nature of the interactions in solution, but it can be very useful in predicting the properties of solutes in various mixtures. It also shows that much of the anomalous behavior of aqueous solutes is related to the peculiar packing density of liquid water.

5.5. Other Theoretical Approaches

As follows from the above discussion, the continuum dielectric and scaled-particle (or "hard-sphere") theories (SPT) rapidly reach their limit of usefulness with regard to the chemical interpretation of solvation phenomena. Molecular models can be much more descriptive, but their usefulness is also restricted for two reasons, namely, the lack of accurate short-range potentials operating within the ion–solvent complex and the difficulty of statistical-mechanical treatments of many-body interactions. Recent progress in the theoretical description of ionic solvation has thus followed two main paths: detailed quantum-mechanical calculations of interaction energies between ions and several water molecules, and computer simulation (Monte Carlo and molecular dynamics) of the behavior of a large collection of solvent molecules containing one or a few ions.

5.5.1. Quantum-Mechanical Calculations

As the name implies, quantum-mechanical calculations of ion–solvent interactions refer to the computation of the charge distributions and potential energies for an ion and one (or a few) solvent molecule. Their objective is to provide potential-energy surfaces for the ion–molecule interactions within an ion–solvent cluster under the least possible approximation in the wave functions and in the truncations of computational procedures. Because the magnitude of the computational problems rapidly increases with the number of electrons that must be taken into account, such calculations are usually performed with the ions of lighter elements (e.g., Li^+, Na^+, K^+, F^-, Cl^-) and relatively simple solvent molecules. These calculations are therefore incapable of predicting free energies or enthalpies for the solvation of ions in liquid solvents. Instead, these results pertain to the energies of ion–molecule interactions in the gas phase, and the latter provide so far the most crucial test of the theoretical predictions.

Following the recent expansion of computational methods, the area of quantum-mechanical calculations has developed into a highly specialized one. In practice, computations are performed at several levels of approximation and it is difficult for anyone to judge the quality of these calculations, except from the agreement between calculated and experimental values. Fortunately, several authoritative monographs[267,268] provide a quantitative assessment of the various computations methods. These may be divided into two broad categories; "*ab initio*" and "semiempirical," the former involving the least assumptions (Hartree–Fock approximation and limitations inherent to the size of the basis set used in representing the molecular orbitals). Although a discussion of these methods is outside the scope of this book, some of the important achievements of quantum-mechanical calculations must be mentioned.

The ion–solvent stabilization energy has been computed by the *ab initio* method for the stepwise addition of water molecules ($n = 1$–10) for several alkali and halide ions.[269] These results are in rather remarkable agreement with the gas-phase solvation data discussed earlier in Section 3.3. For the same hydration process qualitative agreement has also been obtained in the case of the polyatomic NH_4^+ ion.[270] A number of computational results are also available for the interaction of one (or a few) water molecule with larger ions, (e.g., Me_4N^+, $Me_2N^+Et_2$,[267]) although in these cases experimental gas-phase data are not yet available. Several computations with other solvent molecules (e.g., ammonia, formaldehyde, formamide[267]) should provide a basis for the comparison of short-range solvation effects in these solvents. Certainly one of the great merits of the quantum-mechanical calculations is the possibility of specifying the most stable configuration of solvent molecules in an ion–solvent cluster. Hence, subtle differences between anion and cation solvation or between first and second solvation shells can be singled out, and much progress is to be expected in this area. The quantum-mechanical calculations also supply accurate descriptions of the short-range part of ion–atom radial distribution functions, which will be helpful in interpreting scattering data.

For calculations of larger ion–solvent aggregates or with solvents of greater molecular complexity, semiempirical methods (CNDO, INDO) have also been used quite extensively.[271] Recent results of such calculations may be found for cation–water interactions with varying geometry of the solvent cage[272] and for cation–dimethylacetamide and cation–methyl acetate interactions.[273] Calculations on anion and cation solvation by formic acid,[274] methanol,[275] and dimethylsulfoxide[276] have also been reported. Although the charge distributions and absolute energy values obtained in these calculations may differ from those calculated by *ab initio* techniques, they can still be useful in predicting trends within series of ions or solvent molecules.

5.5.2. Monte Carlo and Molecular Dynamics Methods

The basic principles involved in Monte Carlo calculations and molecular dynamics computer simulation have been discussed earlier in relation to the structural features of liquid water. The objective of these methods applied to solvation studies is to give a complete equilibrium and kinetic description of ensembles consisting of one or a few ions and the largest possible number of solvent molecules, using effective potentials given by electrostatic theory or derived from quantum-mechanical calculations. In principle, these computer simulation methods can provide a complete thermodynamic and kinetic (e.g., rotational and translational correlations) representation of ionic solvation. Several recent examples illustrate this rather nicely.

A Monte Carlo study[277] of the stepwise hydration of Li^+, Na^+, K^+, F^- and Cl^-, using an intermolecular potential from *ab initio* calculations (under the pairwise additivity assumption), shows remarkable agreement with the gas-phase data of the enthalpies and free energies of hydration. Monte Carlo and molecular dynamics investigations with larger solvent clusters (typically 200 water molecules)[278-279] yield ion–solvent radial distribution functions that compare favorably with experimental X-ray and neutron scattering data. At this stage any discrepancy between the calculated and experimental parameters probably still reflects the inadequacies of the potential functions used, but the methods are evidently capable of providing a realistic picture of ionic solvation. It may be remarked, however, that such calculations provide evidence of an "experimental" nature rather than tractable solvation theories. These "experiments" describe the condensation, from a noninteracting state, of model solvent molecules and ions to form a model solution. Results of such "experiments" are in themselves important, since they allow a verification of statistical-mechanical theories based on the same models.

5.5.3. The Origin of Hydrophobic Hydration

Although the existence of structural effects in ionic hydration is now well established, the quantitative treatment of structure-breaking and hydrophobic contributions is still in a primitive state. Some progress is being made on hydrophobic hydration and interaction,[280,281] but one recent development on the origin of hydrophobic hydration is worth mentioning. It is well known that hydrophobic hydration is accompanied by negative ΔH_h^θ and ΔS_h^θ. Since hydrophobic hydration is generally considered an unfavorable event, the positive contribution to ΔG_h^θ is assumed to come from the entropy:

$$\Delta G_h^\theta = \Delta H_h^\theta - T\Delta S_h^\theta$$

Similarly, the attraction between two hydrophobic solutes (hydrophobic interactions and bonding) is also considered to be entropy driven. Recently this interpretation has been challenged, e.g., by Shinoda and Fujihara[282] and Patterson and Barbe.[283] This matter has been clarified very elegantly by Frank and Lumry,[284] who considered the transfer of argon from hexane to water in two steps. For this they assumed that hydrazine could be considered, as a first approximation, equivalent to unstructured water. The experimental data are given for various transfer functions as

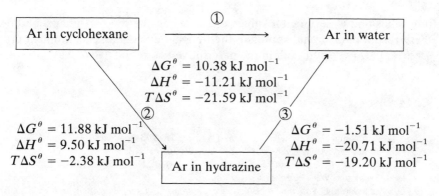

Therefore the structural contribution to hydrophobic hydration (step 3) is largely compensated ($\Delta H^\theta \simeq T\Delta S^\theta$) and the positive sign of ΔG^θ in step 1 is largely due to an unfavorable nonstructural enthalpy (step 2). On the other hand, the temperature dependence of ΔG^θ is primarily controlled by the structural terms. Similarly, the attraction between two hydrophobic solutes is in fact enthalpy driven. Therefore the structural hydration contribution can be quite large for many thermodynamic functions, e.g., heat capacities, but it is often the small residual nonstructural part that gives the sign to ΔG^θ.

5.6. Concluding Remarks

The principal aim of solvation theories consists of understanding the energetics, structure, and dynamics of the solvation phenomena and of providing tools for the prediction of the influence of solvation effects on the physical properties of electrolyte solutions. The brief survey of theoretical approaches presented above clearly shows that at the moment no single solvation theory meets these various goals.

The molecular description of the solvation phenomena appears to be best achieved through a combination of quantum-mechanical calculations and computer simulation techniques. However, the extension of such methods for predicting thermodynamic data for electrolytes in, for example, different solvents is difficult to perform, so the general application appears remote. On the other hand, continuum theories, which can be contained in relatively simple analytical expressions, must remain primitive approximations, since they cannot account for the molecular behavior of the solvent in the first solvation layer. Improvements of these theories can be achieved if a chemical model description of the first coordination sphere is given, as in the work of Beveridge and Schmielle.[244] However, until these approaches are tested for a variety of solvents and thermodynamic properties, little can be said of their true potential. Clearly, however, many recent and current studies are providing increasingly accurate data for many physical and thermodynamic properties of electrolytes in several typical solvents. Together with continuous progress in establishing experimental values of single-ion thermodynamic properties, these recent data should stimulate theoretical developments, and much progress may be expected both in the descriptive and predictive areas.

References

1. J. D. Bernal and R. H. Fowler, *J. Chem. Phys.* **1**, 515 (1933).
2. H. S. Frank and M. W. Evans, *J. Chem. Phys.* **13**, 507 (1945).
3. R. W. Gurney, *Ionic Processes in Solution*, McGraw-Hill, New York (1954).
4. H. S. Frank and W.-Y. Yen, *Discuss. Faraday Soc.* **24**, 133 (1957).
5. O. Ya. Samoilov, *Structure of Electrolyte Solutions and the Hydration of Ions*, English Translation Consultants Bureau Enterprises, New York (1965).
6. F. Franks, ed., *Water, a Comprehensive Treatise*, Plenum Press, New York (Vol. 1, 1972; Vol. 2, 1973; Vol. 3, 1973; Vol. 4, 1975; Vol. 5, 1975; Vol. 6, 1979).
7. A. Ben-Naim, *Water and Aqueous Solutions*, Plenum Press, New York (1974).
8. J. F. Coetzee and C. D. Ritchie, eds., *Solute–Solvent Interactions*, Marcel Dekker, New York (Vol. 1, 1969; Vol. 2, 1976).
9. S. Petrucci, ed., *Ionic Interactions*, Academic Press, New York (1971).
10. T. Erdey-Gruz, *Transport Phenomena in Aqueous Systems*, Wiley, New York (1974).
11. P. Kruss, *Liquids and Solutions*, Marcel Dekker, New York (1977).
12. J. E. Desnoyers, *Phys. Chem. Liq.* **7**, 63 (1977).
13. J. E. Desnoyers and C. Jolicoeur, *Modern Aspects of Electrochemistry*, Vol. 5, J. O'M. Bockris and B. E. Conway, eds., Plenum Press, New York (1969), p. 1.
14. J. I. Lagoswski, ed., *The Chemistry of Non-Aqueous Solvents*, Academic Press, New York (Vol. 1, 1966; Vol. 2, 1967; Vol. 3, 1970; Vol. 4, 1976; Vol. 5A, 1978; Vol. 5B, 1978).
15. T. C. Waddington, ed., *Non-Aqueous Solvent Systems*, Academic Press, New York (1965).
16. G. J. Janz and R. P. T. Tomkins, eds., *Non-Aqueous Electrolytes Handbook*, Academic Press, New York (Vol. 1, 1972; Vol. 2, 1973).
17. R. Bates, in *Solute–Solvent Interactions*, Vol. 1, J. F. Coetzee and C. D. Ritchie, eds., Marcel Dekker, New York (1969), Chap. 2.
18. R. G. Pearson, *J. Chem. Educ.* **45**, 581, 643 (1967).
19. M. J. Kamlet and R. W. Taft, *J. Am. Chem. Soc.* **98**, 377, 2336 (1976); T. Yokoyama, R. W. Taft, and M. J. Kamlet, *ibid.* **98**, 3233 (1976).
20. C. V. Krishnan and H. L. Friedman, in *Solute–Solvent Interactions*, Vol. 2, J. F. Coetzee and C. D. Ritchie, eds., Marcel Dekker, New York (1976), Chap. 9.
21. E. Price, in *The Chemistry of Non-Aqueous Solvents*, Vol. 1, J. I. Lagoswski, ed., Academic Press, New York (1966), Chap. 2.
22. F. Franks, ed., *Water, a Comprehensive Treatise*, Vol. 1, Plenum Press, New York (1972).
23. D. Eisenberg and W. Kauzmann, *The Structure and Properties of Water*, Oxford University Press, New York (1969).
24. W. A. P. Luck, ed., *Structure of Water and Aqueous Solutions*, Verlag Chemie, Weinheim (1974).
25. J. L. Kavanau, *Water and Solute–Water Interactions*, Holden-Day, San Francisco, California (1964).
26. F. Franks, in *Water, a Comprehensive Treatise*, Vol. 1, F. Franks, ed., Plenum Press, New York (1972), Chap. 1.
27. E. M. Arnett and D. R. McKelvey, in *Water, a Comprehensive Treatise*, Vol. 1, F. Franks, ed., Plenum Press, New York (1972), Chap. 6.
28. J. Timmermans, *Physico-Chemical Constants of Pure Organic Compounds*, Elsevier Publishing, New York (Vol. 1, 1950; Vol. 2, 1965).

29. P. R. Philip and C. Jolicoeur, *J. Solution Chem.* **4**, 105 (1975).
30. *Lange's Handbook of Chemistry*, 12th ed., J. A. Dean, ed., McGraw-Hill, New York (1979), pp. 10–125.
31. R. C. Wilhoit and B. J. Zwolinski, *J. Phys. Chem. D.* **2**, Supplement 1 (1973).
32. R. Zana, G. Perron, and J. E. Desnoyers, *J. Solution Chem.*, **8**, 729 (1979).
33. S. W. Benson, *J. Am. Chem. soc.* **100**, 5640 (1978).
34. H. S. Frank, in *Water, a Comprehensive Treatise*, Vol. 1, F. Franks, ed., Plenum Press, New York (1972), p. 515.
35. T. R. Chay and H. S. Frank, *J. Chem. Phys.* **57**, 2910 (1972).
36. A. Cabana and C. Jolicoeur, *Can. Spectrosc.* **12**, 1 (1967).
37. G. Nemethy, in *Structure of Water and Aqueous Solutions*, W. A. P. Luck, ed., Verlag Chemie, Weinheim (1974), Chap. 2, p. 74.
38. J. A. Pople, *Proc. R. Soc. London* **A205**, 163 (1951).
39. L. Pauling and R. E. Marsh, *Proc. Nat. Acad. Sci. USA* **38**, 112 (1952).
40. M. S. John, J. Grosh, T. Ree, and H. Eyring, *J. Chem. Phys.* **44**, 1465 (1966).
41. A. H. Narten, M. D. Danford, and H. A. Levy, *Discuss. Faraday Soc.* **43**, 97 (1967).
42. E. Kalman, E. Lengyel, G. Palinkas, L. Haklik, and E. Eke, in *Structure of Water and Aqueous Solutions*, W. A. P. Luck, ed., Verlag Chemie, Weinheim (1974), Chap. 5, p. 366.
43. A. H. Narten, *J. Chem. Phys.* **56**, 5681 (1972).
44. H. Lidner, Ph.D. dissertation, University of Karlsruhe, Karlsruhe, 1970.
45. G. E. Walrafen, in *Structure of Water and Aqueous Solutions*, W. A. P. Luck, ed., Verlag Chemie, Weinheim (1974), p. 151.
46. G. E. Walrafen, *J. Chem. Phys.* **48**, 244 (1968).
47. W. A. Senior and R. E. Verrall, *J. Phys. Chem.* **73**, 4242 (1969).
48. W. A. P. Luck, *Ber. Bunsenges. Phys. Chem.* **69**, 626 (1965).
49. W. A. P. Luck and W. Ditter, *Z. Naturforsch.* **24b**, 482 (1969).
50. J. Paquette and C. Jolicoeur, *J. Solution Chem.* **6**, 403 (1977).
51. W. A. P. Luck, ed., *Structure of Water and Aqueous Solutions*, Verlag Chemie, Weinheim (1974), Chap. 3, p. 247.
52. P. Philip and C. Jolicoeur, *J. Phys. Chem.* **77**, 3071 (1973).
53. H. E. Hallam, in *Structure of Water and Aqueous Solutions*, W. A. P. Luck, ed., Verlag Chemie, Weinheim (1974), Chap. 3, p. 286.
54. M. Falk and H. R. Wyss, *J. Chem. Phys.* **51**, 5727 (1969).
55. J. R. Scherer, M. K. Go, and S. Kint, *J. Phys. Chem.* **78**, 1304 (1974).
56. W. A. P. Luck and W. Ditter, *J. Mol. Struct.* **1**, 261 (1967).
57. W. A. P. Luck and W. Ditter, *Ber. Bunsenges, Phys. Chem.* **72**, 365 (1968).
58. J. A. Barker and R. O. Watts, *Chem. Phys. Lett.* **3**, 144 (1969).
59. G. N. Sarkisov, G. G. Malenkov, and V. G. Sashevskii, *Zh. Strukt. Khim.* **14**, 6 (1973).
60. G. C. Lie, E. Clementi, and M. Yoshimine, *J. Chem. Phys.* **64**, 2314 (1976).
61. A. Rahman and F. H. Stillinger, *J. Chem. Phys.* **55**, 3336 (1971); **57**, 1281 (1972); **60**, 1545 (1974).
62. C. G. Swain and R. F. Bader, *Tetrahedron* **10**, 183 (1969).
63. J. C. Owicki, B. R. Lentz, A. T. Hagler, and H. A. Scheraga, *J. Phys. Chem.* **79**, 2352 (1975).
64. J. E. Desnoyers and P. R. Philip, *Can. J. Chem.* **50**, 1094 (1972).
65. B. E. Conway and J. O'M. Bockris, *Modern Aspects of Electrochemistry*, Butterworths, London (1954), Chap. 2.
66. P. R. Philip and J. E. Desnoyers, *J. Solution Chem.* **1**, 353 (1972), and references cited therein.
67. J. L. Fortier, P. R. Philip, and J. E. Desnoyers, *J. Solution Chem.* **3**, 523 (1974).
68. H. L. Friedman and C. V. Krishnan, in *Water, a Comprehensive Treatise*, Vol. 3, F. Franks, ed., Plenum Press, New York (1973), Chap. 1.
69. F. J. Millero, in *Water and Aqueous Solutions*, R. A. Horne, ed., Wiley-Interscience, New York (1971).

70. C. V. Krishnan and H. L. Friedman, *J. Phys. Chem.* **74**, 2356 (1970).
71. F. Kawaizumi and R. Zana, *J. Phys. Chem.* **78**, 627, 1094 (1974).
72. C. Shin and C. M. Criss, *J. Solution Chem.* **7**, 205 (1978).
73. Y. S. Choi and C. M. Criss, *Discuss. Faraday Soc.* **64**, 204 (1977).
74. J. O'M. Bockris and A. K. N. Reddy, *Modern Electrochemistry*, Vol. 1, Plenum Press, New York (1970).
75. J. Padova, in *Modern Aspects of Electrochemistry*, Vol. 7, B. E. Conway and J. O'M. Bockris, eds., Plenum Press, New York (1972), Chap. 1.
76. N. Desrosiers, G. Perron, J. G. Mathieson, B. E. Conway, and J. E. Desnoyers, *J. Solution Chem.* **3**, 789 (1974).
77. F. Franks and D. S. Reid, in *Water, a Comprehensive Treatise*; Vol. 2, F. Franks, ed., Plenum Press, New York (1973), Chap. 5.
78. W. Y. Wen and S. Saito, *J. Phys. Chem.* **68**, 2639 (1964).
79. C. Jolicoeur and J. Boileau, *J. Solution Chem.* **3**, 889 (1974).
80. P. Kebarle, *Modern Aspects of Electrochemistry*, Vol. 9, J. O'M. Bockris and B. E. Conway, eds., Plenum Press, New York (1974).
81. P. Kebarle, *Annu. Rev. Phys. Chem.* **28**, 445 (1977).
82. R. Zana and E. Yeager, *J. Phys. Chem.* **70**, 954 (1966); **71**, 521 (1967).
83. B. E. Conway, *J. Solution Chem.* **7**, 721 (1978).
84. T. C. Waddington, *Trans. Faraday Soc.* **62**, 1482 (1966).
85. Y. Pointud and J. Juillard, *J. Chem. Soc. Faraday Trans. 1* **73**, 1048 (1977).
86. M. Y. Schrier, P. J. Turner, and E. E. Schrier, *J. Phys. Chem.* **79**, 1391 (1975).
87. R. B. Cassel and W. Y. Wen, *J. Phys. Chem.* **76**, 1369 (1972).
88. G. Perron, N. Desrosiers, and J. E. Desnoyers, *Can. J. Chem.* **54**, 2163 (1976).
89. C. Jolicoeur and J. C. Mercier, *J. Phys. Chem.* **81**, 1119 (1977).
90. J. G. Mathieson and B. E. Conway, *J. Solution Chem.* **3**, 455 (1974); *J. Chem. Soc. Faraday Trans. 1* **70**, 752 (1974).
91. I. Davidson, G. Perron, and J. E. Desnoyers, *Can. J. Chem.* **59**, 2212 (1981).
92. B. E. Conway and R. E. Verrall, *J. Phys. Chem.* **70**, 3952 (1966).
93. S. Cabani, G. Conti, and E. Matteoli, *J. Solution Chem.* **8**, 11 (1979).
94. H. Schneider, in *Solute–Solvent Interactions*, Vol. 2, J. F. Coetzee and C. D. Ritchie, eds., Marcel Dekker, New York (1976); Chap. 11.
95. W. F. Furter, ed., *Thermodynamic Behavior of Electrolytes in Mixed Solvents, Adv. Chem. Ser.* **155** (1976).
96. J. E. Desnoyers, G. Perron, J. P. Morel, and L. Avekidian, in *Chemistry and Physics of Aqueous Gas Solutions*, W. A. Adams, ed., Electrochemical Society, Princeton, New Jersey (1975), p. 172.
97. G. Perron, D. Joly, J. E. Desnoyers, L. Avedikian, and J. P. Morel, *Can. J. Chem.* **56**, 552 (1978).
98. R. H. Stokes and R. Mills, *Viscosity of Electrolytes and Related Properties*, Pergamon Press, New York (1965).
99. R. L. Kay, D. F. Evans, and M. Matesich, in *Solute–Solvent Interactions*, Vol. 2, J. F. Coetzee and C. D. Ritchie, eds., Marcel Dekker, New York (1976), Chap. 10.
100. R. L. Kay, in *Trace Inorganics in Water*, R. A. Baker, ed., *Adv. Chem. Ser.* **73**, 4 (1968).
101. R. Zwanzig, *J. Chem. Phys.* **38**, 1603 (1963); **52**, 3625 (1970).
102. C. M. Criss, *J. Phys. Chem.* **75**, 2532 (1971).
103. J. E. Desnoyers and G. Perron, *J. Solution Chem.* **1**, 199 (1972).
104. D. E. Irish and M. H. Brooker, in *Advances in Infrared and Raman Spectroscopy*, Vol. 2, R. J. H. Clark and R. E. Hester, eds., Heyden, London (1976), Chap. 6.
105. R. E. Verrall, in *Water, a Comprehensive Treatise*, Vol. 3, F. Franks, ed., Plenum Press, New York (1973), Chap. 5.
106. T. H. Lilley, in *Water, a Comprehensive Treatise*, Vol. 3, F. Franks, ed., Plenum Press, New York (1973), Chap. 6.

107. W. A. P. Luck, ed., *Structure of Water and Aqueous Solutions*, Verlag Chemie, Weinheim (1974), Chap. 3, p. 248.
108. D. E. Irish, in *Structure of Water and Aqueous Solutions*, W. A. P. Luck, ed., Verlag Chemie, Weinheim (1974), Chap. 3, p. 334.
109. M. C. R. Symons, in *Electron–Solvent and Anion–Solvent Interactions*, L. Kevan and B. Webster, eds., Elsevier, Amsterdam (1976), p. 311.
110. D. E. Irish, in *Physical Chemistry of Organic Solvent Systems*, A. K. Covington and T. Dickinson, eds., Plenum Press, New York (1973), p. 433.
111. H. R. Wyss and M. Falk, *Can. J. Chem.* **48**, 607 (1970).
112. K. A. Hartman, *J. Phys. Chem.* **70**, 270 (1966).
113. D. W. James and R. F. Armishaw, *Aust. J. Chem.* **28**, 1179 (1975), and references cited therein.
114. R. E. Hester and R. A. Plane, *Spectrochim. Acta* **23A**, 2289 (1967).
115. G. Brink and M. Falk, *Can. J. Chem.* **48**, 3019 (1970).
116. G. E. Walrafen, *J. Chem. Phys.* **55**, 768 (1971).
117. P. Dryjanski and Z. Kecki, *Rocz. Chem.* **44**, 1141 (1970).
118. D. M. Adams, M. J. Blandamer, M. C. R. Symons, and D. Waddington, *Trans. Faraday Soc.* **67**, 611 (1971).
119. J. F. Coetzee and W. R. Sharpe, *J. Solution Chem.* **1**, 77 (1972).
120. C. C. Addison, D. W. Amos, and D. Sutton, *J. Chem. Soc. A*, 2285 (1968).
121. S. Minc, Z. Kecki, and T. Gulik-Krzywicki, *Spectrochim. Acta* **19**, 353 (1963).
122. E. G. Hohn, J. A. Olander, and M. C. Day, *J. Phys. Chem.* **73**, 3380 (1969).
123. M. S. Greenberg and A. I. Popov, *J. Solution Chem.* **5**, 653 (1976).
124. H. L. Yeager, J. Fedyk, and R. J. Parker, *J. Phys. Chem.* **77**, 2407 (1973).
125. R. M. Moravie and J. Corset, *J. Chim. Phys.* **74**, 707 (1977).
126. B. S. Krumgalz, V. M. Ryabikova, S. Kh. Akopyan, and V. I. Borisova, *Russ. J. Phys. Chem.* **48**, 1538 (1974).
127. R. D. Green, J. S. Martin, W. B. McG. Cassie, and J. B. Hyne, *Can. J. Chem.* **47**, 1639 (1969).
128. A. Allerhand and P. von Ragué Schleyer, *J. Am. Chem. Soc.* **85**, 1233 (1963).
129. I. D. Kuntz, Jr. and C. J. Cheng, *J. Am. Chem. Soc.* **97**, 4852 (1975).
130. W. A. P. Luck and W. Ditter, *J. Mol. Struct.* **1**, 261 (1967).
131. W. A. P. Luck and W. Ditter, *Ber. Bunsenges. Phys. Chem.* **72**, 365 (1968).
132. D. Williams and W. Millet, *Phys. Rev.* **66**, 6 (1944).
133. K. W. Bunzl, *J. Phys. Chem.* **71**, 1358 (1967).
134. W. A. P. Luck, *Ber. Bunsenges. Phys. Chem.* **69**, 69 (1965).
135. W. A. P. Luck, ed., *Structure of Water and Aqueous Solutions*, Verlag Chemie, Weinheim (1974), pp. 278–279.
136. J. D. Worley and I. M. Klotz, *J. Phys. Chem.* **45**, 2868 (1966).
137. C. Jolicoeur, N. D. The, and A. Cabana, *Can. J. Chem.* **49**, 2008 (1971).
138. P. R. Philip and C. Jolicoeur, *J. Phys. Chem.* **77**, 3071 (1973).
139. C. Jolicoeur, J. Paquette, and M. Lucas, *J. Phys. Chem.* **82**, 1051 (1978).
140. P. Philip and C. Jolicoeur, *J. Solution Chem.* **4**, 3 (1975).
140. P. Philip and C. Jolicoeur, *J. Solution Chem.* **4**, 3 (1975).
141. G. E. Walrafen, *J. Chem. Phys.* **40**, 3249 (1964).
142. R. F. Armishaw and D. W. James, *J. Phys. Chem.* **80**, 501 (1976).
143. D. A. Draegert and D. Williams, *J. Chem. Phys.* **48**, 401 (1968).
144. G. E. Walrafen, *J. Chem. Phys.* **47**, 5258 (1967).
145. D. W. James, R. F. Armishaw, and R. L. Frost, *J. Phys. Chem.* **80**, 1346 (1976).
146. R. E. Hester and R. A. Plane, *Inorg. Chem.* **3**, 768 (1964).
147. I. Nakagawa and T. Shimanouchi, *Spectrochim. Acta* **20**, 429 (1964).
148. D. E. Irish and T. Jarv, *Discuss. Faraday Soc.* **64**, 95 (1977).
149. W. F. Edgell, J. Lyford, R. Wright, W. Risen, and A. Watts, *J. Am. Chem. Soc.* **92**, 2240 (1970).

150. C. P. Nash, T. C. Donnelly, and P. A. Rock, *J. Solution Chem.* **6**, 663 (1977).
151. T. J. V. Findlay and M. C. R. Symons, *J. Chem. Soc. Faraday Trans. 2* **4**, 820 (1976).
152. D. E. Irish, in *Ionic Interactions from Dilute Solutions to Fused Salts*, Vol. 11, S. Petrucci, ed., Academic Press, New York (1971), p. 188.
153. D. E. Irish and A. R. Davis, *Can. J. Chem.* **46**, 943 (1968).
154. J. T. Bulmer, T. G. Chang, P. J. Gleeson, and D. E. Irish, *J. Solution Chem.* **12**, 969 (1975).
155. M. Smyrl and J. P. Devlin, *J. Chem. Phys.* **61**, 1596 (1974).
156. G. Ritzhaupt and J. P. Develin, *J. Chem. Phys.* **62**, 1982 (1975).
157. D. E. Irish and A. R. Davis, *Can. J. Chem.* **46**, 943 (1968).
158. A. K. Covington and K. E. Newman, *Modern Aspects of Electrochemistry*, Vol. 12, J. O'M. Bockris and B. E. Conway, eds., Plenum Press, New York (1977), p. 41.
159. E. von Goldammer, *Modern Aspects of Electrochemistry*, Vol. 10, J. O'M. Bockris and B. E. Conway, eds., Plenum Press, New York (1975), p. 1.
160. H. G. Hertz, in *Water, a Comprehensive Treatise*, Vol. 3, F. Franks, ed., Plenum Press, New York (1973), p. 301.
161. P. Laszlo, *Angew. Chem.* **17**, 254 (1978).
162. M. D. Zeidler, in *Water, a Comprehensive Treatise*, Vol. 2, F. Franks, ed., Plenum Press, New York (1973), p. 529.
163. I. R. Lantzke, in *Physical Chemistry of Organic Solvent Systems*, A. K. Covington and T. Dickinson, eds., Plenum Press, New York (1973), p. 483.
164. R. G. Bryant, *Annu. Rev. Phys. Chem.* **29**, 167 (1978).
165. A. I. Popov, in *Solute–Solvent Interactions*, Vol. 2, J. F. Coetzee and C. D. Ritchie, eds., Marcel Dekker, New York (1976), p. 271.
166. H. G. Hertz and M. D. Zeidler, *The Hydrogen Bond*, Vol. 3, P. Shuster, G. Zundel, and C. Sandorfy, eds., North-Holland, Amsterdam (1976), p. 1029.
167. C. Deverell, *Progress in N.M.R. Spectroscopy*, Vol. 4, J. W. Emsley, J. Feeney, and L. H. Sutcliffe, eds., Pergamon Press, New York (1969), p. 235.
168. J. A. Pople, W. G. Schneider, and H. J. Bernstein, *High Resolution Nuclear Magnetic Resonance*, McGraw-Hill, New York (1959).
169. A. Abragam, *The Principles of Nuclear Magnetism*, Oxford University Press, Oxford (1961).
170. C. P. Slichter, *Principles of Magnetic Resonance*, Harper and Row, New York (1963).
171. J. C. Hindman, *J. Chem. Phys.* **44**, 4582 (1966).
172. Z. Luz and G. Y. Yagil, *J. Phys. Chem.* **70**, 554 (1966).
173. F. Fister and H. G. Hertz, *Ber. Bunsenges, Phys. Chem.* **71**, 1032 (1967).
174. D. W. G. Smith and J. G. Powles, *Mol. Phys.* **10**, 451 (1966).
175. J. F. Coetzee and W. R. Sharpe, *J. Phys. Chem.* **75**, 3141 (1971).
176. J. N. Shoorley and B. Alder, *J. Chem. Phys.* **23**, 805 (1955).
177. J. C. Hindman, *J. Chem. Phys.* **36**, 1000 (1962).
178. S. Ormondroyd, E. A. Phillpott, and M. C. R. Symons, *Trans. Faraday Soc.* **67**, 1253 (1971).
179. H. Fukui, K. Mivra, T. Ugai, and M. Abe, *J. Phys. Chem.* **81**, 1205 (1977).
180. J. W. Stockton and J. S. Martin, *J. Am. Chem. Soc.* **94**, 6921 (1972).
181. J. W. Akitt, *J. Chem. Soc. Dalton Trans.* **42** (1973).
182. J. W. Akitt, *Discuss. Faraday Soc.* **64**, 102 (1977).
183. J. Davies, S. Ormondroyd, and M. C. R. Symons, *Trans. Faraday Soc.* **67**, 3465 (1971).
184. C. H. Langford and T. R. Stangle, *J. Am. Chem. Soc.* **91**, 4014 (1969).
185. T. R. Stengle, Y. C. E. Pan, and C. H. Langford, *J. Am. Chem. Soc.* **94**, 9037 (1972).
186. M. S. Greenberg, R. L. Bodner, and A. I. Popov, *J. Phys. Chem.* **77**, 2449 (1973).
187. R. H. Erlich and A. I. Popov, *J. Am. Chem. Soc.* **93**, 5620 (1971).
188. G. E. Maciel, J. K. Hancock, L. F. Lafferty, P. A. Mueller, and W. K. Musker, *Inorg. Chem.* **5**, 554 (1966).
189. J. D. Halliday, R. E. Richards, and R. E. Sharpe, *Proc. R. Soc. London Ser. A* **313**, 45 (1969).
190. E. G. Bloor and R. G. Kidd, *Can. J. Chem.* **50**, 3926 (1972).
191. A. K. Covington, T. H. Lilley, K. E. Newman, and G. A. Porthouse, *J. Chem. Soc. Faraday Trans. 1* **69**, 963 (1973).

192. E. von Goldammer and H. G. Hertz, *J. Phys. Chem.* **74**, 3734 (1970).
193. H. G. Hertz and M. D. Zeidler, *Ber. Bunsenges. Phys. Chem.* **68**, 621 (1964).
194. O. W. Howarth, *J. Chem. Soc. Faraday Trans 1* **12**, 2303 (1975).
195. L. Endom, H. G. Hertz, B. Thul, and M. D. Zeidler, *Ber. Bunsenges. Phys. Chem.* **71**,1008 (1967).
196. H. G. Hertz, *Ber. Bunsenges, Phys. Chem.* **71**, 979, 999 (1967).
197. J. E. Enderby and G. W. Neilson, in *Water, a Comprehensive Treatise*, Vol. 6, F. Franks, ed., Plenum Press, New York (1979), p. 1.
198. A. H. Narten and H. A. Levy, in *Water, a Comprehensive Treatise*, Vol. 1, F. Franks, ed., Plenum Press, New York (1972), p. 311.
199. D. I. Page, in *Water, a Comprehensive Treatise*, Vol. 1, F. Franks, ed., Plenum Press, New York (1972), Chap. 9.
200. R. F. Kruh, *Chem. Rev.* **62**, 319 (1962).
201. J. D. Bernal, *Trans. Faraday Soc.* **33**, 27 (1937).
202. G. W. Brady and J. T. Krause, *J. Chem. Phys.* **27**, 304 (1957).
203. A. N. Narten, F. Vaslow, and H. A. Levy, *J. Chem. Phys.* **58**, 5017 (1973).
204. R. Triolo and A. H. Narten, *J. Chem. Phys.* **63**, 3624 (1975).
205. W. Bol, G. Gerrits, and C. van Panthaleon van Eck, *J. Appl. Crystallogr.* **3**, 486 (1970).
206. A. K. Soper, G. W. Neilson, and J. E. Enderby, *J. Phys. Chem.* **10**, 1793 (1977).
207. I. R. Lantzke, in *Physical Chemistry of Organic Solvent Systems*, A. K. Covington and T. Dickinson, eds., Plenum Press, New York (1973), p. 405.
208. M. C. R. Symons, *Annu. Rep. Chem. Soc.* **73**, 91 (1976).
209. E. M. Kozower, *J. Am. Chem. Soc.* **80**, 3253 (1958); **82**, 2188 (1960).
210. M. J. Blandamer and M. F. Fox, *Chem. Rev.* **70**, 59 (1970).
211. M. Smith and M. C. R. Symms, *Discuss. Faraday Soc.* **24**, 206 (1957).
212. J. E. Wertz and J. R. Bolton, *Electron Spin Resonance*, McGraw-Hill, New York (1972).
213. P. W. Atkins, *Adv. Mol. Relaxation Processes* **2**, 121 (1972).
214. A. Hudson and G. R. Luckhurst, *Chem. Rev.* **69**, 191 (1969).
215. D. Kivelson, *J. Chem. Phys.* **33**, 1094 (1960).
216. *Biological Applications of Electron Spin Resonance*, H. M. Swartz, J. R. Bolton, and D. C. Borg, eds., Wiley-Interscience, New York (1972).
217. V. M. Stockhavsen, *Ber. Bunsenges. Phys. Chem.* **77**, 338 (1973).
218. L. Burlamacchi, G. Martini, and E. Tiezzi, *J. Phys. Chem.* **74**, 3980 (1970).
219. M. Rubinstein, A. Baram, and Z. Luc, *Mol. Phys.* **20**, 67 (1971).
220. M. M. Iannuzzi and P. H. Rieger, *Inorg. Chem.* **14**, 2895 (1975).
221. K. M. Sancier, *J. Phys. Chem.* **72**, 1317 (1968).
222. D. C. McCain and R. J. Myers, *J. Phys. Chem.* **72**, 4115 (1968).
223. W. B. Lewis and L. O. Morgan, in *Transition Metal Chemistry*, Vol. 4, R. L. Corlin, ed., Marcel Dekker, New York (1967), p. 33.
224. N. S. Angerman and R. B. Jordan, *J. Chem. Phys.* **54**, 837 (1971).
225. R. N. Rogers and G. E. Pake, *J. Chem. Phys.* **33**, 1107 (1960).
226. T. E. Eagles and R. E. D. McClung, *Can. J. Chem.* **53**, 1492 (1975).
227. B. Kwert and D. Kivelson, *J. Chem. Phys.* **64**, 5206 (1976).
228. N. M. Atherton, M. R. Manterfield, B. Oral, and F. Zorlu, *J. Chem. Soc. Faraday Trans. 1* **73**, 430 (1977).
229. C. Jolicoeur and H. L. Friedman, *J. Solution Chem.* **3**, 15 (1973).
230. C. Jolicoeur and H. L. Friedman, *J. Phys. Chem.* **75**, 165 (1971).
231. C. Jolicoeur and H. L. Friedman, *Ber. Bunsenges Phys. Chem.* **75**, 248 (1971).
232. C. C. Mills and E. L. King, *J. Am. Chem. Soc.* **92**, 3017 (1970).
233. T. E. Gough, in *Physical Chemistry of Organic Solvent Systems*, A. K. Covington and T. Dickinson, eds., Plenum Press, New York (1973), p. 461.
234. I. M. McKinney, *Electroanal. Chem.* **10**, 97 (1976).
235. M. Eigen and L. de Maeyer, *Techniques of Organic Chemistry*, Vol. 8, A. Weissberger, ed., Wiley, New York (1963), Part 2, Chap. 18.

236. S. Petrucci, ed., *Ionic Interactions*, Academic Press, New York (1971), Chap. 7, p. 39.
237. G. W. Liesegang, M. M. Farrow, F. Arce-Vaguez, N. Purdie, and E. M. Eyring, *J. Am. Chem. Soc.* **99**, 3240 (1977).
238. J. B. Hasted, *Aqueous Dielectrics*, A. D. Buckingham, ed., Chapman and Hall, London (1973).
239. R. Pottel, in *Water, a Comprehensive Treatise*, Vol. 3, F. Franks, ed., Plenum Press, New York (1973), Chap. 8.
240. R. H. Stokes, *J. Am. Chem. Soc.* **86**, 979 (1964).
241. D. S. V. Jain, *Indian J. Chem.* **4**, 466 (1965).
242. R. M. Noyes, *J. Am. Chem. Soc.* **84**, 513 (1962).
243. I. Padova, *J. Chem. Phys.* **56**, 1606 (1972).
244. D. L. Beveridge and G. W. Schmielle, *J. Phys. Chem.* **79**, 2562 (1975).
245. M. H. Abraham and J. Liszi, *J. Chem. Soc. Faraday Trans. 1* **74**, 2858 (1978).
246. H. S. Frank, *J. Chem. Phys.* **23**, 2023 (1955).
247. J. E. Desnoyers, R. E. Verrall, and B. E. Conway, *J. Chem. Phys.* **43**, 243 (1965).
248. L. A. Dunn, *J. Solution Chem.* **3**, 1 (1974).
249. B. E. Conway, R. E. Verrall, and J. E. Desnoyers, *Z. Phys. Chem. (Leipzig)* **230**, 157 (1965).
250. J. W. Cobble, unpublished results.
251. D. D. Eley and M. G. Evans, *Trans. Faraday Soc.* **34**, 1093 (1938).
252. A. D. Buckingham, *Discuss. Faraday Soc.* **24**, 151 (1957).
253. J. S. Muirhead-Gould and K. J. Laidler, *Trans. Faraday Soc.* **63**, 944 (1967).
254. S. Goldman and R. G. Bates, *J. Am. Chem. Soc.* **94**, 1476 (1972).
255. J. O'M. Bockris and P. P. S. Saluja, *J. Phys. Chem.* **76**, 2298 (1972).
256. G. Somsen, *Recl. Trav. Chim. Pays Bas* **85**, 526 (1966).
257. L. Weeda and G. Somsen, *Recl. Trav. Chim. Pays Bas* **86**, 263 (1967).
258. D. D. Eley, *Trans. Faraday Soc.* **55**, 1281, 1421 (1939).
259. H. Reiss, H. L. Frisch, and J. L. Lebovitz, *J. Chem. Phys.* **31**, 369 (1959); **32**, 119 (1960).
260. R. A. Pierotti, *J. Phys. Chem.* **69**, 281 (1965).
261. M. Lucas, *Bull. Soc. Chim. Fr.* **9**, 2994 (1969).
262. A. Feillolay and M. Lucas, *J. Phys. Chem.* **76**, 3068 (1972).
263. M. Lucas and A. Feillolay, *J. Phys. Chem.* **75**, 2330 (1971).
264. W. L. Masterton, D. Bolocofsky, and T. P. Lee, *J. Phys. Chem.* **75**, 2809 (1971).
265. S. K. Shoor and K. E. Gubbins, *J. Phys. Chem.* **73**, 498 (1969).
266. N. Desrosiers and J. E. Desnoyers, *Can. J. Chem.* **54**, 3800 (1976).
267. P. Schuster, W. Jakubetz, and W. Marius, *Topics in Current Chemistry*, Vol. 60, Springer-Verlag, New York (1975), p. 1.
268. E. Clementi, *Determination of Liquid Water Structure, Coordination of Numbers for Ions and Solvation for Biological Molecules*, Lecture Notes in Chemistry, Vol. 2, Springer-Verlag, New York (1976).
269. H. Kistenmacher, H. Popkie, and E. Clementi, *J. Chem. Phys.* **61**, 799 (1974).
270. A. Pullman and A. M. Armbruster, *Chem. Phys. Lett.* **56**, 558 (1975).
271. J. A. Pople and D. L. Beveridge, *Approximate Molecular Orbital Theory*, McGraw-Hill, New York (1970).
272. R. E. Burton and J. Daly, *Trans. Faraday Soc.* **66**, 1281 (1970).
273. P. V. Kostetsky, V. T. Ivanov, Yu. A. Orchinnikov, and G. Shchembelov, *FEBS Lett.* **30**, 205 (1973).
274. B. M. Rode, *Chem. Phys. Lett.* **25**, 369 (1974).
275. M. Salomon, *Can. J. Chem.* **53**, 3194 (1975).
276. M. S. Goldenberg, P. Kruus, and S. K. F. Luk, *Can. J. Chem.* **53**, 1007 (1975).
277. M. R. Mruzik, F. F. Abraham, D. E. Schreiber, and G. M. Pound, *J. Chem. Phys.* **64**, 481 (1976).
278. J. Fromm, E. Clementi, and R. O. Watts, *J. Chem. Phys.* **62**, 1388 (1975).

279. K. Heinznger and P. C. Vogel, *Z. Naturforsch.* **31a**, 463 (1976).
280. A. Ben-Naim, *Water and Aqueous Solutions*, Plenum Press, New York (1974).
281. Z. Elkoshi and A. Ben-Naim, *J. Chem. Phys.* **70**, 1552 (1979).
282. K. Shinoda and M. Fujihara, *Bull. Chem. Soc. Jpn.* **41**, 2612 (1968).
283. D. Patterson and M. Barbe, *J. Phys. Chem.* **80**, 2435 (1976).
284. R. Lumry, in *Bioenergetics and Thermodynamics: Model Systems*, A. Braibanti, ed., D. Reidel, Boston (1980), p. 405.

2

Ionic Interactions and Activity Behavior of Electrolyte Solutions

B. E. CONWAY

1. Introduction

1.1. Scope of the Chapter

In the spirit of this treatise the present account of ionic interaction behavior in electrolyte solutions will cover the subject in both a general and a historical way rather than give a synoptic account only of recent work. Thus, first, the principles and main results of the classical Debye–Hückel theory will be presented with a commentary that distinguishes approximations arising from the model from those of a mathematical kind for various conditions, together with a discussion of the semiempirical approximations for various conditions and extensions that give an account of the specificity of ionic interactions.

The relation of the Debye–Hückel theory to modern statistical-mechanical treatments of ionic solutions in the "primitive" and other levels of modeling will then be examined. The role of the solvent's dielectric properties, especially with regard to hydration effects in ionic activity behavior at appreciable concentrations will be critically discussed in the light of the results of the statistical-mechanical calculations on hard-sphere models, taken

---

B. E. CONWAY • Chemistry Department, University of Ottawa, Ottawa, Ontario, Canada K1N 9B4.

up to concentrations in the $1 \sim 2\,M$ range. Extensions to cases where effective interaction potentials more complex than the hard-sphere repulsion, plus long-range coulombic interaction (primitive model) are considered, e.g., in the Friedman approach to "Hamiltonian" models, will be covered.

Most considerations will be limited to the thermodynamic behavior of aqueous electrolytes at ordinary temperatures, where comparisons between theoretical results and reliable experimental data can be most satisfactorily made. In many nonaqueous media ion pairing complicates the interpretation of experimental behavior.

Connections to both older and more modern treatments of ion distribution in double layers at charged interfaces are noted.

1.2. Historical Aspects of Electrolyte Behavior

The anomalous behavior of electrolytes as solutes in polar solvent media was early recognized in the work of Ostwald and of Arrhenius.[1] The concept and the phenomenon of spontaneous dissociation through a solvolytic reaction, e.g.,

$$HA + S \rightarrow HS^+ + A^- \tag{1}$$

as in acid dissociations in a solvent S (e.g., water), or the direct production of free solvated ions from a crystal lattice already containing ions, e.g.,

$$M^{z+}A^{z-} + (m + a)S \rightarrow M^{z+}(mS) + A^{z-}(aS) \tag{2}$$

were not well understood before the ideas of Clausius and of Arrhenius were proposed and, somewhat later, reluctantly accepted.

Appreciable "dissociation" was recognized on account of the unusual osmotic properties of electrolytes, which required, according to van't Hoff, the introduction of a factor i which defined the ratio of observed osmotic pressure π to that expected π_i for ideal solution behavior: "$i = \pi/\pi_i$." For simple $1:1$ electrolytes in water $i \simeq 2$ indicates dissociation (or dissolution) into 2 g ions per mole of salt.

Attempts to relate the degree of dissociation α at a concentration c to the observed ratio of equivalent conductance Λ_c at c and the infinite dilution value Λ_∞ were made by Arrhenius, who supposed

$$\alpha = \Lambda_c/\Lambda_\infty \tag{3}$$

i.e., that Λ_c differed from Λ_∞ only on account of incomplete dissociation of the electrolyte solute.

Considered in relation to the formal dissociation of a salt MA,

$$MA \underset{K}{\overset{\alpha}{\rightleftharpoons}} M^+ + A^-$$

$$(1 - \alpha)c \qquad \alpha c \qquad \alpha c \tag{4}$$

to a degree α, Arrhenius's ratio for α led to the well-known Ostwald dilution law, which enabled Λ_c to be expressed as a function of c:

$$K\left(\frac{\Lambda_\infty}{\Lambda_c}\right) - K = \frac{\Delta_c}{\Delta_\infty} c \qquad (5)$$

This allows a linear test plot of Λ_c as $f(c)$ to be made.

For electrolytes for which $i \simeq 2$ ($1:1$ type) or $i \simeq 3$ (for $1:2$ or $2:1$ types) Eq. (5) is *not* obeyed, but rather Λ_c is diminished from the infinite dilution value Λ_∞ in proportion to $c^{1/2}$, as found in the accurate work of Kohlrausch.

For such electrolytes Λ_c differs from Λ_∞, mainly, not because $\alpha < 1$, but because of long-range free ion–ion interactions. Correspondingly, for such electrolytes the colligative properties (osmotic pressure, elevation of bp, depression of fp) differ from the ideal values not so much because of incomplete dissociation, but rather on account of the long-range coulombic interactions between the ions of the electrolyte.

For what are now called weak electrolytes, the dissociation factor α is the main factor ($\alpha \ll 1$) determining both the conductance behavior and the thermodynamic behavior, so that the Ostwald dilution law and the Arrhenius ratio, giving α, are approximately correct.

The long-range coulombic factor determines the thermodynamic properties of dilute, strongly dissociated ($\alpha \to 1$) electrolytes and is the basis for explaining deviations of such systems from ideality. Nevertheless, short-range effects are always significant in all but the most dilute solutions—firstly, in regard to recognition of the finite sizes of ions in close encounters and in ionic distribution and, secondly, in cases where incomplete dissociation, or ion association, is a significant but not dominant aspect of the behavior of the electrolyte. In this regard some distinction has to be made between effects that arise from the chemically distinguishable situations of (1) incomplete dissociation in, e.g., a solvolytic reaction and (2) association into ion pairs of a fraction of already independent ions which have arisen in an ionic crystal dissolution process† or from a solvolytic reaction such as in Eq. (1). Spectroscopic (e.g., Raman, IR, or NMR) methods must sometimes be used to distinguish between these situations.[2,3]

In thermodynamic treatments of ionic solution behavior, and in analysis of thermodynamic properties of electrolytes, it is usually necessary to be able to distinguish long-range interaction effects giving rise to the nonideality of ionic solutions from short-range, incomplete dissociation or ion-pair association effects.

Additionally, at high concentrations of electrolytes, ion solvation behavior gives rise to large apparent deviations from ideality due to the lowered solvent activity caused by the strong ion–solvent interactions (dielectric polarization

---

† The distinction between "potential" and "intrinsic" electrolytes is implied here, or the difference between "ionogens" and "ionophores" (Fuoss).

or ion–dipole interactions) normally involved in all solutions of electrolytes in polar media.

The thermodynamic treatment of the behavior of solutions of electrolytes in terms of their activities or activity coefficients is based on an evaluation of both the long-range ion–ion interactions and the ion–solvent interaction effects that begin to be important at moderate or high concentrations, while special quasichemical or electrostatic model approaches are required for treating the short-range effects leading to ion pairing.

We shall begin the treatment of activity behavior with some formal matters concerned with the significance and definitions of the activity and osmotic coefficients that characterize the thermodynamic behavior of electrolyte solutions.

2. Formal Thermodynamic Relations for Activity and Osmotic Coefficients in Relation to the Chemical Potential μ

2.1. Chemical Potential and Activity

For reference purposes in this volume and in order to provide a basis for material that follows in later sections of this chapter, it will be useful to summarize a number of general thermodynamic relations basic to the treatment of binary solutions—in this case electrolytes.

In most work on binary solutions of *non*electrolytes concentration is expressed as a mole fraction X, using "mole fraction statistics," or "volume" or "site fraction statistics" for polymer solutions.[4,5] Thus, for a substance i,

$$\mu_i = \mu_i^0 + RT \ln X_i \tag{6}$$

holds for an ideal solution. The activity coefficients for this type of concentration scale are referred to as "rational" activity coefficients, and the standard state, as Eq. (6) implies, is pure i, $(X_i = 1)$. In some cases volume fraction statistics have been used in the discussion of electrolyte solution behavior.[6]

For electrolyte solutions, where "pure i" as the solute salt is an inconvenient and practically inaccessible standard state in most cases, a molar or molal concentration scale for solute i is usually employed (c for molar concentration, m for molal concentration). However, the activity of the *solvent* is usually expressed on the mole fraction scale; i.e., for infinite dilution of solute salt the solvent activity a_1 is taken as *unity*.

The standard state for the solute salt is chosen in relation to an equation for the activity a_i similar to Eq. (6) but written as

$$\mu_i = \mu_i^0 + RT \ln a_i \tag{7}$$

where a_i may be on a molality (m_i) or a molarity (c_i) scale. If the concentration scale is chosen as molality (this has obvious experimental advantages, since a liter of solution involved on the molarity scale can contain a varying number of moles of water, depending on the electrostriction caused by the ions of the solute and their intrinsic volume in solution), the standard state is chosen as "hypothetical unit molality." The corresponding activity coefficient is written as

$$\gamma_i = a_i/m_i \tag{8}$$

with $\gamma_i \to 1$ as $m_i \to 0$. Ideal solution behavior is approached as $m_i \to 0$, with γ_i tending to unity. Then

$$\mu_{i,m} = \mu_{i,m}^0 + RT \ln m_i\gamma_i \tag{9}$$

$$\mu_{i,m} = \mu_{i,m}^0 + RT \ln m_i + RT \ln \gamma_i \tag{10}$$

The last term of Eq. (10) is, of course, to be recognized as the "nonideal" contribution to the chemical potential of i when the concentration is m_i molal and is often referred to as the "excess" partial molal free energy. It is this quantity, $RT \ln \gamma_i$, that theoretical treatments of ionic solutions aim to calculate, or derivatives of it with respect to T or P which give the negative of the excess entropy or the excess volume. Also, the concentration dependence of γ_i is a critical quantity involved in experimental and theoretical evaluations of the thermodynamic behavior of electrolytes.

The significance of the word *hypothetical* in the definition of the standard state requires further comment. The standard state solution has the same stoichiometric composition as an actual $1\,m$ salt solution, but its "hypothetical" nature arises because all the real interactions in such a $1\,m$ solution are imagined to be absent. Hence γ_i must be taken as unity for the hypothetical $1\,m$ standard state, the same as its value at infinite dilution.

Experimentally, of course, a $1\,m$ solution will usually have an activity coefficient greater or less than unity, depending on the type of ions and solvent, and only coincidentally will the activity coefficient equal unity.

The hypothetical $1\,m$ standard state is such that Eqs. (8) and (9) and (10) for γ_i and μ_i, respectively, hold at all temperatures and pressures. Then, since the partial molal volume $\bar{V}_i$ is the derivative $(\partial\mu_i/\partial P)_T$ and the partial molal enthalpy $\bar{H}_i$ is $[\partial(\mu_i/T)/\partial(1/T)]_P$, it follows that the standard state defined above requires that both $\bar{V}_i$ and $\bar{H}_i$ have the *same* values in the standard state as at infinite dilution.

On the molarity scale the equation analogous to Eq. (9) is

$$\mu_{i,c} = \mu_{i,c}^0 + RT \ln c_iy_i \tag{11}$$

with the molar activity coefficient $y_i \to 1$ as $i \to 0$. Similarly, for mole fractions [cf. Eq. (6)]

$$\mu_{i,x} = \mu_{i,x}^0 + RT \ln Xf_i \tag{12}$$

where f is the "rational" or mole fraction activity coefficient.

2.2. Relations between the Activity Coefficients on Various Scales

The relations between γ_i, y_i, and f_i follow from the respective three concentration scales and the densities d and d_0 of the solvent and solution, respectively. Following the equations conveniently derived by Stokes and Robinson,[7] it is easy to show that

$$\gamma_\pm = \frac{d - 0.001cW_2}{d_0} y_\pm = \frac{c}{md_0} y_\pm \tag{13}$$

$$y_\pm = (1 + 0.001mW_2)\frac{d_0}{d} \gamma_\pm = \frac{md_0}{c} \gamma_\pm \tag{14}$$

$$f_\pm = \gamma_\pm(1 + 0.001\nu W_1 m) \tag{15a}$$

$$f_\pm = y_\pm \frac{d + 0.001c(\nu W_1 - W_2)}{d_0} \tag{15b}$$

where $\nu = \nu_+ + \nu_-$ is the number of moles of ions formed from 1 mol of ionized solute salt, W_1 and W_2 are the molecular weights of the solvent and the solute, respectively, d and d_0 are the respective densities of the solution and the solvent, and m and c are the molal or molar concentrations referred to earlier.

It is also useful to note the interconversions between c and m that can be made by means of the equations

$$c = \frac{md}{1 + 0.001mW_2} \quad \text{and} \quad m = \frac{c}{d - 0.001cW_2} \tag{16}$$

2.3. Chemical Species Referred to in Activity or Chemical Potential Relations for Ions of Electrolytes

The chemical species whose concentrations are measured by c or m are the actual ions, irrespective of their extent of solvation; i.e., the concentration of the solvent in a solution corresponding to a salt concentration c or m is counted as the total solvent, irrespective of what fraction of it is bound up in the solvation shells of the ions. It will be seen later that this is a point of some importance when the activity behavior of salts at high concentrations is considered, i.e., when an appreciable fraction of the available solvent is bound in the hydration shells of the ions. In the theoretical, e.g., Debye–Hückel, treatments it is the interaction between the *solvated* ions that are considered.

2.4. Individual Ionic and Mean Salt Activity Coefficients

The equations written earlier in this section have referred to individual ions i. However, it is well known (see Guggenheim[8]) that activities or activity coefficients of individual ions are not accessible to experimental measurement because no true thermodynamic way exists or is feasible for the μ_i quantities for the individual types of ions to be measured in a solution.

Nevertheless, "individual" ionic activity coefficients have received some attention,[9-13] and the significance of such quantities will be discussed later, especially in relation to the specificities of individual ion–solvent interactions† and to the case of ions of a salt having very different radii.

Experimentally, only a (geometric) "mean" activity coefficient is accessible to measurement and it is denoted by $\gamma_\pm$, $y_\pm$, or $f_\pm$, corresponding to the concentration scale involved. The relation of the mean activity coefficient, e.g., $y_\pm$ for the c scale, to the individual activity coefficients is conveniently obtained as follows.

The activity coefficient y_i is defined, as shown above, in terms of the chemical potentials μ_i of the ions of an electrolyte. For a salt of charge type z_+, z_- corresponding to a stoichiometry ν_+, ν_-, i.e., $M_{\nu_+}^{z_+}A_{\nu_-}^{z_-}$, the total μ is given by

$$\mu = \nu_+\mu_{M^{z+}}^0 + \nu_-\mu_{A^{z-}}^0 + \nu_+RT \ln a_{M^{z+}} + \nu_-RT \ln a_{A^{z-}} \tag{17}$$

in terms of the activities a of the indicated ions. Introducing the ionic activity coefficients $y_{M^{z+}}$ and $y_{A^{z-}}$ and the ionic concentrations $c_{M^{z+}}$ and $c_{A^{z-}}$ leads to

$$\mu = \nu_+\mu_{M^{z+}}^0 + \nu_-\mu_{A^{z-}}^0 + RT \ln (c_{M^{z+}})^{\nu_+} + RT \ln (c_{A^{z-}})^{\nu_-}$$
$$+ RT \ln (y_{M^{z+}})^{\nu_+}(y_{A^{z-}})^{\nu_-} \tag{18}$$

where the last term is the nonideal free energy of the system associated with coulombic ionic interactions and any other effects giving rise to nonideality [ion–solvent size ratio,[20] solvation effects[21] (see below), etc.]. In sufficiently dilute solutions this last term in Eq. (18) is accounted for primarily by long-range coulombic interactions as treated at various levels of the Debye–Hückel theory (depending on the approximations made) or in more modern treatments of this problem.[17-19,22-32]

The last term of Eq. (18) enables us to introduce the definition of the *mean* activity coefficient $y_\pm$ of the electrolyte on the molar concentration scale,

$$y_\pm = (y_{M^{z+}}^{\nu_+} y_{A^{z-}}^{\nu_-})^{1/(\nu_+ + \nu_-)} \tag{19}$$

normally used in treating the experimental activity coefficients of electrolytes.

2.5. Relations between Solvent and Solute Activities

While Eqs. (17) and (18) suffice formally to represent the thermodynamic behavior of electrolytes in solution at any concentration, no account is explicitly given by these equations of the activity behavior of the *solvent* as

---

† For review articles on this latter topic the reader is referred to references 3, 14, and 15 and to a paper by the author (reference 16) as well as to references 17–19 for more modern treatments of the problem.

a function of the salt concentration. This can be introduced through the Gibbs–Duhem equation

$$n_1 d\mu_1 + n_2 d\mu_2 = 0 \tag{20}$$

which relates differentially the chemical potential μ_1 of solvent 1 to the chemical potential μ_2 of solute 2 (an electrolyte in the present case) through the composition defined by n_1 and n_2, where n_1 is the number of moles of solvent associated with n_2 moles of solute. Equation (20) is important because it implies a reciprocal variation of the solvent chemical potential with a changing solute chemical potential, i.e., with changing composition, $d\mu_1 = -(n_2/n_1) d\mu_2$. Since μ_1 and μ_2 involve the activity coefficients of components 1 and 2, Eq. (20) can be written in the form

$$d \ln a_1 = -\frac{n_2}{n_1} d \ln a_2 \tag{21}$$

Applied to a solution at molality m in a solvent s of molecular weight W_1, Eq. (21) gives

$$-\frac{1000}{W_1} d \ln a_s = \nu m d \ln \gamma_\pm m \tag{22}$$

where $\nu = \nu_+ + \nu_-$ [see Eq. (18)] so that the solvent activity a_s is related to the solute activity $\gamma_\pm m$. It is useful to introduce the osmotic coefficient ϕ, characterizing a_s, through the definition

$$\nu m \phi = -\frac{1000}{W_1} \ln a_s \tag{23}$$

Then Eq. (22) may be transformed to

$$(\phi - 1)\frac{dm}{m} + d\phi = d \ln \gamma_\pm \tag{24}$$

which, upon integration, gives the well-known relation

$$\ln \gamma_\pm = (\phi - 1) + \int_0^m (\phi - 1) d \ln m \tag{25}$$

since $\phi \to 1$ as $m \to 0$, and $\ln \gamma_\pm \to 0$. Alternatively,

$$\nu m \phi = -\frac{1000}{W_1} \ln a_s = \int_0^m \nu m d \ln \gamma_\pm m \tag{26}$$

so that

$$\phi = 1 + \frac{1}{m} \int_0^m m d \ln \gamma_\pm \tag{27}$$

Thus a relation is established between the activity coefficient $\gamma_\pm$ of the electrolyte *solute* and the *solvent activity* through ϕ. This relation will be important in considering later the effects of solvent activity change on solute activity coefficients associated with ion–solvent interactions, especially hydration in the case of aqueous solutions. The properties of electrolyte solutions are also conveniently treated directly in terms of the osmotic coefficients,[24] especially when activity data are obtained from isopiestic vapor pressure or fp depression experiments, e.g., in the various works of Stokes.

At this point it will be useful to distinguish the principal factors associated with the solvation of ions of an electrolyte solution that influence the electrolyte activity coefficient in addition to the main long-range coulombic effect. They are the following:

1. The distance-of-closest-approach effect, which is dependent on the average effective distance of solvated ions (not necessarily the sum of the solvation radii of the free ions or the sum of the unsolvated ion radii).
2. Mutual salting-out and dielectric polarization effects due to solvent binding (related to the dielectric saturation effect treated by Hückel).
3. The mutual salting-out effect due to the salting-out of "ion cavities" in the solvent.
4. The overlap of solvation cospheres at appreciable concentrations and ultimate modification of the dielectric constant relevant to ion–ion interactions.

In the material that follows we shall examine the significance of each of these effects, in addition to the coulombic one, and their relation to the ion specificity of the activity coefficient behavior of electrolytes.

2.6. Historical Perspective on Treatments of Ionic Interactions in Solution

Apart from theories of incomplete dissociation, which were mentioned in Section 1.2, the earliest theory of ion interaction was Milner's (1912),[33] who treated the problem by an elaborate procedure consisting of the numerical summation of the interaction energies for a large system of configurations of the ions in solution. No clear functional dependence of interaction energy or corresponding nonideal chemical potential on concentration arose from this work, although it has undoubted historical significance. Lattice theories of ionic interaction played a role in the early theoretical approaches; e.g., Ghosh[34] was one of the first to propose such a representation and deduced a $c^{1/3}$ law for the concentration dependence of equivalent conductance, a relation that is not substantiated by experiment. Bjerrum[35] also showed that a $c^{1/3}$ plot held for the concentration dependence of osmotic coefficients for a number of salts and was improved when hydration effects were allowed for. These approaches anticipated the resurrection of these ideas by Frank and Thompson,[36] Desnoyers and Conway,[37] and more recently by Pytkowicz *et al.*[38-40] for relatively concentrated solutions.

The general ionic strength principle was formulated in 1921 by Lewis and Randall[41] and the square-root relationship in this variable, $I = \frac{1}{2}\sum_1^i m_i z_i^2$, was established for the natural logarithms of activity coefficients. The square-root relation for equivalent conductance was established much earlier by Kohlrausch.[42] In 1922 Brønsted[43] formulated an extension of the empirically observed, limiting $c^{1/2}$ or $m^{1/2}$ laws in osmotic (or activity) coefficients and gave, on the basis of theoretical and empirical arguments, the expressions

$$\phi = 1 - \alpha' m^{1/2} - \beta m \qquad (28)$$

and

$$\ln \gamma = -3\alpha' m^{1/2} - 2\beta m \qquad (29)$$

where α' is equivalent to a Debye–Hückel parameter (see below) and β is an ion-specific interaction parameter, an important aspect of Brønsted's work,[42] which was followed up later[44,45] (Section 7). These approaches, it is to be noted, were formulated before the advent of the Debye–Hückel theory[46] in 1923, based on the Poisson–Boltzmann equation as a distribution function (see below).

In the literature of electrolyte solutions it seems to have been rarely recognized that the Poisson–Boltzmann equation was already used as a distribution function for cation/anion concentrations, or corresponding charge densities, by Chapman in 1913[47] in his quantitative development of the theory of the double layer at charged interfaces, following Gouy's model.[48] In the double-layer problem, the Poisson–Boltzmann equation is used to obtain an expression for the field $d\psi/dx$ as a function of a distance x normal to a charged plane in terms of the cation and anion concentrations, the potential ψ_x as $f(x)$, and the surface charge density σ on the plane. While the Debye–Hückel theory (see Section 3.2) seeks to obtain by a similar procedure the solution for the effective electric potential due to a smoothed charge distribution around a spherical central charge, the earlier double-layer treatment of the charged plane by Chapman sought to obtain the electric potential profile near the plane and the differential relation between σ and potential, giving the diffuse-layer capacitance. It is to be stressed that the starting procedures in these two treatments are very similar; also, the problems and approximations in primitive and more developed theories are closely analogous, as has become apparent especially in modern theoretical work over the past 10 years or so. Thus improvements to primitive theories of ion distribution in both problems have gone along rather similar lines.

3. The Debye–Hückel Evaluation of the Activity Coefficient of an Electrolyte in Solution

3.1. Introduction

Accurate theories of solutions are difficult to formulate owing to the problem of developing and quantitatively evaluating a satisfactory partition function. Generally, for pure liquids a "disordered-solid" approach is preferred to a "compressed-gas" model, at least for temperatures not too much above the melting point. A satisfactory theory of a simple solution must employ statistical-mechanical averaging procedures to represent how the particles are distributed positionally and, in certain cases, orientationally with respect to one another and the solvent, and what is their average energy of interaction at some defined concentration. In the case of ionic solutions the pairwise interaction energy between $++$, $+-$ and $--$ ions is a basic quantity and is given in the simplest analysis by the Coulomb potential $z_+z_-e^2/\varepsilon r$ experienced by the ions of charge z_+e and z_-e at a distance r in a solvent of dielectric constant ε. No orientation distributions are normally involved. Because the Coulomb potential is both numerically strong and long range ("r^{-1}") in character, the interaction effects that arise between ions lead to large deviations from ideal solution behavior (and concentration-dependent equivalent conductance) are already significant in dilute solutions ($0.001 \sim 0.05\ m$). Advantageously, since the interaction effects are large, the dominance of the Coulomb term has enabled theories of ionic solution behavior to be developed that are much more accurate and easier to formulate than theories of binary solutions of nonelectrolytes, where shorter-range and often weaker van der Waals and dipole–dipole interactions predominate (Table 1).

The most successful early theory of ionic solutions (see the historical introduction) was that of Debye and Hückel.[46] By the clever device of treating the ion distribution in terms of a diffused charge "ionic atmosphere" centered around a reference ion and relating the average charge density in it at some particular position and for a defined concentration to the electric potential by means of Poisson's equation and Boltzmann's distribution law, they avoided the difficulties that arise in rigorous statistical-mechanical averaging procedures for this kind of problem. For limitingly low ionic concentrations an *accurate* equation for the activity coefficient or excess free energy results by the use of entirely analytical procedures.

For completeness in this chapter the main features of the derivation of the activity coefficient by Debye and Hückel's method will be presented with appropriate commentary, since other equally important methods, to be discussed subsequently in this chapter, require frequent reference to the original Debye–Hückel (DH) treatment and model.

As in other treatments of ionic solutions, the aim of the DH theory is to calculate the partial molar nonideal or excess free energy G^{ex} and express

Table 1
Comparison between Interaction Potentials in Nonelectrolyte and Electrolyte Solutions

Solution	Interaction potentials	Orientation correlation
Binary, nonpolar	van der Waals, $U \propto r^{-6}$ or "6:12" potential	No
Binary nonpolar/polar or polar/polar	Polarization by dipoles or dipole–dipole interaction + van der Waals; $U \propto \mu_1\mu_2/r^3$ and r^{-6} or "6:12" potential	Yes, especially in hydrogen-bonded solutions
Ionic in polar solvent (ion–ion)	Coulombic + short-range repulsion: $$U = z_+z_-e^2/\varepsilon r \begin{cases} U = \infty, r \leq a \\ U \to \infty, r \to a \end{cases}$$ for $r > a$ Attenuated by ε	No (for spherical ions)
Ionic in polar solvent (ion–solvent)	Ion–dipole: $U \propto -ze\mu \cos\theta/r^2$ + van der Waals or "6:12" potential at short range + long-range Born polarization $[-(ze)^2/2r]$ $(1 - 1/\varepsilon)$	Yes, determined by orientation of dipole (angle θ) and hydrogen bonding in the solvent + ion–quadrupole interaction effects

it through Eq. (10) in terms of an activity coefficient γ: $G^{ex} = RT \ln \gamma$. For electrolyte solutions at reasonable dilutions ($<0.5\,M$) coulombic attractive interactions dominate the deviations from ideality, so that G^{ex} is negative and γ is consequently less than unity.

The principal phenomenological features of the thermodynamic behavior of electrolytes in strongly dissociating solvent media, which have to be accounted for by successful theories of electrolyte solutions, are as follows:

1. Correct limiting law slopes for $\ln \gamma_\pm$ vs $c^{1/2}$ relations.
2. Limiting law slopes that are proportional to the valence factor z_+z_- of the electrolyte and correctly dependent on the dielectric constant ε of the solvent and temperature.
3. Positive deviations from the limiting law lines as $f(c^{1/2})$, as c becomes appreciable ($>0.005\,M$, depending on the valence type of the salt).
4. Turnup of $\ln \gamma_\pm$ vs $c^{1/2}$ relations toward positive values of $\ln \gamma_\pm$ at high concentrations ($>1 \sim 2\,M$), when ion–solvent interaction effects begin to dominate the thermodynamic behavior of the solution.
5. Correct and consistent relations between the themodynamic and conductance behavior of the electrolyte as $f(c)$.

3.2. Debye and Hückel's Evaluation of Activity Coefficients as Functions of Concentration

The model involved in the DH theory is one in which the distribution of ions in solution about a given reference ion is considered in terms of either a point-charge distribution (limiting law case) or a distribution of finite-sized ions that can approach each other to a distance a, the so-called distance of closest approach. The charge distribution is treated in terms of Poisson's equation through the space-charge density ρ. The latter really refers to a continuous charge distribution so that the fact that the real distribution is one of discrete charges introduces a difficulty with regard to the model. This will be discussed further below. However, in sufficiently dilute solutions the distribution of cations and anions about the reference ion can be regarded as a quasicontinuous one, with an excess of ions of charge opposite to that of the central reference ion in the charge distribution about the central ion. This charge distribution is commonly referred to as the "ionic atmosphere" of the central ion.

The Poisson equation defines the divergence of the gradient of electric potential in terms of ρ and the bulk dielectric constant ε as

$$\nabla^2\psi \equiv \text{div grad } \psi = -4\pi\rho/\varepsilon \tag{30}$$

where ε is independent of the field and $\nabla^2\psi = \partial^2\psi/\partial x^2 + \partial^2\psi/\partial y^2 + \partial^2\psi/\partial z^2$ in Cartesian coordinates. It is useful to treat the problem in a spherical coordinate system defined by the radial function r and the Eulerian angles θ and ϕ; thus

$$\nabla^2\psi = \frac{1}{2}\frac{\partial}{\partial r}\left(r^2\frac{\partial\psi}{\partial r}\right) + \frac{1}{r^2\sin\theta}\frac{\partial}{\partial\theta}\left(\sin\theta\frac{\partial\psi}{\partial\theta}\right) + \frac{1}{r^2\sin^2\theta}\left(\frac{\partial^2\psi}{\partial\phi^2}\right) \tag{31}$$

In the case of the equilibrium ionic atmosphere (no net movement of ions) ψ is independent of the angles θ, and ϕ depends only on r and the concentration, etc., so that from Eq. (31)

$$\nabla^2\psi = \frac{1}{r^2}\frac{\partial}{\partial r}\left(r^2\frac{\partial\psi}{\partial r}\right) = -\frac{4\pi\rho_r}{\varepsilon} \tag{32}$$

where ρ_r is a function of r through the potential ψ, which is a function of r.

The equation relating ψ and ρ_r is a purely electrostatic one based on Coulomb's law and is therefore subject to the principle of superposition of potentials; i.e., if charges are changed, corresponding changes of potential (and vector changes of field) must arise in an additive manner. In the treatment of the ionic atmosphere, it is supposed that the local concentrations of cations n_+ and anions n_- about the reference ion are determined by the Boltzmann distribution; i.e., in an element of volume δV at a distance r from the reference ion the local numbers of cations and anions at r are

$$n_{+,r}\delta V = n_+ \exp\left(-z_+ e\psi_r/kT\right)\delta V \tag{33a}$$

and

$$n_{-,r}\delta V = n_- \exp\left(-z_- e\psi_r/kT\right)\delta V \tag{33b}$$

where n_+ and n_- are the bulk stoichiometric concentrations of the cations and anions. The signs of z_+ and z_- are recognized in the exponentials so that $n_{+,r} < n_+$ and $n_{-,r} > n_-$ if the central ion is a cation, and vice versa if it is an anion. The total charge in the volume element is $z_+en_{+,r}\delta V + z_-en_{-,r}\delta V$ and its density is this quantity divided by δV; hence

$$\rho_r = z_+en_+\exp\left(-z_+e\psi_r/kT\right] + z_-en_-\exp\left(-z_-e\psi_r/kT\right) = -\varepsilon\,\nabla^2\psi/4\pi \tag{34}$$

At this point an important difficulty arises, since ρ_r is an *exponential* function of ψ_r, a situation that is inconsistent with the superposition principle referred to above. Expressed in another way, the potential of average force involved in the Boltzmann distribution function is not identical with the mean electrostatic potential in the Poisson equation. The combined Poisson–Boltzmann equation is thus internally inconsistent. This is a well-known problem in the DH treatment and will be referred to subsequently.

Substitution for ρ_r in the Poisson equation thus gives a nonlinear second-order differential equation in ψ that does not have an explicit exact solution. This difficulty was overcome by Debye and Hückel by expanding the exponentials in ψ_r in a Taylor series, retaining only the linear term. Then, from Eq. (34),

$$\rho_r = z_+en_+(1 - z_+e\psi_r/kT) + z_-en_-(1 - z_-e\psi_r/kT) \tag{35}$$

For a salt of stoichiometry $M_{\nu_+}^{z_+}A_{\nu_-}^{z_-}$ at a bulk molar concentration c

$$\nu_+c = n_+, \qquad \nu_-c = n_- \quad \text{and} \quad \nu_+z_+ = \nu_-|z_-|$$

so that

$$z_+en_+ + z_-en_- = 0$$

Hence

$$\rho_r = -(z_+)^2e^2n_+\psi_r/kT - (z_-)^2e^2n_-\psi_r/kT \tag{36}$$

and

$$\nabla^2\psi = \frac{-4\pi\rho_r}{\varepsilon} = \sum_1^i \frac{4\pi n_iz_i^2e^2}{\varepsilon kT}\psi_r = \kappa^2\psi_r \tag{37}$$

say, where the summation is made for ions of type i in the general case, and a parameter κ^2 is introduced equal to

$$\frac{4\pi e^2}{\varepsilon kT}\sum_1^i n_iz_i^2 \tag{38}$$

The result for ρ_r not only eliminates the algebraic difficulty of solving the Poisson–Boltzmann equation for ψ_r but frees the treatment from the inconsistency in the original nonlinear Poisson–Boltzmann equation, since ψ_r is now a linear function of ρ_r after making the linear expansions. Calculation of $z_\pm e\psi_r$

energies shows that the expansion itself is, however, only justified for very dilute solutions where $z_{\pm}e\psi_r < kT$. The situation is obviously worse when $|z_{\pm}| > 1$. Thus this requirement is met less for higher-valent salts than for $1:1$ salts, or, alternatively, the result is only justified for correspondingly more dilute solutions. Also, $z_{\pm}e\psi_r < kT$ will not apply to interactions at short distances†.

In the case of a symmetrical electrolyte $z_+ = z_-$, $=z$, say, $n_+ = n_-$, the expansion of Eq. (34) leads to

$$\rho_r = 0 - 2n_+ze \frac{ze\psi_r}{kT} + 0 - \frac{n_+ze}{3}\left(\frac{ze\psi_r}{kT}\right)^3 + 0 - \cdots \tag{39}$$

where the coefficient of the second-order term in $(ze\psi_r/kT)^2$ is conveniently zero. For this case, therefore, it is interesting to note that the linear expansion approximation is adventitiously more satsifactory than in the case of an unsymmetrical electrolyte where the second-order term from Eq. (34) is

$$\frac{z_+en_+}{2}\left(\frac{z_+e\psi_r}{kT}\right)^2 + \frac{z_-en_-}{2}\left(\frac{z_-e\psi_r}{kT}\right)^2 \neq 0 \tag{40}$$

since $z_+ \neq |z_-|$.

Proceeding now with the solution of the linearized equation for ψ_r, it can be shown to be of the standard form

$$\frac{d^2y}{dr^2} = \kappa^2 y \tag{41}$$

This follows by writing $(1/r^2)(d/dr)(r^2\,d\psi/dr) = \kappa^2\psi$ as

$$\frac{1}{r}\frac{d}{dr}\left(r^2\frac{d\psi}{dr}\right) = r\frac{d^2\psi}{dr^2} + 2\frac{d\psi}{dr} \qquad [=\kappa^2(r\psi)] \tag{42}$$

with the abbreviation $\psi_r \equiv \psi$.

Noting that

$$\frac{d^2(r\psi)}{dr^2} = \frac{d}{dr}\left(r\frac{d\psi}{dr} + \psi\right) = r\frac{d^2\psi}{dr^2} + 2\frac{d\psi}{dr} \tag{43}$$

which is the right-hand side of the original second-order differential equation, it is clear that

$$\kappa^2(\psi r) = \frac{d^2(\psi r)}{dr^2} \tag{44}$$

This has the general solution

$$\psi r = A e^{-\kappa r} + B e^{\kappa r} \tag{45}$$

---

† This may be readily seen if $(ze)^2/\varepsilon r$ is evaluated, for example, in the case of a bi-bivalent salt at, say, $r = 1.0$ nm. The interaction energy ($\varepsilon = 80$) is 6.69 kJ mol^{-1}, in comparison with $RT = 2.51$ kJ mol^{-1} at 300 K.

from which ψ is explicitly obtained after examination of the following boundary conditions: (1) As $r \to \infty$, ψ_r must tend to zero (since ψ_r results from the excess charge near the reference ion, which must tend to zero as $r \to \infty$). Now $e^{\kappa r}$ increases with r more rapidly than does r itself, so that $e^{\kappa r}/r \to \infty$ as $r \to \infty$; hence $B = 0$. (2) In very dilute solutions ψ_r must be due only to the central ion and hence has a value $z_+ e/\varepsilon r$ (for a central cation). Therefore

$$\psi_r = A \frac{e^{-\kappa r}}{r} = \frac{z_+ e}{\varepsilon r} \tag{46}$$

Hence $A = (z_+ e/\varepsilon)e^{-\kappa r}$ and the relation for ψ_r is limitingly

$$\psi_r = \frac{z_+ e}{\varepsilon} \frac{e^{-\kappa r}}{r} \tag{47}$$

or, generally, for a j type of reference ion

$$\psi_{r,j} = \frac{z_j e}{\varepsilon} \frac{e^{-\kappa r}}{r} \tag{48}$$

A more important relation for finite ion sizes, which will hence apply less limitingly to very dilute solutions, may be obtained as follows. It takes into account the fact that the charge distribution, according to Eqs. (34) or (35), cannot extend to zero distance r for ions of finite size but only up to some common contact distance a (Figure 1). This is similar to the Stern[49] modification of the Gouy–Chapman theory of the double layer for finite-size ions (Figure 1c).

For the treatment of a solution of finite-size ions

the total charge in the solution from $r = a$ to ∞ is $\int_a^\infty 4\pi r^2 \rho_r \, dr$ and this must be equal and opposite to $z_j e$.

Hence

$$\int_a^\infty 4\pi r^2 \sum \frac{n_i z_i^2 e^2 \psi_r}{kT} dr = -z_j e \tag{49}$$

and with the definition of κ^2 the integral becomes

$$\int_a^\infty \kappa^2 \varepsilon r^2 \psi_r \, dr = -z_j e \tag{50}$$

The Poisson–Boltzmann equation will still apply, but with a different second limit. Thus, with $\psi_r = A e^{-\kappa r}/r$ again, the integral is

$$\int_a^\infty \kappa^2 \varepsilon r A \, e^{-\kappa r} \, dr \qquad (=-z_j e) \tag{51}$$

Integrating by parts,

$$z_j e = [-\varepsilon A \, e^{-\kappa r}(\kappa r + 1)]_a^\infty = A \, e^{-\kappa a}(1 + \kappa a) \tag{52}$$

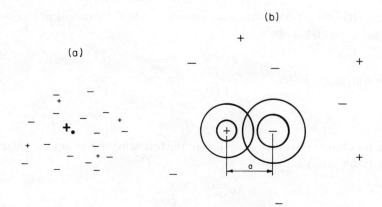

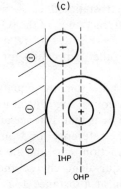

Figure 1. Cutoff distance for the ionic atmosphere charge distributions at an ion in an electrolyte solution or at a charged plane (double layer) according to Stern: (a) point-charge model (limiting law); (b) with distance of closest approach a that may involve hydration shell overlap; (c) corresponding model of Stern and Grahame for ion distribution at a charged interface, with inner (IHP) and outer (OHP) Helmholtz planes at distances of closest approach for anions and hydrated cations.

Hence

$$A = \frac{z_j e\, e^{\kappa a}}{\varepsilon(1 + \kappa a)} \tag{53}$$

and

$$\psi_r = \frac{z_j e}{\varepsilon} \frac{e^{-\kappa r}}{r} \frac{e^{\kappa a}}{1 + \kappa a} \tag{54}$$

which at very low concentrations, where $\kappa a \ll 1$, becomes identical with the Eq. (48) for the point-charge case. ψ_r is thus the point charge result [Eq. (48)] modified by the factor $e^{\kappa a}(1 + \kappa a)$.

Returning to the simpler solution for ψ_r for the remainder of the calculation, we now require the energy associated with the system of interacting charges, where the net potential at any point r from the reference ion j is given by the solution deduced above. ψ_r is made up of composite quantities owing to (1) the central j ion and (2) all the other ions in the atmosphere (a quantity we shall call ψ_{atm}). It is clear that it is the latter quantity that is required in order to evaluate the electrostatic energy of the typical ion j with respect to its interactions with all the other ions. By the principle of superposition, again, we may write

$$\psi_r = \psi_{j,r} + \psi_{atm} \qquad (55)$$

so that with $\psi_{j,r} = z_j e / \varepsilon r$

$$\psi_{atm,r} = \frac{z_j e}{\varepsilon r} e^{-\kappa r} - \frac{z_j e}{\varepsilon r} = \frac{z_j e}{\varepsilon r} (e^{-\kappa r} - 1) \qquad (56)$$

Again, for dilute solutions $e^{-\kappa r} \doteq 1 - \kappa r$, the following important but elegantly simple result arises:

$$\psi_{atm} = -\frac{z_j e}{\varepsilon} \kappa \qquad (57)$$

This potential characterizes the average electrostatic effect of the ionic atmosphere, and the electrostatic free energy of interaction g_e of the reference ion with its atmosphere is therefore $z_j e \psi_{atm}$, i.e., $-(z_j e)^2 \kappa / \varepsilon$ (a related and similar conclusion would have resulted had the fuller expression for ψ_r for ions of finite size been taken).

The significance of κ may now be considered. From the equation for ψ_{atm} and the resulting averaged energy of interaction of the central ion with the atmosphere ions, it is seen that the quantity $1/\kappa$ is that distance at which an ion of equal but opposite charge to that of the reference ion would give rise to the same electrostatic mean energy of interaction, since the interaction energy can be written $-(z_j e)^2 / \varepsilon (1/\kappa)$. This equivalent charge therefore behaves as an "image" charge of the central ion at a distance $1/\kappa$. $1/\kappa$ is thus the effective "radius" of the ionic atmosphere" and its significance is similar to the characteristic "lattice length" of the diffused charge lattice model, which gives rise to a $c^{1/3}$ relation for $\ln \gamma_\pm$ (see below). Also, since ρ_r is $-\sum (n_i z_i^2 e^2 / kT) \psi_r$, and with the definition of κ^2, ρ_r may be represented as

$$\rho_r = \frac{-\varepsilon \kappa^2 \psi_r}{4\pi} \qquad (58)$$

Then with the point-charge solution for ψ_r

$$\rho_r = \frac{-z_+ e}{4\pi} \frac{\kappa^2 e^{-\kappa r}}{r} \qquad (59)$$

Differentiating ρ_r with respect to r and equating the result to zero for a maximum in ρ_r gives

$$\frac{d\rho_r}{dr} = -\frac{z_+ e \kappa^2}{4\pi} \frac{-r\kappa e^{-\kappa r} - e^{-\kappa r}}{r^2} = 0 \qquad (60)$$

Therefore

$$-r\kappa = 1$$

or

$$-r = 1/\kappa \qquad \text{for } \rho_{max} \tag{61}$$

Hence $-1/\kappa$ is also the *imaginary* distance at which the space-charge density would be a maximum. However, from Eq. (34) or (59) for ρ_r it is clear that the real ρ_r must decrease continuously with r for positive values of r and real values of κ.

The original quantity desired from the calculation was the nonideal molar chemical potential contribution $RT \ln \gamma_{\pm}$. Hence g_e must be evaluated for the whole system of ions and the result differentiated with respect to the number of ions constituting the electrolyte solution system in order to obtain μ_e, the electrostatic chemical potential contribution corresponding to g_e and the nonideal free energy $RT \ln \gamma_{\pm}$.

Various ways of obtaining the results have been proposed: (1) the "Guntelberg–Müller[50] charging process" in which the work required to charge the central ion j against the atmosphere potential is calculated; (2) the "Debye charging process" in which the work of charging all ions simultaneously from a hypothetically uncharged state (but with the same spatial distribution maintained as when the ions were fully charged) is evaluated (in both cases the result is differentiated appropriately to obtain the chemical potential); and, finally, (3) a simple and intuitive method suggested by Robinson and Stokes,[7] namely, that the electrostatic chemical potential contribution is simply $[(z_j e)^2/2\varepsilon]\kappa$, since $z_j e$ multiplied by $(-z_j e/\varepsilon)\kappa$ gives an energy twice times that associated with all the ions, the factor 2 entering, since atmosphere ions can just as well be counted as reference ions in the overall summation of interaction energies (a similar factor enters any lattice energy calculation in terms of the energy of pair interactions and the coordination number).

The procedure for evaluating $\gamma_{\pm}$ may be illustrated by reference to the first method. The energy of charging the ion j with the ionic atmosphere maintained is given by

$$g_e = -\int_0^1 (\lambda e)^2 z_j^2 \frac{\kappa}{\varepsilon} d\lambda = \frac{-(z_j e)^2 \kappa}{3\varepsilon} \tag{62}$$

for n_j ions per cubic centimeter, and the total electrical work is $n_j g_e$, i.e., say,

$$-n_j(z_j e)^2 \kappa/3\varepsilon = \Delta G_e \tag{63}$$

To obtain the corresponding chemical potential contribution μ_e it is noted that

$$\mu_e = \left(\frac{\partial \Delta G_e}{\partial n_j}\right)_{T,P} = \frac{-(z_j e)^2 \kappa}{2\varepsilon} \tag{64}$$

taking into account the fact that n_j appears both directly in the expression for ΔG_e above and in κ. Then, with the definition of the nonideal chemical potential contribution in terms of $\gamma_\pm$,

$$\mu_e = kT \ln \gamma_j = -(a_j e)^2 \kappa / 2\varepsilon$$

or

$$\ln \gamma_j = -(z_j e)^2 \kappa / 2\varepsilon kT \tag{65}$$

is obtained as a limiting law (the so-called Debye–Hückel "limiting law"). Alternatively, for the finite-ion-size case,

$$-\ln \gamma_j = \frac{(z_j e)^2}{2\varepsilon kT} \frac{\kappa}{1 + \kappa a} \tag{66}$$

The concentration term in κ can be expressed in terms of molarity through the ionic strength I, i.e.,

$$\kappa = \left(\frac{8\pi e^2 N}{1000\varepsilon kT} \right)^{1/2} I^{1/2} \tag{67}$$

where I is $\frac{1}{2}\sum_1^i c_i z_i^2 = c/2 \sum_1^i \nu_i z_i^2$, N is Avogadro's number, and c is the molarity. Then for the overall $+-$ salt

$$\log \gamma_\pm = \frac{-A(z_+ z_-)I^{1/2}}{1 + \mathring{a}BI^{1/2}} \tag{68}$$

where $\mathring{a}$ is the distance of closest approach usually expressed in Å units. Numerically, A is given by

$$A = \left(\frac{2N}{1000} \right)^{1/2} \frac{e^3}{2.303k^{3/2}} \frac{1}{(\varepsilon T)^{3/2}} = \frac{1.8246 \times 10^6}{(\varepsilon T)^{3/2}} \tag{69}$$

and

$$B = 10^{-8} \left(\frac{8\pi e^2 N}{1000k} \right)^{1/2} \frac{1}{(\varepsilon T)^{1/2}} = \frac{5.029 \times 10}{(\varepsilon T)^{1/2}} \tag{70}$$

In Eq. (68) we have assumed the connection between γ_+, γ_-, and $\gamma_\pm$.[†] This follows from the definition of $\gamma_\pm$ given in Eq. (19) [see Eq. (19), p. 117] by substituting the individual values of $\ln \gamma_j$ and $\ln \gamma_i$ from the result of the DH calculation. Thus for the general electrolyte $M_{\nu_+}^{z^+} A_{\nu_-}^{z^-}$

$$\log \gamma_\pm = \frac{\nu_+ \log \gamma_+ + \nu_- \log \gamma_-}{\nu_+ + \nu_-} \tag{71}$$

Then, using the limiting law result,

$$\log \gamma_\pm = \frac{\nu_i A(z_+)^2 I^{1/2} + \nu_- A(z_-)^2 I^{1/2}}{\nu_+ + \nu_-} \tag{72}$$

[†] The calculation of *individual* ionic activity coefficients, their significance, and their specific values in relation to the hydration of cations and anions will be dealt with in Section 5.6.

Noting that $\nu_+z_+ + \nu_-z_- = 0$ and hence substituting for ν_+ or ν_- with the inclusion of the $+$ or $-$ signs in z's, the limiting law result is obtained as

$$\log \gamma_\pm = \frac{-z_+z_-(z_+ - z_-)}{z_- - z_+} AI^{1/2} = -A|z_+z_-|I^{1/2} \tag{73}$$

which has the same form as the equations for the individual ion activity coefficients.

A tabulation of DH constants is given in Table 2.

A very recent paper by Clarke and Glew[51] should be noted. It gives a useful calculation and tabulation of DH limiting law slopes for aqueous solutions between 273 and 423 K.

3.3. Numerical Solutions and Expanded Exponential Terms

The main source of inaccuracy of the DH treatment in its limiting law forms (i.e., including the case for finite ion size) is the linearization of the Boltzmann exponential terms, which leads to inconsistencies with experimental data already at quite low concentrations. Gronwall et al.[52,53] made a calculation in which the exponentials were expanded in a power series for $e\psi/kT$ in terms of a characteristic distance $s = e^2/\varepsilon kT$ (see the Bjerrum distance for ion association in a 1:1 salt solution) and a, the distance of closest approach. More recently Card and Valleau[31] used a DH exponential distribution and showed that it is in quite good agreement with exact Monte Carlo and hypernetted chain statistical-mechanical calculations (see Section 4.2.).

An accurate numerical solution of the Poisson–Boltzmann equation for the case of ions in solution (DH problem) was given by Guggenheim.[54] The first procedure for a numerical solution of the Poisson–Boltzmann relation was proposed by Müller in 1927 and employed by him for symmetrical electrolytes. However, by the use of modern digital computing techniques this type of problem is now entirely tractable,[54] including the case for unsymmetrical charge types.

The achievement of an exact numerical solution for the Poisson–Boltzmann equation raises again the question discussed by Onsager[55] and referred to above, concerning the superposition inconsistency, which becomes significant particularly in the case of unsymmetrical electrolytes. The probability of occupation of two sites in the solution, one by a cation (or anion) and the other by the anion (or cation) of the salt, differs from that determined by random occupation by the Boltzmann exponential factor in $\pm ze\psi/kT$. This factor should, of course, be independent of which ion is chosen to be the "central reference ion" in the treatment of Debye and Hückel. This requires that the quantity be the same for the two types of ions. This condition is satisfied in the DH approximation, where the Boltzmann distribution functions are expanded only to the first linear term in a Taylor series, but *not* in the accurate solution of the Poisson–Boltzmann equation such as that evaluated

Table 2
The Debye–Hückel Equation for Aqueous Solutions[a]

Values of the constants A and A' in the equations

$$\log f_{\pm} = \frac{-A\Gamma_i^{1/2}}{1 + B\Gamma_i^{1/2}} \quad \text{and} \quad \log f_{\pm} \frac{-A'c_i^{1/2}}{1 + B'c_i^{1/2}}$$

where the latter equation applies only for a single electrolyte and the former equation is general for single or mixed electrolyte solutions of ionic concentration Γ_i or molarity c_i (where Γ equal to $2I_{(m)}$, is the ionic strength).

Values of the constants $B/\mathring{a}$ and $B'/\mathring{a}$ in the equations

$$\frac{B}{\mathring{a}} = \frac{\kappa \times 10^{-8}}{\Gamma_i^{1/2}} \quad \text{and} \quad \frac{B'}{\mathring{a}} = \frac{\kappa \times 10^{-8}}{c_i^{1/2}}$$

where $\mathring{a}$ is the mean distance of nearest approach of the ions (measured in Å).

The valency factors w, w', and w'' are given by

$$w = \frac{1}{\nu}\sum_1^0 \nu_i z_i^2, \quad w' = \frac{1}{\nu 2^{1/2}}\left(\sum_1^Q \nu_i z_i^2\right)^{3/2}, \quad w'' = \left(\frac{1}{2}\sum_1^Q \nu_i z_i^2\right)^{1/2} = \frac{w'}{w''}$$

where Q is the number of types of ions derived from the electrolytic dissociation and ν is the number of ions resulting from this.

For binary electrolytes the factor w'' reduces to z ($=z_1 = |z_2|$).

T (K)	A	A'	$B/\mathring{a}$	$B'/\mathring{a}$
273	0.3446 w	0.4870 w'	0.2994	0.3244 w''
278	0.3466	0.4902	0.2299	0.3251
283	0.3492	0.4938	0.2305	0.3260
288	0.3519	0.4977	0.2311	0.3268
291	0.3538	0.5002	0.2315	0.3274
293	0.3549	0.5019	0.2318	0.3278
298	0.3582	0.5065	0.2325	0.3288
323	0.3777	0.5341	0.2366	0.3346
353	0.4083	0.5774	0.2429	0.3434
373	0.4325	0.6116	0.2476	0.3501

[a] Errors in the limiting slopes A and A' do not exceed ±0.1%; errors in B and B' do not exceed ±0.05%. From H. S. Harned and B. B. Owen, *Physical Chemistry of Electrolytic Solutions*, Reinhold, New York (1943); see also B. B. Owen and S. R. Brinkley, *Ann. N.Y. Acad. Sci.* **51**, 753 (1949) and Table III.2 therein. Revised DH parameters based on an improved empirical equation for the dielectric constant of water have been given recently by Pitzer and Bradley [*J. Phys. Chem.* **83**, 1599 (1979)] and a comprehensive tabulation of DH slopes for aqueous solutions between 273 and 423 K have just been published by Clarke and Glew.[51]

by Guggenheim.[54] The differences between results calculated for unsymmetrical electrolytes when either the cation or anion is taken as the reference ion can become appreciable, as was shown in Guggenheim's calculation.

The error arising from the inconsistency involving the potentials will also be retained in any other extended form of the DH theory for higher concentrations (e.g., that of Gronwall, LaMer, and Sandved[52]) that expresses terms

to powers higher than unity. In the linear expansion for symmetrical electrolytes the situation is satisfactory because the second-order term in ψ_r^2 has,
as shown above, a coefficient of zero and third-order terms may reasonably
be neglected. For unsymmetrical electrolytes this favorable algebra no longer
exists (see p. 125), so that for a better approximation in the Taylor series the
second-order term in any fuller treatment would be retained. However, for
all such terms involving ψ_r to a power greater than unity, the inconsistency in
the superposition principle will arise.

This matter is manifested in another way in a difference between the details
of the results of the "charging" processes of Debye and Hückel and of
Guntelberg and Müller in the calculation of the overall electrostatic free
energy of ionic interaction. The apparent "error" in the latter method to
which Onsager[55] has referred and which was discussed by Gronwall, LaMer,
and Sandved,[52] essentially arises for the same reason as do the other intrinsic
difficulties, i.e., because of the inconsistency in using the same potential in
the Poisson equation and Boltzmann distribution terms.

3.4. Other Distribution Functions

A modified distribution function for ionic solutions was proposed by
Eigen and Wicke.[56] This treatment took into account excluded volume, a
factor that becomes significant at higher concentrations. The treatment refers
essentially to concentrations in terms of site fractions, a type of procedure
used earlier by Stern[49] (and Glueckauf[6] in a different way) in his treatment
of the double layer and ionic adsorption. The site fraction term is analogous
(for three-dimensional solutions) to the configurational term $\theta/1 - \theta$ in Lang-
muir's isotherm. As discussed by Stokes and Robinson,[7] the solution for the
potential becomes

$$\psi_r = \frac{z_j e}{\varepsilon} \frac{e^{\kappa' a}}{1 + \kappa' a} \frac{e^{-\kappa' r}}{r} \tag{74}$$

where κ' is a function analogous to κ and is given by

$$\kappa' = \left[\frac{4\pi e^2}{\varepsilon k T} \sum_1^i n_i z_i^2 \left(1 - \frac{n_i}{N_i}\right) \right]^{1/2} \tag{75}$$

and N_i is the number of sites available to n_i ions in a unit volume. When $n_i \ll N_i$
for very dilute solutions, $\kappa' \equiv \kappa$ and the simpler theory given earlier applies.

Bagchi[57] proposed that the distribution function should be of the
"Fermi–Dirac" form,

$$n_i' = n_i/[1 + \exp (z_i e \psi_j / kT)] \tag{76}$$

but it is unclear if this can lead to any advantageous result, e.g., the limiting
law result is not apparently predicted. Neither of these distribution functions
have been taken up in later work.

3.5. The Solute–Solvent Size Ratio in the Thermodynamics of Electrolyte Solutions

An interesting point connected with the properties of the solvent is the extent to which nonideality in mixing arises from the difference in the sizes of solute ions and solvent molecules. The size factor can be allowed for by use of "volume fraction" statistics.[5,6] This is an effect that can arise in nonelectrolyte solutions,[58] but it may be significant with special electrolytes whose ions are large, e.g., tetraalkylammonium or tetraphenylboron salts. Based on volume fraction statistics equations, Conway and Verrall[20] evaluated these effects in terms of a nonelectrostatic (NE) activity coefficient contribution $\ln \gamma_{NE}$ given by

$$\ln \gamma_{NE} = \ln \frac{N_2 + N_1}{rN_2 + N_1} + (r - 1)\frac{N_2}{N_2 + n_1/r} \tag{77}$$

or, as an approximation,

$$\ln \gamma_{NE} \doteq (r - 1)^2 \frac{N_r}{rN_2 + 1} \tag{78}$$

where r is the relative size of solute ion and solvent molecules, and the composition of the solution is defined by N_r moles of r-mer ion in N_1 mol of solvent. When $r \gg 1$, say 5–10, the deviation of γ_{NE} from unity is appreciable and increases with concentration. It gives positive deviations from the limiting law; however, it is to be noted that in testing the significance of such effects as have been mentioned above and in reference 20 for *non*electrolyte solutions, it is found[59-61] that they are less than are expected on the basis of statistical lattice models of binary solutions. This is probably because a liquid solution is not really a regular lattice in which solvent particles can be substituted by solute particles in various well-defined configurations; also, the hole properties of real solvents at finite temperatures must be taken into account.

Glueckauf[6] used a volume fraction statistics approach in his treatment of the activity behavior of ionic solutions in which a $c^{1/2}$ relation was shown to go over to a $c^{1/3}$ one at elevated concentrations (see also reference 73 later).

3.6. Specific Ion Interaction Approaches

3.6.1. Brønsted's Equation

Various attempts to provide a better account, albeit empirically, of the dependence of activity coefficient behavior of salts on the specific pair of +– ion types involved than that given by the DH theory have been made. Already, before the appearance of the DH theory, Brønsted (1922)[62] had proposed a relation of the form

$$1 - \phi = \alpha' m^{1/2} + \beta m \tag{79}$$

for the osmotic coefficient of $1:1$ electrolytes. In this equation α' is recognized as the "limiting law" slope and β is introduced as a *salt-specific* interaction parameter, applicable to the dependence of ϕ on m above the limiting law region of concentrations, for which ϕ depends on $m^{1/2}$.

Of course, in the "finite-ion-size" calculation of the DH theory [Eq. (66)] some account of ion-specific behavior is built in through recognition of the distance of closest approach factor a in the term $\kappa a/1 + \kappa a$, which depends in some way on the effective sum of the crystal ionic radii and hydration shell thicknesses.[3,16]

3.6.2. Guggenheim's Treatments

The specific ion interaction approach of Brønsted[62] was taken up in more detail by Guggenheim in important early papers[63,64] and in a more recent one (1955) with Turgeon.[44] Parameters were derived that represent the salt-specific activity and osmotic behavior of electrolytes, not treated in the DH theory except indirectly through the distance of closest approach factor a, up to an ionic strength of $0.1\ m$ for $1:1$, $2:1$, and $1:2$ electrolytes. Pitzer's further extensions[45,65] of the specific interaction treatments allow representations of activity behavior up to $\sim 1.0\ m$.

The thermodynamic formulas involve so-called specific ion interaction coefficients $\beta_{M,X}$ for each combination of a cation M^{z+} and an anion X^{z-}. Assigned values of β have the advantage that they can also be used for representing the thermodynamic properties of *mixed* electrolytes. Tabulations of $\beta_{M,X}$ have been made[66] for numerous electrolytes in aqueous solutions from various determinations.[64] The paper by Guggenheim and Turgeon[44] was concerned mainly with presentation of a revised series of β values (appropriate to the molality scale rather than the mole fraction scale previously used), with applications to acid–base equilibria.

First the basic equations involving the β terms must be given. For the stoichiometric activity coefficient, γ_{MX}

$$\ln \gamma_{MX} = -\alpha' z_M |z_X| \frac{I^{1/2}}{1 + I^{1/2}} + \frac{2\nu_+}{\nu_+ + \nu_-} \sum_{X'} \beta_{X',M} m_{X'} + \frac{2\nu^-}{\nu_+ + \nu_-} \sum_{M'} \beta_{M',X} m_{M'} \quad (80)$$

for any electrolyte or mixture. For a single electrolyte, omitting the subscripts M and X, $\ln \gamma_{MX}$ becomes

$$\ln \gamma_{\pm} = -\alpha' z_+ |z_-| \frac{I^{1/2}}{1 + I^{1/2}} + 2\nu\beta m \quad (81)$$

where $\nu = \nu_+ + \nu_-$ and α' is the limiting law DH coefficient. Alternatively, for $\log \gamma_{\pm}$ the relation is usually written as

$$\log \gamma_{\pm} = -A z_+ |z_-| \frac{I^{1/2}}{1 + I^{1/2}} + Bm \quad (82)$$

where, obviously, $A \equiv \alpha'/\ln 10$ and $B = 2\nu\beta/\ln 10$.

Correspondingly, the osmotic coefficient ϕ is given by

$$1 - \phi = \tfrac{1}{3}\alpha'z_+|z_-|I^{1/2}\sigma(I^{1/2}) - \nu\beta m \tag{83}$$

for the single electrolyte, where the function

$$\sigma(y) \equiv \frac{3}{y^3}[1 + y - (1 + y)^{-1} - 2\ln(1 + y)] \tag{84}$$

and $y \equiv I^{1/2}$ in Eq. (83). Tables of the function σ have been given.[67]

Also, at 273.1 and 298.1 K the values of the DH parameter α' for aqueous solutions are

273.1 K: $A = 0.4883\ \mathrm{kg}^{1/2}\ \mathrm{mol}^{-1/2}$, $\tfrac{1}{3}\alpha' = 0.374\ \mathrm{kg}^{1/2}\ \mathrm{mol}^{-1/2}$

and at

298.1 K: $A = 0.5085\ \mathrm{kg}^{1/2}\ \mathrm{mol}^{-1/2}$

Some values of β are given in Table 3 (column a refers to fp data at 273.1 K, column b refers to emf data at 298.1 K, and column c refers to isopiestic data, measured relative to NaCl). It is convenient to remember that the limiting law coefficient for $-\ln \gamma$ is three times that for $1 - \phi$ as a function of $m^{1/2}$.

It should be mentioned that the term $I^{1/2}/(1 + I^{1/2})$ in Eq. (80) allows for restriction in the ion distribution corresponding to the DH a term for

Table 3
Guggenheim and Turgeon's[44] Values for β for Some 1:1 Electrolytes in Water

Electrolyte	Column a 273.1 K β (kg mol^{-1})	Column b 298.1 K β (kg mol^{-1})	Column c 298.1 K β (kg mol^{-1})
HCl	0.25	0.27	
HBr		0.33	
HI			0.36
HClO$_4$			0.30
HNO$_3$	0.16		
LiCl	0.20		0.22
LiBr	0.30		0.26
LiI			0.35
LiClO$_3$	0.25		
LiClO$_4$	0.35		0.34
LiNO$_3$	0.23		0.21
LiOAc	0.19		0.18

Table 3 (cont.)

Electrolyte	Column a 273.1 K β (kg mol^{-1})	Column b 298.1 K β (kg mol^{-1})	Column c 298.1 K β (kg mol^{-1})
NaF			0.07
NaCl	0.11	0.15	(0.15)
NaBr	0.20		0.17
NaI			0.21
NaClO$_3$	0		0.10
NaClO$_4$	0.05		0.13
NaBrO$_3$			0.01
NaIO$_3$	−0.41		
NaNO$_3$	−0.04		0.04
NaOAc	0.26		0.23
NaCNS			0.20
NaH$_2$PO$_4$			−0.06
KF	0.05		0.13
KCl	0.04	0.10	0.10
KBr	0.06		0.11
KI			0.15
KClO$_3$	−0.19		−0.04
KClO$_4$	−0.55		
KBrO$_3$			0.07
KIO$_3$	−0.43		−0.07
KNO$_3$	−0.26		−0.11
	−0.31		
RbCl			0.06
RbBr			0.05
RbI			0.04
RbNO$_3$			−0.14
RbOAc			0.26
CsCl			0
CsBr			0
CsI			−0.01
CsNO$_3$			−0.15
CsOAc			0.28
AgNO$_3$			−0.14

finiteness sizes of the ions (in their treatment (see Section 3.2.) this term has the form $I^{1/2}/(1 + \mathring{a}BI^{1/2})$, where $\mathring{a}B$ is on the order of unity). The introduction of the β (or B) parameters allows for specificity of ion interactions in a different way ($\mathring{a}B = 1$ is taken). (The DH "a" value is taken in Å and written $\mathring{a}$, as in Eq. (68).)

The interaction coefficients β in Table 3 give, as was mentioned, a good representation of $\ln \gamma_{\pm}$ and ϕ as a function of concentration up to ~0.1 m. They take care of effects of specific differences in sizes, shapes, polarizabilities, and probably hydration of ions. These are also the specific properties that determine extents of ion association that were not specifically treated in Guggenheim and Turgeon's work.[44] However, provided that degrees of association are small, the formulas of Guggenheim adequately deal with association through the B parameter. Limitingly, $B \simeq 0.1 \pm 0.2$ kg mol^{-1} is characteristic of situations where association is unimportant, while for $B < -0.1$ kg mol^{-1} association is important.[68]

3.6.3. Treatment of Pitzer et al. [45,65]

In recent years Pitzer and coworkers[45,65] have made important contributions to the modern treatment of ion–ion interactions in electrolyte solutions, especially through the specific ion interaction approach. It will be remembered that in earlier years Pitzer, with Latimer and Slansky,[69] developed one of the first scales of individual ionic standard free energies and entropies of ions in aqueous solution, using extrapolation methods based on the Born equation. The resulting data led to a clear appreciation of the individuality of ion–water interactions and the differences between anion and cation behavior, e.g. in their hydration.[3,16]

The treatment of Pitzer[45] and others develops the specific ion interaction approach of Guggenheim[63,64] and Guggenheim and Turgeon[44] [see Section 3.6.2. and Eqs. (80)–(82)]. For data at relatively high concentrations, where β is not necessarily a constant, Eq. (83) with Eq. (84) can be solved for β giving, in terms of ϕ,

$$\beta = m^{-1}[\phi - 1 + \tfrac{1}{3}\alpha|z_+z_-|I^{1/2}\sigma(I^{1/2})] \quad \text{(see Eq. 83)} \tag{85}$$

or, in terms of $\gamma_{\pm}$,

$$\beta = (2m)^{-1}[\ln \gamma_{\pm} + \alpha|z_+z_-|I^{1/2}(1 + I^{1/2}) \tag{86}$$

as discussed by Scatchard[70] and by Pitzer and Brewer.[71] A quantity $B = 2\beta/2.303$ [for Eq. (86) written in Briggsian logarithms] is the apparent ion interaction coefficient and can be plotted versus m and is little dependent on m above ~0.75 m, as shown in Figure 2.

For the primitive model (hard-core ions, continuum solvent) an equation for $\phi - 1$ can be obtained in terms of the pairwise distribution functions for like-charged ions, $g_{++} = g_{--}$ (see Section 4):

$$\phi - 1 = \frac{e^2}{24\varepsilon kT}cz^2\int_a^{\infty}[g_{++}(r) - g_{+-}(r)]4\pi r\, dr + c\frac{\pi a^3}{6}[g_{++}(a) + g_{+-}(a)] \tag{87}$$

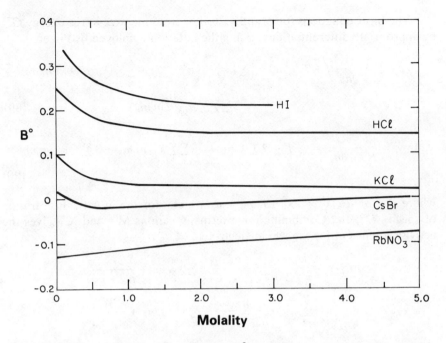

Figure 2. The apparent ion interaction coefficient B^0 for several $1:1$ electrolytes as a function of molality (from *Thermodynamics*, K. S. Pitzer and L. Brewer, revised from the 1st ed. by G. N. Lewis and M. Randall, copyright 1961 by McGraw-Hill; reproduced with permission by McGraw-Hill).

In the above equation $g_{ij}(a)$ is the value for $r = a + \delta r$, i.e., a limiting value of $g_{ij}(r)$ just as r approaches a from $r > a$. The radial distribution functions have been calculated by various authors: the Monte Carlo approach (see Section 4.4.2) applied by Card and Valleau[31] is exact in principle and susceptible only to computer errors in execution (see reference 72); computations of results from other approaches have been compared with those from the Monte Carlo method by Rasaiah *et al.*[32] (see Section 4). The first term in Eq. (87), of course, corresponds to the coulombic interactions as in the DH treatment, while the second is a short-range term, sensitive to specific ion interactions; a is a distance of closest approach corresponding to the conditions

$$U_{ij} = \infty, \quad r < a \text{ (hard-core potential)} \tag{88a}$$

$$U_{ij} = z_i z_j e^2 / \varepsilon r, \quad r > a \text{ (coulombic potential)} \tag{88b}$$

The $g_{ij}(a)$ terms have an interesting dependence on ionic strength, as shown in Figure 3, and represent the ion contact probabilities, $g_{++}(a)$ or $g_{--}(a)$ for like ion encounters being, of course, less than unity.

The basic equations for ϕ and $\gamma_\pm$ follow from the excess free energy G^{ex} by appropriate differentiations. Using the notation employed by Pitzer,

$$\phi - 1 = \frac{\partial G^{ex}}{\partial n_w} \Big/ RT \sum_i m_i = \left(\sum_1^i m_i \right)^{-1} \left((If' - f) + \sum_i \sum_j (\lambda_{ij} + I\lambda'_{ij})m_i m_j \right.$$
$$\left. + 2 \sum_i \sum_j \sum_k \mu_{ijk}m_i m_j m_k \right) \tag{89}$$

$$\ln \gamma_i = \frac{1}{RT}\frac{\partial G^{ex}}{\partial n_i} = \frac{z_i^2}{2}f' + 2\sum_j \lambda_{ij}m_j + \frac{z_i^2}{2}\sum_j\sum_k \lambda'_{jk}\, m_j m_k + 3\sum_j\sum_k \mu_{ijk}m_j m_k \tag{90}$$

where $f' = df/dI$, $\lambda'_{ij} = d\lambda_{ij}/dI$, $m_i = n_i/n_w$, etc. for the molal concentration of species i, j, etc. Combining ionic terms in cations M^{z+} and X^{z-} gives the activity coefficient for the whole electrolyte $M_{\nu M}X_{\nu X}$ as

$$\ln \gamma_{MX} = \frac{|z_M z_X|}{2}f' + \frac{2\nu_M}{\nu_M + \nu_X}\sum_j \lambda_{M_j}m_j + \frac{2\nu_X}{\nu_M + \nu_X}$$
$$+ \sum_j \lambda_{X_j}m_j + \frac{|z_M z_X|}{2}\sum_j\sum_k \lambda'_{jk}m_j m_k + \frac{3\nu_M}{\nu_M + \nu_X}$$
$$+ \sum_j\sum_k \mu_{M_{jk}}m_j m_k + \frac{3\nu_X}{\nu_M + \nu_X}\sum_j\sum_k \mu_{X_{jk}}m_j m_k \tag{91}$$

where $f(I)$ is a function of ionic strength I and expresses the effects of the long-range electrostatic forces, while λ terms are interaction coefficients for short-range i–j encounters, related to the behavior of β and g_{ij} shown in Figures 2 and 3.

For a pure single electrolyte MX these equations are written

$$\phi - 1 = \frac{|z_M z_X|}{2}\left(f' - \frac{f}{I} \right) + \frac{m}{\nu}[2\nu_M \nu_X(\lambda_{MX} + I\lambda'_{MX}) + \nu_M^2(\lambda_{MM} + I\lambda'_{MM})$$
$$+ \nu_X^2(\lambda_{XX} + I\lambda'_{XX})] + \frac{6\nu_M \nu_X m^2}{\nu}(\nu_M \mu_{MMX} + \nu_X \mu_{MXX}) \tag{92}$$

and

$$\ln \gamma = \frac{|z_M z_X|}{2}f' + \frac{m}{\nu}[2\nu_M \nu_X(2\lambda_{MX} + I\lambda'_{MX}) + \nu_M^2(2\lambda_{MM} + I\lambda'_{MM})$$
$$+ \nu_X^2(2\lambda_{XX} + I\lambda'_{XX})] + \frac{9\nu_M \nu_X m^2}{\nu}(\nu_M \mu_{MMX} + \nu_X \mu_{MXX}) \tag{93}$$

where $\nu = \nu_M + \nu_X$, in the usual way. Interaction terms in μ for triple ion encounters, MMM or XXX, have been omitted on account of their low probabilities, at least in aqueous solutions.

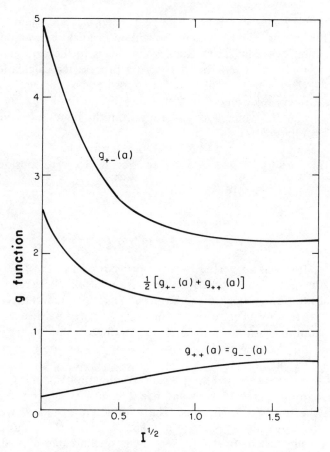

Figure 3. Dependence of radial distribution functions g at hard-core contact for the primitive model for a 1:1 electrolyte versus ionic strength. Also shown is the $g_{+-}(a) + g_{++}(a)$, which appears in the expression for the ion interaction coefficient (from Pitzer[65]).

Three types of function were considered for F, the electrostatic energy term:

$$F^\phi_{\mathrm{DHO}} = -A_\phi[I^{1/2}/(1 + bI^{1/2})] \qquad (94)$$

$$F^\phi_{\mathrm{DHC}} = -A_\phi I^{1/2}\sigma(bI^{1/2}) \qquad (95)$$

$$F^\phi_G = -A_\phi(I^{1/2}/4)\{1 + [3/(1 + bI^{1/2})^2]\} \qquad \text{for } bI^{1/2} < 2.0 \quad (96)$$

$$= \tfrac{1}{3}A_\phi(2I/b)^{1/3} \qquad \text{for } bI^{1/2} > 2.0 \quad (97)$$

where

$$A_\phi = \tfrac{1}{3}A_\gamma = \tfrac{1}{3}(2\pi N_0 d_w/1000)^{1/2}(e^2/\varepsilon kT)^{3/2} \qquad (98)$$

Here the A's are the DH coefficients, as usual.

In Eq. (94) DHO stands for "Debye–Hückel osmotic," and Eq. (94) is based on the use of a DH (Poisson–Boltzmann) radial distribution function in the equation of statistical mechanics that leads to the osmotic pressure. DHC refers to the Debye–Hückel charging process for calculation of the activity coefficient, which is converted to an osmotic coefficient by means of the Gibbs–Duhem relations [Eq. (20) or (21)]. The third form of F is that due to Glueckauf,[73] which goes smoothly from the limiting law in $I^{1/2}$ to a $I^{1/3}$ relation ("lattice result"; see Section 5.5) at high I. The quantity b is empirically evaluated for best fit for each of the forms for F.

Two relations for the second virial coefficient, β^ϕ, were considered:

$$\beta_I^\phi = \beta^0 + \beta^1/(1 + \alpha'I^{1/2})^2 \tag{99}$$

and

$$\beta_{II}^\phi = \beta^0 + \beta^1 \exp{(-\alpha'I^{1/2})} \tag{100}$$

The first relation was suggested by an improved form of the DH treatment made by Pitzer[45,65] and based on expansion of the Poisson–Boltzmann exponentials to three terms (cf. the Gronwall et al.[52b,53] and Guggenheim[54] numerical expansions). The second form was considered because of its simple form and limitingly useful properties: (1) its finite value at zero I, (2) its rapidly changing value with $I^{1/2}$ at low I, and (3) its smooth approach to a limiting value at high I.

Tests were made for the six possible combinations of the forms of F^ϕ and β^ϕ above for eight 1:1 salts and five 2:1 or 1:2 salts up to 2 M, with respect to the osmotic coefficients. The parameters b and α' were varied systematically, but with each held at the same value for all the substances for which data were tested. Best representations of the experimental behavior were achieved by taking $b = 0.12$ and $\alpha' = 2$ (standard deviations in fit ~0.0015).

Taking the electrostatic term F^ϕ for the osmotic function in the form

$$F^\phi = -A_\phi[I^{1/2}/(1 + 1.2I^{1/2})] \tag{101}$$

the Gibbs energy may be obtained by usual procedures as

$$G = -A_\phi(4I/1.2)\ln{(1 + 1.2I^{1/2})} \tag{102}$$

using $b = 1.2$, mentioned above. Then for the activity coefficient

$$\ln \gamma = -A_\phi[I^{1/2}/(1 + 1.2I^{1/2})] + (2/1.2)\ln{(1 + 1.2I^{1/2})} \tag{103}$$

It is considered that the form for F^ϕ and the Gibbs excess free energy are more convenient than the older DH forms, although the activity coefficient function is more complex. New calculations of the DH parameter A_ϕ have been made with an improved empirical equation for the dielectric constant of water.[74] The specific ion interaction approach lends itself to applications to mixtures (see Harned's rule). Detailed equations were developed by Pitzer[65] and others for this type of problem, and extensive tabulations of β^0, β^1, and the third virial coefficient C are available[71,65] for a wide range of salts including electrolytes of special interest such as tetra-n-alkylammonium halides and sulfonic acids and sulfonates. Useful data for 2:1 and 3:1 electrolytes are also available.[65] Applications to ionic mixtures, such as sea water, were treated and the osmotic coefficients calculated.

Cases for ion interaction involving an association equilibrium were also treated with special reference to phosphoric and sulfuric acids. As was mentioned earlier, the Guggenheim equation can take such effects into account provided that the degrees of ionization are large or those of association small.

3.6.4. Hückel's Term

Hückel[75] added a term, linear in m, to the DH result and attributed ion-specific activity behavior at elevated concentrations to ion–solvent interaction as manifested in the dielectric decrement δ ($\varepsilon_c = \varepsilon_0 + \delta c$ or $\varepsilon_m = \varepsilon_0 + \delta' m$).

Further discussion of this approach will be given in Section 5 on ion–solvent interaction effects to which Hückel's treatment pertains more closely, while modern statistical-mechanical approaches (76–87) will be treated next.

4. Statistical-Mechanical Treatments of Ionic Interactions in Solution

4.1. Introduction: Problems with the Debye–Hückel Theory

The DH theory and various improvements on its type of approach have dominated the treatment of the thermodynamics of ionic solutions for many years. The basis of such approaches has been the development of analytical expressions (as given in Section 3.2) representing the behavior of a well-defined model of ion distribution and ionic interactions in solution. The averaging processes characteristic of rigorous statistical-mechanical procedures are replaced by a model—the ionic atmosphere—that averages the behavior and distribution of the ions and is associated with an average electrical potential, ψ_{atm}, and a diffused space charge. The ionic atmosphere model is treated at various levels of mathematical rigor that corresponds to well-defined approximations in the DH theory, e.g., linearization of exponentials in the Poisson–Boltzmann terms for ion charge distributions or series expansion of such terms to second-, third-order terms, etc., and linearization of $\exp[-\kappa a]$ terms for $\kappa a \ll 1$. Additionally, empirical terms are added to account for the activity behavior at high concentrations, based on ion–solvent interaction or corresponding changes of solvent dielectric properties;[75] these empirical modifications amount to a change of the original DH model.

Discussions of the quantitative validity of the DH treatment have revolved around two aspects of the original theory and subsequent improvements of it: (1) the extent to which the *model itself* is a valid representation of the interaction situation in ionic solutions and (2) the extent to which the *mathematical approximations* of various kinds limit the validity of the treatment of the model to certain definable upper limits of concentration, determined by the valence type of the salt, the temperature and the dielectric constant of the solvent.

At concentrations above ~ 1.0 m *both* the model *and* various mathematical aspects of the treatment of its properties must be called into question. Surprisingly, the DH ionic atmosphere model seems[31,65] to work quite well up to ~ 1.0 m for $1:1$ salts (1) if allowance is made for the cut off in the ion distribution in the ionic atmosphere, corresponding to the effective distance of closest approach, a, of hydrated ions (cf. the double-layer case,[49] Figure 1c), and (2) if the Boltzmann exponential terms are expanded to third-order terms or if a numerical (computer) solution is made of the Poisson–Boltzmann equation without approximations (Guggenheim[54]).

This satisfactory behavior of the treatment is achieved notwithstanding three questions of some significance: one, the validity of the ionic atmosphere as a model for real discrete ion charge distribution that has been raised by Frank and Thompson,[36] who put an upper limit to the concentration at which discreteness of ion charge distribution in the ionic atmosphere corresponds to physical reality as ~ 0.003 m (Fuoss mentions a rather higher figure of ~ 0.01 m). This is a problem of "fine-grainedness" in the ionic atmosphere—cf. the discreteness of charge problems in double-layer theory.

Two, how, and the extent to which, specific ion–solvent interaction (solvation effects) must be introduced into the treatment if the thermodynamic behavior of electrolytes is to be accounted for up to quite high concentrations (>1 m). This question will be dealt with in some detail in later sections of this chapter. And third is the question of the superposition principle, dealt with earlier.

While some of the limitations in the DH theory, e.g., regarding the superposition problem (see p. 124), were already mentioned by Fowler in his monograph "Statistical Thermodynamics" in 1929 and Mayer's classic paper had appeared in 1950,[30] the first most useful statistical-mechanical examination of the DH theory, as such, was made by Kirkwood and Poirier[88] in 1954. For the case of point-charge ions ($a \to 0$) the validity of the linearized Poisson–Boltzmann equation for the first power of a parameter ξ representing the extent of charging of ions, in the expression for the potential of mean force in the interaction between a pair of ions, was demonstrated. The validity of the DH limiting law was thus shown in an unambiguous way.

When the finite sizes of ions are taken into account, the linearized integral equation for the potential of average force for a pair of ions has an oscillatory solution at high ionic concentrations ($\kappa a > 1.003$). This corresponds to a $+-$

alternation of charge density in the neighborhood of each ion. This oscillatory behavior of the local charge density corresponds to the expectations of behavior of a lattice model of ionic solutions. The lattice characteristics appear at sufficiently high concentrations, and the nonlattice behavior (limiting law, ionic atmosphere) at sufficiently high dilutions.

When the finite-ion-size case is considered in a symmetrical way for both ions of a pair, the linearized integral equation for the potential of average force cannot be converted to a Poisson–Boltzmann equation but, rather, gives oscillatory solutions at high ionic strengths. This aspect of the treatment of Kirkwood and Poirier[88] is one of the most interesting features of their results and is consistent with conclusions of later workers, e.g., Rasaiah, Card and Valleau, and Frank and Thompson, regarding the periodicity of local charge density and the relation of this behavior to lattice models.

4.2. Modern Statistical-Mechanical Treatments

4.2.1. General Aspects

Modern statistical-mechanical theories of ionic solutions seek to avoid the somewhat artificial aspects of the DH theory (the modelistic problems) and the various degrees of approximation required to improve the fit of that theory to experiment for other than limitingly dilute solutions. Advanced approaches are based on rigorous statistical-mechanical treatments of solutions developed in terms of the grand partition function and canonical ensembles.

The pairwise interaction potential $U(r)$ is the basic quantity of interest that determines the properties of the system. By various approximation methods the correlation function [written generally here as $g(r)$] for a pair of interacting particles separated by a distance r can be estimated at various concentrations and used to calculate thermodynamic functions for the solution. The procedures, using various approximations on numerical methods, may be illustrated in terms of the scheme (Rasaiah[24]) below:

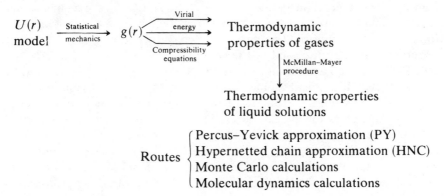

The calculations are based on the following general integral equation for any excess function $X_{(ex)}$ for a solution of interacting ions i and j:

$$X_{(ex)} = \sum_i \sum_j c_i c_j \int_0^\infty X_{i,k}(r) g_{i,j}(r) 4\pi r^2 \, dr \tag{104}$$

where $X_{i,j}(r)$ is a weight function depending on the property concerned.[76] The significance of this relation is that each i–j pair at a specified separation r makes a contribution $X_{i,j}(r)$ to the excess property $X_{(ex)}$. Since the influence of i on j through their mutual interaction must normally go to zero as $r \to \infty$, $X_{i,j}(r) \to 0$ as $r \to \infty$, as in the DH theory.

In particular, for the excess energy U^{ex} per unit volume the relation becomes

$$U^{ex} = \frac{1}{2} \sum_{i=1}^\sigma \sum_{j=1}^\sigma c_i c_j \int_0^\infty \frac{\partial(\beta u_{ij})}{\partial \beta} g_{ij}(r) 4\pi r^2 \, dr \tag{105}$$

where σ is the number of solute species, the c's are the indicated concentrations of i and j, and β, as usual, is equal to $1/kT$. (Here β is not to be confused with the specific interaction coefficient referred to in Guggenheim's and Pitzer's treatments referred to in an earlier section).

The main contributing interaction potential in $X_{i,j}(r)$ or U_{ij} and the longest-range one is the Coulomb potential $e_i e_j / \varepsilon r$. Introduction of the Coulomb potential function $f_X e_i e_j$ leads to the term, analogous to that in Eq. (104),

$$f_X e_i e_j \int_0^\infty [g_{i,j}(r) - 1] 4\pi r^2 \, dr \tag{106}$$

A general problem with the long-range Coulomb potential is that this integral diverges unless $g_{i,j}(r) - 1$ goes to zero fast enough at large r compared with the increase of $4\pi r^2 \, dr$, the spherical volume integral. (Note that $g_{i,j}(r)$ normally goes to unity for any real interaction potential as $r \to \infty$, since the interaction forces are then zero and there is no specific correlation between i and j). The evaluation of the integral in Eq. (106) thus depends rather critically on the form of $g_{i,j}(r) - 1$ as a function of r as $r \to \infty$ i.e., the asymptotic form of $g_{i,j} - 1$ is important with regard to convergence, or otherwise, of the integral as has been discussed by Friedman.[19,76]

There are some indications that the $++$ and $+-$ pair distribution or charge-density functions become oscillatory at some critical concentration, e.g., as shown by Poirier[23] and Rasaiah.[24] For water this behavior arises at $\sim 6\,M$ or can be as low as $2\,M$ according to various model calculations.[77] These oscillations are not predicted by the DH treatment or by the various modifications of it. It appears that this interesting behavior may be connected with the onset of applicability of a quasilattice model for the behavior of electrolytes, which can lead to a cube-root law in concentration for the excess free energy[36,37] (see p. 186). The details of such calculations and the problems involved have been treated or reviewed rather fully in various publications by Friedman *et al.*[17,19,25] and will not be repeated here.

The nature of solvation effects has not received particular attention in these theories and is left (see below), rather, as a residual interaction to be empirically introduced after terms for other interactions have been treated as accurately as possible.[25,26] Since the solvation effect is always significant in determining activity and osmotic coefficients at appreciable or high concentrations, it is unfortunate that no satisfactory attempts have been made to calculate *a priori* such solvation interaction terms in $RT \ln \gamma_{\pm}$ in the framework of modern statistical-mechanical treatments of ionic solutions.

At this point it may be useful to illustrate the experimental dependence of the osmotic and activity coefficients for a series of electrolytes with a common (Cl^-) anion and on a thermodynamic hydration property of the individual,[16] noncommon ions. This is shown in Figure 4 for the case of the individual standard ionic entropy of hydration of the noncommon ions. This function is chosen because it is well known that the entropy of hydration provides a useful measure of the local electrostatic and structural influence of ions on the solvent, which result in a change of its activity at a given temperature.

4.2.2. McMillan–Mayer Treatment

A major departure from the classical DH approach for the evaluation of thermodynamic properties of electrolyte solutions was made by Mayer[30]† based on McMillan and Mayer's[78] treatment of solutions in 1945. The approach is based on advanced statistical mechanics, with a specially defined type of interaction potential in the correlation function, $g_{i,j}(r)$, which can jointly take into account various interaction effects, including hydration.

Normally, for ideal gases or at *large* interparticular distances in dense fluids, e.g., solvents, $g_{i,j}(r) = 1$. However, the determination of $g_{i,j}(r)$ for relatively *small* values of r is one of the main problems in theories of liquid-state solutions.

When the fluid is a low-density gas, $g_{i,j}(r)$ can be specified simply as

$$g_{i,j}(r) = \exp\left[-U_{i,j}(r)/kT\right] \qquad (107)$$

where $U_{i,j}(r)$ is the interaction energy of particle i with particle j at a separation distance r at temperature T. Equation (107) is, of course, the well-known Boltzmann factor. At appreciable particle densities, however, the simple Eq. (107) is inadequate, since other i, j particles and the solvent molecules are close enough to a given particle to affect the distribution of i with respect to

† Unfortunately, despite laborious mathematical development, no attempt to calculate activity coefficient values for any electrolytes was made.

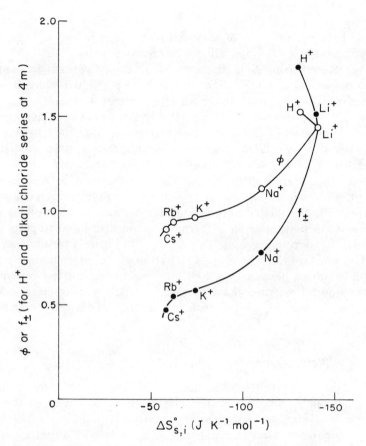

Figure 4. Experimental osmotic and activity coefficients plotted as functions of the hydration entropy of the noncommon ions in a series of salts in aqueous solution at 298 K.

j. For example, for small r the hydration of particle i, if it is an ion, will grossly distort its distribution with regard to the position of j (also hydrated) in comparison with that predicted by Eq. (107) for interactions in a dilute gas phase. Not only are other interaction energies involved in $U_{i,j}(r)$ at small r (e.g., in concentrated solutions), but excluded volume effects become significant, as discussed by Eigen and Wicke,[56] for electrolyte solutions.

The essence of the McMillan–Mayer theory of solutions is that some of the problems outlined above can be avoided if an "effective" pair interaction potential $\bar{U}_{i,j}(r)$ is considered. Suppose we consider that a correlation function $g_{i,j}^0(r)$ exists for a pair of i, j *hydrated* ions at quasi-infinite dilution so that this function contains all the effects of a real molecular solvent, including local hydration of the ions and longer-range dielectric effects. The simple Boltzmann function, Eq. (107), will not, however, apply, since the fluid is a dense one rather than a dilute gas. However, an *effective* interaction potential

$\bar{U}_{i,j}^0(r)$ can be defined for which

$$g_{i,j}^0(r) = \exp\left[-\bar{U}_{i,j}^0(r)/kT\right] \tag{108}$$

This function enables the real system of the solvated ions in a fluid to be treated *as though* it were a system of particles in a low-density gas, since the factors that make Eq. (107) inapplicable to the dense fluid situation have been incorporated into the *effective* potential $\bar{U}_{i,j}^0(r)$. This approach enables the calculation of ion-pair correlation functions to be made for an ionic solution at finite concentrations as if the solvent molecules were absent, provided that the effective ionic interaction potentials, rather than the real ones, are employed. Thus all the effects associated with the presence of the solvent are absorbed into the defined "effective potential." This is, in fact, the approach implicitly involved in the DH theory which treats interactions between "solvated" ions. The McMillan–Mayer treatment thus allows the calculation of solution thermodynamic properties as through the system of distributed ions were hypothetical dilute plasma.

In practice, since at finite salt concentrations the activity of water in a salt solution differs from that at infinite dilution, a variety of "effective potentials" would have to be used. This problem is circumvented by using, in fact, one set of effective potentials but taking the solution to be at an elevated pressure of $1\,\text{atm} + \pi$, the osmotic pressure, so that the solvent activity is caused to remain constant. Thus the thermodynamic properties at various concentrations c are calculated at various pressures $1\,\text{atm} + \pi(c)$ and converted to their 1-atm values by standard procedures.[31]

4.2.3. Evaluation of the Effective Potentials

In the McMillan–Mayer approach the problem of calculating thermodynamic properties of the solution becomes centered on the representation and calculation of the correct form for the effective interionic potentials. This, of course, is a major problem, especially for more concentrated ionic solutions. Limiting cases are relatively easily identified; thus at very short distances a core repulsive potential must always be operative and serves to define the "hard-sphere" radii of the particles. At sufficiently long distances the interaction potential must be the Coulomb potential $e_i e_j/\varepsilon r$ or $z_i e_i z_j e/\varepsilon r$ for a z_i, z_j-valent salt, where ε is the dielectric constant of the pure solvent (see Section 5). This is the term that must give rise to the observed DH limiting law behavior for excess functions for electrolytes at sufficiently high dilutions.

$\bar{U}_{i,j}^0(r)$ is thus written, in the simplest analysis, as

$$\bar{U}_{i,j}^0(r) = \bar{U}_{\text{core}}^0(r) + z_i z_j e^2/\varepsilon r + \bar{U}_{\text{other}}^0(r) \tag{109}$$

where the last term covers any remaining interaction effects such as those associated with solvation of the ions. The so-called "primitive model" or "restricted primitive model" is concerned with the case where only the first two terms on the right-hand side of Eq. (109) are taken into account. More detailed treatments[24-28] consider the factors, including ion solvation, that are contained in $\bar{U}^0_{\text{other}}(r)$. We consider first the primitive model case, which has been treated by several computational procedures.

4.3. Primitive Model Calculations

In the so-called primitive model such effects as hydration and mutual and "cavity" salting-out effects (see below) are omitted and only the effects due to the first two terms on the right-hand side of Eq. (109) are treated. The core potential can be treated in the simplest analysis as a step function: $U_{\text{core}}(r) = \infty$ for $r < r_i + r_j$ and $\bar{U}^0_{\text{core}}(r) = 0$ for $r > r_i + r_j$, the sum of the ionic radii.†

The primitive model has been treated by several methods, with surprisingly successful results. One of the most successful and direct ones obtains the results by means of a statistical computer calculation, well known in other problems of physics and of liquids[79,80]—the Monte Carlo method.

4.3.1. Monte Carlo Method

In applying the Monte Carlo method to statistical-mechanical problems,[79] a system of several hundred to several thousand particles is chosen as a basis for simulation of the system's behavior. Larger samples may give more accurate results but are prohibitive in computer time and expense. The Monte Carlo approach is essentially a "game" in which the positions of the particles are progressively changed in such a way that successive configurations have relative probabilities that converge to the Boltzmann exponent $\exp(-\Delta U/RT)$, where ΔU is the difference of potential energies the two configurations successively involved. The change of configurations and related probabilities is what is carried out relatively easily by means of a modern fast computer. Quantities of interest are then averaged over this sequence of configurations. Convergence (Figure 5) to some significant average value of a property may be attained over ~2000 configuration changes, depending on the number of particles chosen for the modeling calculation.

4.3.2. Monte Carlo Calculation of Ionic Solution Properties

The primitive model, it will be useful to reiterate, treats an ionic solution as a distribution of hard spheres influencing each other entirely through their coulombic force, corresponding to energies of interaction U_{ij} defined (see Eqs. 88a and 88b) by

$$U_{ij} = z_i z_j e^2 / \varepsilon r_{ij} \qquad (r_{ij} \geq a) \qquad (110a)$$

† This sum is analogous to the $\mathring{a}$ parameter in the DH treatment for finite-sized ions.

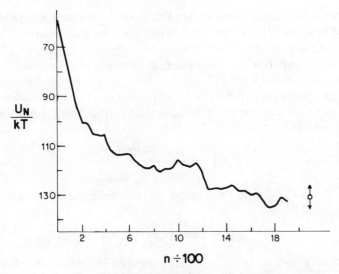

Figure 5. Convergence of a Monte Carlo evaluation of U to the equilibrium region. This shows the configurational energy of a 200-ion system, 1.968 M, starting from a configuration generated by placing the ions in a nonoverlapping but otherwise random manner. The equilibrium average energy obtained in the rest of the run is shown along with its standard deviation ($\bigcirc$). The line segments join points that are averages over the preceding 50 configurations and n is the number of configurations (from Card and Valleau[31]).

and feeling each other at short range by the core repulsion

$$U_{ij} = \infty \quad \text{(hard-sphere repulsion) for } r_{ij} < a \qquad (110b)$$

where a is an effective contact distance, as in the DH theory, presumably corresponding to some overlap of solvation shells. (Strictly, ions having real nonspherical solvation shells are unlikely to exhibit "hard-sphere" repulsive interactions. In fact, cases for various "softer" interactions have been treated[24-28]).

It is the aim of Monte Carlo calculations to provide an "exact" (within definable limits of computational accuracy) treatment of the given model and thus predict the thermodynamic properties of a solution behaving according to the defined model. Such predictions can then be compared with those of the DH theory up to various concentrations and with those of other statistical-mechanical treatments of the DH model, e.g., by the integral equation methods.[27-29]

The Monte Carlo approach was first investigated in work by Poirier[23] with Shaw[81] and applied more systematically in independent work in Russia.[72] Related studies have been made on plasmas where the charge interaction problems are similar but without, of course, a solvent dielectric present.

A thorough Monte Carlo study of the primitive model was published by Card and Valleau,[31] and since its conclusions are particularly illuminating and the accuracy of the evaluation of convergent configurational energies appears to be superior to that obtained in other calculations,[72] we shall present aspects of this work in some detail.

If P is a property expressible as a function of the configuration q of the particles in a classical system, its canonical ensemble average may be written (see reference 31) as

$$\langle P \rangle = \int \frac{P(q) \exp [-\beta U(q)] \, dq}{Q} \tag{111}$$

where $\beta = 1/kT$, $U(q)$ is the energy of the configuration q, and Q is the configurational integral

$$Q = \int \exp [-\beta U(q)] \, dq \tag{112}$$

$\langle P \rangle$, according to Eq. (111) may be estimated by means of Monte Carlo procedures,[79,80] which enable a Markov chain of configurations to be generated for which the relative frequencies of the configurations in the limit distribution are appropriately weighted by the Boltzmann factor $\exp(-\beta U)$, e.g., Figure 5.

The required thermodynamic averages $\langle P \rangle$ involve the sum U of the potentials u_{ij} for N particles of the system in a volume V:

$$U = \sum_{i}^{N-1} \sum_{j=i+1}^{N} u_{ij} \tag{113}$$

The configurational energy U for, say, a 200-ion system ($\equiv 1.968 \, M$) converges towards the "equilibrium" value after about 1800 successive configurations (after some chosen initial random one) having progressively more favorable Boltzmann probabilities that have been generated in the chain (Figure 5).

For the simple $+-$ ion model there are only two distinct pair correlation functions:

$$g_{++}(r) = g_{--}(r) \qquad \text{to be denoted by } g_A(r) \tag{114}$$

and

$$g_{+-}(r) \equiv g_{-+}(r) \qquad \text{to be denoted by } g_B(r) \tag{114b}$$

The Poisson–Boltzmann equations in the DH theory (for the finite-ion-size case) give for these g's exponential functions in

$$F(r) = (e^2/\varepsilon kT)\{\exp [\kappa (a - r)]/(1 + \kappa a)r\} \tag{115}$$

where κ^{-1}, the ionic atmosphere reciprocal radius, is defined from

$$\kappa^2 = (4\pi e^2/\varepsilon kT)\{[N_+(z^+)^2 + N_-(z^-)^2]/V\} \tag{116}$$

In order to minimize errors the potential ψ derived from a linearized Poisson–Boltzmann equation is used as an exponential factor in the correlation functions g_{ij}. These g's, written in exponential form, are the basis of what Card and Valleau call the DHX calculation. Further comment on this type of calculation in relation to the role of hydration effects in the behavior of $1:1$ electrolytes is given later.

The correlation functions are then written as

$$g_{A,DHX}(r) = \exp[-F(r)] \qquad r \geq a \qquad (117)$$

$$g_{B,DHX}(r) = \exp[F(r)] \qquad r \geq a \qquad (118)$$

$$g_{A,DHX}(r) = g_{B,DHX}(r) = 0 \qquad r < a \text{ (hard-sphere repulsion)} \qquad (119)$$

They are then used in the McMillan–Mayer formalism. The values of the parameters were as follows. a was taken as 0.425 nm, $T = 298$ K, and ε as 78.5.

The (reduced) configuration energy per ion is then obtained as

$$\frac{U_{DHX}}{NkT} = 0.25 \, \kappa^2 \int_a^\infty [g_{A,DHX}(r) - g_{B,DHX}(r)] r \, dr \qquad (120)$$

and the osmotic coefficient as

$$\phi_{DHX}^{-1} = (\pi a^3 N/3V)[g_{A,DHX}^{(a)} + g_{B,DHX}^{(a)}] + U_{DHX}/3NkT \qquad (121)$$

The $g_{A,DHX}^{(a)}$ and $g_{B,DHX}^{(a)}$ are the pair correlations at contact $(r_i + r_j = a)$ and were estimated from the corresponding $g(r)$ functions for a series of small positive separations Δr from $r = a$.

The pair correlation functions $g_A(r)$ and $g_B(r)$ were also calculated from the data generated in the Monte Carlo calculations. $g_A(r)$ and $g_B(r)$ as functions of r were evaluated for a "0.00911 M" solution for $N = 200$, 64, and 16 ions. For $N = 200$ and $N = 64$ the agreement with the respective $g_{DHX}(r)$ functions was excellent $(r = 5\text{–}50 \text{ Å})$. At higher concentrations, "0.1038 M" and "0.425 M", deviations of g_{DHX} from the Monte Carlo values become small but significant, whereas at "1.968 M" they are appreciable, indicating failure of the DHX at this concentration $(r = 0.42\text{–}1.8 \text{ nm})$. While the latter discrepancies are interesting, the striking feature of these Monte Carlo calculations of Card and Valleau is that the exponential distribution function† is still

---

† It should be mentioned that by DHX in Card and Valleau's treatment they meant a theory in which the potential ψ resulting from the *linearized* Poisson–Boltzmann equation, for finite ion size, is inserted back in *exponential* factors to obtain g_{ij}. Then these g_{ij} (g_{+-} and g_{++} or g_{--}) are used in the McMillan–Mayer formalism to derive the osmotic pressure. This means that "repulsive force" contributions to the osmotic pressure are taken into account at least approximately and not discarded through the usual device of using a "charging process" to obtain the electrostatic contributions to the nonideal chemical potential of the electrolyte. Thus the rise in the DHX activity or osmotic coefficient at high concentrations is not due only to the coulombic contributions when the behavior is expressed in the McMillan–Mayer system.

remarkably satisfactory up to ~"0.425 M" and excellent at "0.1038 M." (For practical purposes it is to be noted that the Monte Carlo results are to be regarded as *exact* for the model as defined—the so-called "primitive" model—there are no approximations arising in the model itself except other additional factors, which will be considered later, e.g., in reference to the treatments of Rasaiah and Friedman and with regard to Figure 6).

Activity coefficients in the McMillan–Mayer system (constant T and fixed chemical potential of the solvent, μ_{H_2O}) were derived from the Monte Carlo results and related to the conventional activity coefficient system (constant T, P). Agreement between the Monte Carlo results (Figures 7 and 8) for $\phi - 1$ or $\ln \gamma_\pm$ in the primitive model and those given by the exponential DH theory are remarkably good, even up to ~2.0 M for 1:1 electrolytes. This agreement,

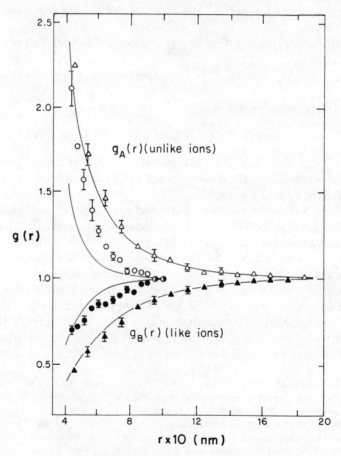

Figure 6. Pair correlation functions $g_B(r)$ and $g_A(r)$ for unlike and like ion configurations for 0.425 M ($\triangle$) and 1.968 M ($\bigcirc$, $\bullet$) for a sample N of 200 ions. The solid line curves represent the DHX behavior (from Card and Valleau[31]).

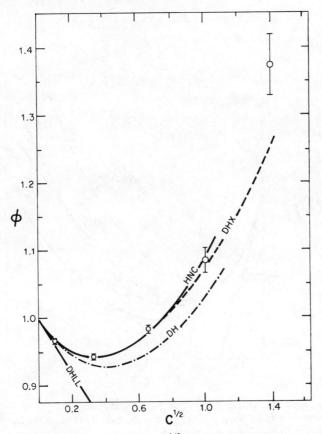

Figure 7. Osmotic coefficient as a function of $c^{1/2}$. The error bars shown correspond to $\pm 3\sigma$ (where σ is the standard error). The HNC line is from the HNC work of Rasaiah and Friedman. In the original, common version of the DH theory the pair correlation functions were linearized; DH theory shows the result of using these linearized functions. The graph also shows the limiting law (DHLL) (from Card and Valleau[31]).

it may be noted, is achieved without introduction of any ion hydration parameter in this type of calculation. Good agreement is also found with the results of Rasaiah and Friedman[27,28] using the hypernetted chain (HNC) approach (Figures 7 and 9). Some comparisons with other results from the integral equation methods [Percus–Yevick, (PY) and HNC] are shown in Table 4 for ϕ data and Table 5 for McMillan–Mayer and conventional $\ln \gamma_{\pm}$ data.

The authors have noted that the functional behavior of their results for the Monte Carlo treatment of the primitive model differ appreciably from those of Vorontsov-Vel-yaminov *et al.*[72] obtained similarly (apart from somewhat different initial parameters). These differences, e.g., in $\ln \gamma_{\pm}$ as a function of concentration ($\ln \gamma_{\pm}$ values are too low), are attributed to an inaccurate sampling procedure in the computer calculations. Discrepancies in

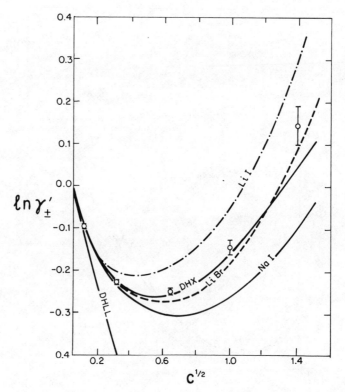

Figure 8. Mean ionic activity coefficient $\gamma'_{\pm}$ at fixed T and P. Error bars show 3σ as in Figure 7. The lines represent DHX for $a = 0.425$ nm and the experimental results for three salts in aqueous solution at 298 K. DHLL as in Figure 7.

$\phi - 1$ are attributed to poorly defined values for the $g(r)$ function rather than unsatisfactory U values in the Russian work. The reliability of the Card and Valleau calculations[31] is indicated by the good agreement they give with the HNC results (see below).[32]

Some detailed improvements to the primitive model can be made by relieving the "hard-sphere" repulsion relation in the core term and replacing it by a more realistic interaction potential function, such as is used in the "6–12" Lennard–Jones and Devonshire theory of molecular interactions. More generally, $\bar{U}^0_{COR} = D_{ij}r^{-n}$, where D_{ij} is a characteristic constant for the i–j repulsive interaction and n is a power index (12 in the Lennard-Jones and Devonshire model). Ramanathan and Friedman[26] have used a "soft" r^{-9} function for the COR repulsion, with advantageous results.

4.3.3. Integral Equation and Other Methods

For electrolyte solution problems, in addition to the Monte Carlo calculations,[31] the so-called integral equation approaches have been profitably used,

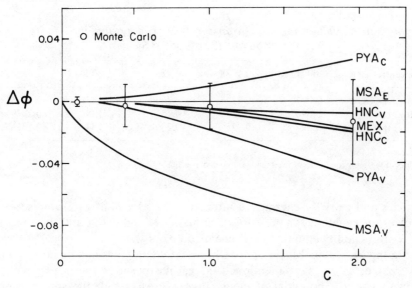

Figure 9. Comparison of results for deviation $\Delta\phi$ as a function of electrolyte concentration C, calculated by various methods (see text for abbreviations) (from Rasaiah, Card, and Valleau[32]).

e.g., by Carley[29] and by Rasaiah and Friedman.[27,28] From the point of view of structural aspects of solutions, emphasis has been placed on evaluation of radial distribution functions $g_{ij}(r)$ for the spatial relation of one particle, i, with respect to another, j, influenced by the interaction energy u_{ij} between the pair and the temperature. Information in $g_{ij}(r)$ is often more useful than that in some mean electrostatic potential (as in the DH theory). Generally there are three routes from $g_{ij}(r)$ functions to evaluate thermodynamic properties: One involves the energy of the system in terms of number densities of

Table 4
Values of the Osmotic Coefficient ϕ from Monte Carlo Evaluation of g(a) and U/NkT (Extrapolated to $1/N \to 0$) and the DHX Prediction in Comparison with Integral Equation Results[a]

c	ϕ	ϕ_{DHX}	ϕ^{PY} [Carley[29]]	ϕ^{HNC} [Carley[29]]	ϕ^{HNC} [Rasaiah and Friedman[27,28]]
0.00911	0.9704 ± 0.0008	0.9701	0.985	0.970	0.970
0.1038	0.9454 ± 0.0011	0.9451	0.952	0.946	0.945
0.425	0.9850 ± 0.0023	0.9791	0.976	0.984	0.980
1.000	1.085 ± 0.0058	1.0768	(1.08)	(1.09)	1.090
1.968	1.373 ± 0.014	1.2561	(1.34)	(1.41)	—

[a] From Card and Valleau.[31]

Table 5
Comparison of Values for ln γ± from the DHX Theory and the Monte Carlo
Results in the McMillan–Mayer System.[a]

Concentration c	0.00911	0.1038	0.425	1.000	1.968
ln γ± (DHX)	−0.0970	−0.2285	−0.2600	−0.1448	+0.1375
ln γ± (MC)	−0.0967	−0.2280	−0.2511	−0.128	+0.2879
ln γ± (MC)[b]	−0.097	−0.228	−0.250	−0.144	+0.145

[a] From Card and Valleau.[31]
[b] Values of ln γ'± for conditions of fixed temperature and pressure.

particles and pairwise energies of interaction u_{ij} as well as $g_{ij}(r)$; the second involves an equation for pressure in terms of u_{ij} and $g_{ij}(r)$, and the third is the compressibility equation (see scheme on p. 145).

In the McMillan–Mayer approach the solute particles in a liquid solution are treated, as we have mentioned, as gaslike particles interacting with a pairwise *effective* potential of mean force $\bar{U}_{ij}$, as though the solvent were not present, but with the effective potential itself taking account of interactions as though the solvent were present. The next step is to derive approximate pair correlation functions in terms of $\bar{U}_{ij}$. A complex cluster diagram analysis has to be made for g_{ij} functions for electrolyte solution problems.

Currently, approximations for g_{ij} functions are made in various ways through the direct correlation function c_{ij} in the integral equation approach, defined by the Ornstein–Zernicke[82] equation

$$g_{ij} - 1 = c_{ij} + \Sigma_s \rho_s \int (g_{is} - 1) c_{sj} \, ds \qquad (122)$$

The direct correlation function reflects, in its long-range behavior, the long-range part (e.g., coulombic) of u_{ij} very simply. The asymptotic behavior of c_{ij} is in fact

$$c_{ij}(r) \rightarrow -\frac{1}{kT} u_{ij}(r) \qquad \text{as} \quad r \rightarrow \infty \qquad (123)$$

Various approximate integral equation treatments, based on the Ornstein–Zernicke equation, have been made: Thus, if c_{ij} is chosen as

$$c_{ij} = g_{ij} - 1 - \ln g_{ij} - \beta u_{ij} \qquad (124)$$

the so-called HNC approximation arises. This name originates from the diagram notation[82] in the cluster expansion treatment of the pair correlation function, where only diagrams having a certain chainlike configuration are retained. The diagrams are topological representations of mathematical terms and summing over types of diagrams provides the integration operations required.[30,83]

Another procedure allows

$$c_{ij} = g_{ij}[1 - \exp(\beta U_{ij})] \tag{125}$$

which corresponds to the PY approximation,[84] first derived by a collective coordinate method.

Another integral equation treatment, which has been developed for treating the primitive model of electrolyte solutions, is the so-called mean spherical approximation (MSA). For this the exact condition $g_{ij} = 0$ for $r_{ij} < a_{ij}$ applies, corresponding to hard-sphere contact, while for $r_{ij} > a_{ij}$ it is approximated by its asymptotic behavior

$$c_{ij} = \beta \frac{z_i z_j e^2}{\varepsilon r_{ij}} \tag{126}$$

which applies strictly when $r_{ij} \to \infty$. Exact solutions can be obtained for this case, as were derived by Waisman and Lebowitz for the restricted primitive model.[85]

4.3.4. Comparison between Results of Various Statistical-Mechanical Methods

Card and Valleau,[31] as mentioned in Section 4.3.2, made a comparison between their Monte Carlo (MC) results and those obtained by the HNC method by Rasaiah and Friedman[27,28] and found excellent agreement up to $c \sim 1\ M$. However, the latter authors concluded that the analogue of the HNC equation is superior to that of the PY (Percus–Yevick[84]) equation as treated by Allnatt[86] ("PYA").

A more recent (see reference 31) detailed comparison between the MC, PYA, and HNC results (Figure 9) has been given by Rasaiah, Card, and Valleau,[32] who obtained new data with the HNC and PYA equations for the same concentrations as those of the MC work and other parameters; concentrations were extended to $2\ M$ to compare with the results of the MC calculations. Also, a treatment by Anderson and Chandler[87] [the "mode expansion method" (MEX)] has been applied to electrolyte solutions in the restricted primitive model and the results compared with those of the MC method. In the MEX treatment a reference system of uncharged hard spheres is considered, with the coulombic interaction introduced as a "perturbation." MEX theory expresses the effect of these interactions on the free energy as an infinite series of terms ("modes") that involve Fourier transforms of the electrostatic perturbation potential and of the correlation functions of the reference system.

We shall not quote here all the details of the discussion of these comparisons, but differences and similarities in the results of the various calculations are conveniently seen in Tables 6 and 7 for the relative configurational energy U/NkT and the osmotic coefficient ϕ (see also Figure 10). The latter

Table 6
Comparison of MC Results for $U/N\kappa T$ and CONTACT with Results from the HNC, PYA, and MSA Theories, with $a = 4.25$ Å, $\varepsilon = 78.5$, $\partial\varepsilon/\partial T = 0$, and $T = 298$ K

C (mol liter^{-1})	$-U/N\kappa T$			
	MC	HNC	PYA	MSA
0.00911	0.1029 ± 0.0013	0.1014	0.1014	0.0992
0.10376	0.2739 ± 0.0014	0.2712	0.2712	0.2675
0.42502	0.4341 ± 0.0017	0.4295	0.4285	0.4264
1.0001	0.5516 ± 0.0016	0.5447	0.5418	0.5405
1.9676	0.6511 ± 0.0020	0.6460	0.6376	0.6362

C (mol liter^{-1})	Contact			
	MC	HNC	PYA	MSA
0.00911	0.0044 ± 0.0007	0.0041	0.0041	0.0017
0.10376	0.0359 ± 0.0011	0.0357	0.0351	0.0203
0.42502	0.1217 ± 0.0045	0.1228	0.1194	0.0867
1.001	0.2777 ± 0.0045	0.2741	0.2595	0.2191
1.9676	0.5625 ± 0.0088	0.5668	0.5240	0.4878

Table 7
Comparison of MC Results for the Osmotic Coefficient with the HNC, PYA, MSA, and MEX Theoretical Treatments, with $a = 4.25$ Å, $\varepsilon = 78.5$, and $T = 298$ K

C (mol liter^{-1})	ϕ_ν^{MC}	ϕ_ν^{HNC}	ϕ_c^{HNC}	ϕ_ν^{PYA}
0.00911	0.9701 ± 0.008	0.9703	0.9705	0.9703
0.10376	0.9445 ± 0.0012	0.9453	0.9458	0.9452
0.42502	0.9774 ± 0.0046	0.9796	0.9800	0.9765
1.0001	1.094 ± 0.005	1.0926	1.0906	1.0789
1.9676	1.346 ± 0.009	1.3514	1.3404	1.3114

C (mol liter^{-1})	ϕ_c^{PYA}	ϕ_ν^{MSA}	ϕ_E^{MSA}	ϕ^{MEX}
0.00911	0.9705	0.9687	0.9709	0.9707
0.10376	0.9461	0.9312	0.9454	0.9452
0.42502	0.9844	0.9446	0.9806	0.9787
1.0001	1.1076	1.039	1.097	1.091
1.9676	1.386	1.2757	1.3595	1.342

[a] From Rasaiah, Card, and Valleau.[32]

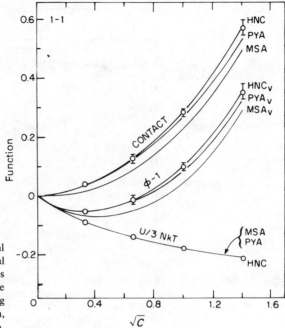

Figure 10. MC and theoretical values for the configurational energy U contact and for $\phi - 1$ as a function of the square root of the electrolyte concentration C, using the virial theorem (from Rasaiah, Card, and Valleau[32]). See Eq. (127).

is expressed either as ϕ_v, by applying the virial theorem, which gives ϕ in terms of the pair functions:

$$\phi_v - 1 = \langle U \rangle / 3NkT + \text{CONTACT} \tag{127}$$

where

$$\text{CONTACT} = (\pi Na^3/3V)[g_A(a_+) + g_B(a_+)] \tag{128}$$

or ϕ is expressed as ϕ_c, from the compressibility equation treatment of Rasaiah and Friedman. ϕ from the MSA calculation is referred to as ϕ_E, and a is the contact diameter of the ions $[g(a^+) = \lim_{\delta r \to 0} g(a + \delta r)]$, as previously.

Agreement between predictions of HNC and MC calculations for both the like and unlike ion pair correlation functions as functions of r is also excellent up to $c = 1 \cdot 968\,M$, but there is an interesting charge-inversion region in g_{+-} and $g_{++}\,(= g_{--})$ over the region $r \doteq 9$ to $r \doteq 15$ Å. This represents the onset of oscillatory charge distribution characteristic of lattice behavior, a phenomenon predicted by Kirkwood and Poirier[88] that had been the subject of some controversy in earlier years (Figure 11). A cube-root behavior in ϕ as a function of c is clearly predicted by MC, HNC, and MSA calculations (Figure 12), corresponding to quasilattice behavior in the range $c^{1/3} \simeq 0.1$ to $c^{1/3} \simeq 0.6$. It should be noted, however, that if some variable is a function of c, such as $c^{1/2}$ or $c^{1/3}$, it is often difficult to distinguish the functionality in c unless a wide range of c values can be covered in the test plots.

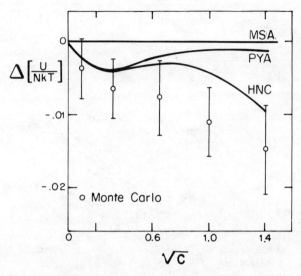

Figure 11. Excess energy results. The graph shows differences from the MSA predictions and thus displays, on an enlarged scale, only the differences between the various results. The MC error bars are again plus or minus three standard errors. The onset of oscillatory behavior is seen.

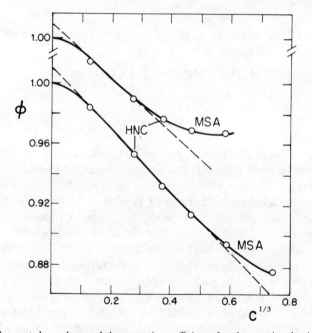

Figure 12. Cube-root dependence of the osmotic coefficients for the restricted primitive model with diameters $a = 0.387$ nm (upper curve) and $a = 0.241$ nm (lower curve). The ordinates of the upper curve are displaced by 0.03. In this range of concentration the MC, MSA, and HNC results all agree excellently and are presumably accurate results for the restricted primitive model.

4.3.5. Charge Effects and Ion Size

An interesting conclusion from the MSA osmotic coefficients ϕ_E is that for sufficiently large ions of equal size (e.g., $a = 6.33$ Å) the charge effect on ϕ_E as a function of c is almost *eliminated* at a concentration as low as ~ 1 M, that is, the ϕ_E behavior as a function of c is almost independent of the charges if the ions are large enough. This conclusion probably reflects the real situation with some tetraalkylammonium salt solutions, which behave, in some respects, like neutral molecule solutes in water, e.g., with regard to structure promotion. Results are shown in Figure 13.

5. Ion–Solvent Interactions in the Activity Behavior of Electrolytes

5.1. Introduction

The basis of early treatments[41] of the activity behavior of electrolytes did not involve any role of ion–solvent interactions, since the deviations from ideality were treated in terms of interactions between already hydrated

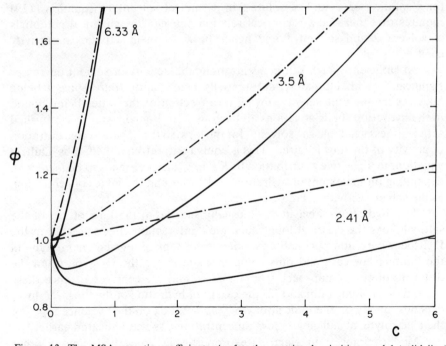

Figure 13. The MSA osmotic coefficients ϕ_E for the restricted primitive model (solid line) compared with the values ϕ^0 they would have were the charges removed from the ions (dashed line). The sets of curves are labeled according to the ionic diameters.

(solvated) ions; i.e., the reference state for activity coefficients is defined as the infinite dilution of *solvated* ions ($\gamma_\pm \to 1$) (see Section 3). Solvent properties then only enter into the expressions for activity coefficients through the bulk dielectric constant ε of the solvent medium involved in the long-range ion–ion interaction terms. Since hydrated (solvated) ions are the defined species between which interactions occur, the relevant value of ε involved in expressions for the nonideal free energy contribution $RT \ln \gamma_\pm$ is that for the pure solvent rather than ε_s, that for the solution. This is because any decrease of ε to ε_s on account of the presence of the ions is due to the local dielectric saturation effects[89–93] associated with ion–solvent interactions within the solvation shells or, more generally, the solvation cospheres[93] of the ions. A more complex situation arises, of course, at concentrations $> \sim 2\ m$, where virtually no free bulk solvent, uninfluenced by ions, remains; the activity behavior of salts under these conditions will be discussed in more detail later (see Section 5.3).

An effect of changes of solvent activity, due to ion–solvent interactions, on solute salt activity arises indirectly for quite general and fundamental principles, as follows from the Gibbs–Duhem equation

$$n_1 d\mu_1 + n_2 d\mu_2 = 0 \tag{129}$$

for a solution of n_2 mol of solute 2 in n_1 mol of solvent 1. Equation (129) requires that there be a reciprocal relation between the chemical potentials of solvent 2 and solvent 1 and hence between their activities or activity coefficients (p. 118).

At sufficiently high salt concentrations the activity of solvent decreases significantly and, ultimately, substantially below unity (pure solvent being taken as having unit activity) owing to a fraction of the water being bound in the hydration shells of cations and anions. This bound water of diminished activity is less available as "solvent" for the ions, so the apparent concentration or activity of the ions is raised. This is equivalent, through the Gibbs–Duhem equation, to a positive contribution to $RT \ln \gamma_\pm$; i.e., $\gamma_\pm$ tends to be *increased*, depending on the extent of hydration of the ions and the ionic concentration, as the solvent activity decreases.

This hydration effect in $\gamma_\pm$ is usually specific to the type of ion of the salt, whereas the classical long-range ionic interaction effects treated in the DH theory are nonspecific for the identity of ions of a given charge type in the limiting law case and only somewhat specific to the ion type when the distances of closest approach (contact condition) are considered, since these depend on (1) ionic radii and (2) the extent of hydration of the ions. $RT \ln \gamma_\pm$ depends mainly on the ionic strength $\sum_i c_i z_i^2 / 2$, i.e., on the "valence type" of the electrolyte, at sufficiently low concentrations, as was indicated earlier.

Modern interest in electrolyte properties is concerned, among other things, with the specificities of ion–solvent interaction both with respect to the types of individual ions involved and to the nature of the solvent, e.g., in relation to the accessibility of its principal dipole and/or any lone pair electrons in the molecule. These ion- and solvent-specific interactions are revealed,[3] on the one hand, in the spectroscopic behavior of electrolyte solutions (see the many papers on IR, Raman, and NMR spectroscopic studies of electrolytes) and, on the other, in the thermodynamic properties of electrolyte solutions where (1) thermodynamic properties at infinite dilution or in some standard state usually reveal ion-specific behavior or (2) the electrolyte activity coefficients at appreciable concentrations have characteristic values determined to a significant extent, at sufficiently high concentrations, by the ion–solvent interactions.

In some of the modern theories of electrolytes[24-28] the latter effects are treated in a more or less empirical way in terms of the so-called Gurney parameters, after more rigorous *a priori* expressions have been used for other types of ionic interaction potential (see Section 4).

5.2. Dielectric Constant and Hydration Effects

5.2.1. Role of Dielectric Saturation at Ions

It is well known that the field near ions, in the region $3 \sim 5$ Å from their periphery, is sufficient to cause appreciable dielectric saturation effects. This question was treated theoretically by Webb[92] in terms of the Debye–Langevin model and in a more satisfactory way for water by Booth,[90] based on the Kirkwood–Onsager theory for associated dielectrics. Experimental evidence for this dielectric saturation effect follows from the observation[89] of appreciable molar dielectric decrements δ for most salts in aqueous solutions and has been discussed by Schellman.[94] The observed dielectric constant of an ionic solution is a mole fraction weighted average of the dielectric constant of "free water" and the effective dielectric constant of water in the tightly bound region of hydration cospheres of cations and anions. The values of δ correspond to primary hydration numbers[89] on the order of $2 \sim 6$, with a local value of the dielectric constant of ~ 6 (see reference 95).

The question of whether this local dielectric saturation effect at ions, which is concentration dependent to an extent determined by δ ($\varepsilon_c = \varepsilon_{c=0} + \delta c$), would lead to a deviation from the general dependence of $\ln \gamma_{\pm}$ on the square root of ionic strength was discussed by Debye and Pauling.[96] They considered the electric potential near an ion in two separate cosphere regions, one with low ε, corresponding to a degree of solvent dipole orientation saturation, and another further out, with a normal value of ε. A complex expression for the potential of the central ion in its ionic atmosphere was obtained, but upon expansion in powers of concentration it can be shown that this expression gives the limiting law as a leading term. They did not, however, discuss the significance of deviations from the limiting law, which can already arise[97] at low but significant concentrations in this treatment.

5.2.2. Relation between Dielectric Saturation, Local Hydration of Ions, and the Contact Distance a

A more explicit treatment of this problem was given by Frank,[97] who considered (1) the nature of deviations from the limiting law that the Debye–Pauling approach[96] leads to and (2) the relation of local dielectric saturation effects to the significance of the DH a parameter corresponding to close encounters between ions in the "finite-ion-size" treatment: $\ln \gamma_\pm = z_+ z_- e^2 \kappa / 2\varepsilon kT \, (1 + \kappa a)$.

The model is shown in Figure 14: the ion has a hard-sphere radius of r, a dielectric constant ε_1, and a cosphere of strongly bound solvent up to a radius R. This region was also assumed to have a dielectric constant ε_1 (this is probably strictly incorrect, as the ion's dielectric constant will probably be less than that of the solvent molecules in the region $r \to R$, depending on the radius of the ion and its charge). Beyond R in the bulk solution the dielectric constant is ε_2.

The free energy G_i of a central reference ion in relation to its interaction with other ionic atmosphere ions is calculated, giving

$$G_i = \frac{(ze)^2}{2\varepsilon_1 R} - \frac{(ze)^2}{2\varepsilon_2 R} - \frac{(ze)^2 \lambda}{2\varepsilon_1 \Delta_1} \frac{1 + \beta E}{1 + \lambda r/\Delta_1 - (1 - \lambda r/\Delta_1)\beta E} \qquad (130)$$

where $\lambda = (4\pi e^2/kT) \sum_1^i c_i z_i^2$, $\Delta = \varepsilon^{1/2}$ so that λ/Δ for either region is the DH κ for that region, $E = \exp[(2\lambda/\Delta_1)(r - R)]$, and $\beta = [1 + \lambda R/(\Delta_1 + \Delta_2)]/[1 + \lambda R/(\Delta_2 - \Delta_1)]$. For the DH model ε_1 (for hydration shells) is not distinguished from ε_2, so that the usual result, $G_i = -(ze)^2 \kappa/2\varepsilon(1 + \kappa r)$, is recovered (here r is the DH $\mathring{a}$ parameter, see below).

A value of a_{eff} is defined so that $G_1 = -(ze)^2 \kappa/ze(1 + \kappa r_{eff})$ is the expression for the free energy, using a value r_{eff}, which is the value r would have to have in a system of uniform dielectric constant to give the corresponding actual value of G_i.

Equation (130) is similar to that of Debye and Pauling, except that the free energy is calculated and is $ze/2$ times their expression for the potential of the ion.

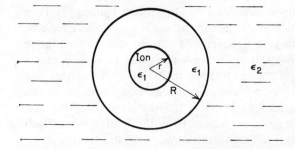

Figure 14. Model used in Pauling and Debye's[96] and Frank's[97] treatments of activity coefficients in relation to the dielectric constant of the solution. ε_1 is the dielectric constant of the ion solvate entity; ε_2 is the dielectric constant of the solution solvent.

The ratio of G_i to the limiting law value $-(ze)^2\kappa/2\varepsilon$ was computed for various values of ε_1, r, and R and is shown in Figure 15, while Figure 16 shows plots of G_i versus $c^{1/2}$. Under certain conditions *negative* deviations from the limiting law are predicted (Figure 16), depending on the choice of ε_1 and $R - r$. Experimentally, of course, such behavior† is never observed. A Bjerrum ion association correction does not offer a solution to this disagreement. However, by taking r equal to or somewhat less than R, the predicted large deviations from the limiting law are avoided. Then the "hard-sphere" contact distance between an anion and a cation is not the sum of the crystallographic radii but is a larger value, as is in fact indicated by the values of the DH $\mathring{a}$ parameter required to fit $\ln \gamma_{\pm}$ data as a function of $c^{1/2}$ in the finite-ion-size treatment.

A value of r somewhat less than R, due to the penetration of anion and cation hydration cospheres, was visualized by Frank[97] and corresponds to the empirical experimental situation that $\mathring{a}$, for a given electrolyte, is usually greater than sums of the ionic radii of the anions and cation involved, but is less than the sum of probable hydration radii (Table 8). This situation corresponds to the solvent-shared type of ion pair but is transient.

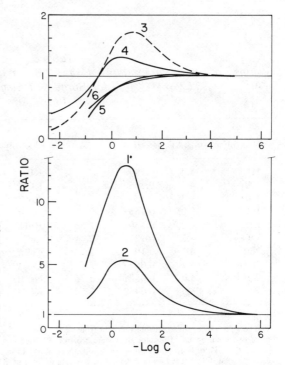

Figure 15. Ratio of G_i to the limiting law value as a function of $\log c$, where c is in mol liter^{-1}: curve 1, $\varepsilon_1 = 4$, $a = 2\,\text{Å}$, $R = 5\,\text{Å}$; curve 2, $\varepsilon_1 = 4$, $a = 3\,\text{Å}$, $R = 5\,\text{Å}$, curve 3, $\varepsilon_1 = 25$, $a = 3\,\text{Å}$, $R = 10\,\text{Å}$; curve 4, $\varepsilon_1 = 25$, $a = 2\,\text{Å}$, $R = 5\,\text{Å}$; curve 5, $\varepsilon_1 = 49$, $a = 3\,\text{Å}$, $R = 10\,\text{Å}$; curve 6, $\varepsilon_1 = 25$, $a = 3\,\text{Å}$, $R = 5\,\text{Å}$ (from Frank[97]).

† Apparent negative deviations arise when strong ion pairing occurs, but this is a situation not considered in the limiting law treatment, which applies, of course, to fully dissociated electrolytes.

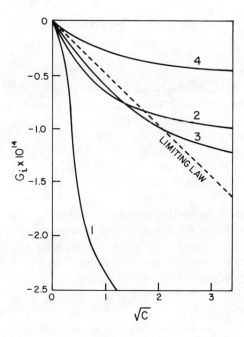

Figure 16. G_i as a function of $c^{1/2}$, in ergs per ion $\times 10^{14}$, c in mol liter^{-1}: curve 1, $\varepsilon = 4$, $a = 3$ Å, $R = 5$ Å; curve 2, $\varepsilon_1 = 25$, $a = 3$ Å, $R = 10$ Å; curve 3, $\varepsilon_1 = 25$, $a = 2$ Å, $R = 5$ Å; curve 4, $\varepsilon_1 = 49$, $a = 3$ Å, $R = 5$ Å (from Frank[97]).

Table 8

Values of the Parameters a and B in the DH Equation $\log \gamma_\pm = -Al^{1/2}/[1 + 35.57a(\varepsilon T)^{-1/2}l^{1/2}]$ for Some 1:1 Electrolytes at 298 K (Valid up to 1 M)[a]

Electrolyte	$\mathring{a}$ (Å)	B	$(r_+ + r_-)$ (Å)	Electrolyte	$\mathring{a}$ (Å)	B	$(r_+ + r_-)$ (Å)
HI	5.0	0.197		KBr	3.84	0.0282	3.28
HBr	4.4	0.165		KCl	3.8	0.0202	3.14
HCl	4.3	0.133		RbCl	3.6	0.010	3.29
LiI	5.05	0.165	2.77	RbBr	3.55	0.101	3.43
LiBr	4.3	0.130	2.56	RbI	3.5	0.0085	3.65
LiCl	4.25	0.121	2.41	CsCl	3.0	0	3.46
NaI	4.2	0.100	3.13	CsBr	2.93	0	3.61
NaBr	4.1	0.0687	2.91	CsI	2.87	0	3.82
NaCl	4.0	0.0521	2.76	NaOH	3.24	0.0460	
KI	3.94	0.0462	3.50	KOH	3.7	0.1294	

Notes:
1. The values of $\mathring{a}$ are somewhat different from these given by Stokes and Robinson[7.21] in their treatment of hydration effects. The data given here are from Harned and Owen, *Physical Chemistry of Electrolytic Solutions*, Third Edition, Reinhold Publishing, New York (1958).
2. Somewhat different values of a and B fit the data up to 4 M, if higher. Hückel terms are included in the equation for $\log \gamma_\pm$.
3. Fourth columns give sums of cation and anion crystal radii.
4. Compare Table 11 for individual ionic contact radii.

From consideration of values of r and free energies of close interaction between hydrated anions and cations, Frank deduced a series of standard free energies of ion association for the alkali halide series, which were consistent with the known, very low degrees of association.

Returning to the practical question of what value of the dielectric constant is to be used in expressions for ln $\gamma_\pm$ in relation to local hydration and dielectric saturation at ions, it is evident that the normal value $\varepsilon = \varepsilon_2$ suffices. However, at appreciable concentrations ($> \sim 2\text{--}3\ m$) virtually no "free" water having a normal dielectric constant remains, so that the "solution" dielectric constant is then the relevant quantity. Thus, in a treatment by Scatchard[98] for high concentrations (see Section 5.3 below), the concentration dependence of ε is introduced, using experimental data of Wyman,[99] into the expression for the DH parameter κ. However, at such concentrations the whole basis of the DH ionic atmosphere model is to be called into question and is probably no longer applicable.[36]

5.2.3. Dielectric Constant and the Activity Behavior at High Concentrations

Despite the generalities of the DH treatment for dilute solutions in terms of ionic strength and its extensions to solutions of somewhat higher concentrations by taking account of finite ion size and higher terms in the expansion of exponentials, it was apparent, even in early work, that the often ion-specific behavior of electrolytes at higher concentrations might be explained in terms of hydration of the ions. Perhaps surprisingly, this question was already discussed by Bjerrum[100] in 1919 (four years before the publication of the DH theory) in terms of the "removal" of water from its solvent function by strong hydration association with the ions, and Brønsted's principle of specific interaction of ions had appeared in 1920 to account for his observations of specific ion effects in studies of the solubility changes of complex cobalt salts brought about by changes of the activity coefficients of the ions of such salts by added simple electrolytes.

These effects were introduced in a general way, as an extension of the DH theory, by Hückel,[75] by adding a positive term, Cc, *linear* in concentration c to the expression for ln $\gamma_\pm$. This took account of the decrease of the dielectric constant of ionic solutions with increasing concentration (see reference 10), which is due to the local dielectric saturation in primary hydration cospheres of ions and is associated with mutual salting-out effects[98] among the ions.

Hückel suggested an empirical treatment of ion hydration effects in activity coefficient behavior, taking into account two factors:

(1) The ions are crowded out near other ions owing to strong hydration forces at given ions; this is a kind of "excluded volume" or mutual salting-out effect (see Section 5.3.2).

(2) Since the ions cause solvent orientation and dielectric saturation effects in their primary hydration shells, the average dielectric constant of the solution ε_c will become changed from ε_0 as the salt concentration c increases to some lower value ε_c ($\varepsilon_c = \varepsilon_0 - \delta c$). Hückel showed that $\ln \gamma_\pm$ could be written as

$$\ln \gamma_\pm = \frac{z_+ z_- e^2}{2\varepsilon \kappa T} \frac{\kappa a}{1 + \kappa a} + f(\kappa) \tag{131}$$

where κ is the usual DH reciprocal radius of the ionic atmosphere. $f(\kappa)$ was taken as $C(c_i z_i^2)$, where C is a positive constant dependent on the extent to which ions depress the dielectric constant ε_0 of the solvent. It is well known that this depression of ε_0 is directly connected with ion–solvent interactions,[89,94] especially orientation of solvent molecules in the fields of the ions. An empirical equation, similar to Hückel's, was also proposed by Harned.

The Hückel term constant in Cc is given by

$$C = \frac{0.4343\, e^2}{\pi \varepsilon^2 k T} \left(\frac{\nu_+ \delta_+}{a_+} + \frac{\nu_- \delta_-}{a_-} \right) \tag{132}$$

where δ_+ and δ_- are dielectric decrement terms and a_+ and a_- are ion size parameters. C is thus a parameter specific to the types of ions and related to their interactions with the solvent. Values of C for various electrolytes were deduced from accurate conductivity measurements.

The first term in the expression for $\ln \gamma_\pm$ involving the ratio $\kappa a/(1 + \kappa a)$, it is to be noted, can only go asymptotically to a line parallel to the $c^{1/2}$ axis, with increasing $c^{1/2}$ in κ; it cannot lead to a turn up of $\ln \gamma_\pm$ at high c with a slope of sign different from that predicted by the limiting law. However, when the term in Cc is taken into account and is positive, it is seen [eq. (131)] that $\ln \gamma_\pm$ can be eventually increased above the normal negative values predicted by the DH law and the slope $d \ln \gamma_\pm/dc$ reversed in sign. This is as found experimentally. At sufficiently high concentrations $\log \gamma_\pm$ may exceed 0, i.e., $\gamma_\pm > 1$, for some salts. The predictions for various dielectric decrements δ of salts in solution are shown in Figure 17 and can be compared with the experimental behavior, discussed by Stokes and Robinson,[21] illustrated in Figure 18. It is clear, qualitatively, that the behavior of $\ln \gamma_\pm$ at high concentrations can be accounted for in terms of ion hydration through the parameter δ. It will be seen later (Section 5.3) that a related method, introducing hydration numbers, gives similar results. However, at this point it will be useful to remark that the hydration effect *per se* is *not the only factor* that can give rise to this kind of behavior at high κ (see the calculations of Card and Valleau[31] and Rasaiah et al.,[32] Section 4.2).

While the term in Cc is related to the specific hydration of ions, as are the factors discussed in Section 5.2.2 above, it is to be stressed that it arises

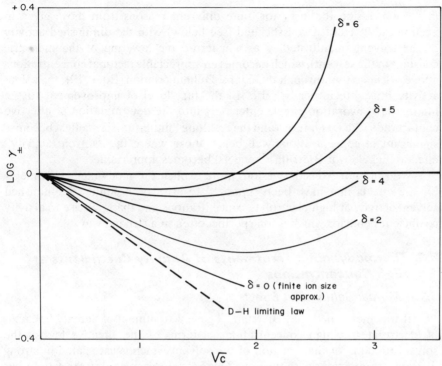

Figure 17. Predictions for $\log \gamma_{\pm}$ as a function of $c^{1/2}$ according to the Hückel equation for various values of the mean dielectric decrement δ.

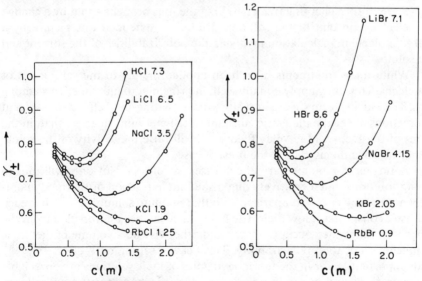

Figure 18. Some experimental values of $\log \gamma_{\pm}$ for various salts as a function of molal concentration at high concentrations, according to Stokes and Robinson.[21]

in the expression for $\ln \gamma_{\pm}$ for quite different reasons from those given in Section 5.2.2: Here it is associated (see below) with the diminished activity of the solvent (manifested[89] as a macroscopic lowering of the dielectric constant of the solvent), which becomes an appreciable factor at concentrations above $\sim 1\ m$ so that through the Gibbs–Duhem relation [Eq. (129)] the solute activity must "reciprocally" rise. In the first level of improvement to the limiting law, hydration effects enter only into the determination of effective contact radii (the DH a term) and the resulting "finite-ion-size" effect becomes significant at concentrations well below those where the "solvent activity" effect (Hückel[75] or Bjerrum[35] effects) becomes appreciable.

We shall now consider various more explicit thermodynamic treatments for the effects that have been supposed to arise on account of diminished solvent activity at high electrolyte concentrations and that give rise to activity coefficient behavior which is ion specific, often in a striking way.

5.3. Thermodynamic Treatments of Activity Coefficients at High Concentrations

5.3.1. Mutual Salting-Out Effects

It was mentioned that Bjerrum[35] gave a treatment of activity behavior of electrolytes at high concentrations in terms of the effective loss of the solvent function of that fraction of the solvent which is associated in strong hydration interactions with the ions. This behavior is related to the salting-out effects that ions exert on other solutes by raising their activity coefficients. Debye and MacAulay[101] treated this problem in terms of the changed self-energy (Born charging free energy) of the ions brought about by a change of dielectric constant of the solution. Butler[102] extended this picture by a more "molecular" calculation involving the polarizabilities of the solvent and the solute.

While these treatments have been applied mainly to the salting out of nonelectrolytes, they apply equally well, in principle, to mutual effects among ions. The Hückel term Cc arises really from a "salting-out" effect; ion–solvent interactions lower the dielectric constant of the medium, an effect that, in the terms of the Debye–MacAulay theory[101] will raise the activity of the solute i.e., of the ions themselves in the present case.

A further complementary effect arises as follows. Not only will ions in strong solutions find a relatively diminished mole fraction of thermodynamically normal solvent in comparison with that in dilute solutions,[21,98] but each ion itself constitutes a low dielectric hole. Such cavities are repelled from the other ions, since they occupy what would otherwise be a volume of dielectric of much greater total polarizability. This "cavity salting-out" effect[37,103,104] is an important factor in the thermodynamics of ionic solutions at appreciable concentrations and arises in addition to the effects associated with strong ion–solvent interactions in the solvation cosphere of ions.

5.3.2. Scatchard's Treatment

A comprehensive theoretical treatment of the activity behavior of strong electrolytes at elevated concentrations was made by Scatchard,[98] with special reference to the alkali halides. The treatment is based, among other factors, on terms arising from salting out among the cations and anions, an effect that is related to their radii r_1 and r_2, which are effective for salting out, and their volumes V_1 and V_2. Also, a concentration-dependent dielectric constant ε_c is taken into account. The equation derived[98] by Scatchard for the stoichiometric activity coefficient $\gamma_\pm$ for a salt with ions labeled 1 and 2 is[†]

$$
\ln \gamma_\pm = -\ln\left(1 + \frac{\nu_s N_s}{N_0}\right) + \frac{\varepsilon^2 N}{2RT\varepsilon_0}\left[\frac{z_1 z_2 \kappa}{1 + \kappa a} + \left(\frac{\kappa}{1 + \kappa a} + X\right)z_1 z_2 V_s m d_0\right.
$$
$$
\left. + \frac{4\nu_1\nu_2}{\nu_s}\left(\frac{V_2 z_1^2}{r_1} + \frac{V_1 z_2^2}{r_2}\right)m d_0\right] + \frac{Am d_0(2 + \nu_s m d_0)}{RT(1 + V_s m d_0)^2} \tag{133}
$$

where N_s is the number of moles of electrolyte, N_0 is the number of moles of solvent, V_1 and V_2 are the molal volumes of the ions, and V_0 is the molal volume of solvent. The parameters r_1 and r_2 are the radii of the ions that are effective for "salting out." The other symbols are defined by the following equations, where a is the sum of the ionic radii effective in ionic collisions (mean distance of closest approach) and a_{12}, a_{10}, and a_{20} are the mutual cohesive energy densities:

$$m d_0 = N_s/V_0 N_0$$

$$\nu_s = \nu_1 + \nu_2$$

$$V_s = \nu_1 V_1 + \nu_2 V_2$$

$$\varepsilon_c = \varepsilon V_0 N_0/V$$

$$\kappa^2 = -\frac{8\pi N^2 \varepsilon^2 z_1 z_2 \nu_s N_s}{1000 RT\varepsilon_c V} = -\frac{8N^2\varepsilon^2 z_1 z_2 \nu_s m d_0}{1000 RT\varepsilon} \quad \text{(signs in } z_1, z_2\text{)}$$

$$\tag{134}$$

$$X = \frac{\kappa}{1 + \kappa a} - Y$$

$$Y = \left(1 + \kappa a - \frac{1}{1 + \kappa a}2\ln(1 + \kappa a)\right)/\kappa^2 a^3$$

$$A = 2\nu_1\nu_2 V_1 V_2(2a_{12} - a_{10} - a_{20})/\nu_s$$

The significance of the terms on the right-hand side of Eq. (133) is as follows: The first term is simply the expression for conversion of the rational activity coefficient $f_\pm$ to the stoichiometric activity coefficient $\gamma_\pm$. The first

---

[†] The extended terms of Gronwall, LaMer, and Sandved[52b] are omitted.

two terms within the square bracket are essentially those of Debye and Hückel for the ion–ion interactions, except that κ is defined for a concentration-dependent dielectric constant ε_c rather than for a constant dielectric coefficient ε. Whereas Hückel assumed (cf. reference 89) a linear variation of the dielectric constant with the molarity of electrolyte, Scatchard employed the relation $\varepsilon_i = \varepsilon V_0 N_0 / V$, found by Wyman[99] for nonelectrolyte solutions. The influence of this difference in κ turns out to be small. The effect is proportional to the number of ions per unit volume of solvent. The third term within the brackets is the salting-out or charge-molecule interaction term and corresponds to the Debye and MacAulay expression[101] for this effect. These terms are related to those of the Hückel extension[75] of the theory. The final term outside the square brackets is introduced to represent any nonelectrolyte molecule–molecule interactions or the departure of purely nonelectrolyte solutions from the ideal solution (see reference 20).

Scatchard's equation gives a good account of the osmotic and activity coefficients of the alkali halides in concentrated aqueous solutions at ordinary temperatures.

5.3.3. Stokes and Robinson's Treatment

A different and thermodynamically more formal approach was made by Stokes and Robinson,[21] who related the activity coefficients of salts at high concentrations to the lowering of activity of the solvent and the degree of hydration of the ions. Their treatment is related to that of Bjerrum[35] but is developed in more explicit terms. In Stokes and Robinson's treatment the extent of binding of water molecules to ions is introduced through a hydration number n, whereas in Scatchard's treatment the ion–solvent interactions are considered in terms of their electrostatic salting-out effects.

The basis of Stokes and Robinson's method[21] is to calculate the properties of the electrolyte solution in two thermodynamically equivalent ways: (1) to consider the solute ions as "unsolvated" and (2) to consider that a total of n mol of solvent are associated in a quasipermanent† way with ν_1 mol of cations and ν_2 mol of anions per mole of salt; i.e., the ion–solvent interactions are associated with exoenergetic energies of interaction $\gg kT$. Thermodynamic quantities associated with mode (1) of describing the solution will be primed ('), while those associated with mode (2) (the normal basis of discussing properties of electrolyte solutions) will be unprimed.

The free energy G of the solution may then be written in two ways for ν_1 mol of cations and ν_2 mol of anions dissolved in S moles of solvent s:

$$G = S\mu_s + \nu_1\mu_1 + \nu_2\mu_2 \tag{135a}$$

---

† Kinetically, for many ions this is now known to be unrealistic, e.g., as indicated by various spectroscopic techniques, especially NMR; cf. the dynamical model of ion solvation of Samoilov.[105]

and

$$G = (S - n)\mu_s + \nu_1\nu_1' + \nu_2\mu_2' \tag{135b}$$

Then each potential term is expressed in terms of the natural logarithm of the mole fraction of the respective species and its rational activity coefficient, giving, after some algebraic rearrangements,

$$\nu_1(\mu_1^0 - \mu_1^{0\prime})RT + \nu_2(\mu_2^0 - \mu_2^{0\prime})/RT + n\mu_s^0/RT + n \ln a_s$$
$$+ \nu \ln \left(\frac{S + \nu - n}{S + \nu}\right)\nu_1 \ln f_1 + \nu_2 \ln f_2 = \nu_1 \ln f_1' + \nu_2 \ln f_2' \tag{136}$$

If conditions for infinite dilution are considered next, $S \to \infty$, i.e., $\ln a_s \to 1$, and all the activity coefficients become unity so that the logarithmic terms become zero. Hence the sum of the first three terms of the left-hand side of Eq. (136), i.e., involving the standard chemical potential terms, are also zero. Then, at the same time, introducing mean activity coefficients $f_\pm$ and $f_\pm'$,

$$\ln f_\pm' = \ln f_\pm + \frac{n}{\nu} \ln a_s + \ln \left(\frac{S + \nu - n}{S + \nu}\right) \tag{137}$$

In practice, it is more useful when applying this equation to experimental data to express it in terms of the mean molal (stoichiometric) activity coefficient $\gamma_\pm$. Then, using the relations $f_\pm = \gamma_\pm(1 + 0.001\nu W_s m)$ and $S = 1000/W_s m$, where W_s is the molecular weight of solvent, the result

$$\ln f_\pm' = \ln \gamma_\pm + \frac{n}{\nu} \ln a_s + \ln[1 + 0.001 W_s(\nu - n)m] \tag{138}$$

is obtained.[21] Alternatively, introducing the osmotic coefficient ϕ in order to substitute for a_s ($\ln a_s = (-\nu W_s m/1000)\phi$), the relation for $\ln f_\pm'$ is

$$\ln f_\pm' = \ln \gamma_\pm - 0.001 W_s n m\phi + \ln[1 + 0.001 W_s(\nu - n)] \tag{139}$$

Since ϕ or a_s can be calculated if $\gamma_\pm$ is known over the range of compositions up to that considered or, similarly, $\gamma_\pm$ can be calculated if ϕ data and a_s are known, the above equations enable the rational mean ionic activity coefficient of solute salt, assumed solvated with n mol of solvent per mole of salt, to be expressed in terms of the conventional activity coefficients.

In order to test Eq. (138) or (139) along the lines mentioned above, $\ln f_\pm$ is replaced by the DH expression, since this applies to interactions between "solvated" rather than free ions. Stokes and Robinson then wrote

$$\log \gamma_\pm = -\frac{A(z_1 z_2)I^{1/2}}{1 + \beta a I^{1/2}} - \frac{n}{\nu} \log a_s - \log[1 + 0.001 W_s(\nu - n)m] \tag{140}$$

where A, B, and a are the DH constants, and I the ionic strength. Thus a two-parameter (n and a)† equation results that can be tested against experimental data for $\ln \gamma_\pm$ as a function of concentration, at high concentrations.

The only criticism of the applicability of Eq. (140) is that the finite "ion size" form of the DH result is used for $\ln f'_\pm$, whereas at the high salt concentrations for which tests of Eq. (140) are to be made it is unlikely that such a simple form will be applicable; at least a form in which higher terms of the expansion of $\exp \pm ze\psi/kT$ terms has been made should be used. Although Eq. (140) gives a good account of the variation of $\ln \gamma_\pm$ with concentration for a variety of electrolytes with ion- (or salt-) specific values of both n and a, as expected, it is to be noted, as mentioned earlier, that if an exponential form (exponentials evaluated rather than linearized) of the DH theory had been used for the long-range ion interaction effects, smaller hydration effects in causing a turn up of $\ln \gamma_\pm$ at higher concentrations would have been required to secure agreement with experiment. Thus the DHX calculation considered by Card and Valleau[31] and the Monte Carlo and HNC calculations already give these effects *without* the introduction of hydration in the primitive model or of any hydration parameters.

Some comparisons between the results of Stokes and Robinson's treatment and experiment are given in Table 9, which lists the values of n and a required to fit the experimental behavior (Figure 18) of a selection of electrolytes.

Stokes and Robinson also proposed that the association of hydration water with ions in concentrated solutions was analogous to behavior represented by the Brunauer, Emmett, and Teller (BET) type of adsorption isotherm for multilayer adsorption at surfaces. They showed, in fact, that the water activity a_s in such solutions could be related to the molality m of the solution and to a parameter n_1, the number of water molecules in a monolayer hydration cosphere around ions, when complete, and a constant l related to the energy of "adsorption" E of the water molecules adjacent to the ion, where $l = \exp (E - E_L)/RT$ and $-E_L$ is the heat of vaporization of pure water.

Their relation, based on the BET isotherm approach, has the form

$$m = \frac{55.51(1 - a_s)}{a_s}\left(\frac{1}{ln_1} + \frac{l-1}{ln_1} a_s\right) \tag{141}$$

where $s \equiv H_2O$ in this case. This equation gives a surprisingly good fit for the dependence of water activity in strong aqueous electrolyte solutions to the salt molality m over a wide range of m values.

---

† It is assumed that n is independent of salt concentration. This is only true (as noted in reference (21)) for concentrations below those at which an appreciable overlap of solvation cospheres begins.

Table 9
Constants of the Two-Parameter Equation of Stokes and Robinson[21]
Giving Best Fits to the Experimental Activity Coefficients of a Selec-
tion of 1:1 and 2:1 Salts

Electrolyte	n	a (Å)	Range fitted (m)	Average difference in γ
HCl	8.0	4.47	0.01–1.0	0.001
HBr	8.6	5.18	0.1–1.0	0.001
HI	10.6	5.69	0.1–0.7	0.002
HClO$_4$	7.4	5.09	0.1–2.0	0.001
LiCl	7.1	4.32	0.1–1.0	0.001
LiBr	7.6	4.56	0.1–1.5	0.001
LiI	9.0	5.60	0.1–1.0	0.003
LiClO$_4$	8.7	5.63	0.2–1.0	0.003
NaCl	3.5	3.97	0.1–5.0	0.002
NaBr	4.2	4.24	0.1–4.0	0.001
NaI	5.5	4.47	0.1–1.5	0.002
NaClO$_4$	2.4	4.04	0.2–4.0	0.0015
KCl	1.9	3.63	0.1–4.0	0.002
KBr	2.1	3.85	0.1–4.0	0.0025
KI	2.5	4.16	0.1–4.0	0.001
RbCl	1.2	3.49	0.1–1.5	0.001
RbBr	0.9	3.48	0.1–1.5	0.001
RbI	0.6	3.56	0.1–1.5	0.005
MgCl$_2$	13.7	5.02	0.1–1.4	0.001
MgBr$_2$	17.0	5.46	0.1–1.0	0.002
MgI$_2$	19.0	6.18	0.1–0.7	0.001
CaCl$_2$	12.0	4.73	0.1–1.4	0.001
CaBr$_2$	14.6	5.02	0.1–1.0	0.0005
CaI$_2$	17.0	5.69	0.1–0.7	0.0005
SrCl$_2$	10.7	4.61	0.1–1.8	0.001
SrBr$_2$	12.7	4.89	0.1–1.4	0.0015
SrI$_2$	15.5	5.58	0.1–1.0	0.001
BaCl$_2$	7.7	4.45	0.1–1.8	0.001
BaBr$_2$	10.7	4.68	0.1–1.5	0.001
BaI$_2$	15.0	5.44	0.1–1.0	0.0025
MnCl$_2$	11.0	4.74	0.1–1.4	0.001
FeCl$_2$	12.0	4.80	0.1–1.4	0.002
CoCl$_2$	13.0	4.81	0.1–1.0	0.001
NiCl$_2$	13.0	4.86	0.1–1.4	0.0015
Zn(ClO$_4$)$_2$	20.0	6.18	0.1–0.7	0.001

5.4. Solvation Effects in the Statistical-Mechanical Treatments

In the simplest primitive model calculations specific ion solvation is not taken into account. Improvements of various kinds make some allowance for solvation effects, which become more significant at appreciable concentrations.

5.4.1. Ion Size Effects and Hydration in the Repulsive Term

We have mentioned earlier that various repulsive functions other than the hard-sphere, discontinuous one have been used. The HNC approach, used by Rasaiah *et al.*,[27,28,32] is readily adaptable to modifications of the short-range interaction potential term and has been used for the calculation of osmotic coefficients and their comparison with experiment.

For alkali halides in aqueous solution at 298 K (Figure 19a) it is found that the HNC calculation for various distances of closest approach a_{+-} of oppositely charged ions gives only a qualitatively good account of the thermodynamic behavior as a function of the square root of ionic concentration. Here, in the primitive model, the ion size parameter a_{+-} is the sum of the Pauling crystallographic radii. Although the calculated results (Figure 19b)

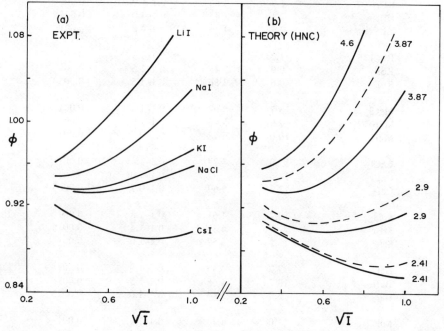

Figure 19. Experimental (a) and theoretical (b) (calculated by the HNC method) osmotic coefficients as a function of $I^{1/2}$. Theoretical curves are for various a values (from Rasaiah[24]). ——, $r_+ = r_-$; - - -, $r_-/r_+ = 3.6$.

have essentially the right form in relation to experiment, it is found[24] that choice of a_{+-} values required to fit the experimental data at low concentrations leads, in accurate calculations, to too large values of ϕ at higher concentrations. The introduction of unequal radii (but having the correct sum $r_+ + r_-$ to fit the data at low concentrations) serves only to increase the discrepancy between accurate theoretical calculations and the experimental behavior. Hence some improvement of the primitive model is required.

A simple modification that reduces the discrepancy between theory and experiment at higher concentrations but gives self-consistent results for a family of alkali halides is the so-called square-well or square-mound model.[27,106] The *short*-range potential $U^*_{i,j}(r)$ for the close approach of two unlike ions is now defined by the conditions

$$U^*_{i,j}(r) = u_{i,j} \qquad \text{if } a_{i,j} < r < b_{i,j} \qquad (142)$$

where $a_{i,j}$ is the hard-sphere ion contact distance based on Pauling ionic radii and $b_{i,j}$ is a hydrated-ion contact radius differing from $a_{i,j}$ by the diameter of a water molecule: $b_{i,j} = a_{i,j} + 2.76$ Å.

Also,

$$U^*_{i,j}(r) = \infty \qquad \text{if } r < a_{i,j} \qquad (143)$$

and

$$U^*_{i,j}(r) = 0 \qquad \text{if } r > b_{i,j} \qquad (144)$$

The specification of $b_{i,j}$ in relation to $a_{i,j}$ implies an approach of two *hydrated* ions of unlike charge to a configuration within the solvent, where their separation between centers is specified by $a_{i,j} < r < b_{i,j}$, with the acquisition of extra energy $U^*_{i,j}(r)$ (energy "mound"[24]) that is associated with the displacement of some water between the ions when they come into contact. The extent of the displacement of water will depend on the ionic charges and the crystal radii of the ions. It is unlikely, of course, that all the hydration water will be lost in such close $+-$ ion encounters. (This is analogous to the situation introduced empirically in the DH treatment for finite-size ions). In fact, in the study by Hamann, Pearce, and Strauss[107] on ion-pairing equilibria as a function of pressure the (positive) volume of ion pairing is found to be appreciably less than would correspond to complete loss of electrostriction of the two ions. The pair is thus normally a solvent-shared ion pair. Presumably a similar situation applies here in the model for short-range interactions.

The requirements of an improved calculation are indicated by data plotted in Figure 20; here a quantity $\Delta\phi$, defined as the difference between the experimental ϕ values for various salts and the values of ϕ calculated on the primitive model for ionic radii equal to the Pauling ion hard-sphere radii, is related to the concentration. It is evident that in this plot the deviations are connected with hydration effects, as reflected, e.g., in the parameters of the more classical treatment of Stokes and Robinson.[21] Using the square-well

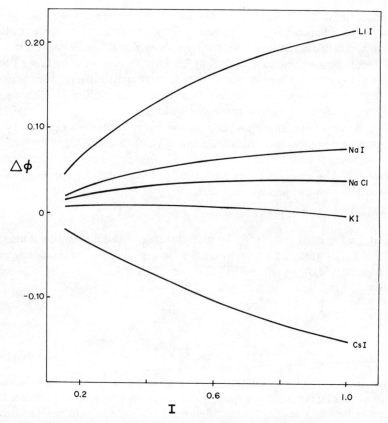

Figure 20. Differences $\Delta\phi$ of experimental and calculated ϕ values for several alkali halide salts in water. The theoretical values of ϕ are derived from the primitive model with Pauling ionic radii (redrawn from Reference 106).

(mound) model, it is clear that what is required for a closer agreement between calculation and experiment is a positive $U_{i,j}^*(r)$ when $\Delta\phi > 0$, and a negative value when $\Delta\phi < 0$. It is found that a better account of the variation of the osmotic coefficients with concentration at higher concentrations is given by introduction of finite values of the $U_{i,j}^*(r)$ parameter ($i = +, j = -$), and this is the only adjustable parameter in the calculations.

The osmotic coefficients were calculated for LiBr, NaBr, KBr, and CsBr by the HNC method, using the sum of Pauling ionic radii for $a_{i,j}$ together with $U_{++}^*(r) = 0$, $U_{--}^*(r) = 0$, and $U_{+-}^*(r)$ taken with a selection of appropriate values for the respective salts. The magnitudes of $U_{+-}^*(r)$ are reasonable and large positive values correspond to strong hydration of either or both of the ions, e.g., for Li^+ in Li^+ in LiBr.

A remaining problem with this square-well, hydration shell penetration model is that it is likely that a sharp cutoff as implied by the conditions defined by Eqs. (142)–(144) is unrealistic except for strongly hydrated ions. Interpenetration (cospheres overlap) of the secondary hydration shell regions, in which structure-changed[108] solvent is involved, is a likely complication (see reference 45). It is in fact indicated[24] by the observed positive signs of $d[U^*_{+-}(r)]/dT$ values.

The $U^*_{+-}(r)$ parameter, introduced by Rasaiah, turns out to have similar effects in calculated results to those of the GUR A parameter treated by Friedman and co-workers (to be discussed below), which takes account of the overlap of Gurney solvation cospheres.

5.4.2. The "OTHER" Term in the Effective Potential and Its Relation to Ion–Solvent Interaction

For appreciable concentrations ($> 0.1 \sim 1.0\, m$) two terms connected with solvation of the ions must be included in Eq. (109), i.e., in the $\bar{U}_{\text{OTHER}}(r)$ term. The first is the cavity (CAV) polarization term, arising from the mutual salting out of the ion envelopes in the solution dielectric, discussed in Section 5.3. This term is written, according to Levine and co-workers[103,104] [see reference 37 and Eq. (151) below]

$$\text{CAV}_{i,j}(r) = \frac{1}{2\varepsilon r^4}\frac{\varepsilon - \varepsilon_c}{2\varepsilon + \varepsilon_c}(z_i^2 er_j^3 + z_j^2 er_i^3) \tag{145}$$

where ε_c is the dielectric constant of the ion cavity (~ 2) and ε, as before, is the dielectric constant of the solvent. Since $\varepsilon \gg \varepsilon_c$ for most solutions, $\text{CAV}_{i,j}(r)$ is always a repulsive term.

The final term is the so-called Gurney (GUR) term associated with interpenetration of the Gurney solvation cospheres[93] of the ions according to the model shown in Figure 21, as treated by Ramanathan, Krishnan, and Friedman.[25,26] Especially, but not exclusively, in more concentrated solutions there will be a tendency for the ion-influenced regions of solvent, i.e., the polarization cospheres of the ions, to overlap with one another. This can be represented by a process such as that shown in Figure 21, where a volume element V_{mu} (the mutual overlap volume) is returned to the solvent with a certain free energy (or, in the general case, change of the property X) change. The GUR energy term is represented by a function

$$\text{GUR}(r) = A_{i,j}\frac{V_{\text{mu}}}{V_s}(r_i + R_j, r_j + R_j, r) \tag{146}$$

where V_s is the molar volume of the solvent, $A_{i,j}$ the Gurney free-energy parameter for the system of two ions i and j at a separation r, and R_i and R_j are the respective radii of the cospheres of ions i and j. From the geometry of overlapping spheres of radii a and b the function in the brackets of Eq. (146) can be written[25,26]

$$V_{\text{mu}}(a, b, r) = \pi[-(4r)^{-1}(a^2 - b^2) + \tfrac{2}{3}(a^3 + b^3) - \tfrac{1}{2}r(a^2 + b^2) + r^3/12] \tag{147}$$

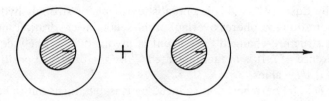

Like-ions with hydration co-spheres

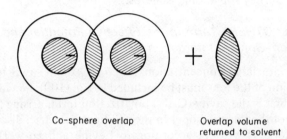

Co-sphere overlap   Overlap volume
                          returned to solvent

Figure 21. Gurney cosphere overlap model for calculation of A_{ij}, etc., parameters as treated by Friedman in various papers (e.g., References 25 and 26).

which, for two ions of identical radii, e.g., in $++$ or $--$ interactions, becomes, writing $a = b = x$, say,

$$V_{mu}(a, b, r) = \pi[\tfrac{4}{3}x^3 - rx^2 + r^3/12] \tag{148}$$

Usually (see references 25 and 26) the cosphere radii are taken to correspond to the radius of the ion plus the thickness of a single water molecule layer coordinated to the ion. For some ions this is probably too small, e.g., with divalent ions and possibly with R_4N^+ ions, where a structure-enhanced region of relaxationally less mobile water may extend further than one water molecule from the "periphery" of the ion.[3]

An outline of the functional dependence on r of the various terms in the effective potential is shown in Figure 22 according to Ramanathan et al.[25] for a tetraalkylammonium salt. The magnitude of $\pm |A_{ij}|$ is ~0.4–0.8 kJ mol^{-1} for univalent ions, depending on the nature of the ions (Table 10). Gurney parameters for other functions such as the excess enthalpy, entropy, or volume may also be introduced into equations for those respective functions. Comparisons of some computed and experimental ϕ values are shown in Figures 23 and 24 (see reference 26).

In solutions of moderate concentrations the $+-$ Gurney interactions will predominate, due to weighting caused by the attractive ion-pair interaction potential, over the $--$ or $++$ contributions. In more concentrated solutions both the $--$ and $++$ interactions will become of increasing importance, but the $+-$ interactions will, of course, also increase in probability.

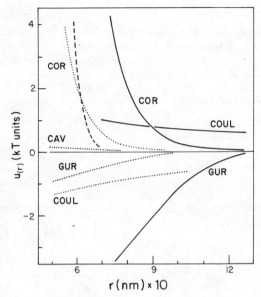

Figure 22. Functional dependence of r (nm) of the various terms in the interionic potential function for a model of aqueous Et_4NCl solutions that fits the osmotic coefficient behavior. The terms of U_{+-} are shown as dotted lines, and those of U_{++} as solid lines. The dashed line is the COR term in U_{+-}, with an exponential form in the distance of approach. The CAV term is omitted for clarity (from Ramanathan, Krishnan, and Friedman[25]).

Hitherto the GUR potential term has been evaluated empirically as a "catch-all" term, assuming that the other terms are expressed relatively accurately as functions of r in the equations or conditions given earlier. It is unfortunate that this aspect—an important one—of the modern theory of electrolyte solutions is left in an empirical condition, since the ion-specific characteristics of the excess G function, and those of other functions, arise primarily from the specificities of individual ion solvation[21,16] (including the solvation-dependent a parameter), which are of great current interest, thermodynamically and spectroscopically.

A further difficulty is that not only are the $A_{i,j}$ parameters empirical, but more than one set can be assigned to fit the experimental $\ln \gamma_{\pm}$ data for a given salt (see reference 26), e.g., as illustrated in Table 10.

5.4.3. A Priori Calculation of Gurney Parameters

The *a priori* calculation of the GUR terms has not been attempted for the case of three-dimensional electrolytes in the context of McMillan–Mayer types of calculation. However, Levine and Rozenthal[104] gave an excellent and detailed calculation of the energy of ion pairing for $+-$ ion pairs that bears a close relation to the above problem. They considered the interaction energy as a function of a reduced interionic separation parameter for two ions of unlike charge with separated and with overlapping solvent coordination shells. The calculations are very complex but lead to potential profiles and effective dielectric constants as a function of ion separation for the electrostatic interaction. It seems that it would be useful to extend this type of approach

B. E. CONWAY

Table 10

Two Sets of A_{ij} Parameters That Fit the Excess Free Energy Data for Aqueous Alkali Halides (Values in cal mol^{-1}; for Upper Right Half of Table All $A_{--} = 0$, for Lower Left Half All $A_{++} = 0$)[a]

Ions	Li$^+$	Na$^+$	K$^+$	Rb$^+$	Cs$^+$	F$^-$	Cl$^-$	Br$^-$	I$^-$
Li$^+$	0	—	—	—	−200	—	50	40	45
Na$^+$	—	−50	—	—	—	−50	−47	−50	−54
K$^+$	—	—	−100	—	—	−10	−80	−86	−95
	—	—	—	−100	—	—	—	—	—
Rb$^+$	—	—	—	—	—	0.0	−90	−102	−120
Cs$^+$	−100	—	—	—	−100	0.0	−110	−120	−135
F$^-$	—	−5	0.0	12	0.0	—	—	—	—
	—	—	—	—	—	−200	—	—	—
Cl$^-$	100	−14	−67	−84	−110	—	—	—	—
	—	—	—	—	—	—	−88	—	—
Br$^-$	90	−18	−75	−96	−120	—	—	—	—
	—	—	—	—	—	—	—	−75	—
I$^-$	95	−25	−82	−114	−135	—	—	—	−60

[a] From Ramanathan and Friedman.[26]

to the improved and less empirical evaluation of excess thermodynamic properties of electrolytes at elevated concentrations.

From an inspection of the calculations of Levine and Rozenthal[104] in some detail, it would seem that it is somewhat of an oversimplification to represent the Gurney cosphere overlap free energy simply in terms of a characteristic interaction parameter A_{ij} in Eq. (146), constant for a given pair of ions, independent of the mutual volume V_{mu} of overlap, i.e, independent of the interionic distance r in the overlap configuration.

Very difficult questions of the variation of the local dielectric constant or orientation of solvent dipoles must arise[104] as two like or unlike ions approach each other with an overlap of their solvation cospheres. Probably the correct calculation for pairs of ions of unlike charge is more difficult than that for pairs of like charge (see references 109 and 110).

In certain situations, e.g., in double layers at charged interfaces, a related problem arises, but the calculations involve mainly interactions of ions of a given *like* charge that are accumulated in two dimensions at a surface (e.g., a polarized electrode or charged colloidal particle) of opposite charge. The lateral interactions in such an ad-layer are then determined (1) by coulombic and core repulsions, (2) cavity repulsions, and (3) repulsions associated with overlap of the Gurney cospheres, since in most cases the two-dimensional concentrations of ions adsorbed at charged interfaces are quite large and comparable with three-dimensional concentrations in ordinary electrolyte solutions in the range 2–4 M. Evaluation of the total interaction potential

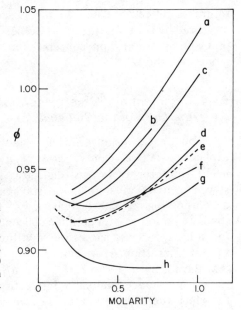

Figure 23. Osmotic coefficients calculated for models of aqueous NaCl, using GUR and CAV terms, compared with the experimental behavior (concentration in mol liter^{-1}): (a) all $A_{ij} = 0$; (b) all $A_{ij} = 0$ and without CAV; (c) $A_{+-} = -20$; (d) $A_{+-} = -50$; (e) $A_{++} = 50$, $A_{+-} = -47$, $A_{--} = 0$; (f) exponential curve; (g) $n = 15$, all $A_{ij} = 0$ and without CAV; (h) primitive model ($n = \infty$) (from Ramanathan and Friedman[26]) n is the exponent of repulsive effects in the COR term.

enables an adsorption isotherm to be set up for the ad-ion species (see references 109 and 110). The free energy of adsorption ΔG^0_{ads} in such an expression is a function of ion coverage, θ, due to the lateral ion–ion interactions in the interphase.

Conway and Dhar[109] and Conway[110] made the first model calculation of the GUR overlap energies for *like*-charged ions at interfaces on the basis of (1) overlap of Born polarization cospheres and (2) (in order to introduce known ion specificities) overlap of polarization cospheres in which dielectric

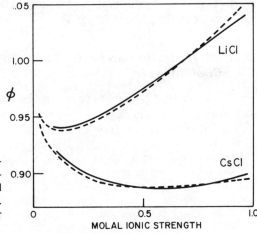

Figure 24. Computed (- - -) and experimental (——) osmotic coefficients compared in two typical cases, LiCl and CsCl (from Ramanathan and Friedman[26]). Data given in the McMillan–Mayer system.

saturation effects, related to the observed bulk dielectric decrements as measured by Hasted, Ritson and Collie,[89] were involved. ΔG_{ads}^0 as a function of two-dimensional ion concentration at the interface was calculated for models (1) and (2).[110]

5.4.4. Comment on Hückel's and Stokes and Robinson's Results in the Light of the Statistical-Mechanical Calculations

The treatments of Hückel[75] and Stokes and Robinson[21] sought to explain deviations from a DH function for $\log \gamma_\pm [= -A(z_+z_-)I^{1/2}/1 + \mathring{a}BI^{1/2}]$ in terms of factors (dielectric decrement or solvation numbers and solvent activity) related to ion solvation. In particular, the turn up from decreasing $\log \gamma_\pm$ with increasing $I^{1/2}$ to increasing (eventually positive in some cases) $\log \gamma_\pm$ at high ionic concentrations is semiempirically accounted for in an apparently successful way through parameters characterizing ion–solvent interaction. However, the problem involved in these approaches is that the function $-A(z_+z_-)I^{1/2}/1 + aBI^{1/2}$ is inappropriate for accounting for the long-range coulombic interactions at the appreciable concentrations for which the added solvation-related terms or parameters are intended to apply. Thus Card and Valleau[31] showed that an exponential DH distribution function ("DHX") would already account for a turn up of $\log \gamma_\pm$ versus $I^{1/2}$ for 1:1 electrolytes† at about $1 \sim 2\ M$ *without* introduction of any hydration term. Similarly, the Monte Carlo and HNC results for the primitive model in which no ion–hydration interactions are introduced automatically give (see diagrams in reference 31) a similar curved relation for $\log \gamma_\pm$ versus $I^{1/2}$, with a minimum.‡ Hence a main added term for ion-specific hydration effects (although they are still significant) is not apparently needed to account for the minimum in the $\log \gamma_\pm$ vs. $I^{1/2}$ relation. Residual hydration effects are, however, accounted for in the GUR term in the nonprimitive model, which takes into account more complex contributions in the "effective potential," as treated by Ramanathan *et al.*[25,26]

5.5. Ion Hydration and the "Cube-Root Law" for Activity Coefficients

The idea that properties of electrolyte solutions, especially at appreciable concentrations, might depend on the *cube root* of concentration was explored by a number of authors. It originates from the view that an electrolyte solution

---

† For unsymmetrical or 2:2 electrolytes the DHX calculation is less satisfactory.

‡ It should be noted that this conclusion is made on the basis of calculated results in the McMillan–Mayer system. It is the opinion of the authors of this work (J. P. V. *et al.*, private communication) that a similar conclusion would apply to results corrected to regular conditions. Presumably the behavior indicated from the calculations originates from the effect of the COR repulsive term at elevated concentrations.

can be represented as a quasilattice of positive and negative ions disordered by the thermal fluctuations in the solution at finite temperatures. Then the average interionic distance is proportional to the inverse cube root of concentration so that the coulombic electrical lattice interaction energy of such a system would be related to $c^{1/3}$. Ghosh[34] appears to have been the first to have proposed such a representation, and he deduced that the equivalent conductivity should be a linear function of $c^{1/3}$. This is not substantiated by experiment, but as Stokes and Robinson have shown,[21] $\ln \gamma_{\pm}$ over an appreciable range of intermediate concentrations plots out almost linearly in $c^{1/3}$ rather than in $c^{1/2}$ (as it must at "limiting law" concentrations).

A $c^{1/3}$ plot was also considered by Bjerrum[35] for osmotic coefficients for a number of salts, and in a later paper[100] he showed that allowance for hydration effects extended the validity of such a plot to concentrations of several moles per litre.

More recently this topic was taken up by Frank and Thompson,[36] who discussed the question, "What is the concentration beyond which the DH ionic atmosphere model itself (in distinction to various approximations) might be considered to break down?" They regarded the very low concentration of $10^{-3} M$ as this limit, but the higher limit of $0.02 M$ given by Fuoss and Onsager[111] seems more realistic. Beyond such a limit, the concept of a countercharge, conjugate to that on the central reference ion, diffusely distributed in spherical-shell elements of the solvent about the central ion, ceases to have validity. The countercharge is too localized[36] for the model to be valid any more. Under such conditions a disordered lattice model becomes a realistic alternative and its reality may to some extent be supported by the oscillatory behavior of the pair distribution functions for $+-$ and $++$ or $--$ ion arrangements in the solution, or the charge density ρ as a function of distance, which is found in statistical-mechanical calculations.[23,24]

Rasaiah[112] working with Frank, showed that the emf of a $Zn/Hg|ZnSO_4$, $PbSO_4(s)|Pb/Hg$ cell was linear in $m^{1/3}$ over an appreciable range of concentrations down to $m^{1/3}$ ($ZnSO_4$) ~ 0.06 and discussed the basis of a $m^{1/3}$ plot for purposes of extrapolation to evaluate the cell emf E^0.

Lattice models involving partial "long-range order" have been treated recently in papers by Pytkowicz *et al.*[38–40] The degree of order is found to be larger at higher concentrations and lower temperatures, as is expected on the basis of thermal fluctuations in relation to energies of coulombic interaction between ions. We shall not discuss Pytkowicz's treatment any further here, as it has been presented by him in detail in a chapter in another recent monograph.[40]

Unfortunately, tests of the validity of cube-root relations for the excess free energy encounter the difficulty that at the appreciable concentrations at which such behavior might be manifested and demonstrable in a plot based on experimental data, complications due to hydration of the ions which raise $\ln \gamma_{\pm}$ become significant and often large (see reference 21). Such effects were discussed in Section 5.3.3. This point was evidently appreciated by Bjerrum.[100]

Some newer (at that time) evidence for the reality of a cube-root dependence of ln $\gamma_\pm$, originating from the ion–ion electrostatic interactions, was given by Desnoyers and Conway.[37] These authors overcame the difficulty mentioned above concerning the hydration effects in $RT \ln \gamma_\pm$ by subtracting from this term estimates of the nonideal free-energy contributions due to (1) the main hydration effect, which lowers a_{H_2O}, thus raising $RT \ln \gamma_\pm$, and (2) the cavity salting-out effect (see Section 5.4.2 and below).

Formally, the experimental activity coefficient γ can be separated into three main component terms (expressed as rational activity coefficients):

$$\log \gamma = \log f_c + \log f_h + \ln f_{so} \qquad (149)$$

where f_c is the coulombic contribution to be evaluated, f_h is that associated with direct hydration effects in which a certain fraction of the solvent molecules are removed from their solvent role, and f_{so} is a term that involves the cavity salting-out effect. There can also be a further contribution associated with nonideality of the solution arising on account of the different *sizes* of solute ions and solvent molecules.[25,64] For simplicity, this will not be included in Eq. (149); it was briefly discussed in Section 3.5.

The f_h contribution can be evaluated using the relation derived by Stokes and Robinson [see Eq. (140)];

$$\log f_h = \frac{0.018 \, h \, m \, \phi}{2.303} - \log [1 - 0.018(n - \nu)m] \qquad (150)$$

Here the osmotic coefficient ϕ is the experimentally known quantity and n is the hydration number parameter evaluated in reference 21 (Section 5.3.2).

The mutual salting out (cavity effect) was evaluated using the expression

$$\Delta U(r) = z_i e^2 \bar{V}_i / 8\pi N_A \varepsilon r^4 \qquad (151)$$

where $\bar{V}_i$ is the partial molar volume of the ion, N_A is Avogadro's number, and r is the distance of the ion cavity being "salted out" from a reference ion. $\Delta U(r)$ adds to the electric Coulomb potential $z_i e \psi_j / kT$ and determines the distribution of ions with respect to each other. ψ_j is taken as the DH result for finite ion size:

$$\psi_j = \frac{z_j e}{\varepsilon} \frac{e^{-\kappa r}}{r} \frac{e^{\kappa a}}{1 + \kappa a}$$

The total mean ionic salting-out contribution $\log f_{so}$ is found to be given by the following two equations, one for 1–1 electrolytes and the other for 2–1 electrolytes:

$$\log f_{so} = \frac{e^2 c}{4605 \varepsilon kT} \left(\frac{1}{3} \frac{\bar{V}_A}{r_{h(M)}} + \frac{\bar{V}_M}{r_{b(M)}} + \frac{1}{3} \frac{\bar{V}_M}{r_{h(A)}} + \frac{\bar{V}_A}{f_{h(A)}} \right) \qquad (152)$$

where c is the molar concentration. For a $2:1$ salt (e.g., $CaCl_2$) the mean salting-out activity coefficient contribution is

$$\log f_{so} = \frac{e^2 c}{4605 \varepsilon k T} \left(\frac{\bar{V}_A}{r_{h(M)}} + \frac{4 \bar{V}_M}{r_{h(M)}} + \frac{(4/5)\bar{V}_M}{r_{h(A)}} + \frac{2 \bar{V}_M}{r_{h(A)}} \right) \qquad (153)$$

In these equations the subscript M refers to the cation and A to the anion; r_h terms indicated for M and A ions are their respective hydration radii, and $\bar{V}$ their partial molar volumes.

Evaluation of $\log f_h$ and $\log f_{so}$ then enables $\log f_c$ to be evaluated, and its dependence on the cube root of the concentration tested. Results for most 1–1 electrolytes give *linear* relations over the concentration range $m^{1/3} = 0.45–1.0$, supporting the suggestion of Frank and Thompson that a disordered lattice model is preferable to an ionic atmosphere model at intermediate and high concentrations. Of course, the cube-root relation must always go over to the square-root limiting law equation at sufficiently high dilutions. Some results for alkali metal halides, plotted in terms of $c^{1/3}$, are shown in Figure 25 (from reference 37). Similar plots in $c^{1/3}$ were demonstrated by Glueckauf and Mayorga and Pitzer[6,113,114] (see below).

Glueckauf[6,115] also investigated the relation between hydration and electrostatic effects in the dependence of $\log \gamma_\pm$ on concentration, using a volume fraction statistics approach (see references 20 and 59–61). He considered that systematic deviations from the equation for $\log \gamma_\pm$ based on the finite-ion-size DH term plus a term (see reference 21) for hydration effects and a statistical term could be represented by an extended electrostatic free

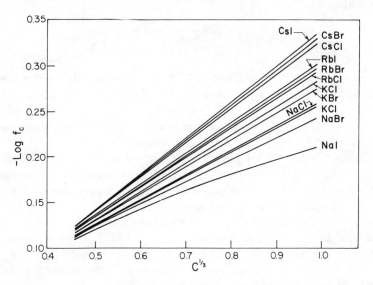

Figure 25. Plot of $\log f_c$ (i.e., after allowance for solvation effects) for some alkali halide salts as a function of the cube root of the concentration (from Reference 37).

energy term $X(m)$. This term was supposed to have a value ~ 0 at $0.1\ m$ but become rapidly more significant for concentrations $>0.1\ m$, assuming the hydration parameter (see reference 26) remains constant (except at very high concentrations). $X(m)$ takes the form

$$X(m) = -\frac{0.06\ m^2}{(1 + m/2)^2} \tag{154}$$

The mean stoichiometric activity coefficient is then given by

$$\log \gamma_{\pm} = -\frac{0.509 m^{1/2}}{1 + am^{1/2}} - \left(\frac{0.06m}{1 + m/2}\right)^2 + \log \gamma_{vf} \tag{155}$$

where $RT \ln \gamma_{vf}$ is a nonideal free-energy term in volume fraction statistics involving the degree of hydration of the ions and their sizes (see references 25 and 67) and a is a DH finite-ion-size parameter.

It was found that the first two terms of Eq. (155) produce a curve with a section *linear* in $m^{1/3}$ over an appreciable range of concentrations. For alkali halides and HCl in water the equation

$$\log \gamma_{\pm} = -Km^{1/3} + \log \gamma_{vf} \tag{156}$$

gives an excellent fit of the experimental values of $\log \gamma_{\pm}$ over the concentration range $m = 0.2 \sim 0.5$ to $2 \sim 4\ m$, depending on the electrolyte.

It was deduced on the basis of an ionic (solution) lattice model that the electrostatic contribution in $\log \gamma_{\pm}$ [the $-Km^{1/3}$ term in Eq. (156)] would be of the magnitude

$$\log \gamma_{\pm,el} \sim -2.3 \left[\frac{4Ae^2}{3\varepsilon kT}\left(\frac{N}{1000}\right)^{1/3}\right]c^{1/3} = -0.385c^{1/3} \tag{157}$$

where A is a Madelung constant ($=1.75$) for the quasilattice. The constants of the term in $c^{1/3}$ vary between 0.225 and 0.305 for HCl and the alkali halides in water, which probably reflects some solution structure effects[108] due to the specific hydration properties of the various ions and some variation of the hydration parameter in the $\log \gamma_{vf}$ term.

The conclusions of Glueckauf[6,115] support the early view of Bjerrum, that a cube-root relation for $\log \gamma_{\pm}$ may apply at elevated ionic concentrations.

Finally, it should be mentioned that the accurate primitive model calculations[31,32] also give results for the osmotic coefficient which, over an intermediate range of concentrations ($c^{1/3} \simeq 0.1$ to $0.7\ m$), give a linear relation between ϕ and $c^{1/3}$. Deviations arise when $c^{1/3} > \sim 0.7\ M$. Care must be exercised, however, in considering the significance of any linear region in ϕ vs $c^{1/3}$ plots if it arises only over a relatively short range in $c^{1/3}$, as various functions of c may also plot out almost linearly (p. 161).

5.6. "Individual" Ionic Activity Coefficients and Specificity of Ion Solvation

A table of 26 individual ionic activity coefficients, reported as preliminary values, was published in 1923 by Lewis and Randall (see revised edition in reference 41). Individual ionic thermodynamic properties[15,16] cannot, however, be defined or measured experimentally, as was emphasized in the well-known papers of Guggenheim (p. 117).[8]

Individual ion activity coefficients are a little more difficult to visualize than other ionic thermodynamic properties (see reference 16), such as $\bar{S}_i^0$, $\bar{C}_{p,i}^0$, or $\bar{V}_i^0$, since by their nature they involve interactions between *two* different kinds of ions as well as between like kinds. However, from the formal point of view of the DH theory it is possible, in principle, to formulate the energy of interaction of a cation i with all other ions of types i and j in an ionic atmosphere and the corresponding energy for an anion j with the atmosphere. For ions of a given charge such interactions will only differ in dilute solutions on account of the finite, ion-specific effective sizes of the solvated cations and anions.

Normally, for the mean activity coefficient at concentrations above limiting law dilutions it is necessary to express $\ln \gamma_\pm$ in terms of the empirical DH parameter $\mathring{a}$, the distance of closest approach of one ion to another. This parameter therefore normally involves *both* types of ions in an electrolyte. Any attempt to evaluate "individual" ionic activity coefficients therefore necessarily involves a procedure for estimating "individual" $\mathring{a}$ values. Since the $\mathring{a}$ value for a salt is usually neither the sum of the crystallographic radii nor that of the separate ionic hydration radii, it is difficult to evaluate individual γ_i's, except in the limiting law case, which is hardly profitable from the point of view of significance of individual activity coefficients, except for their dependence on the charge of the ion.

Kielland[12] had discussed this question and quotes a method due to Bonino and Centola[116] for the evaluation of individual $\mathring{a}_i$ values, using the relation

$$10^8 \mathring{a}_i / z_i = 0.9 \times 10^8 r_i / 10^{24} \alpha_i + 2 \qquad (158)$$

or from the ionic conductance λ_i^0

$$10^8 \mathring{a}_i = 182 z_i / \lambda_i^0 \qquad (159)$$

Alternatively, using an empirical form due to Brull,[117]

$$10^8 \mathring{a}_i = 216 z_i^{1/2} / \lambda_i^0 \qquad (160)$$

where α_i is the polarizability ("deformability"[12]) and r_i the crystal ionic radius. It would appear that Eq. (158) does not include hydration effects, while Eq. (159) involves the whole Stokes solvation radius† of the ion.

Keilland gave a tabulation of $\mathring{a}_i$ values, using these formulas, and calculated a system of individual ionic activity coefficients, using the DH relation:

$$\log f_i = \frac{-0.358z_i^2 I^{1/2}}{1 + 10^8 \mathring{a}_i 0.2325I^{1/2}} = \log \gamma_i + \log (1 + 0.018m_i) \qquad (161)$$

Data were given for a large variety of ions, grouped according to the similarity or identity of their $\mathring{a}_i$ values as estimated by the formulas given above. Use of the a_i values calculated from the equations of references 116 or 117 implies "$\mathring{a}_i$" values, characteristic of individual free ions, that would be different from the normal a values for $+-$ ion contacts in the DH equation for $\ln \gamma_\pm$ values (see Table 8). A selection of $\mathring{a}_i$ values from reference 12 is given in Table 11.

In a paper by Frank[118] the individualities of $\ln \gamma_\pm$ behavior for salts at appreciable concentrations were related in a qualitative way to the individual ionic structure-making or structure-breaking properties the cations or anions exhibit in their interactions with water as solvent, as treated in the work of

Table 11

Individual $\mathring{a}_i$ Values for Some Ions Derived by Keilland[12] Using the Principles of References 116 and 117[a]

Ion	$\mathring{a}_i$ (Å)	Ion	$\mathring{a}_i$ (Å)
H_{aq}^+	9	Mg^{2+}	8
Li^+	6	Ca^{2+}	6
Na^+	4.5	Sp^{2+}	5
K^+	3	Ba^{2+}	5
Rb^+	2.5		
Cs^+	2.5	F^-	3.5
Tl^+	2.5	Cl^-	3
Ag^+	2.5	Br^-	3
NH_4^+	2.5	I^-	3

[a] Compare Table 8 for the $\mathring{a}$ values required to fit log $\gamma_\pm$ data for salt pairs of ions according to the DH relation for finite ion size.

---

† Estimates of Stokes' radii r_s of ions from λ_+^0 or λ_-^0 should not now be made simply on the basis of a formula $z_i e/300 = 6\pi \eta r_s (\lambda_\pm^0/F)$, but on one in which an allowance for the dielectric relaxation friction component (Fuoss, Zwanzig effect) is made through an effective friction coefficient of $6\pi \eta r_i + 2e^2 \tau (\varepsilon_0 - \varepsilon_\infty)/\varepsilon^2 3r_i^3$, where r_i is the ionic radius, ε_0 the static dielectric constant, ε_∞ the optical dielectric constant, and τ the dielectric relaxation time of the solvent. This is equivalent to an effective Stokes radius of $r_i + e^2 \tau (\varepsilon_0 - \varepsilon_\infty)/\varepsilon^2 9\pi \eta r_i^3$. It does not appear, however, that such a relation includes the ion-specific effects that are known to be significant in empirically evaluated r_s data and in the applicability of Walden's rule for various solutions.

Frank and Evans[119] and Frank and Wen. [108] The special behavior that may arise with large R_4N^+ cations coupled with small anions in various tetraalkyl-ammonium salts was stressed. Here the large difference in size of the cation and anion and the special hydrophobic interactions between the larger R_4N^+ ions and water tend to make the individual cation activity coefficient appreciably different from the anion activity coefficient at a given R_4N^+ salt concentration.

In a further paper[119] Frank has explored more quantitatively the significance and evaluation of individual ionic activity coefficients in relation to hydration. Following Bjerrum,[100] the hydration contributions to $\ln \gamma_\pm$ are written

$$\ln \gamma_{+,h} = \frac{3h_+ + h_- - 2}{55.51} m \qquad (162)$$

and

$$\log \gamma_{-,h} = \frac{3h_- + h_+ - 2}{55.51} m \qquad (163)$$

where h_+ and h_- are the "hydration numbers" of the cation and anion, respectively. These relations give the standard result for hydration effects in $\ln \gamma_{\pm,h}$:

$$\ln \gamma_{\pm,h} = \frac{2(h_+ + h_-)^{-2}}{55.51} m \qquad (164)$$

Also, the ratio $\gamma_{+,h}/\gamma_{-,h}$ can be obtained:

$$\ln (\gamma_{+,h}/\gamma_{-,h}) = \frac{2h_+ - h_-}{55.51} m \qquad (165)$$

which characterizes the extents to which $\ln \gamma_+$ and $\ln \gamma_-$ can differ owing to hydration. For $0.1\,M$ aqueous HCl taking $h_+ = 8$ and $h_- = 0$ (see reference 119), γ_+/γ_- is deduced as ~ 1.05, as inferred from literature pH values.

In R_4N^+ iodide solutions γ_{I^-} seems to be larger, and sometimes much larger, than $\gamma_{R_4N^+}$. This is attributed to the unusual structure-promoting properties of R_4N^+ cations in water coupled with the structure-breaking behavior of I^-. These effects are associated with the extra work terms that are involved when R_4N^+ ions are imagined to be introduced at some finite $R_4N^+I^-$ salt concentration into water, the structure of which is modified by the I^- ions, or vice versa. The cavity salting-out effects (see p. 181) will also differ appreciably for the two types of ions owing to their appreciable difference in volume.

For considering numerical data for single-ion activities it was suggested[119] that a new quantity be defined, the mean ionic activity coefficient deviation, denoted by $\bar{\delta}_\pm$ and given by

$$\bar{\delta}_\pm = (\gamma_+^{\nu_+}/\gamma_-^{\nu_-})^{1/\nu} \qquad (166)$$

When combined with the definition of the mean activity coefficient $\gamma_\pm = (\gamma_+^{\nu_+}\gamma_-^{\nu_-})^{1/\nu}$, the single-ion activity coefficients become accessible through the identities

$$\gamma_+ \equiv (\gamma_\pm\bar{\delta}_\pm)^{\nu/2\nu_+} \tag{167}$$

and

$$\gamma_- \equiv (\gamma_\pm\bar{\delta}_\pm)^{\nu/2\nu_-} \tag{168}$$

Experimental emf data for a cell based on KI/AgI, Ag, and $R_4N^+I(R \equiv$ Me, Et, n-Pr, and n-Bu)/AgI, Ag electrodes (see reference 120) were considered and $\bar{\delta}_\pm$ data derived.

It is to be concluded that individual ionic activity coefficients can have some useful but extrathermodynamic significance, especially when strong or special solvation interactions are involved that are manifested in the excess properties for nondilute solutions and also when the cation and coanion differ appreciably in size.

Other approaches for the evaluation of individual ionic activity coefficients have been proposed by Tamamushi,[10] using a combination of the DH activity coefficient equations with an ionic mobility equation, and by Goldberg and Frank[11] with regard to evaluation of liquid junction potentials at very short times after the formation of electrolyte boundaries (see the question of liquid junction potentials in individual ionic activity coefficient evaluation discussed in references 119 and 120).

A general idea of the individuality of ionic activity coefficients can be derived from the differences of $\ln \gamma_\pm$ at a given concentration for pairs or series of salts with a common anion or cation, using data for intermediate or high concentrations, as has been discussed in other ways in earlier sections.

Some advantages may result from the calculation of individual ionic activity coefficients or these may be combined to give mean activity coefficients for "salt" species in mixtures (see references 65 and 71). Thus Whitfield[13], using equations of the type treated by Pitzer (Section 3.6.3), has calculated individual ionic activity coefficients of 29 ions in sea water.

We have referred to a nonequilibrium method for the possible estimation of individual ionic activity coefficients based on the initial diffusion of ions at a junction.[10] Ultrasensitive methods may be developed in the future that will enable this area of activity coefficient evaluation to be explored further.[121]

6. Relation to Conductance Behavior

Studies of the properties of electrolyte solutions by thermodynamic and conductivity experiments have gone hand in hand for many years. The ionic atmosphere model has to be used in a consistent way when dealing with the dynamical problem of ion mobility through modifications of the model in which a nonequilibrium, steady-state perturbed atmosphere situation is treated. The interaction forces that give rise to thermodynamic nonideality in the chemical potential expression are manifested as net actual forces in the dynamical situation of net ion transport in conductance, so the theories of conductance have to be consistent with the thermodynamic representation of ion interaction behavior.

This has been the basis of the older treatments of Debye, Hückel, and Onsager, of Falkenhagen and Kelbg, and also of the more modern treatments of ion conductance. A full account of conductance behavior of electrolytes and its theoretical treatment is given in Chapter 3 by J.-C. Justice in this volume.

7. Short-Range Ionic Interactions

7.1. Introduction

The DH theory and more modern treatments of ionic solutions recognize the finite size of anions and cations and the cutoff effective contact distance a in the ion or charge-density (ρ) distribution for this critical codistance of closest approach. This closest distance of approach factor a does not imply in these theories any finite time of residence of a $+-$ pair of ions in the closest approach configuration; it is simply a boundary condition.

However, with ions of sufficient charge density and/or in solvents of low dielectric constant, a significant degree of association $1-\alpha$ manifested in kinetic (conductance and diffusion) and thermodynamic (activity and osmotic coefficients) behavior does arise and is formally represented by an association equilibrium constant. For the above conditions, ion pairs are significant chemical species, with their own kinetic, spectroscopic, and thermodynamic properties.

The role of short-range ion association effects in electrolyte solutions was early (1926) recognized by Bjerrum.[122]

7.2. Definitions

First, it should be noted that, in certain cases, association between ions M^+ and A^- of a salt M^+A^- must be distinguished from incomplete dissociation. Thus, in the case of a potential electrolyte ("ionogen") such as HCl or CH_3COOH, dissociation by an acid/base proton transfer reaction with a solvent S will first occur to give free solvated ions $SH^+{\cdot}nS$ and $Cl^-{\cdot}n'S$, for the HCl case. These ions may be in equilibrium with solvated ion pairs $[SH^+{\cdot}nS][Cl^-{\cdot}n'S]$, with partly solvated pairs involving less than $n + n'$ solvent molecules, and with a small fraction of unreacted molecules. In certain cases successive stages of desolvation down to the so-called contact ion pair M^+A^- can be distinguished, where M^+ and A^- are a cation and an anion, respectively.

Other ligands, besides molecules of the solvent, may be involved in the ion pair, particularly in the case of those ions that form stable Werner complexes with donor molecules. The gegen-anion itself may be a ligand and the "ion pair" is then, in the limiting case, a complex ion capable itself of further association with, e.g., a gegen-cation. In all these cases short-range interactions are involved and their effects must be distinguished from those arising on account of long-range ionic interactions treated in the DH type of theory, treated earlier.

An important distinguishing feature of the behavior of ion pairs is that they can move at least for a significant period of time as a single kinetic entity and they may have characteristic absorption spectra in some cases.

Historically, most electrolytes were regarded by early workers as incompletely dissociated and only on account of the van't Hoff i factors being near 2 for simple 1–1 salts in water and because of the failure of the Ostwald dilution law did it become clear that dissociation of most electrolytes was virtually complete, at least at all except the highest concentrations in water. For an equilibrium of the form

$$M^+A^- \overset{K}{\rightleftharpoons} M^+ + A^- \tag{169}$$

proceeding to an extent α when the stoichiometric concentration of salt M^+A^- is C, K is given by

$$K = \frac{\alpha^2 c}{1 - \alpha} \tag{170}$$

neglecting activity coefficient effects. According to Arrhenius, α could be expressed as the equivalent conductance ratio Λ_c/Λ_0, it being assumed that Λ_c differed from the infinite dilution value Λ_0 only on account of the limited dissociation. Insertion of the conductance ratio for α in the expression for K gives the well-known dilution law

$$K = \frac{\Lambda_c^2}{\Lambda_0^2} c \Big/ \left(1 - \frac{\Lambda_c}{\Lambda_0}\right) \tag{171}$$

Rearrangement gives

$$\Lambda_0 K \frac{1}{\Lambda_c} - K = \Lambda_c c / \Lambda_0 \tag{172a}$$

or

$$\Lambda_0^2 K \frac{1}{\Lambda_c} - \Lambda_0 K = \Lambda_c c \tag{172b}$$

a relation that can be tested by plotting a suitable function involving the experimental quantities Λ_c and c. The fact that this relation was *not* obeyed by most salts but *was* obeyed by the "weak" acids provided final evidence against the Arrhenius view of incomplete but appreciable dissociation, except in limiting cases where $1 - \alpha$ was less than but near unity, e.g., for "weak" electrolytes such as CH_3COOH or Hg_2Cl_2 in water. Bjerrum suggested that the conductance behavior could empirically be treated better in terms of a conductance ratio

$$\Lambda_c/\Lambda_0 = f\alpha \tag{173}$$

where f was essentially an interionic attraction factor and α a true degree of dissociation. When $\alpha \to 1$, f was the main factor, leading to Λ_c/Λ_0 being less than unity; when $\alpha \ll 1$, interionic attractions would be small and $f \to 1$, so that the conductance ratio would be mainly determined, as in the classical theory, by the concentration-dependent α.

Strictly speaking, ion-pair formation must be distinguished chemically from incomplete dissociation, though phenomenologically, in conductance or thermodynamic measurements, the effects arising in experimental measurements are difficult to assign to one or the other of these types of "association." IR, Raman, and NMR spectroscopic methods are, however, able in some cases to make the required distinction.

7.3. Thermodynamics and Energetics of Ion-Pair Equilibria

7.3.1. The Equilibrium Constant

In a modern approach to the problem of incomplete dissociation the Bjerrum f factor must be treated in terms of the relevant activity coefficient quotient, so that for the dissociation of M^+A^- considered above

$$\bar{K} = \frac{a_{M^+}a_{A^-}}{a_{M^+A^-}} = \frac{\alpha_m^2 m}{1 - \alpha_m} \gamma_\pm^2/\gamma_{M^+A^-} \tag{174}$$

where $\gamma_\pm$ is the mean molal activity coefficient for the free ions at concentrations $\alpha_m m$, α_m is the degree of dissociation at a total concentration of salt m, and $\gamma_{M^+A^-}$ is the activity coefficient (usually unity) for the undissociated molecule or the associated ions. $\bar{K}$ in Eq. (174) is a true thermodynamic dissociation constant and differs from K considered earlier in Eq. (170) by proper inclusion of activity coefficients in the mass action term (the latter is derived, of course, from the condition of equality of chemical potentials of M^+, A^-, and M^+A^- at equilibrium).

Since $\bar{K}$ is $\exp(-\Delta G^0/RT)$, the entropy and enthalpy change associated with ion association can usually be evaluated by general procedures (see below) based on determinations of $\bar{K}$ as a function of T, where both α and $\gamma_\pm$ may vary with T.

7.3.2. Bjerrum Definition of Association

The first *a priori* treatment of ion association was given by Bjerrum[122] and formed the basis of early work on incomplete dissociation. Related to his treatment, there is a general problem in regard to defining association of molecular species, whatever is the force field operating between them, i.e., if it is an attractive one: To define association it is necessary to specify within what distance particles are interacting with a certain minimum energy. The aggregate concerned is energetically specified by this critical energy. Normally this energy is a function of interionic distance. The stronger the energy of interaction and the more short range the interaction forces, the easier it is to define a particular associated species within a short critical distance. In Bjerrum's treatment[122] the Coulomb force field is involved with an energy cutoff limit of $2kT$ for the interaction corresponding to a distance q given by Coulomb's law:

$$2kT = z_+ z_- e^2 / \varepsilon q \qquad (175)$$

$2kT$ is a useful limit (see below) for defining association and is convenient in the solution of the integral that results when the evaluation of the equilibrium constant K for association is made (see below). Within a spherical envelope of radius q the ions involved in the pair are regarded as associated. In fact, at a distance q, where the energy is $2kT$, there is a minimum probability of finding an ion of opposite charge to the central ion. Within the shell of radius q, two ions are attracted together with energies $> 2kT$ so that the probability of their existence as a pair increases. There is obviously a cutoff distance (a) $(<q)$ determined by the closest-approach condition, which defines ions actually in some state of physical contact. Outside the shell at q there is also a geometrically increasing probability of finding gegen-ions because of the properties of the spherical $\int 4\pi r^2 \, dr$, but the energy of interaction of such ions with the central ion is $<2kT$ and rapidly diminishes with r as $r > q$.

In the Bjerrum equation, Eq. (175), ε was taken as the usual bulk solvent value; this is a factor that may be questioned (see below). For aqueous solutions, accepting this assumption about ε, it is easily seen that q is on the order of 0.45 nm, i.e., physically corresponding to a reasonable value of the contact distance a. However, for organic solvents with $\varepsilon = 10$, say, q is much larger (4.5 nm) than any physically reasonable value of a, so there is a sphere of influence of radius $q \gg a$ within which kinetic association obtains. In any case, for water there is a problem with the Bjerrum relation, since between the associated ions ε is going to be <80 owing to dielectric saturation, e.g., with a solvent-separated ion pair where $a > r_+ + r_-$, the sum of the crystallographic ionic radii. Although the Bjerrum theory was proposed as early as 1926, it has continued to be used until recent times (e.g., Brown and Prue,[123] 1955), with some success. Hence it will be desirable to discuss the treatment in more detail and to then describe some recent modifications of the theory.

7.3.3. Relation to Debye–Hückel Theory

In terms of the ionic atmosphere DH model of electrolyte solutions the electric potential ψ_r is given through the Poisson–Boltzmann equation by

$$\nabla^2 \psi_r = -\frac{4\pi}{\varepsilon} \sum_1^i c_i z_i e \exp\left(\frac{-z_i e \psi_r}{kT}\right) \tag{176}$$

This relation does not give a simple result for ψ_r, and hence for the coulombic interaction, when $z e \psi_r > kT$, as it usually is *near* to an ion. However, Bjerrum avoided this difficulty by noting that near the ion ionic atmosphere effects will not be important in determining the short-range local attraction, so that the local potential due to a central ion i can be written to a satisfactory approximation simply as

$$\psi_i = z_i e / \varepsilon r \tag{177}$$

It is also interesting to examine the relation between the Bjerrum condition for ion association and the DH approximation $z_+ e \psi_r / kT < 1$ (or $\ll 1$). For a limiting case where $z_+ e \psi_r = kT$ it is seen that ψ_r is equal to $z_+ e / 2\varepsilon q = \psi_q$. Of course, the ionic atmosphere model will hardly be applicable to give ψ at the short distance q in aqueous media ($\varepsilon \approx 80$), as noted by Bjerrum. The point of interest with regard to the conditions of applicability of the DH theory is that when $z_- e \psi_r > kT$, as may physically arise with high charge-density ion salts and/or at short interionic distances or in low dielectric constant solvents, not only is there a limitation on the analytical solution of the Poisson–Boltzmann equation but also significant ion-pairing can arise at distances less than q, corresponding to the potential ψ_q. When ε is small, this q can be quite large, as will be discussed later.

Generally speaking, however, the interaction energy condition $|z_+ z_- e^2 / 2\varepsilon q| > 2kT$ implies a larger electrostatic energy of interaction between cation 1 and anion 2 than that corresponding to the average condition for which the DH approximation is justified.

7.3.4. Relation to the Degree of Association and the Ion Association Equilibrium Constant

The degree of association $1 - \alpha$ is obtained by integrating the number of ions in all shells within q to the distance of closest approach a. This gives the time average (fractional) number of ions (<1) that are effectively associated with the given central ion, and this is therefore a quantity to be identified with the degree of association, and the number of counter-ions j in a spherical-shell element of thickness dr at a distance r from the central ion i will be

$$n_j = n_j^0 \exp\left(\frac{-z_j e z_i e}{kT \varepsilon r}\right) 4\pi r^2 \, dr \tag{178}$$

where n_j^0 is the bulk concentration of j ions. It is from this equation that it becomes apparent that a minimum in the distribution function for n_j arises, as seen from the tabulation of approximate figures shown in Table 12 for a 1:1 electrolyte. The minimum arises at ~ 0.357 nm for water ($\varepsilon = 78.3$, 298 K); the number of ions of like charge obviously increases monotonically as r increases, owing to the repulsion effect.

The rationalization of the choice of a distance q, within which the ionic interaction energy is numerically $>2kT$, is seen as follows:

If the probability of finding ions of opposite charge is written [see Eq. (178)] as $P = r^2 \exp(-x/r)$, where $x = z_i z_j e^2/\varepsilon kT$, P has a minimum corresponding to

$$\frac{dP}{dr} = x + 2r = 0 \tag{179}$$

i.e., at $r = -x/2$ or at a distance $q = |z_i z_i| e^2/2\varepsilon kT$ for which the coulombic interaction energy is $2kT$. Hence the origin of Eq. (175).

Let $y = z_i z_j e^2/\varepsilon kTr\, (= x/r)$. Hence $r = z_i z_j e^2/\varepsilon kTy$ and

$$\frac{dr}{dy} = -\frac{1}{y^2}\frac{z_i z_j e^2}{\varepsilon kT}$$

Then

$$\int_a^q \exp\left(\frac{z_i z_j e^2}{\varepsilon kTr}\right) r^2\, dr = -\left(\frac{z_i z_j e^2}{\varepsilon kT}\right)^3 \int_b^2 e^{-x} x^{-4}\, dx \tag{180}$$

and $4\pi n_j^0$ times this integral is the number of ions associated, i.e., $1 - \alpha$. That is,

$$1 - \alpha = 4\pi n_j^0 \frac{z_i z_j e^2}{\varepsilon kT} Q(b) \tag{181}$$

Table 12
Numerical Illustration of Distribution of Ions in Bjerrum Ion-Pairing Treatment

r (nm)	$\exp[e^2/\varepsilon kTr]$	$4\pi r^2\, dr \times 10^{25}$ ($\delta r = 0.01$-nm shells)	Number of ions in successive 0.01-nm shells	
			+-	++
0.2	35.57	0.50	$1.77\, n_j^0$	$0.001\, n_j^0$
0.2	17.36	0.79	$1.37\, n_j^0$	$0.005\, n_j^0$
0.3	10.78	1.13	$1.22\, n_j^0$	$0.01\, n_j^0$
0.357	7.39	1.60	min. $1.18\, n_j^0$	$0.02\, n_j^0$
0.4	5.95	3.14	$1.20\, n_j^0$	$0.03\, n_j^0$
0.6	3.28	4.52	$1.48\, n_j^0$	$0.14\, n_j^0$

where $Q(b)$ is a definite integral that can be tabulated. b is simply the value of y when $r = a$ and 2 is the value of y when $r = q$.

Then from the law of mass action for *dissociation*, neglecting differences of activity coefficients [Eq. (174)] from unity, and in dilute solutions where $\alpha \to 1$

$$\frac{1}{K} = \frac{1 - \alpha}{c_j} = \frac{4\pi N}{1000} \frac{z_i z_j e^2}{\varepsilon k T} Q(b) \tag{182}$$

where

$$\frac{c_j}{1000} N = n_j^0 \tag{183}$$

7.3.5. Problems with the Bjerrum Model: Other Treatments

One of the main problems with the Bjerrum calculation is the question of the physical significance of the $1 - \alpha$ values that can be calculated. All ions within a critical distance are calculated, but for low dielectric constant solvents this distance would include, as explained earlier, some ions not physically in contact; thus in a 96% dioxane–water mixture q is 8.0 nm.

A serious problem is the question of the local value of ε. This will be particularly important for a short-range interaction such as that involved between a solvated cation and anion in, for example, water as the solvent, for which $q = 0.357$ nm. This matter has been considered by Rosseinsky[124] and Pankhurst.[125]

For distances between ions less than 0.5 nm dielectric saturation effects become important and the Bjerrum type of expression for K is then written by Rosseinsky[124] in two terms (cf. Conway, Desnoyers and Smith[126]):

$$K = \frac{4\pi N}{1000} \int_a^{0.5 \text{ nm}} \exp\left(\frac{U(r)}{kT}\right) r^2 \, dr + \int_{0.5 \text{ nm}}^{d} \exp\left(\frac{z^2 e^2}{\varepsilon_0 r k T}\right) r^2 \, dr \tag{184}$$

where $U(r)$ is calculated on the basis of the electrostatic attraction between the ions forming the pair, taking account of the field − and hence the distance dependence of ε. In the second integral the normal value ε_0 of the dielectric constant is used.

While this treatment is an improvement on the Bjerrum analysis, there is the complication that as the pair of ions become close to each other, the dielectric saturation in regions of solvent not along the line of centers may be quite different from that obtaining at each ion separately at a corresponding radial distance from the center. The field vectors will combine in a complex way and make ε difficult to evaluate locally. In the case of rigidly oriented water molecules around a pair of ions, Levine and Wrigley[127] have made a much more sophisticated mathematical treatment and report somewhat different conclusions. At long distances from an ion pair the pair will appear to the solvent as an uncharged entity, but at short distances local fields at one side of one of the ions will obviously be far from negligible, as may be readily evaluated by the principle of superposition of fields. This matter is closely connected with the question of evaluation of the local profile of the dielectric constant about an ion pair (cf. Levine and Rozenthal[128]).

The second main difficulty in the Bjerrum theory is a technical one, that of the divergent distribution function, Eq. (178). It is avoided in Bjerrum's treatment by the arbitrary cut off at the minimum of the distribution referred to previously. A more satisfactory convergent function was treated first by Fuoss[129] in 1934. Later (1959) he made an improved treatment in which only ions really in contact are regarded as associated[130] (cf. Denison and Ramsey[131]) along the following lines (Robinson and Stokes[132]). The cations are treated as charged particles of finite radius a and the anions, for the purposes of the calculation, are taken as point charges. This is equivalent to allowing anions to be located at any point on a "sphere of closest approach" having a volume $V = \frac{4}{3}\pi a^3$. From the DH potential ψ_j for such a situation, i.e., for a distribution of ions of finite size, where

$$\psi_j = \frac{z_j e}{\varepsilon} \frac{e^{\kappa a}}{1 + \kappa a} \frac{e^{-\kappa r}}{r} \tag{185}$$

ψ_j becomes (a more complete argument involving the local field strength at the surface of the cation is given in the original derivation by Fuoss[130])

$$\psi_{j,a} = \frac{z_j e}{\varepsilon a} \frac{1}{1 + \kappa a} \tag{186}$$

for $r < a$, so that the potential energy of the anion with respect to the central ion is

$$U = -\frac{|z_i z_j| e^2}{\varepsilon a (1 + \kappa a)} \tag{187}$$

In point of fact, locally at $r = a$, $\psi_{j,a}$ is usually no longer given at all exactly by the DH solution owing to dielectric saturation effects and because (see the earlier discussion in relation to hydration and the Debye–Pauling treatment) $ze\psi_{j,a} \not< kT$. However, the assumption made enables a final result to be obtained that has some advantageous features over the treatment of Bjerrum.

The solution may be considered as containing n_+ cations per unit volume, of which n'_+ are free ions and n''_+ are associated in ion pairs. A $1:1$ electrolyte will be considered so that a similar distribution of anions in the solution and in pairs will be involved.

If we imagine that quantities δ_{n^+} and δ_{n^-} of ions of each kind are added to the solution, the probability of an anion remaining free is proportional to $1 - n'_+ \nu$ and the probability of it entering into ion-pair formation is $n'_+ \nu \exp[-U/kT]$, since the interaction energy U facilitates pair formation.

For electroneutrality an equal fraction of the added cations, δn_+, must also form ion pairs so that the proportion of bound added cations to unbound added anions is given by

$$\frac{\delta n_+''}{\delta n_-'} = \frac{2n_+'\nu \exp(-U/kT)}{1 - n_+'\nu} \tag{188}$$

The expression for U may be separated into a constant term (b) and a concentration-dependent term. Then, dividing by kT,

$$-U/kT = b - b\kappa a/(1 + \kappa a) \tag{189}$$

The second term is the logarithm of the square of the DH activity coefficient $f_\pm^2$, i.e.,

$$b\kappa a/(1 + \kappa a) = z_i z_j e^2 \kappa/(1 + \kappa a) = -\ln f_\pm^2 \tag{190}$$

where

$$b = |z_i z_j| e^2/\varepsilon k T a \tag{191}$$

Then

$$\exp(-U/kT) = f_\pm^2 b \tag{192}$$

Performing an integration on the differential equation corresponding to that written above in terms of δ's, assuming that the solution is sufficiently dilute so that $1 - n_+'\nu \doteq 1$ and taking into account the molar activity coefficient $y_\pm$ ($y_\pm \doteq$ rational activity coefficient $f_\pm$ at high dilution) gives for the concentration of ion pairs

$$n_\pm'' = (n_+')^2 \nu y_\pm^2 \exp(|z_i z_j| e^2/\varepsilon k T a) \tag{193}$$

The corresponding expression for the equilibrium constant in terms of molar concentrations is then, for $n_\pm \ll n_+'$,

$$\frac{1}{K} = \frac{1 - \alpha}{\alpha^2 y_\pm^2 C} = \frac{4\pi N a^3}{3000} \exp(|z_i z_j| e^2/\varepsilon k T a) \tag{194}$$

where $1 - \alpha$, the fraction associated, is $n_\pm''/n_+'$ and α is $1 - n_\pm''/n_+'$. The relation for $1/K$ fits data for tetra-*iso*-amylammonium nitrate in dioxane–water mixtures and the results for lanthanum ferricyanide (Davies and James[133] and Dunsmore and James[134]) can be accounted for accurately with $a = 0.775$ nm.

In the treatment of Denison and Ramsey[131] the free energy of association is evaluated for a 1:1 salt simply as $Ne^2/\varepsilon a$, giving an equilibrium constant for association of

$$-\ln K = e^2/a\varepsilon k T \tag{195}$$

which is satisfactory when the association energy is numerically $\gg kT$, so that effectively associated species are only the contact pairs. This derivation was based on consideration of a Born–Haber type of cycle involving the charging of the ions and the energy of separation of ions in the ion pairs *in vacuo* and in the solvent medium. From the expression for K given above, the entropy and heat changes associated with the ion-pair formation are readily obtained in terms of $d\varepsilon/dT$, and experimentally from measurements of K as a function of T. Shedlovsky's treatment[135] of the conductance data is usually used.

A relation of a different form was proposed by Gilkerson,[136] where the equilibrium constant can be expressed in a form related to that given by Fuoss,[130] discussed above, which is as follows:

$$1/K = \frac{4\pi N a^3}{3000} \frac{\exp{(z_i z_j e^2/\varepsilon k T a - E_s)}}{kT} \tag{196}$$

The term E_s allows for interaction between ions and solvent dipoles and relates to the difference of solvation energy of the ions and the ion pair. This expression was used by Sadek and Fuoss[137] in their studies on conductance of tetrabutylammonium bromide in various mixed solvent systems involving CCl_4, MeOH, EtOH, and $PhNO_2$.

The expression of Gilkerson,[136] based on Kirkwood's[138] "zeroth-order" partition function for a particle in solution, takes into account the free volume V_0 available to the ions and ion pairs, the a parameter, and the solvation energies. The introduction of V_0 and E_s enables solvent-specific K values to be obtained for a given electrolyte.

In the work of Accascina[139] the disssociation constants K of a number of tetraalkylammonium picrates and nitrates in almost *iso*dielectric solvents (ethylene chloride, $\varepsilon = 10.23$; ethylidine chloride, 10.00; and *o*-dichloroben-zene, 9.93) were determined from conductance measurements using the relations of Shedlovsky.[135] The K values differ by factors far in excess of any differences that could be predicted with Bjerrum's treatment, taking reasonable a values. Use of the treatment involving the solvent dipole moments gave a good account of the differences of the dissociation constants in the three solvents and also the dependence of the association effect on εT, which is usually not given at all quantitatively by Bjerrum's theory.

7.4. Successive Stages of Ion Pairing

A problem of a different kind arises when successive ion-pair equilibria arise. This can occur in two ways: (1) Successive dehydration may lead to several kinetically and thermodynamically distinguishable types of ion pairs, from the case where both the cation and anion remain hydrated and a loose hydrate contact pair is formed, to the Fuoss "contact" ion pair where no hydrate water remains between the ions; (2) in some cases one or more anions may become associated with a given cation, particularly if the latter is di- or trivalent. Charged or neutral ion aggregates can then be formed. A related case arises in solvents of low ε where ion pairs are initially formed, followed by ion triplets $+-+$ or $-+-$ at higher concentrations. The latter types of ion aggregate are indicated by anomalous increases of equivalent conductance at moderate or high concentrations in nonaqueous solvents[140,141] after initial decreases of Λ_c with c.

Successive ion-pair equilibria involving varying degrees of solvation may be represented by

$$M_{aq}^{z_+} + A_{aq}^{z_-} \rightleftharpoons [M_{aq}^{z_+}, A_{aq}^{z_-}] \rightleftharpoons [M^{z_+}H_2O\,A^{z_-}]_{aq} + xH_2O \rightleftharpoons$$

loose hydrate   solvent-separated
complex   ion pair

$$(MA^{(z_++-z_-)+})_{aq} + (x-1)H_2O, \text{ etc.} \tag{197}$$

contact ion pair

A small distinction between the various ion pairs can be made on the basis of whether they are "outer-sphere" species (one or two solvent molecules or ligands involved between the pair of ions) or "inner-sphere" complexes corresponding to the contact situation (contact ion pairs).

Conductance measurements detect ion pairs in which there is any kind of kinetically significant association. Spectral methods, on the other hand, are particularly sensitive to specific short-range interactions such as are involved in the replacement of ligands or in contact pair formation. Distinctions may also be made on the basis of comparisons of spectral effects for given pairs of ions in the UV and visible spectra.[142]

In certain cases clear evidence for the significant existence of both inner- and outer-sphere ion pairs with a given type of ion has been obtained. In a spectrophotometric study of the association of the SO_4^{2-} ion with the aquopentammine cobaltic cation, the replacement of an inner coordination sphere water ligand by the SO_4^{2-} ion was found to be a slow enough process to be followed in the visible spectrum:

$$[Co(NH_3)_5H_2O]^{3+} + SO_4^{2-} \rightleftharpoons [Co(NH_3)_5SO_4]^+ + H_2O \tag{198}$$

Immediate spectroscopic effects of mixing $[Co(NH_3)_5H_2O]^{3+}$ with SO_4^{2-}-containing solution indicated formation of an outer-sphere complex with SO_4^{2-}, which was followed by further slow changes in the visible spectrum corresponding to the replacement of the inner-sphere H_2O ligand by SO_4^{2-}, as in Eq. (198).

Important evidence for such successive hydration equilibria in ion-pair formation is afforded by ultrasonic absorption studies on bivalent metal sulfates, e.g., in the well-known work of Eigen,[143] Eigen and Tamm,[144] and Atkinson and Kor.[145] The ion-pair equilibria are sensitive to local adiabatic changes of pressure owing to the changes of solvation and local volume. A periodic mechanical vibration signal, as in an ultrasonic wave, can therefore suffer absorption of energy at characteristic frequencies (compare dielectric relaxation and rf absorption) corresponding to the reciprocals of the relaxation times of any energy-absorbing processes. The absorption curves for $2:2$

electrolytes show two maxima that were attributed to specific interactions with exchange of hydrate water between the ions after they had approached sufficiently near to each other by random diffusional motion. The types of process envisaged are analogous to those in the general case considered above and may be represented as follows:

$$M_{aq}^{2+} + SO_{4aq}^{2-} \underset{k_{-1}}{\overset{k_1}{\rightleftharpoons}} [M^{2+} \cdot 2H_2O \cdot SO_4^{2-}]_{aq}$$

$$\underset{k_{-2}}{\overset{k_2}{\rightleftharpoons}} [M^{2+} \cdot H_2O \cdot A^{2-}]_{aq} \underset{k_{-3}}{\overset{k_3}{\rightleftharpoons}} [M^{2+} \cdot A^{2-}]_{aq} \qquad (199)$$

The high-frequency absorption process was not very dependent on the cation and was therefore attributed to relaxation associated with process 2 involving hydrate water molecules but where the ions remain at least separated by one water molecule. The lower-frequency absorption depended on the cation and corresponded to process 3 involving the formation of a water-free contact pair. The rate constants for the latter process for some $2:2$ salts are shown in Table 13. The association equilibrium constant K $(=k_1/k_{-1})$ is seen to be about 0.1–0.2 for all of the cations in the above series of salts; i.e., appreciable association is involved.

Since the successive removal of electrostricted water and changes of local field intensity will result from the successive equilibria shown above, it is to be expected that there will be substantial pressure effects on ion association equilibria. If K values are known as a function of pressure, the corresponding volume changes in the equilibria can be evaluated from well-known thermodynamic principles.

Studies of this kind have been conducted, for example, by Fisher[146] on $MgSO_4$ and by Hamann et al.[147] on $La[Fe(CN)_6]$. The (positive) volume changes (7.3 cm^3 mol^{-1} for $MgSO_4$ and 8 cm^3 mol^{-1} for $La[Fe(CN)_6]$) that can be deduced from the pressure effects are substantially smaller than would correspond to elimination of all electrostriction upon charge-pair formation, so that appreciable association of the pair with water *between* the ions and/or residually around the pair is indicated. Alternatively, a can be considered to be independent of pressure and the variation of K with pressure then attributed to the increase of ε and density of the solvent with pressure, an effect that would lead to the possibility of an explanation of the observed behavior in terms of Bjerrum's theory. Both effects are probably significant.

Table 13
Kinetic Data for 2:2 Ion-Pair Equilibrium Processes

	BeSO$_4$	MgSO$_4$	MnSO$_4$	NiSO$_4$	ZnSO$_4$
k_3 (sec^{-1})	1×10^{-2}	1×10^5	4×10^6	1×10^4	3×10^7
k_{-3} (sec^{-1})	1.3×10^{-3}	8×10^5	2×10^7	1×10^5	$>10^8$

7.5. Ion Association and the Wien Effect

The field-dissociation or "second" Wien effect (Onsager,[148] Wilson[149]) provides further information on ion-pair formation. The effects are particularly significant with $2:2$ electrolytes and polymeric electrolytes (Patterson and Bailey[150]). In the latter case, the high-field enhancement of conductance is due to increased dissociation of site-localized counter-ions and the conductivity does not immediately return to a normal (low-field) value after application of a high-field pulse (10^4 V cm^{-1}). The effects are on the order of 13% or more. With $2:2$ electrolytes the Wein dissociation effects are larger than those predicted theoretically.

According to the theory of Onsager,[151] the second Wien effect arises because of modifications of the ionic atmosphere (first Wien effect), which, as a secondary effect, enhances the degree of dissociation of any ions combined in pairs. Essentially, a kinetic theory of electrolytic dissociation of pairs and recombination of ions is developed in terms of Brownian motion. The probabilities of collision leading to association, compared with the probability of dissociation against or with the direction of the field, will be modified in comparison with the situation at zero field. The increase of the dissociation constant due to the field is measured by the quantity

$$b = 2\beta q = \frac{z_1 w_1 + z_2 w_2}{w_1 + w_2} z_1 z_2 \frac{Xe^3}{2\varepsilon k^2 T^2} = \frac{z_1^2 z_2^2 (\Lambda_1 + \Lambda_2)}{z_2 \Lambda_1 + z_1 \Lambda_2} 9.636 \frac{V}{\varepsilon T^2}$$

(200)

in the relation

$$k_{(x)} = K_{x=0} = K_{x=0}[1 + 2\beta q + (4\beta q)^2/2!3! + \cdots$$

(201)

where z_1 and z_2 in the relation for b are the valences of ions 1 and 2, Λ's are the equivalent conductances, e the electronic charge, X the field in esu, and V the field in V cm^{-1}. For $1:1$ salts b reduces to $9.636\, V/\varepsilon T^2$, q is the Bjerrum distance, and

$$\beta = |Xe(z_1 w_1 - z_2 w_2|/kT(w_1 + w_2)$$

(202)

where kTw are the diffusion constants of the ions and zew are the velocities of electrolytic migration at unit esu field and w's are the hydrodynamic friction coefficients of the ions.

7.6. Experimental Methods for the Study of Ion Association

Three main types of methods are available:

1. Those based on thermodynamic studies of salt solutions.
2. Those based on studies of conductivity.
3. Spectroscopic methods of various kinds.

In the first class of measurements there is the general difficulty that the nonideality of electrolyte solutions is characterized thermodynamically by two principal causes. The first arises from long-range interactions (and solvation effects at high concentrations) such as those treated in the DH and other theories discussed earlier; the second cause arises from short-range ion association effects. It is a general problem in the thermodynamic methods of approach to separate the nonideality contributions arising from long-range effects from those arising from true association between the ions. This problem was in principle recognized by Bjerrum.

7.6.1. Thermodynamic Methods

Among the thermodynamic methods, the procedure based on the determination of emf's of reversible cells is the most precise. It has been employed principally for the determination of the dissociation constants of weak electrolytes and can be used for any salts where the cations and anions can form reversible half-cells. The method was first employed by Harned and Ehlers[152] in a study of the dissociation of acetic acid in aqueous solution. The measurements involved the cell

$$Pt/H_2/HA, m_1; MA, m_2; MX, m_3/AgX/Ag$$

and the emf is dependent on the activity of hydrogen and chloride ions and on the dissociation equilibrium constant, as shown in Eqs. (203) and (204):

$$E = E^0_{AgCl} - (RT/F) \ln m_{H^+} m_{Cl^-} \gamma_{H^+} \gamma_{Cl^-} \qquad (203)$$

and for the general case $Cl^- \equiv A^-$

$$K = m_{H^+} m_{A^-} \gamma_{H^+} \gamma_{A^-} / m_{HA} \gamma_{HA} \qquad (204)$$

The electrolyte solution used in the cell is comprised of the acid HA, a salt of the acid, MA, and another salt of M with a halide ion to which the silver electrode is reversible. In effect, the cell measures the extent of dissociation of the acid as determined by the ratio of concentrations of the salt and the acid according to the general principles of buffering. Very precise measurements of the dissociation equilibrium constant K are obtained by this type of method.

In the case of dissociation of salts a cell is set up, for example, with hydrogen, the acid HA, and ions of the salt $M^{z^+}A^{z^-}$, together with a halide ion electroactive with respect, for example, to Ag/AgX. The equilibria involved in such a cell are as follows, with AgCl:

$$H^+ + A^{z^-} \rightleftharpoons HA^{z-1} \text{ and } M^{z^+} + A^{z^-} \rightleftharpoons (MA)^{z^+ - z^-} \qquad (205)$$

as well as $AgCl + e \rightleftharpoons Ag + Cl^-$ and the solubility equilibrium of AgCl. The activity of protons is determined by the acid dissociation involving HA, and the activity of A^{z^-} ions is determined by the ion association between M^{z^+} cations and A^z anions, as shown in the right-hand equilibrium of Eq. (205). Successive approximations are usually required in order to evaluate the dissociation equilibrium constant.

Another difficulty is that in mixed electrolytes such as are involved in the system shown in Eq. (205) there can arise specific interactions that give rise to activity coefficients not necessarily identical with those of the isolated electrolytes alone in the solvent. Electrolyte mixture treatments must then be applied.[152] In a number of cases successive association processes are involved, for example, with some complexes, and then rather special procedures are required to evaluate the equilibrium constants of the individual processes. The details of the procedures required under such circumstances have been described by Monk.[153]

A second type of thermodynamic method is based on interpretations of solubility products of sparingly soluble salts in the presence of various added electrolytes. This method was first employed by Davies. The principle of this method is that if it is found that the activity coefficients of the species involved in the solubility equilibrium are to be the same in any solution of the same ionic strength, independently of the nature of the added salt, then some allowance for incomplete dissociation of the sparingly soluble salt must usually be made. Provided that the solutions are sufficiently dilute, it can be assumed that the DH relation accounts adequately for the activity coefficients, so that any further deviations of solubility from those expected on the basis of the ion activity product must be attributed to incomplete dissociation, dependent on the type of added ion. Again, the method requires the use of successive approximations for the ionic strength and the activity coefficients in order to obtain the required association constant for the sparingly soluble species.

7.6.2. Conductance Methods

Methods based on conductivity are the most precise and are applicable to quite dilute solutions, in fact, even in the region of concentration where the limiting Onsager law for conductance applies; however, they depend on the accuracy with which a theoretically based equation for the free ion conductivity can be formulated. When ion association is significant, the conductivity is affected by the degree of association in two ways: first, at any concentration, the equivalent conductivity is less than would be expected simply from the concentration and the equivalent conductance at infinite dilution; second, the apparent dependence of the conductivity on, for example, the square root of the concentration differs from that which would be expected on the basis of complete dissociation, since the dissociation factor α enters into the slope of the Onsager relation as well. This may be written

$$\Lambda_c = \alpha[\Lambda_0 - b(\alpha c)^{1/2}] \tag{206}$$

which may be rearranged to the form

$$\alpha = \frac{\Lambda_c}{\Lambda_0 - b(\alpha c)^{1/2}} \tag{207}$$

An approximate value of the dissociation constant α may be inserted equal to the Arrhenius conductance ratio Λ_c/Λ_0. This value of α is substituted in the concentration-dependent term in the right side of Eq. (207) and a revised value of α is obtained from Λ_c and Λ_0. This revised value is then inserted again in the denominator of Eq. (207) and a series of successive approximations for α are thus obtained; this operation is continued until a constant value of the degree of dissociation is calculated. Usually this can be reached after three or four approximations. The above method, due to Davies, is based on the use of the limiting Onsager law for conductance. In a more recent treatment of Fuoss and Onsager (see Chapter 3 in this volume) the main assumptions of the Debye–Onsager theories are unchanged, but the ions are considered as spheres of finite size. The equation for completely dissociated 1:1 electrolytes is then

$$\Lambda_c = \Lambda_0 - bc^{1/2} + Qc \log c + Jc \tag{208}$$

Fuoss and Kraus have applied the Fuoss–Onsager equation (see Chapter 3) to the problem of ion association, expressing the conductance by

$$\Lambda_c = \Lambda_0 - b(\alpha c)^{1/2} + Q(\alpha c) \log (\alpha c) + J(\alpha c) - K(\alpha c)f_\pm^2 \tag{209}$$

The treatment is obviously more complex than that of Davies and gives satisfactory results for studies of some salts in nonaqueous solvent mixtures. Further details of this method are discussed below, as rather precise α values can be obtained. An alternative procedure is that of Shedlovsky,[135] and the relevant conductance equations are as follows:

$$\Lambda_c = \alpha \Lambda_0 - b(\Lambda_c/\Lambda_0)(\alpha c)^{1/2} \tag{210}$$

where α is the fraction dissociated, $b = 8.15 \cdot 10^5 \Lambda_0/(\varepsilon T)^{3/2} + 82/(\varepsilon T)^{3/2}$, and K is the dissociation constant (see below). The second equation is

$$1/\Lambda_c S(z) = C\Lambda_c f_\pm^2 S(z)/K\Lambda_0^2 + 1/\Lambda_0 \tag{211}$$

where

$$S(z) = 1 + z + z^2/2 + z^3/8 + \cdots \quad \text{and} \quad z = b(c\Lambda_c)^{1/2}/\Lambda_0^{3/2} \tag{212}$$

These enable Λ_0 and K to be evaluated by plotting $1/\Lambda_c S(z)$ (ordinate) *vs.* $C\Lambda_c f_\pm^2 S(z)$. The slopes give $1/K\Lambda_0^2$ and the intercept $1/\Lambda_0$. This method is perhaps the most direct for the evaluation of K.

In most cases of significant ion association it is fairly easy to establish the significance of the association by inspection of the slopes of the equivalent conductivity–concentration plots. The limiting slopes can usually be established theoretically by reference to experimental viscosity and dielectric constant data, so that deviations from the theoretical limiting slopes can be evaluated.

It has been noted that a significant conductance contribution can indirectly arise from the ion pairs themselves. They can come together in one direction, rotate as a dipole in the cell field, and dissociate after rotation. This is equivalent to a small net transport, related to a, which will add to the free ion mobility. It is a Grotthuss type of effect. Thus

$$\oplus \quad \ominus \rightarrow \oplus \ominus \rightarrow \ominus \oplus \rightarrow \ominus \quad \oplus$$

7.6.3. Spectroscopic Methods

Spectroscopic methods fall into two classes: one in which spectroscopy is simply used in a spectrophotometric way to evaluate, in suitable cases, the relative concentrations of paired and unpaired ion species (then the results are applied in a thermodynamic way to evaluate K) and another in which direct spectroscopic information is obtained on the ion-pair state, giving information about its type of bonding and/or the participation of solvent molecules.

First, referring to spectrophotometric methods, these are used to evaluate the concentration, usually of the nonassociated species or of the associated species, if one of the species has an absorption spectrum different from that of the other. For acid dissociation processes, for example, this method is often easy to employ in the case of organic acids, where the acid molecules and conjugate base ions have characteristically different spectra. Changing equilibrium among the species, e.g., with changing temperature is often indicated by the existence of an isosbestic point in the absorption profiles. Examples are the nitrophenols and picric acid. The method may also be used for a number of inorganic ions, for example, nitrate when absorption spectrometry in the UV must be employed.

Among the spectroscopic methods one of the most important is that involving the use of Raman spectra. Here the Raman spectrum is very often able to detect the undissociated species uniquely, even in some cases of inorganic acids or salts in solution. New techniques of ion laser Raman excitation, providing high exciting intensities, enable this type of method to be extended to concentrations lower than those previously studied, for example, in aqueous HCl or nitric acids.

In the case of the iodide ion, examination of the cation dependence of the UV spectrum in solvents of low polarity provides a means of evaluating cation–I^- association equilibria as studied by Blandamer, Gough, and Symons.[154] By a comparison between the energies for the first absorption maximum for alkali metal iodides in the gas phase and in various nonpolar, weakly polar, and polar solvents (tetrahydrofuran, t-butanol) and in relation to the spectra of I^- in solutions of tetraalkylammonium salts, it is concluded that I^- is involved in solvent-shared ion pairs. Correlation with the Kosower Z values[155] of the solvents were made in relation to the possible direction of charge transfer in the ion pairs. The electronic transition responsible for the observed spectrum, it was concluded, is one in which the dipole moment of the pair is decreased. In the case of polar solvents the absorption spectrum of I^- (and Cl^- and Br^-) has been interpreted in terms of transitions involving "charge transfer to solvent," the excited state being determined by the proximity and orientation of adjacent solvent molecules.

NMR studies have been useful in the investigation of the ionization of acids, for example, in the work of Hood, Redlich, and Riley.[156] The principle of this type of method is that the proton chemical shift associated with the acid will be different, depending on the environment of the protons. A related approach has been employed by Nancollas and Park,[157] who investigated the proton magnetic resonance in some metal complexes of ligands related to EDTA and EGTA. Complex formation leads to a redistribution of electrons in the ligand structures, with consequent changes of proton shielding. Inductive "deshielding" and long-range shielding effects in the ligands are distinguished. In a number of other cases other nuclei can be used that are involved specifically in ion-pair formation with anions of the solution. A case involving a heavy nucleus, ^{205}Tl, has been studied by Figgis[158] in the presence of halide ions in terms of formation of a series of Tl (III) halide complexes.

The presence of ionic solutes in water can also bring about changes of the proton relaxation rate, the effect being largest for paramagnetic solutes, and smallest for diamagnetic solutes. Any interconversion of diamagnetic and paramagnetic ions in solution can therefore be readily followed. Various investigations have been concerned with the study of spin–lattice relaxation measurements of this kind, for example, with respect to the problem of identifying the complexes present in solutions of nickelous cyanide, and also information about the stability of the nickel-hexacyano complex ion was obtained.

In certain favorable cases, e.g., with aromatic anion radicals, electron spin resonance can be employed in the study of ion pairing. For example, pairing between a diamagnetic cation and an organic radical ion can be investigated by reference to the hyperfine structure generated by the interaction between the unpaired electron and the nucleus of the cation, as in the work of Atherton and Weissmann[159] and Blandamer, Gough, and Symons.[154]

The linewidth of the resonance absorption generally depends on the lifetimes of the radical in its two possible environments: free in solution or associated with the cation. If the interchange process is sufficiently rapid, the energy levels of the various spin states in the radical cease to be precisely definable, owing to the operation of the uncertainty principle. Also, if the radical exists in two equivalent states, to each of which the countercation may become associated, then the cation can migrate between the two anion sites.

Linewidth alternation may then occur if the rate constant for the interchange process is of such a value that the lifetime of each configuration is of the same order of magnitude as the reciprocal of the differences between the hyperfine coupling constants.

Effects of these types have been investigated by Gough and Symons[154] for processes involving the association of durosemiquinone cations with various alkali metal and tetra-n-hexylammonium ions. In the case of Li^+ very little effect of the cation on the total spin density on the durene ring is indicated. In the case of association with Na^+ (in t-pentanol), alternating linewidths arise in a marked way. In this case the hyperfine splitting of the Na indicates that the migration of Na^+ across the durosemiquinone occurs intramolecularly, without complete separation of the cation and anion. Solvent dependence of the sodium splitting is observed, however, so that the structure of the ion pair is sensitive to solvent composition, and is presumably not simply a contact pair, as is the case with the tetra-n-hexylammonium ion.

In nonaqueous solutions the complementary use of alkali metal ion NMR studies with IR spectroscopic measurements has been particularly profitable, e.g., in the work of Popov.[160] In a number of cases solvent-shared ion pairs can be clearly distinguished from contact ion pairs. The alkali metal ion NMR spectrum is sensitive to whether it has a shared solvent molecule as a neighbor in the pair or another ion, the anion, in a Fuoss-type contact pair. Similarly, the solvent IR spectrum in the solvent-shared situation in the pair is characteristic of the associated ion species.

7.6.4. Data on Ion Pairing

A summary of some thermodynamic data for ion pairs and ion-pairing reactions is given in Tables 14 and 15.[161]

7.7. Problems with Successive Ion Association Steps

In a number of ion association equilibria it is possible for several steps to be involved, as we have discussed earlier in this chapter. This directly makes some difficulties with regard to interpretations employing such procedures as those of Bjerrum[122] or Fuoss.[130]

7.8. The Fuoss–Onsager Equation† and the Evaluation of Degrees of Ion Association

A conductance equation taking into account the finite size of ions has been given by Fuoss and Onsager[162,163] and should be applicable up to $\kappa a \not> 0.2$; a further revised form, taking into account higher-order terms in

† For a more detailed discussion of the Fuoss–Onsager equations see Chapter 3.

Table 14
Thermodynamic Data for Some Ion Association Reactions[a]

Reaction[b]	K_A	ΔG^0 (kcal mol^{-1})	ΔH^0 (kcal mol^{-1})	ΔS^0 (cal K^{-1} mol^{-1})	References
$Pb^{2+} + I^-$	83	-2.62	-0.3	7.8	l
$Pb^{2+} + NO_3^-$	15.1	-1.62	-0.57	3.5	a
$Pb^{2+} + Cl^-$	41	-2.20	1.25	11.6	l
$Pb^{2+} + Br^-$	72	-2.53	0.3	9.4	l
$Cd^{2+} + Cl^-$	91	-2.67	0.98	12.2	l
$Cd^{2+} + Br^-$	141	-2.93	-0.32	8.8	l
$Cd^{2+} + I^-$	282	-3.34	-2.05	4.3	l
$Tl^+ + OH^-$	6.67	-1.12	0.37	5.1	m
$Tl^+ + Cl^-$	4.76	-0.93	1.43	-1.7	m
$Tl^+ + Br$	7.67	-1.2	-2.45	-4.2	n
$Tl^+ + NO_3^-$	2.15	-0.45	-0.65	-1.0	n
$Tl^+ + CNS^-$	6.25	-1.09	-2.96	-6.4	m
$Tl^+ + N_3^-$	2.44	-0.53	-1.33	-2.7	n
$Ag^+ + Cl^-$	2.0×10^3	-4.5	-2.7	6	b
$Ag^+ + IO_3^-$	6.8	-1.13	5.14	20.3	c
$Ca^{2+} + OH^-$	25	-1.91	1.19	10.4	m
$Ca^{2+} + OH^-$	23	-1.86	1.25	11.3	d
$Sr^{2+} + OH^-$	6.7	-1.12	1.15	7.6	d
$Ba^{2+} + OH^-$	4.4	-0.87	1.75	8.8	d
$Mg^{2+} + OH^-$	380	-3.5	—	—	e
$La^{3+} + OH^-$	2×10^3	-4.1	—	—	o
$Fe^{3+} + OH^-$	6.4×10^{11}	-16.1	-3.0	44	f
$V^{3+} + OH^{-1}$ [b]	5×10^{12}	-17.2	-3.7	45.5	g
$U^{4+} + OH^-$	2.5×10^{13}	-18.2	-2.5	52	h
$Ca^{2+} + SO_4^{2-}$	200	-3.15	1.65	16.1	m
$Mg^{2+} + SO_4^{2-}$	234	-3.22	4.55	26.1	p, q
$Mn^{2+} + SO_4^{2-}$	181	-3.07	3.37	22.6	p, q
$Co^{2+} + SO_4^{2-}$	230	-3.21	1.74	16.6	p, q
$Ni^{2+} + SO_4^{2-}$	211	-3.16	3.31	21.7	p, q
$Zn^{2+} + SO_4^{2-}$	240	-3.25	4.01	24.4	p, q
$Mg^{2+} + CH_3 \cdot CO_2^-$	17.6	-1.69	-1.52	0.6	a
$Ca^{2+} + CH_3 \cdot CO_2^-$	17.5	-1.69	0.91	8.7	a
$La^{3+} + CH_2(CO_2)_2^{2-}$	1.0×10^5	-6.79	4.8	39	i
$Gd^{3+} + CH_2(CO_2)_2^{2-}$	2.5×10^5	-7.32	5.1	42	i
$Lu^{3+} + CH_2(CO_2)_2^{2-}$	5.3×10^5	-7.77	5.2	44	i
$La^{3+} + Fe(CN)_6^{3-}$	5.5×10^3	-5.09	2.0	23.9	r
$La^{3+} + Co(CN)_6^{3-}$	5.8×10^3	-5.13	1.33	21.7	j
$Mg^{2+} + EDTA^{4-}$ [c]	3.5×10^8	-11.65	3.14	50.5	k
$Ca^{2+} + EDTA^{4-}$ [c]	3.3×10^{10}	-14.34	-6.45	26.9	k
$Co^{2+} + EDTA^{4-}$ [c]	1.15×10^{16}	-21.9	-4.4	59.7	k
$Ni^{2+} + EDTA^{4-}$ [c]	2.0×10^{18}	-24.96	-8.35	56.7	k
$Zn^{2+} + EDTA^{4-}$ [c]	1.7×10^{16}	-22.12	-5.61	56.3	k
$Al^{3+} + EDTA^{4-}$ [c]	6.9×10^{15}	-21.60	12.58	116.6	k
$Y^{3+} + EDTA^{4-}$ [c]	6.2×10^{17}	-24.26	0.32	83.8	k

Table 15
Thermodynamic Properties of Some Ion Pairs[a,b]

Ion pair	S_g $(MX^{(n-m)+})$	ΔS^0	S^0 $(MX^{(n-m)+})$	$-\Delta S_{hyd}$ $(MX^{(n-m)+})$	$(r_+ + r_-)^{-1}$
CaOH$^+$	53.8	10.8	−4.8	58.6	0.397
SrOH$^+$	56.0	7.6	−4.3	60.3	0.376
BaOH$^+$	58.4	8.8	9.3	49.1	0.347
VOH^{2+}	53.7	45.5	−22	75.7	0.457
FeOH^{2+}	53.9	44	−23.2	77.1	0.469
CrOH^{2+}	53.8	—	−16.4	70.2	0.462
UOH^{3+}	58.5	52	−28.5	86.0	0.413
TlOH	58.8	5.1	33.0	25.8	0.352
TlCl	60.9	−1.7	41.9	19.0	0.308
TlBr	62.8	−4.2	45.5	17.3	0.295
TlNO$_3$	80.4	−1.0	64.4	16.0	0.289
TlCNS	69.8	−6.4	60	9.8	~0.30
TlN$_3$	66.7	−2.7	59.7	7.0	~0.32
AgCl	58.9	6	36.5	22.4	0.326
AgIO$_3$	82.2	20.3	66.0	16.2	~0.29
MgSO$_4$	68.2	26.1	2.0	66.2	0.26
CaSO$_4$	69.5	16.1	7.0	62.5	0.24
MnSO$_4$	70.1	22.6	6.7	63.4	0.252
CoSO$_4$	70.2	16.6	−1	71	0.255
NiSO$_4$	70.2	21.7	4	67	0.256
ZnSO$_4$	70.3	24.4	3.0	67.3	0.257
CH$_3$·CO$_2$Mg$^+$	65.0	0.6	−1.0	66.1	~0.35
CH$_3$·CO$_2$Ca$^+$	66.5	8.7	22.2	44.3	~0.31

[a] From Nancollas.[161]
[b] All S values are in terms of cal K^{-1} mol^{-1}.

Footnotes to Table 14

[a] From Nancollas.[161]
[b] Ionic strength = 1.0.
[c] Ionic strength = 0.1.
References: (a) Nancollas, *J. Chem. Soc.*, 1955, 1358; *ibid.* 1956, 744; (b) Jonte and Martin, *J. Am. Chem. Soc.* **74**, 2053, 1952; (c) Renier and Martin, *ibid.* **78**, 1833, 1956; (d) Gimblett and Monk, *Trans. Faraday Soc.* **50**, 965, 1954; (e) Stock and Davies, *ibid.* **44**, 856, 1948; (f) Milburn, *J. Am. Chem. Soc.* **79**, 537, 1957; (g) Furman and Garner, *ibid.* **72**, 1785, 1950; (h) Betts, *Can. J. Chem.* **33**, 1775, 1955; (i) Gelles and Nancollas, *Trans. Faraday Soc.* **52**, 680, 1956; (j) James and Monk, *ibid.* **46**, 1041, 1950; (k) Staveley and Randall, *Discuss. Faraday Soc.* **26**, 157, 1958; (l) Austin, Matheson, and Parton, *Structure of Electrolyte Solutions*, W. Hamer, ed., John Wiley & Sons, New York, 1959, Chap. 24; (m) Bell and George, *Trans. Faraday Soc.* **49**, 619, 1953; (n) Nair and Nancollas, *J. Chem. Soc.*, 318, 1957; (o) Davies, *J. Chem. Soc.*, 1256, 1951; (p) Nair and Nancollas, *J. Chem. Soc.*, 3706, 1958; (q) Nair and Nancollas, *J. Chem. Soc.*, 3934, 1959; (r) Davies and James, *Proc. Roy. Soc. London* **A195**, 116, 1948.

concentration, was given by Fuoss.[130,164] Fuoss and Onsager[165] also proposed a treatment for dealing with significant ion association in terms of the Fuoss–Onsager equation. Generally, the full equation should be used for such analyses, as estimates of the degree of dissociation α at a given finite concentration will obviously be incorrect if the conductance itself is not properly represented in a function applicable to the contributions of the free ions at such concentrations. In terms of the revised Fuoss–Onsager equation the equivalent conductance Λ_c is given by

$$\Lambda_c = \Lambda_0 - Sc^{1/2}\alpha^{1/2} + Ec\alpha \log{(c\alpha)}Jc\alpha - K_A\Lambda f_{\pm}^2 c\alpha \qquad (213)$$

where K_A is the ion association constant, $f_{\pm}$ the mean ionic activity coefficient, S the Onsager limiting slope, and E and J are the coefficients of the higher terms in the Fuoss–Onsager equation and involve Λ_0, solvent properties (S and E), and the ion size parameter a (in J) (Fuoss[130,164,166]). Some weighted mean data for a (denoted by $\bar{a}$) discussed by Kay[167] in terms of the revised Fuoss–Onsager equation are given in Table 16 below for aqueous solutions at 298 K, based on various earlier experimental data. A method of calculation involving successive approximations for α has also been described by Kay.[167] For KCl, KBr, and KI the revised equation gave $\bar{a}$ values almost constant with temperature in the range 278–328 K.

7.9. Molecular Treatment of Ion Association

The Bjerrum treatment,[122] that of Fuoss,[130,166] and the revised treatment of Rosseinsky,[124] taking into account local electric saturation effects in short-range ionic interactions, are essentially continuous distribution calculations based on a continuum model for the solvent. A related treatment involving an adjustable macroscopic dielectric constant has been considered by Magnusson.[168] Detailed calculations based on a more discrete molecular

Table 16
Some Weighted Mean Data for $\bar{a}$ and K_A Values for Alkali Halides Using the Fuoss–Onsager Equation

Salt (aqueous)	$10^8\bar{a}$	$10^8\sigma\,(\bar{a})$	K_A (approximate)
LiCl	3.29	0.07	0.0
NaCl	3.23	0.05	0.2
KCl	3.11	0.02	0.4
RbCl	2.90	0.02	0.6
CsCl	2.61	0.04	0.8
NaBr	3.45	0.05	0.2
KBr	3.28	0.02	0.4
NaI	4.07	0.05	0.0
KI	3.69	0.15	0.3

picture of the ion–solvent interaction involved in pair formation between hydrated ions have been given in various papers by Levine and co-workers.[127,128,169] The problem of simultaneously considering the short-range ion–ion interactions and the solvent interactions in the region between, and outside, the ions is a very complex one, particularly for water, where the response of a water molecule to an anion field is probably not simply the "mirror image" of the response to a cation field of similar magnitude (see hydrogen-bonding effects). Analogously, the ion/water quadrupole interactions will lead to complicating behavior in regard to orientation of the molecules, and the quadrupole moment induced in each hydrated ion must be taken into account.

In the treatments of Levine *et al.* the two ions involved in the pair are regarded as being of the same size, hydration number, and polarizability; the primary coordination sphere is regarded as consisting of discrete oriented "point-dipole" solvent molecules or a corresponding spherical region of dielectric constant different from that of the bulk. Beyond this region a continuous dielectric medium is assumed with the normal bulk dielectric constant. Both nonoverlapping and overlapping hydrate shells were considered as models for the solvent-separated ion pairs. In the latter case the charge distribution was replaced by one involving charges embedded in an ellipsoid of revolution (a prolate spheroid) having symmetry about the major axis.

Three approaches were made to the calculation of the interaction energy of the pairs of ions in the solvent medium: (1) one based on a charging process of the Guntelberg–Müller type; (2) another based on calculating the work involved in surrounding with dielectric medium the hydrated ions having the required configurations established *in vacuo*, and (3) the third based on determining the force between the ions, using the Maxwell electrostatic stress tensor (Levine and Rozenthal[128]). In the charging process the energy change associated with association that is to be calculated must be referred to the energy of the infinitely separated ions; i.e., their own Born charging energy must be subtracted. At the present time, results of this type of calculation have not been compared in any detail with experiment for a variety of pairs of ions of varying size, but Levine and Wrigley[127] performed some numerical calculations of a semiempirical kind for K^+F^-.

For the case of nonoverlapping hydration shells the two ion–hydrate complexes were represented as spheres of effective dielectric constant ε' ($<\varepsilon$, the bulk solvent dielectric constant). The problem of individual orientations of hydrate water molecules (Vaslow[170]) is thus avoided, but this is an important departure from what the real situation must be, insofar as local orientations must presumably change as the pair of hydrated ions is brought into contact. This is allowed for, in effect, by obtaining ε' as a function of the interionic separation. For separations r of the ions that were not too small, a simplified result was obtained for the interaction energy $U(r)$ of the ions in the pair:

$$U(r) = -\frac{e^2}{\varepsilon r}\left(1 + \frac{10^3\delta}{4\pi N \varepsilon r^3}\right) = \frac{e^2}{\varepsilon r}\left(\frac{10.7 \times 10^{-24}}{r^3} - 1\right) \tag{214}$$

where δ, the dielectric decrement for the ions K^+ and F^- (assumed equal) is taken as -6.5, $\varepsilon = 80$, and N is Avogadro's number. For the model of four coordinated K^+ and F^- considered by Levine and Wrigley[127] the distance of closest approach ($2a$ in their model) varied between 0.618 and 0.818 nm.

More recently the question of apparent association of certain pairs of ions of *like* charge, e.g., tetraalkylammonium and alkali metal cations, has been investigated by Wen and Nara.[171] Here the association effects arise, it is believed, for *non*electrostatic reasons connected with structure formation effects in the water solvent.[3] Thermodynamically, they arise presumably from anomalous positive entropy changes associated with the effects of both ions, in close proximity, on the local water structure. Similar effects arise (but partly for different reasons) in micelle formation between long chain cations or anions (colloidal electrolytes). They are referred to as hydrophobic association effects and have been treated, e.g., by Ben-Naim.[172,173]

Acknowledgment

Grant support from the Natural Sciences and Engineering Research Council, Canada, to the author assisted the preparation and production of this contribution. Discussions in correspondence with Dr. J. P. Valleau are gratefully acknowledged.

References

1. S. Arrhenius, *Z. Phys. Chem.* **1**, 481 (1887); *Philos. Mag.* **26**, 81 (1888); *see also* H. Ostwald, *Z. Phys. Chem.* **2**, 36 (1888); **2**, 270 (1888).
2. E.g., see A. I. Popov, *Pure Appl. Chem.* **41**, 275 (1975).
3. B. E. Conway, *Ionic Hydration in Chemistry and Biophysics*, Elsevier Publishing, Amsterdam (1981).
4. E. A. Guggenheim, *Mixtures*, Clarendon Press, Oxford (1952).
5. M. L. Huggins, *J. Phys. Chem.* **46**, 151 (1942); *Ann. N.Y. Acad. Sci.* **41**, 1 (1942).
6. E. Glueckauf, *Trans. Faraday Soc.* **51**, 1235 (1955).
7. R. H. Stokes and R. A. Robinson, *Electrolyte Solutions*, Butterworths Scientific Publications, London (1955).
8. E. A. Guggenheim, *J. Phys. Chem.* **33**, 842 (1929); **34**, 1540 (1930).
9. H. S. Frank, *Z. Phys. Chem. (Leipzig)* **228**, 364 (1935).
10. J. Tamamushi, *Bull. Chem. Soc. Jpn* **47**, 1921 (1974).
11. R. N. Goldberg and H. S. Frank, *J. Phys. Chem.* **76**, 1258 (1972).
12. J. Kielland, *J. Am. Chem. Soc.* **59**, 1675 (1937).
13. M. Whitfield, *Geochim. Cosmochim. Acta* **39**, 1545 (1975).
14. B. E. Conway and J. O'M. Bockris, *Modern Aspects of Electrochemistry*, Vol. 1, J. O'M. Bockris, ed., Butterworths, London (1954), Chap. 2.

15. J. E. Desnoyers and C. Jolicoeur, *Modern Aspects of Electrochemistry*, Vol. 5, B. E. Conway and J. O'M. Bockris, eds., Plenum, New York (1969), Chap. 1.
16. B. E. Conway, *J. Solution Chem.* **7**, 721 (1978).
17. H. L. Friedman, *Ionic Solution Theory*, Wiley, New York (1972).
18. P. Resibois, *Electrolyte Theory*, Harper and Row, New York (1968).
19. H. L. Friedman, *Modern Aspects of Electrochemistry*, Vol. 6, B. E. Conway and J. O'M. Bockris, eds., Plenum, New York (1971), Chap. 1.
20. B. E. Conway and R. E. Verrall, *J. Phys. Chem.* **70**, 1473 (1966).
21. R. H. Stokes and R. A. Robinson, *J. Am. Chem. Soc.* **70**, 1870 (1948).
22. H. Anderson, *Modern Aspects of Electrochemistry*, Vol. 11, B. E. Conway and J. O'M. Bockris, eds., Plenum, New York (1975), Chap. 1.
23. J. C. Poirier, in *Chemical Physics of Ionic Solutions*, B. E. Conway and R. G. Barradas, eds., Wiley, New York (1966), Chap. 2; see also *J. Chem. Phys.* **21**, 965 (1953); **21**, 972 (1953).
24. J. C. Rasaiah, *J. Solution Chem.* **2**, 301 (1973).
25. P. S. Ramanathan, C. V. Krishnan, and H. L. Friedman, *J. Solution Chem.* **1**, 237 (1972).
26. P. S. Ramanathan and H. L. Friedman, *J. Chem. Phys.* **54**, 1086 (1971).
27. J. C. Rasaiah and H. L. Friedman, *J. Phys. Chem.* **72**, 3352 (1968).
28. J. C. Rasaiah and H. L. Friedman, *J. Chem. Phys.* **48**, 2742 (1968); **50**, 3965 (1969).
29. D. D. Carley, *J. Chem. Phys.* **46**, 3783 (1967).
30. J. E. Mayer, *J. Chem. Phys.* **18**, 1426 (1950).
31. D. N. Card and J. P. Valleau, *J. Chem. Phys.* **52**, 6232 (1979).
32. J. C. Rasaiah, D. N. Card, and J. P. Vallaeu, *J. Chem. Phys.* **56**, 248 (1972).
33. S. R. Milner, *Philos. Mag.* **23**, 551 (1912); **25**, 742 (1913).
34. J. C. Ghosh, *J. Chem. Soc.* **113**, 449 (1918).
35. N. Bjerrum, *Z. Elektrochem.* **24**, 321 (1918).
36. H. S. Frank and P. T. Thompson, *J. Chem. Phys.* **31**, 1086 (1959).
37. J. E. Desnoyers and B. E. Conway, *J. Phys. Chem.* **68**, 2305 (1964).
38. R. M. Pytkowicz, K. Johnson, and C. Curtis, *Geochem. J.* **11**, 1 (1977).
39. R. M. Pytkowicz and D. A. Kester, *Am. J. Sci.* **267**, 217 (1969).
40. R. M. Pytkowicz and K. Johnson, in *Activity Coefficients in Electrolytes*, Vol. 1, R. M. Pytkowicz, ed., CRC Press, Boca Raton, Florida (1979), Chap. 8.
41. G. N. Lewis and J. T. Randall, *J. Am. Chem. Soc.* **43**, 1112 (1921); *Thermodynamics*, McGraw-Hill, New York (1923); *see also* revised edition by K. S. Pitzer and L. Brewer, McGraw-Hill, New York (1961).
42. F. Kohlrausch, *Wied. Ann.* **60**, 315 (1897).
43. J. N. Brønsted, *J. Am. Chem. Soc.* **44**, 877, 938 (1922); **42**, 761 (1920); **45**, 2898 (1923).
44. E. A. Guggenheim and J. C. Turgeon, *Trans. Faraday Soc.* **51**, 747 (1955).
45. K. S. Pitzer, *J. Phys. Chem.* **77**, 268 (1973); *Acc. Chem. Res.* **10**, 371 (1977).
46. P. Debye and E. Hückel, *Phys. Z.* **24**, 185 (1923); **24**, 334 (1923); **25**, 97 (1924).
47. D. L. Chapman, *Philos. Mag.* **25**, 475 (1913).
48. G. Gouy, *J. Phys.* **9**, 457 (1910); *Ann. Phys. (Paris)* **7**, 163 (1917).
49. O. Stern, *Z. Elektrochem.* **30**, 508 (1924).
50. E. Guntelberg, *Z. Phys. Chem.* **123**, 199 (1926).
51. E. C. W. Clarke and D. N. Glew, *J. Chem. Soc. Faraday Trans.* **176**, 1911 (1980).
52. (a) T. H. Gronwall, V. K. LaMer, and L. J. Greiff, *J. Phys. Chem.* **35**, 2245 (1931); (b) T. H. Gronwall, V. K. LaMer, and K. Sandved, *Phys. Z.* **29**, 358 (1929).
53. V. K. LaMer, T. H. Gronwall, and L. J. Grieff, *J. Phys. Chem.* **35**, 2245 (1931).
54. E. A. Guggenheim, *Trans. Faraday Soc.* **55**, 1714 (1959).
55. L. Onsager, *Phys. Z.* **28**, 277 (1927); *Chem. Rev.* **13**, 73 (1933).
56. M. Eigen and E. Wicke, *Naturwissenschaften* **38**, 453 (1951); **39**, 545 (1952); *Z. Elektrochem.* **56**, 551 (1952); **57**, 319 (1953).
57. S. N. Bagchi, *J. Indian Chem. Soc.* **27**, 199 (1950).

58. R. H. Fowler and G. S. Rushbrooke, *Trans. Faraday Soc.* **33**, 1272 (1937).
59. D. H. Everett and M. Penney, *Proc. R. Soc. London* **A212**, 164 (1952).
60. D. H. Trevena, *Proc. Phys. Soc. London* **84**, 969 (1964).
61. K. Shinoda and J. H. Hildebrand, *J. Phys. Chem.* **61**, 789 (1957).
62. J. N. Brønsted, *K. Dan. Vidensk. Selsk. Mat. Fys. Medd.* **4**, 4 (1921); *J. Am. Chem. Soc.* **44**, 877 (1922); **45**, 2898 (1923).
63. E. A. Guggenheim, *Rep. Scand. Sci. Congr.* (Copenhagen, 1929), p. 298.
64. E. A. Guggenheim, *Philos. Mag.* **19**, 588 (1935).
65. K. S. Pitzer, in *Activity Coefficients in Electrolyte Solutions*, Vol. 1, R. M. Pytkowicz, ed., CRC Press, Boca Raton, Florida (1979), Chap. 7.
66. E. A. Guggenheim, *Thermodynamics*, North-Holland Publishing, Amsterdam (1949), p. 318.
67. H. S. Harned and B. B. Owen, *Physical Chemistry of Electrolyte Solutions*, 2nd ed., Reinhold, New York (1950), p. 597.
68. C. W. Davies, *J. Chem. Soc. London* 2093 (1938); *Annu. Rep. Chem. Soc. London* **49**, 30 (1952).
69. W. M. Latimer, K. S. Pitzer, and C. Slansky, *J. Chem. Phys.* **7**, 108 (1939).
70. G. Scatchard, *Chem. Rev.* **19**, 309 (1936).
71. K. S. Pitzer and L. Brewer, Lewis and Randall's *Thermodynamics*, rev. ed., McGraw-Hill, New York (1961).
72. P. N. Vorontsov-Vel'yaminov, A. M. El'yashevich, and A. K. Kron, *Elektrokhimiya* **2**, 708 (1966); **4**, 1430 (1968).
73. E. Glueckauf, *Proc. R. Soc. London* **A310**, 449 (1969).
74. K. S. Pitzer and D. J. Bradley, *J. Phys. Chem.* **83**, 1599 (1979).
75. E. Hückel, *Phys. Z.* **26**, 93 (1925).
76. H. L. Friedman, *J. Electrochem. Soc.* **124**, 421c (1977).
77. C. W. Outhwaite, in *Statistical Mechanics*, Vol. 2, Specialist Periodical Report, K. Singer, ed., Chemical Society, London (1975), p. 188.
78. W. G. McMillan and J. E. Mayer, *J. Chem. Phys.* **13**, 276 (1945).
79. N. Metropolis, A. W. Rosenbluth, M. N. Rosenbluth, A. H. Teller, and E. Teller, *J. Chem. Phys.* **21**, 1087 (1953).
80. W. W. Wood, in *Physics of Simple Liquids*, H. N. V. Temperley, S. S. Rowlinson, and G. S. Rushbrooke, eds., Wiley, New York (1968).
81. J. N. Shaw, Monte Carlo calculation for a system of hard-sphere ions, Ph.D. thesis, Duke University, Durham, North Carolina, 1963.
82. See, e.g., H. L. Frisch and J. L. Lebowitz, *The Equilibrium Theory of Classical Fluids*, W. Benjamin, New York (1964), and R. M. Mazo and C. Y. Mou, in *Activity Coefficients in Electrolyte Solutions*, Vol. 1, R. L. Pytkowicz, ed., CRC Press, Boca Raton, Florida (1979), Chap. 2.
83. J. E. Mayer and M. G. Mayer, *Statistical Thermodynamics*, Wiley, New York (1940).
84. J. K. Percus and G. J. Yevick, *Phys. Rev.* **110**, 1 (1958).
85. E. Waisman and J. L. Lebowitz, *J. Chem. Phys.* **56**, 3086 (1972); **56**, 3094 (1972).
86. A. R. Allnatt, *Mol. Phys.* **8**, 533 (1964).
87. H. C. Anderson and D. Chandler, *J. Chem. Phys.* **53**, 547 (1970); **54**, 26 (1971); **55**, 1497 (1971).
88. J. G. Kirkwood and J. C. Poirier, *J. Phys. Chem.* **58**, 591 (1954).
89. J. B. Hasted, D. M. Ritson, and C. H. Collie, *J. Chem. Phys.* **10**, 1 (1948).
90. F. Booth, *J. Chem. Phys.* **19**, 391, 1327 (1951); **19**, 1615 (1951).
91. K. J. Laidler, *Can. J. Chem.* **37**, 138 (1959).
92. J. C. Webb, *J. Am. Chem. Soc.* **48**, 2589 (1926).
93. R. W. Gurney, *Ionic Processes in Solution*, Dover, New York (1963).
94. J. A. Schellman, *J. Chem. Phys.* **25**, 350 (1956).
95. R. W. Rampolla, R. C. Miller, and C. P. Smyth, *J. Chem. Phys.* **30**, 566 (1959).

96. P. Debye and L. Pauling, *J. Am. Chem. Soc.* **47**, 2129 (1925).
97. H. S. Frank, *J. Am. Chem. Soc.* **63**, 1789 (1941).
98. G. Scatchard, *Phys. Z.* **33**, 32 (1932).
99. J. Wyman, *J. Am. Chem. Soc.* **58**, 1482 (1936).
100. N. Bjerrum, *Medd. Vetenskab. Akad. Nobelinst.* **5**(16) (1919); *Z. Anorg. Chem.* **109**, 275 (1920).
101. P. Debye and J. MacAulay, *Phys. Z.* **26**, 22 (1925).
102. J. A. V. Butler, *J. Phys. Chem.* **33**, 1015 (1929).
103. S. Levine and G. M. Bell, in *Electrolytes*, B. Pesce, ed., Pergamon Press, New York (1962).
104. S. Levine and D. K. Rozenthal, in *Chemical Physics of Ionic Solutions*, B. E. Conway and R. G. Barradas, eds., Wiley, New York (1966), Chap. 8.
105. O. Y. Samoilov, *Structure of Electrolyte Solutions and Hydration of Ions*, Consultants Bureau Translation, New York (1965); *Discuss. Faraday Soc.* **24**, 141 (1957).
106. J. C. Rasaiah, *J. Chem. Phys.* **52**, 704 (1970).
107. S. D. Hamann, P. J. Pearce, and W. Strauss, *J. Phys. Chem.* **68**, 375 (1964).
108. H. S. Frank and W. Y. Wen, *Discuss. Faraday Soc.* **24**, 133 (1957).
109. B. E. Conway and H. P. Dhar, *J. Colloid Interface Sci.* **48**, 73 (1974).
110. B. E. Conway, *Elektrokhimiya* **13**, 822 (1977) (Consultants Bureau Translation, *Sov. Electrochem.* **13**, 695 (1977)).
111. R. M. Fuoss and L. Onsager, *J. Chem. Phys.* **61**, 668 (1957).
112. J. C. Rasaiah, Ph.D. thesis, University of Pittsburgh, Pittsburgh, Pennsylvania (1965).
113. See K. S. Pitzer, *J. Phys. Chem.* **77**, 268 (1973).
114. K. S. Pitzer and G. Mayorga, *J. Phys. Chem.* **77**, 2300 (1973).
115. E. Glueckauf, in *The Structure of Electrolyte Solutions*, W. J. Hamer, ed., Wiley, New York (1959), Chap. 7.
116. G. B. Bonino and G. Centrola, *Mem. R. Accad. Ital. Cl. Sci. Fis. Mat. Nat. Chim.* **4**, 445 (1933).
117. L. Brull, *Gazz. Chim. Ital.* **64**, 624 (1934).
118. H. S. Frank, *Z. Phys. Chem.* (*Leipzig*) **228**, 364 (1965).
119. H. S. Frank, *J. Phys. Chem.* **67**, 1554 (1963).
120. M. A. V. Devanathan and M. J. Fernando, *Trans. Faraday Soc.* **58**, 784 (1962).
121. R. Gomer and G. Tryson, *J. Chem. Phys.* **66**, 4413 (1977).
122. N. Bjerrum, *Koninklinge Dans. Vidensk. Selsk.* **7**(9) (1926); *Selected Papers*, Einar Munksgaard, Copenhagen (1949), p. 108.
123. P. G. M. Brown and J. Prue, *Proc. R. Soc. London* **A232**, 320 (1955).
124. D. R. Rosseinsky, *J. Chem. Soc.* 785 (1962).
125. M. H. Panckhurst, *Aust. J. Chem.* **15**, 383 (1962).
126. B. E. Conway, J. E. Desnoyers, and A. C. Smith, *Philos. Trans. R. Soc. London* **A256**, 389 (1964).
127. S. Levine and H. E. Wrigley, *Discuss. Faraday Soc.* **24**, 43, 73 (1957).
128. S. Levine and D. K. Rozenthal, *Chemical Physics of Ionic Solutions*, B. E. Conway and R. G. Barradas, eds., Wiley, New York (1966), p. 119.
129. R. M. Fuoss, *Trans. Faraday Soc.* **30**, 967 (1934).
130. R. M. Fuoss, *Proc. Nat. Acad. Sci. USA* **45**, 807 (1959).
131. J. T. Denison and J. B. Ramsey, *J. Am. Chem. Soc.* **77**, 2615 (1955).
132. R. A. Robinson and R. H. Stokes, *Electrolyte Solutions*, 2nd ed., Butterworths, London (1959).
133. C. W. Davies and J. C. James, *Proc. R. Soc. London* **195A**, 116 (1948); see also *J. Chem. Soc.*, 1094 (1950).
134. H. S. Dunsmore and J. C. James, *J. Chem. Soc.* 2925 (1951).
135. T. Shedlovsky, *J. Franklin Inst.* **225**, 739 (1938).
136. S. Gilkerson, *J. Chem. Phys.* **25**, 1199 (1956).
137. H. Sadek and R. M. Fuoss, *J. Am. Chem. Soc.* **81**, 4507 (1959).

138. J. G. Kirkwood, *J. Chem. Phys.* **7**, 911 (1939).
139. F. Accascina, *Proc. Nat. Acad. Sci. USA* **39**, 917 (1953).
140. R. M. Fuoss and C. A. Kraus, *J. Am. Chem. Soc.* **55**, 2387 (1933).
141. R. M. Fuoss and C. A. Kraus, *Chem. Rev.* **17**, 27 (1935).
142. J. M. Smithson and R. J. P. Williams, *J. Chem. Soc.*, 457 (1958).
143. M. Eigen, *Discuss. Faraday Soc.* **24**, 25 (1957).
144. M. Eigen and K. Tamm, *Z. Elektrochem.* **66**, 107 (1962).
145. G. Atkinson and S. K. Kor, *J. Phys. Chem.* **60**, 128 (1965).
146. F. H. Fisher, *J. Phys. Chem.* **66**, 1607 (1962).
147. S. D. Hamann, P. F. Pearce, and W. Strauss, *J. Phys. Chem.* **68**, 375 (1964).
148. L. Onsager, *J. Chem. Phys.* **2**, 599 (1934).
149. W. S. Wilson, Ph.D. thesis, Yale University, New Haven, Connecticut (1936).
150. A. Patterson and F. E. Bailey, *J. Polymer Sci.* **33**, 235 (1958).
151. L. Onsager, *Phys. Z.* **32**, 545 (1931); *Chem. Rev.* **13**, 72 (1933)
152. See H. S. Harned and B. B. Owen, *Physical Chemistry of Electrolytic Solutions*, 3rd ed., Reinhold, New York (1958), Chap. 7.
153. C. B. Monk, *Electrolytic Dissociation*, Academic Press, New York (1961).
154. L. Blandamer, T. E. Gough, and M. C. R. Symons, *Trans. Faraday Soc.* **62**, 269 (1966).
155. E. M. Kosower, *J. Am. Chem. Soc.* **80**, 3253 (1958).
156. G. C. Hood, O. Redlich, C. A. Riley, *J. Chem. Phys.* **22**, 2067 (1954).
157. G. H. Nancollas and A. C. Park, *J. Phys. Chem.* **71**, 3678 (1967).
158. B. N. Figgis, *Trans. Faraday Soc.* **55**, 1075 (1959).
159. N. M. Atherton and S. I. Weissmann, *J. Am. Chem. Soc.* **83**, 1330 (1961).
160. A. I. Popov, *Rev. Pure Appl. Chem.* **41**, 275 (1975).
161. G. H. Nancollas, *Quart. Rev. Chem. Soc., London* **14**, 402 (1960).
162. R. M. Fuoss and L. Onsager, *J. Phys. Chem.* **61**, 668 (1957).
163. R. M. Fuoss and L. Onsager, *J. Phys. Chem.* **62**, 1339 (1958).
164. R. M. Fuoss, *J. Am. Chem. Soc.* **80**, 3163 (1958).
165. R. M. Fuoss and L. Onsager, *J. Am. Chem. Soc.* **79**, 3301 (1957).
166. R. M. Fuoss, *J. Am. Chem. Soc.* **81**, 2659 (1959).
167. R. L. Kay, *J. Am. Chem. Soc.* **82**, 2099 (1960).
168. L. B. Magnusson, *J. Chem. Phys.* **39**, 1953 (1963).
169. S. Levine and G. M. Bell, *Electrolytes*, B. Pesce, ed., Pergamon Press, London (1962), p. 77.
170. F. Vaslow, *J. Phys. Chem.* **67**, 2777 (1963).
171. W. -Y. Wen and K. Nara, *J. Phys. Chem.* **71**, 3907 (1967).
172. A. Ben-Naim, *J. Phys. Chem.* **69**, 1922 (1965).
173. A. Ben-Naim, *Hydrophobic Interactions*, Plenum, New York (1980); see also A. Ben-Naim, *Water and Aqueous Solutions*, Plenum, New York (1974).

3

Conductance of Electrolyte Solutions

JEAN-CLAUDE JUSTICE

1. Introduction

1.1. Outline and General Aims of the Review

As an introduction to a review of the present state of the theory of conductance, it seems appropriate to state first to whom this review is addressed, the aims of the author's presentation, the points that will be developed, and the intentions concerning the presentations of these points.

This review is first of all intended to be useful to the nonspecialist in the field who wants to make his way through the forest of recent publications on the subject. No mathematical calculations will be found here such as solutions of integral/differential equations, since these can be found adequately in the original papers. However, enough details will be given so that the meaning of the basic equations may be understood, and with enough generality so that the immediate comparison of one treatment with another can be easily made.

It is sometimes considered that there is some confusion in the field of conductance (this is said at least at meetings, if not in written reports of the discussions at these meetings). Is there any active scientific field that does not look confusing from the outside? Whenever progress is fast and multifold, complexity in approach, in formalisms, and even in conclusions is to be expected. A modern review is thus clearly needed.

JEAN-CLAUDE JUSTICE • Laboratoire d'Electrochimie, Université Pierre et Marie Curie, Place Jussieu, Paris, France.

This chapter will only consider what is commonly called the Onsager approach to conductance, because it is the only kind of treatment that goes beyond the derivation of the limiting law, a necessary condition for conductance theory to be useful.

The recent developments of conductance are based on considerations that refer to the structure of the Onsager continuity equation. The importance of this equation is such that it becomes necessary to center the discussion around it and, consequently, not to take it merely as a starting point. The Onsager equation, as it was used by Onsager,[1,2] Pitts,[3] Fuoss and Onsager,[4] Falkenhagen *et al.*,[5] and others, with some variations, is in fact an approximation of a more general equation. Making a separate presentation of each approximation and to not even mention, much less explain, the variations can lead only to confusion, whereas showing in a precise manner how each equation is a particular case and approximation of a more general formulation can provide a unified view of the overall problem. The price to pay in order to achieve this goal is to begin the approach at a more fundamental level, with the risk of becoming less easily comprehensible to the nonspecialist of statistical mechanics.

It is the belief of the author that this danger can be circumvented if the basic quantities, phenomena, and assumptions are clearly defined, stated, and translated into a simple symbolism. This is possible if the various time average mean quantities that are at the core of the Onsager treatment can be easily interpreted and represented by an adequate system of indices, allowing a complete display to be given of the set of restricting conditions involved in each time-averaging process considered. This will necessitate a generalization of the original Onsager system of indexing, but the result is rewarding, since a simple look allows these conditions to be immediately understood; this is a principal point because the exact definition of the restrictive conditions at each step of the development of the treatment is of paramount importance for a clear and simple understanding of the whole theory in general, as well as for properly setting each particular approximation in the general context.

It will then be found that the general system of diffusion-type equations that characterizes the Onsager approach can be simply derived from a minimum set of assumptions and will ultimately reduce to four systems of hierarchies, which, after combination, will lead to a continuity equation strictly identical to the Ebeling continuity equation. The latter can also be derived after some approximations from the Liouville equation, which is at the basis of statistical-mechanical approaches. Hopefully the advantage of the present derivation lies in the fact that each term at any level of the derivation systematically has an easily comprehensible physical meaning. This may help the nonspecialist to concentrate on the ideas involved instead of dissipating his efforts disentangling mathematical developments, which, after all, are of no immediate physical importance.

1.2. Why Study and Measure Conductance?

To answer this question it must be recalled that the conductance of electrolytes can be accurately measured down to very low concentrations. The response of ions to an externally applied electrical field can be measured with very high precision. A low concentration range is precisely a condition for a theory to be tractable with enough accuracy. There results, therefore, a conjunction of conditions that makes conductance a remarkable tool for the derivation of information on the electrolyte–solvent systems studied. This, of course, explains the large number of experimental and theoretical contributions that has characterized this field over the last 25 years. The importance of understanding the specific behavior of a solute in solution need not be emphasized.

Most important chemical reactions occur in the bulk of a liquid. Reactions between solute particles are greatly influenced, if not totally controlled, by solvation effects. This field of solute–solvent interactions is still relatively unknown owing to its complexity. How many problems would be solved if we knew what were the exact state of the solvent within the solvation cosphere, that is (to use the language of Gurney, which is quite useful for dilute solutions) in the region around the solute particle, where the solvent has properties different from those for its pure state?

The solution of this problem is arduous, since a many-particle dense system is involved. For the most part, only vague ideas exist about the state of the pure solvent, that is, about the organization of solvent particles around a given solvent particle, so it is then difficult to even imagine what the solvent structure is around a foreign particle. Herein lies the major problem in theoretically evaluating thermodynamic transfer quantities such as ΔG^0, ΔH^0, and ΔS^0 at equilibrium, and D_0 and λ_0, the diffusion coefficient and the limiting conductance, respectively, of a solute particle in solution.

The problem of evaluating excess quantities is perhaps still more urgent, since these are the parameters that influence the reaction between two species in solution. For noncharged solutes these quantities are difficult to determine experimentally in dilute solution. This is not the case, however, for electrolyte solutions, so that measuring conductance is clearly the ideal, yet indirect, technique for evaluation. The difficulty is not in the performance of the measurements but, rather, in using a sound theory that relates the excess parameters to the experimental mobility data in order to allow these parameters to be evaluated. It is our belief that this is now possible in a way that may receive general agreement.

Parameters such as the Gurney cosphere excess quantities, as expressed, for instance, in the simplest model used by Rasaiah and Friedman, may be reliably determined. Whenever appropriate comparisons could be made with equilibrium thermodynamic data, confirmation of the theoretical predictions was found. This theoretical aspect will be stressed in the present review to show how it is possible to handle models more elaborate than the usual "restricted primitive" one, which has for long been the only case considered. This can be achieved by disregarding the classical Debye approximation, which, by its nature, eliminates any possibility of building a reliable theory for short-range ionic interactions. Recent developments such as the echo-formulation and strong coupling approximations used in conjunction with former classical theoretical derivations will prove to be one key that allows this significant stage to now become accessible.

1.3. Historical Survey

One of the most important properties of an electrolyte solution is its conductance. Quite early in the history of chemistry Faraday introduced the concept of ions and chose this name for those hypothetical entities that were "moving" in an electrical field and could thus carry the current through the solution. The question was, however, where do these ions come from? It was first thought that they were produced by the action of the electrical field itself, which polarized molecules so much that they dissociated into electrically charged particles, giving birth to anions and cations. In 1887 Arrhenius proposed a simpler but more daring explanation, that a certain fraction of molecules *spontaneously* dissociate in solution into anions and cations, quite independently of the presence of an external field. The consequences of Arrhenius's theory were great indeed, since it settled the concept of ions as stable entities. With the use of the Guldberg and Waage mass action law it became possible to measure this "degree of dissociation."

Assuming that the intrinsic mobility of an ion is a constant and introducing the mass action law for the fraction γ of dissociated electrolyte gives

$$\Lambda = \gamma \Lambda_0 \tag{1}$$

$$\frac{\gamma^2 C}{1 - \gamma} = K_d \tag{2}$$

which leads to the so-called Ostwald dilution law after eliminating γ:

$$\frac{1}{\Lambda} = \frac{1}{\Lambda_0} + \frac{1}{K_d} c \frac{\Lambda}{\Lambda_0^2} \tag{3}$$

A plot of $1/\Lambda$ versus $c\Lambda$ then gives a straight line, which allows Λ_0 and K_d to be easily evaluated. As a consequence, measuring the conductance of solutions became a very precise tool for the determination of important thermodynamic constants of chemical equilibria in solutions.

The remarkable success of this theory was, however, limited by striking exceptions. Kolrausch's measurements showed that there were many electrolyte solutions which did not follow the Ostwald law but which carried current more easily than those which did. It was found that the conductance of these so-called "strong electrolytes" varied almost linearly with the square root of their concentration. This was the case of alkali halides in water, for instance. Clearly a new theoretical explanation had to be found to understand their behavior. Bjerrum[6] in 1908 was among the first to claim that these electrolytes were entirely dissociated at all concentrations ($K_d = \infty$) and that the observed changes of Λ with concentration varied not with γ but, rather, with the mobility of free ions, which decreased with increasing concentration because of coulombic interactions between the ions.

We now know that this view is correct, but the idea seemed so original at the time that it took some years before it was accepted. The only way to convince the skeptics was to evaluate the effect of coulombic interactions. This presented, however, a mathematical difficulty: The $1/r$ dependence of the corresponding coulombic potential energy led to divergences in the solution of the problem according to all the methods of treatment used at the time, all classical integrals that converge for, say, Lennard-Jones potentials used for gases, diverge with the coulombic potential. Debye and Hückel,[7] as we know, were the first in 1923 to circumvent this mathematical difficulty by devising a new approach. It consisted in combining Boltzmann's statistical law and Poisson's electrostatic law and truncating the expansion of the resulting exponential function to the third term. The result was the famous linearized Poisson–Boltzmann equation†

$$\Delta\psi = \kappa^2\psi \tag{4}$$

which allowed the mean potential of interaction ψ and hence the limiting law for electrolytic properties to be evaluated as a function of κ (the reciprocal of the so-called Debye ionic atmosphere radius), which varies exactly as the square root of the electrolyte concentration. The road was then opened to a tremendous amount of work on electrolyte solutions, both for equilibrium as well as for transport properties, and among the latter, conductance.

The well-known Debye–Hückel (1923) result for the activity coefficient $f_\pm$ of symmetrical electrolytes should be specially mentioned,

$$\log f_\pm = -\frac{\kappa q}{1 + \kappa a} \tag{5}$$

(where q is the Bjerrum length and a the ionic diameter) as well as the Debye–Onsager (1927) limiting conductance law,[2]

$$\Lambda = \Lambda_0 - (\alpha\Lambda_0 + \beta)c^{1/2} \tag{6}$$

One could have hoped that at this stage all was unified and clearly understood: Strong electrolytes like KCl in water ($K_d = \infty$ and $\gamma = 1$) followed the Debye–Onsager law, whereas weak electrolytes like acetic acid in water

---

† This equation was first used, however, by Chapman in 1913 to solve the corresponding problem of one-dimensional ion distribution in double layers at charged interfaces.

(K_d small and $\gamma \ll 1$) followed the same law combined with the mass action law; examples are given in Figure 1. For the latter, if the square-root limiting law term could be neglected, the Ostwald dilution law held. But, again, the case of the supposedly strong electrolyte brought about new difficulties: Those

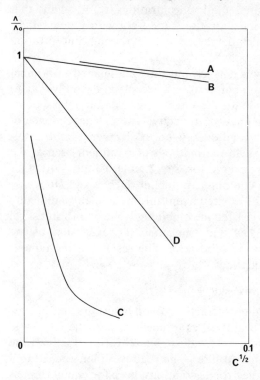

Figure 1. Typical behavior of conductance coefficients $f_\Lambda = \Lambda/\Lambda_0$ as a function of the square root of concentration: curve A is typical of ionophores in water (like KCl); curve B represents the limiting law in water; curve C indifferently represents ionogens (like CH_3COOH) in water or ionophores in H_2O–dioxane 80%; and curve D is the limiting law for KCl in H_2O–dioxane 80%.

electrolytes like sodium chloride, which were at that time duly recognized as intrinsically composed of pure ionic entities, behaved in some solvents like weak electrolytes. However, as time passed and knowledge increased, it was hard, or even impossible, to imagine that ions such as Na^+ and Cl^- could chemically react in any solvent to give neutral molecules of NaCl as CH_3COO^- and H^+ react to give CH_3COOH. Thus those ions that have the electronic structure of rare gases were correctly considered as being chemically inert to each other.

In 1926 Bjerrum[8] was again the first to propose a solution to the puzzling situation. He clearly saw that Debye's starting point, $\Delta\psi = \kappa^2\psi$, was a crude approximation that became less and less valid as coulombic interactions got larger, that is, for more highly charged ions or for lower dielectric constant solvents. However, the mathematical difficulty of proceeding from the original Poisson–Boltzmann equation was such that no calculation was possible at the time. Bjerrum proposed an elegant method of avoiding the calculation: He considered that short-range ion-pair configurations in dilute solutions do not interact with the rest of the ions of the solution; as a consequence, they behave thermodynamically as neutral molecules, *though no chemical bond is involved*.

If R is the cutoff distance, then $1 - \gamma$, the fraction of anions and cations that are within a distance r ($<R$) of each other, can be calculated by the mass action law describing their equilibrium with the remainder of the ions. This gives

$$\frac{1-\gamma}{\gamma^2 c f_{\pm}'^2} = K_A \tag{7}$$

where the activity coefficient $f_{\pm}'$ of the so-called free ions is given by the appropriately modified Debye and Hückel law:

$$\log f_{\pm}' = -\frac{\kappa q \gamma^{1/2}}{1 + \kappa R \gamma^{1/2}} \tag{8}$$

which allows an evaluation of the stoichiometric activity coefficient $f_{\pm}$ of the total electrolyte solute to be made,

$$\log f_{\pm} = \log f_{\pm}' + \log \gamma \tag{9}$$

The constant K_A is given by the simple relation

$$K_A = \frac{4\pi N}{1000} \int_0^R r^2 \exp\left(-\frac{U_{+-}}{kT}\right) dr \tag{10}$$

where U_{+-} is the isolated ion-pair potential energy. In the case of the restricted primitive model (a hard sphere of diameter a in a continuous dielectric solvent), which was the only one considered in the Debye theory, this constant becomes

$$K_A = \frac{4\pi N}{1000} \int_a^R r^2 \exp\left(\frac{2q}{r}\right) dr \tag{11}$$

Observing that the integrand is at a minimum for $r = q$, Bjerrum proposed this distance for R, since it minimizes the consequences of the introduced discontinuity. One then obtains the original Bjerrum association constant. When the solvent dielectric constant or the charge of the ions is such that $q < a$, one must use $R = a$, which leads directly back to the original Debye and Hückel law ($K_A = 0$, $\gamma = 1$).

Initially the Bjerrum concept was mainly tested on excess free energy data of $2:2$ salts in water. Conductance data were not used at that time as a field of application. The reason is probably that the Bjerrum theory originated in 1926 and the Debye–Onsager theory of conductance originated in 1927 and applies only to the limiting law; the Debye–Hückel law for the activity coefficients was more developed. Consequently, there appeared to be no utility in applying such a correction to a treatment that did not even take into consideration the size of the ions. One had to wait until 1953 for a conductance equation which introduced ion size at the same level of approximation as in the Debye–Hückel 1923 law for the activity coefficient. Unfortunately, at this time, the quantitative correction of Bjerrum was considered out-of-date. However, in 1938 Shedlovsky[9] proposed that the mass action law be introduced into the Debye–Onsager equation,

$$\Lambda = \gamma(\Lambda_0 - Sc^{1/2}\gamma^{1/2}) \tag{12}$$

with

$$\frac{1-\gamma}{\gamma^2 cf_{\pm}'^2} = K_A \tag{13}$$

where S is the Onsager limiting law coefficient. After some simple manipulations this set of equations can be rewritten as

$$\frac{1}{\Lambda S(z)} = \frac{1}{\Lambda_0} + \frac{c\Lambda f_{\pm}'^2 K_A}{\Lambda_0^2} S(z) \tag{14}$$

where $S(z)$ can be calculated from experimental data and an estimated value of Λ_0 through

$$S(z) = \{z/2 + [1 + (z/2)^2]^{1/2}\}^2 \tag{15}$$

with

$$z = (S/\Lambda_0^{3/2})(c\Lambda)^{1/2} \tag{16}$$

This constitutes a generalization of the Ostwald dilution law. It allows, through a series of iterations involving numerical calculations and plots of $1/\Lambda S(z)$ versus $c\Lambda f_{\pm}'^2 S(z)$, the parameters Λ_0 and K_A to be evaluated.

At a time when computers were not available to optimize the parameters by automatic least-squares adjustments the Shedlovsky procedure as well as some other variations of this method were quite useful in extracting the conductance parameters. As the theory of conductance developed by introducing higher-order terms beyond those for the limiting law, these graphical iterative methods became more and more complex. They culminated with the well-known Fuoss y–x plot, which allows the numerical evaluation of Λ_0, K_A, and J, where J is the coefficient of the linear concentration term in the conductance equation.[10] These calculations were unavoidable until in 1960 Kay proposed a program of least-squares fitting to be used with a computer. This was very welcome, since by that time experimenters had to devote much more time to processing their data than to make the experimental conductance measurements.

The next theoretical development in conductance, following the Onsager[2] (1927) approach, was achieved in 1953, when Pitts,[3] starting from the 1927 Onsager continuity equation and introducing a set of original boundary conditions, published his equation. Unfortunately, probably owing to a complex formulation, this equation was not tested with experimental data until the work by Stokes[11] in 1961.

In 1957 Fuoss and Onsager[4,12] derived the now well-known equation:

$$\Lambda = \Lambda_0 - Sc^{1/2} + Ec \log c + J(a)c + J_2(a)c^{3/2} \tag{17}$$

where the $J(a)$ function is at a level of approximation equivalent to that for activity coefficients in the original Debye–Hückel equation. This equation is also based on the original continuity equation of Onsager, to which are added boundary conditions that were given by Fuoss and Onsager in 1955 in a very compact formulation.[13] To reduce the drawbacks of the Debye approximation Fuoss and Kraus[14-16] proposed the following modification of the original equation after neglecting the $c^{3/2}$ contribution:

$$\Lambda = \gamma[\Lambda_0 - Sc^{1/2}\gamma^{1/2} + Ec\gamma \log c\gamma + J(a)c\gamma] \tag{18}$$

$$\frac{1-\gamma}{\gamma^2 cf_{\pm}^{\prime 2}} = K_A^F \tag{19}$$

This equation was quite similar to one of Bjerrum's formulations, except for two points: The $J(a)$ function was kept unchanged and the association constant became

$$K_A^F = \frac{4\pi Na^3}{3000} e^b \tag{20}$$

which was supposed to represent the behavior of paired ions upon contact.

In 1967 the present author[17,18] proposed the following equation, which is a mere transposition of the original Bjerrum modification of the Debye–Hückel activity coefficient law to the case of conductances:

$$\Lambda = \gamma[\Lambda_0 - Sc^{1/2}\gamma^{1/2} + Ec\gamma \log c\gamma + J(R)c\gamma + J_{3/2}(R)c^{3/2}\gamma^{3/2}] \tag{21}$$

where γ, K_A, R, and a are related to one another as displayed above in the description of the Bjerrum system of Eqs. (7–11). Again, when $q < a$, the conductance equation reduces to the original Fuoss–Onsager equation, just as Bjerrum's equation reduces to the original Debye–Hückel equation. This last system of equations shows in a simple way how conductance constitutes one way of measuring activity coefficients, since the quantities Λ and $f_{\pm}$ are related to each other by the same intermediary parameter K_A. This justification of using the Bjerrum formulation in a conductance equation was based mainly on the experimental observation that many of the inconsistencies that had accumulated in the literature were then explained.

There was also a strong theoretical reason: Since it was established that the Bjerrum formulation provides an excellent correction to the Debye–Hückel activity coefficient law, which was based on the Debye approximation, the same could be expected to be true for any other theoretical derivation based on the same approximation, as was indeed the case for the Fuoss–Onsager equation (as well as Pitt's equation).

However, given the initial opposition to these views, it seems appropriate to now briefly review and discuss the objection sometimes made to the Bjerrum concept in general and to its application to conductance in particular. Obviously this concept is an approximation grafted onto another approximation, that of Debye, whose limits of validity are known or can at least be estimated. It is interesting to inquire whether the whole approach is nothing but an error added to an error, with no gain, or whether the approximations of both treatments are correlated in such a way that their conjunction leads to an improvement.

1. It is sometimes argued that the value $R = q$ chosen for the cutoff distance has no real physical meaning[19] and that in reality nothing happens at $R = q$, which justifies the use of two radically different properties on each side of this boundary. Such a formulation of the criticism seems definitive. However, there is a better way to present this criticism which shows that an answer may be found: Are the mathematical approximations used on both sides of the boundary in such contradiction as to raise the question of an inconsistency? Outside the sphere $r = R$ the Debye mean potential is used and the corresponding Boltzmann exponential function linearized. Inside the sphere the mean potential is approximated by the direct potential but the Boltzmann exponential function is kept explicitly. Obviously it is better to use the two approximations in their respective domains rather than to use only one of them over the entire space. The result is indeed a discontinuity at $r = R$, but this is the price to be paid whenever two different approximations are used in such a way. One thus sees that this criticism constitutes no real objection. Moreover, it is easy to check that the Bjerrum formulas for $f_\pm$ and Λ are in no way sensitive to the choice $R = q$ without significantly changing the calculated values.

2. One sometimes finds as a criticism the fact that Bjerrum's association constant K_A is valid only for the restricted primitive model and that it cannot be applied to real electrolyte–solute systems. The answer in this case is easy, since the generalization of Eq. (10) to any Hamiltonian model is straightforward since all *specific* interactions occur at short distances.

3. More interesting is the following remark: Why should the fraction $1 - \gamma$ of the anions lose their electrolytic properties and no longer interact with the other ions? The answer is that at high dilutions, most of the nearest neighbors to a Bjerrum pair are at such distances from the pair that they interact with it as with a dipole, that is, in a negligible way. Since the average distance of ions is given with a good approximation by κ^{-1} at high dilutions, one may expect the Bjerrum concept to hold for concentrations such that $\kappa q < 0.5$ (or, even better, $\kappa q \gamma^{1/2} < 0.5$), which is the range of concentration commonly used (at least in conductance work).

4. Changing the functions $J(a)$ and $J_{3/2}(a)$ into $J(R)$ and $J_{3/2}(R)$ has been objected to by Fuoss[20] on arguments based on the short-range boundary condition, which states that for the hard-sphere model

$$\left\{ [\mathbf{v}_{ji}(\mathbf{r}) - \mathbf{v}_{ij}(-\mathbf{r})] \frac{\mathbf{r}}{r} \right\}_{r=a} = 0 \tag{22}$$

This condition takes into account the fact that for hard-sphere collisions the difference of the radial part of the two-particle mean-velocity vectors must vanish. It was then argued that changing a to R was equivalent to considering hard spheres of diameter R for the ions of the solutions. Though this argument seems very strong indeed, it does not, however, stand up to a careful analysis.

Hard-sphere collisions cause the radial part of the $\mathbf{v}_{ji}(\mathbf{r})$ vectors to vanish at $r = a$; but the converse is not true! Stating that the velocity vectors $\mathbf{v}_{ji}(\mathbf{r})$ (or their radial part) are null at any distance R simply means that at this distance this quantity is *unperturbed by the external field*, since it takes its equilibrium value, which is precisely zero. The real problem is then the following: In the cases of highly charged ions or in low dielectric constant solvents, do the radial parts of the two-particle velocity vectors $\mathbf{v}_{ji}(\mathbf{r})$ become negligible below a certain distance R?

A statistical treatment of hard-sphere collisions tells us that these have to vanish at $r = a$, but *for other reasons*, it happens that they already practically vanish at $r = R \sim q$ when $q > a$. The reason for this is in fact the answer to the following more fundamental question. Why do the short-range anion–cation pairs not also carry current when $q > a$? One way to answer this question is to introduce this property early enough in a fundamental treatment and to see whether or not it leads to theoretical evaluations in agreement with less restrictive treatments. This problem will be treated in some detail in the discussion of the Justice–Ebeling[21] strong coupling approximation $(\mathbf{v}_{ji}(\mathbf{r}) = 0$ for $\mathbf{r} < R \sim q)$. We shall see that it leads to practically the same analytical evaluations of the linear contribution of the relaxation field, which is the only contribution that can be calculated from the exact boundary condition. It presents, in addition, the advantage of avoiding mathematical difficulties inherent in the exact treatment. Moreover, one observes that once introduced into the exact echo formulation,[22] the strong coupling approximation becomes a very powerful tool that justifies the whole Bjerrum association mass action law at high dilutions $(\kappa q \gamma^{1/2} < 0.5)$, and not only its first linear concentration contribution.

In parallel to the work of Pitts[3] and Fuoss and Onsager,[4] the school of Falkenhagen has produced a series of contributions to the treatment of conductance that have been reviewed by Falkenhagen, Ebeling, and Hertz,[23] and Falkenhagen, Ebeling, and Kraeft[5] (FEK). The FEK work has many interesting features. It shows how the Onsager approach can be related to the fundamental equation of Liouville and how the boundary condition and the osmotic contribution to the relaxation field can be given very general and simple derivations. In addition, it does not make use of the Debye approximation, so that the evaluation of the $J(a)$ coefficient can be used to test how the Bjerrum approximation fits in the theory.

It should be noted that in 1962 Fuoss, Onsager, and Skinner[24] made the first attempt to avoid the Debye approximation. Unfortunately, the linear coefficient $J(a)$ had then not been discussed thoroughly enough by these authors. It was concluded that it corresponded to an association concept, but its identity with the Bjerrum result had been overlooked.

Other derivations were published by Murphy and Cohen,[25,26] who attacked the problem of unsymmetrical electrolytes without using the Debye approximation, Quint and Viallard,[27] who treated the general problem of mixtures of electrolytes of any kind, and recently by Chen,[28] who investigated the same problem without using the Debye approximation.

We should also mention here the work of Carman,[29] who tried to extend the calculation to higher concentrations, and that of Lee and Wheaton,[30,31] who tried to introduce a new model taking into account Gurney's solvation cospheres. Along the same lines, Fuoss gave several contributions[32-34] in order to try to avoid the restricted primitive model. A recent contribution by Ebeling *et al.*[35] should also be quoted. The problem of changing from the restricted primitive model to more refined ones, closer to reality, is important. Trying to introduce such models *ab initio* in the fundamental equations on which the treatment of conductance is based is an arduous task that, of course, necessitates avoiding the Debye approximation and also leads to mathematical difficulties. We shall see that, when combined with the echo effect, the strong coupling approximation shortcuts this mathematical difficulty, since the direct potential U_{ij} between ions of various kinds in a pure solvent can be formally accounted for in the final conductance equation without specifying any particular model.

2. The Basis of the Onsager Approach

2.1. The Fundamental Quantities

The Onsager theory of electrolyte solution conductance uses time-averaged quantities, the definition of which must be briefly summarized here so that the phenomenological content of each step in the derivation is clear.

The most important quantities are the distribution functions, which are the probabilities per unit volume of ionic species $k, l, \ldots$ of a given subset being at positions $R, S, \ldots$ when other ionic species $i, j, \ldots$ are eventually at positions $P, Q, \ldots$. These distribution functions are designated $g_{iP,jQ,\ldots}^{kR,lS,\ldots}$, where the superscripts designate the species (and their respective positions) described by the distribution function and the subscripts specify any restrictions imposed by the time averaging.

The physical meaning of the distribution functions is clear when one imagines the following Maxwell observer experiment (see Figure 2). Let us suppose that such a microscopic observer records in his notebook the history, for an interval of time τ, of various volume elements $dV_P, dV_Q, \ldots, dV_R, dV_S, \ldots$ centered around points $P, Q, \ldots, R, S, \ldots$ in the solution. By the history of a volume element we simply mean a record of time when an ion of type i enters or leaves the volume element dV_P, and similarly for ions of type j in dV_Q, or of type k in dV_R, and so on.

Such a record can be summarized by the diagram shown in Figure 2. From this diagram it is simple to evaluate the net time τ^{iP} an ion of species i is present in dV_P during the interval τ of the experiment. Let us define g^{iP} by

$$n_i g^{iP} dV_P = \frac{\tau^{iP}}{\tau} \tag{23}$$

as the "unrestricted" distribution function of the ion of type i at P, where n_i is the bulk concentration of the ion in the homogeneous solution. The functions g^{iQ}, g^{kR}, and g^{lS} are defined in a similar way. Obviously this definition leads to $g^{iP} = g^{iQ} = g^{kR} = g^{lS} = 1$.

The following definition will be more useful. Let $\tau^{iP,jQ}$ be the time during which an ion of type i and an ion of type j are simultaneously present in dV_P and dV_Q, respectively. One can now define three new quantities, the unrestricted two-particle distribution function $g^{iP,jQ}$ and the restricted two-particle distribution functions g_{jQ}^{iP} and g_{iP}^{jQ}, by

$$n_i n_j g^{iP,jQ} dV_P dV_Q = \frac{\tau^{iP,jQ}}{\tau} \tag{24}$$

$$n_i g_{jQ}^{iP} dV_P = \frac{\tau^{iP,jQ}}{\tau^{jQ}} \tag{25}$$

$$n_j g_{iP}^{jQ} dV_Q = \frac{\tau^{iP,jQ}}{\tau^{iP}} \tag{26}$$

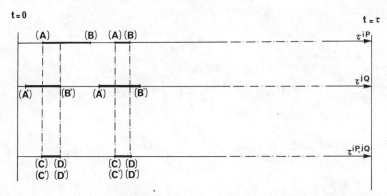

Figure 2. Records of the Maxwellian observer concerning events in volume elements during a given time length τ. The thick segments indicate the presence of an ion of type i in dV_P (first line) and that of an ion of type j in dV_Q (second line), and allow an evaluation of τ^{iP} and τ^{jQ}. The third line is derived from the first two lines to evaluate the time of simultaneous presence $\tau^{iP,jQ}$. (A) and (B) denote the coordinates of ion i at each change of situation with respect to the volume element dV_P for ion i in space P, and similarly for (A') and (B') and ion j; (C)-(D) and (C')-(D') are the coordinates of ion i in space P and of ion j in space Q, respectively, at the beginning and the end of their simultaneous presence. The generalization to higher-order levels of simultaneity, i.e., involving more than two ions, is straightforward.

It follows that for homogeneous solutions

$$g_{jQ}^{iP} = \frac{g^{iP,jQ}}{g^{jQ}} = g^{iP,jQ} = \frac{g^{iP,jQ}}{g^{iP}} = g_{iP}^{jQ} \tag{27}$$

which is an important property of the two-particle distribution functions.

It is now a simple exercise to define the various quantities $g_{kR,ls}^{iP,jQ}$ and their relations to one another. Let us simply mention here as typical examples

$$g_{iP,jQ}^{kR} = \frac{g^{kR,iP,jQ}}{g^{iP,jQ}} \tag{28}$$

$$g_{iP,jQ,kR}^{ls} = \frac{g^{ls,iP,jQ,kR}}{g^{iP,jQ,kR}} \tag{29}$$

which will prove to be useful later. This notation is a generalization of that first used by Debye and Falkenhagen.[36] The advantage of this notation lies in the fact that it is immediately obvious from the subscripts and superscripts what conditions apply to the distribution functions.

The distribution functions are of primary importance in the equations of the Born, Bogoliubov, Green, Kirkwood, and Yvon (BBGKY) hierarchy, which are equivalent to the Liouville equation, which is the basis of the statistical-mechanical study of isolated systems of particles at equilibrium. The BBGKY set of equations constitutes a recursion system. We shall give here only the first three equations as an example from which the recursion relation is obvious:

$$\nabla_P g^{iP} + \frac{1}{kT} \sum_k n_k \int \nabla_P U(iP, kR) g^{iP,kR} \, dV_R = 0 \tag{30}$$

$$\nabla_P g^{iP,jQ} + \frac{1}{kT} g^{iP,jQ} \nabla_P U(iP, jQ) + \frac{1}{kT} \sum_k n_k \int \nabla_P U(iP, kR) g^{kR,iP,jQ} \, dV_R = 0 \tag{31}$$

$$\nabla_P g^{iP,jQ,lS} + \frac{1}{kT} g^{iP,jQ,lS} \nabla_P U(iP, jQ) + \frac{1}{kT} g^{iP,jQ,lS} \nabla_P U(iP, lS)$$

$$+ \frac{1}{kT} \sum_k n_k \int \nabla_P U(iP, kR) g^{kR,iP,jQ,lS} \, dV_R = 0 \tag{32}$$

where $U(iP, jQ)$ is the potential energy of direct interaction between the two particles i and j located at P and Q, respectively.

Making use of the properties of the distribution functions derived above, the original BBGKY set of equations can easily be transformed into a new formulation, the physical meaning of which will become clear by dividing each term by the appropriate distribution function multiplied by $-kT$. It then follows that

$$-kT\nabla_P \ln g^{iP} - \sum_k n_k \int \nabla_P U(iP, kR) g_{iP}^{kR} \, dV_R \equiv \mathbf{F}^{iP} \tag{33}$$

$$-kT\nabla_P \ln g_{jQ}^{iP} - \nabla_P U(iP, jQ) - \sum_k n_k \int \nabla_P U(iP, kR) g_{iP,jQ}^{kR} \, dV_R \equiv \mathbf{F}_{jQ}^{iP} \tag{34}$$

$$-kT\nabla_P \ln g_{jQ,lS}^{iP} - \nabla_P U(iP, jQ) - \nabla_P U(iP, lS)$$

$$- \sum_k n_k \int \nabla_P U(iP, kR) g_{iP,jQ,lS}^{kR} \, dV_R \equiv \mathbf{F}_{jQ,lS}^{iP} \tag{35}$$

where for the system at equilibrium, to which the BBGKY hierarchy applies, the right-hand sides of Eqs. (33)–(35) are identically zero. However, for a system that is not at equilibrium, the right-hand sides are no longer zero and Eqs. (33)–(35) are the definitions of $\mathbf{F}^{iP}$, and so on. To understand the meaning of the $\mathbf{F}^{iP}$ terms we note that each term of each equation now has the dimension of a force, so that each equation constitutes the definition of an internal force acting on an ion i at P, which is zero for an isolated system of particles at equilibrium. The logarithmic terms can be identified with Einstein's fictive diffusion forces and the following terms with the direct forces acting between the particles whose presence at given locations is considered in the time averaging. The last term is the force resulting from all the other particles for which no spatial restrictions were imposed during the time averaging. The BBGKY hierarchy is, in fact, a hierarchy for internal forces that will prove to be very important in the development of the basic equations of the Onsager approach to conductance.

We have introduced the hierarchies of distribution functions and internal forces that are obviously necessary to handle the problems of ionic solutions. However, one now needs to define the time average velocities of ions, since conductance is directly related to the velocity that is imparted to the ions by the various forces (internal and external) acting on them.

In order to define such velocity quantities we shall again call on the microscopic Maxwell observer experiment introduced above. The observer will now be asked to record more information. Defining the volume elements by cubes around each point concerned, we shall require him to list the coordinates of the points A and B (Figure 3) where ions enter and leave each cubic element. This enables us to define the average velocity vectors of an ion i during the time τ^{iP} it remains in the immediate vicinity of point P; thus

$$\mathbf{v}^{iP} = \sum (\mathbf{AB})/\tau^{iP} \tag{36}$$

where the summation is carried out over all the occurrences of an ion i in P during the time τ and $\mathbf{AB}$ is the vector from A to B.

Similarly, we can define the time averaging restricted velocity vector as

$$\mathbf{v}_{jQ}^{iP} = \sum (\mathbf{CD})/\tau^{iP,jQ} \tag{37}$$

Figure 3. Example of one observation of Maxwell's demon concerning
the various positions of an ion of
type i that crosses the volume element dV_P in the solution. A and B
are respectively the entrance and
exit points located on the surface of
the cube. C and D are the extreme
points of the trajectory of the ion i
in (or on) the cube within which the
presence of an ion of type j was
observed in the volume element
dV_Q. If this is the event labeled α
during the observation time τ, the
mean-velocity vectors introduced
are easily calculable from the spatial
and time data recorded by the Maxwellian observer.

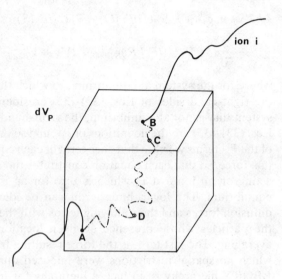

$$\mathbf{v}^{iP} = \sum_{\alpha} \mathbf{A}_{\alpha}\mathbf{B}_{\alpha} \Big/ \sum_{\alpha} (t_{B\alpha} - t_{A\alpha})$$

$$= \sum_{\alpha} (\mathbf{AB})_{\alpha} \Big/ \tau^{iP}$$

$$\mathbf{v}_{jQ}^{iP} = \sum_{\alpha} \mathbf{C}_{\alpha}\mathbf{D}_{\alpha} \Big/ \sum_{\alpha} (t_{D\alpha} - t_{C\alpha})$$

$$= \sum_{\alpha} (\mathbf{CD})_{\alpha} \Big/ \tau^{iP,jQ}$$

where C and D are now the locations of the ion i in the volume element
dV_p at the beginning and end of each occurrence of the simultaneous presence
of an ion i in dV_p and an ion j in dV_Q.

Again the summation is carried out over all such occurrences in the time
τ. The quantity $\mathbf{v}_{jQ}^{iP}$ is now the time average velocity vector of an ion of type
i in the vicinity of P when an ion of type j is simultaneously in the vicinity
of Q. At equilibrium these vector quantities are zero, but this is no longer
true when the system is perturbed, as in the case of conductance, where an
external electrical field $\mathbf{X}$ is imposed.

In order to specify that there is no accumulation or depletion of ions of
type i around P one uses the mass conservation relation

$$\nabla_p g^{iP} \mathbf{v}^{iP} = -\frac{\partial g^{iP}}{\partial t} = 0 \tag{38}$$

Such an expression relates the distribution functions to the velocity vectors.
For the case of single-particle quantities this relation is trivial and of no use
for the conductance problems. More important is the corresponding expression
at the level of the two-particle quantities. The position of an ion i at point P
and of an ion j at point Q can be represented by a simple point in the
six-dimensional space $P + Q$. The velocity vector of this point in the space

$P + Q$ is just the sum of its projections in the spaces P and Q. This is expressed as

$$\mathbf{v}^{iP,jQ} \equiv \mathbf{v}^{iP}_{jQ} + \mathbf{v}^{jQ}_{iP} \tag{39}$$

If one again expresses the mass evolution in this space it follows that

$$-\frac{\partial g^{iP,jQ}}{\partial t} = \nabla_{P+Q} g^{iP,jQ} \mathbf{v}^{iP,jQ} = -\frac{\partial g^{iP}_{jQ}}{\partial t} = -\frac{\partial g^{jQ}_{iP}}{\partial t} \tag{40}$$

where the divergence operator ∇_{P+Q} functions in the six dimensions of $P + Q$. Given the definition of $\mathbf{v}^{iP,jQ}$ and the properties for $g^{iP,jQ}$ above, it follows that

$$-\frac{\partial}{\partial t} g^{iP}_{jQ} = \nabla_P g^{iP}_{jQ} \mathbf{v}^{iP}_{jQ} + \nabla_Q g^{jQ}_{iP} \mathbf{v}^{jQ}_{iP} = -\frac{\partial}{\partial t} g^{jQ}_{iP} \tag{41}$$

This very important relation, commonly referred to as the continuity equation, permits the pair distribution functions to be evaluated and is the basis of the Onsager approach.

It would be easy to define a hierarchy of mass conservation relations by considering higher-order hyperspaces $P + Q + R$; however, we shall see later that approximations can be introduced that make this unnecessary.

We now have at our disposal the fundamental quantities and relations with which it is possible to handle the general problem of irreversible transport processes, in which the usual low-field, steady-state conductance is just a special case!

2.2. Relations between the Fundamental Quantities

Under the action of a uniform electrical field $\mathbf{X}$ an ion of type j with charge e_j immersed in a viscous fluid will move in the direction of the field, with a limiting velocity characterized by the vector $\mathbf{v}^j$. As a consequence, the observed current density is

$$\mathbf{i} = \sum_{j=1}^{s} n_j e_j \mathbf{v}^j \tag{42}$$

for a solution containing s types of ions.

The ionic molar conductance contribution λ^j of the species j is defined by the scalar quantity

$$\lambda^i = Z_j \mathscr{F} \frac{\mathbf{X}}{X} \frac{\mathbf{v}^j}{X} \tag{43}$$

where Z_j is the algebraic valence of the ions j and $\mathscr{F}$ is the Faraday constant. This, in turn, leads to the conductivity χ of the solution:

$$\chi = 10^{-3} \sum_{j=1}^{s} c_j \lambda^j \tag{44}$$

where c_j is now the molar concentration of the ions of species j.

As a result, any practical treatment of conductance must have an analytical expression that gives the quantities λ^j as a function of the various parameters which characterize both the ionic solute and the solvent. Equation (43) shows us that this problem reduces to the evaluation of $\mathbf{v}^j$, which identifies with the one-particle mean-velocity vector $\mathbf{v}^{jQ}$ defined in the previous section.

The Onsager approach is based on the assumption that the limiting velocity imparted to ions is the result of three forces:

1. The external force $e_j\mathbf{X}$.
2. The average interionic forces $\mathbf{F}^{jQ}$ defined by Eq. (33).
3. The force transmitted by the solvent, which is itself subject to a velocity field through the momentum transmitted to it by the moving ions.

The Onsager approach is thus based on the following phenomenological relation:

$$\mathbf{v}^{jQ} = \mathbf{v}_e^{jQ} + \omega_j(e_j\mathbf{X} + \mathbf{F}^{jQ}) \tag{45}$$

where $\mathbf{v}_e^{jQ}$ is called the electrophoretic velocity vector. In the previous section we have seen the relation between $\mathbf{F}^{jQ}$ and the ion-pair distribution function [Eq. (32)]. The velocity vector $\mathbf{v}_e^{jQ}$ is also related to the same quantity through the Oseen relation

$$\mathbf{v}_e^{jQ} = \sum_k n_k \int \mathbf{\chi}(Q, R, \omega_j, \omega_k)(e_k\mathbf{X} + \mathbf{F}_{jQ}^{kR})g_{jQ}^{kR}\,dV_R \tag{46}$$

where $\mathbf{\chi}(Q, R, \omega_j, \omega_k)$ is an electrophoretic coupling tensor that is a function of the viscosity of the supporting medium, of the distance $|\mathbf{r}_R - \mathbf{r}_Q|$, and of the friction coefficient of the ion concerned in dV_R.

The hydrodynamic coupling tensor $\mathbf{\chi}(Q, R, \omega_j, \omega_k)$ is an operator that transforms the force transmitted by an ion of type k to the fluid at point R into a velocity vector belonging to an ion of type j located at a point Q. The summation over k gives the total force transmitted to the element of fluid dV_R and the integral over the whole space leads to the total electrophoretic contribution $\mathbf{v}_e^{iP}$.

The coupling hydrodynamic tensor is given by

$$\mathbf{\chi}(Q, R, \omega_j, \omega_k) = \mathbf{\lambda}(Q, R) + \mathbf{\lambda}'(Q, R, \omega_j, \omega_k) \tag{47}$$

$$\mathbf{\lambda}(Q, R) = (1/8\pi\eta r)[\mathbf{\delta} + (\mathbf{rr}/r^2)] \tag{48}$$

$$\mathbf{\lambda}'(Q, R, \omega_j, \omega_k) = (1/8\pi\eta r^3)(\omega_j^{-2} + \omega_k^{-2})(\mathbf{\delta}/3 - \mathbf{rr}/r^2) \tag{49}$$

where $\mathbf{r} = \mathbf{r}_R - \mathbf{r}_Q$ and $\mathbf{\delta}$ is the unit tensor. These equations result from recent progress in the theory of hydrodynamics.[37] Let us note that when the short-range tensor $\mathbf{\lambda}'$ is neglected, as in most cases so far, the electrophoretic expressions are greatly simplified.

The theoretical problem of conductance thus reduces in the last analysis to the evaluation of distribution functions, since both $\mathbf{F}^{jQ}$ and $\mathbf{F}_{jQ}^{kR}$ also depend on them. The pair distribution functions can be evaluated by solving the mass evolution equation [Eq. (41)], which calls for the use of the two-particle restricted velocity vectors $\mathbf{v}_{iP}^{jQ}$ and $\mathbf{v}_{jQ}^{iP}$. The two-particle velocity vector $\mathbf{v}_{iP}^{jQ}$ is given by the second equation of the velocity hierarchy, which proceeds from Eq. (45):

$$\mathbf{v}_{iP}^{jQ} = \mathbf{v}_{e_iP}^{jQ} + \omega_j(e_j\mathbf{X} + \mathbf{F}_{iP}^{jQ}) \tag{50}$$

where, in turn, $\mathbf{v}_{e_iP}^{jQ}$ is the second equation of the hierarchy of electrophoretic two-particle velocity relations that proceed from Eq. (46),

$$\mathbf{v}_{e_iP}^{jQ} = \sum_k n_k \int \boldsymbol{\chi}(Q, R, \omega_j, \omega_k)[(e_k\mathbf{X} + \mathbf{F}_{jQ,iP}^{kR})]g_{iP,jQ}^{kR} \, dV_R$$

$$+ \boldsymbol{\chi}(P, Q, \omega_i, \omega_j)(e_i\mathbf{X} + \mathbf{F}_{jQ}^{iP}) \tag{51}$$

The first term of the right-hand side of Eq. (51) is the straightforward generalization of the right-hand side of Eq. (45). The second term is the additional contribution due to the presence of one ion of type i at P, which transmits to the solvent at P the force $e_i\mathbf{X} + \mathbf{F}_{jQ}^{iP}$.

After substituting the equations of the BBGKY hierarchy for the mean forces $\mathbf{F}_{iP}^{jQ}$ and $\mathbf{F}_{jQ,iP}^{kR}$ into Eqs. (51) and (50) and using this result in the two-particle mass evolution equation [Eq. (41)], one obtains a continuity equation (52) that refers explicitly to no other unknown quantities but two-, three-, and four-particle distribution functions. This complicated equation, given in Table 1, is identical to the result obtained by Ebeling from considerations starting at the level of the Liouville equation. This continuity equation will permit the problem of evaluating pair-distribution functions to be attacked (see Section 2.4) and solved once certain approximations are introduced. Table 1 summarizes the various sets of hierarchy equations, the combinations of which lead to the Ebeling equation, which is the basis of the treatment of conductance described in the present section.

2-3. Outline of the Original Derivation of Ebeling[38]

For a mixture of M particles (N solute molecules and N_s solvent molecules) let $P_M(\mathbf{r}_1, \mathbf{r}_2, \ldots \mathbf{r}_M, \mathbf{v}_1, \ldots, \mathbf{v}_M, t)$ be the general distribution function in the $6M$-dimensional space phase, where $\mathbf{r}_k$ and $\mathbf{v}_k$ are the location and velocity vectors of the kth particle. Integrating over all velocities and all solvent molecule positions leads to the N-solute-particle space distribution function D_N:

$$D_N(\mathbf{r}_1, \mathbf{r}_2, \ldots, \mathbf{r}_N, t) = P_M \, d\mathbf{r}_{N+1} \cdots d\mathbf{r}_M \, d\mathbf{v}_1 \cdots d\mathbf{v}_M \tag{53}$$

Table 1
Summary of the Four Basic Hierarchical Sets of Equations in the Onsager Treatment of Conductance

1. $\mathbf{v}^{iP} = \mathbf{v}_e^{iP} + \omega_i(e_i\mathbf{X} + \mathbf{F}^{iP})$ (44)

   $\mathbf{v}_{jQ}^{iP} = \mathbf{v}_{e_{jQ}}^{iP} + \omega_i(e_i\mathbf{X} + \mathbf{F}_{jQ}^{iP})$ (49)

2. $\mathbf{F}^{iP} = -kT\nabla_P \ln g^{iP} - \sum_l n_l \int \nabla_P U(iP, lS) g_{iP}^{lS}\, dV_S$ (32)

   $\mathbf{F}_{jQ}^{iP} = -kT\nabla_P \ln g_{jQ}^{iP} - \nabla_P U(iP, jQ) - \sum_l n_l \int \nabla_P U(iP, lS) g_{iP,jQ}^{lS}\, dV_S$ (33)

   $\mathbf{F}_{jQ,kR}^{iP} = -kT\nabla_P \ln g_{jQ,kR}^{iP} - \nabla_P U(iP, jQ) - \nabla_P U(iP, kR) - \sum_l n_l$

   $\times \int \nabla_P U(iP, lS) g_{iP,jQ,kR}^{lS}\, dV_S$ (34)

3. $\mathbf{v}_e^{iP} = \sum_l n_l \int \boldsymbol{\chi}(P, S, \omega_i, \omega_l)(e_i\mathbf{X} + \mathbf{F}_{iP}^{lS}) g_{iP}^{lS}\, dV_S$ (50)

   $\mathbf{v}_{e_{jQ}}^{iP} = \sum_l n_l \int \boldsymbol{\chi}(P, S, \omega_i, \omega_l)(e_i\mathbf{X} + \mathbf{F}_{iP,jQ}^{lS}) g_{iP,jQ}^{lS}\, dV_S + \boldsymbol{\chi}(P, Q, \omega_i, \omega_j)(e_j\mathbf{X} + \mathbf{F}_{iP}^{jQ})$ (51)

4. $\dfrac{\partial}{\partial t} g^{iP,jQ} = \dfrac{\partial}{\partial t} g_{iP}^{jQ} = \dfrac{\partial}{\partial t} g_{jQ}^{iP} = -\nabla_{P+Q} g^{iP,jQ} \mathbf{v}^{iP,jQ}$

   $= -(\nabla_P g_{jQ}^{iP} \mathbf{v}_{jQ}^{iP} + \nabla_Q g_{iP}^{jQ} \mathbf{v}_{iP}^{jQ})$

   $= \nabla_P \Big\{ \omega_i \Big(kT\nabla_P \ln g_{jQ}^{iP} + \nabla_P U(iP, jQ) + \sum_k n_{k'} \int [\nabla_P U(iP, kR)] g_{iPjQ}^{kR}\, dV_R - e_i\mathbf{X} \Big)$

   $+ \boldsymbol{\chi}(P, Q, \omega_j, \omega_i) \Big(kT\nabla_Q \ln g_{jQ}^{iP} + \nabla_Q U(iP, jQ)$

   $+ \sum_k n_k \int [\nabla_Q U(jQ, kR)] g_{iP,jQ}^{kR}\, dV_R - e_j\mathbf{X} \Big)$

   $+ \sum_k n_k \int \boldsymbol{\chi}(P, R, \omega_i, \omega_k) \Big(kT\nabla_R g_{iP,jQ}^{kR} + \nabla_R U(iP, kR) + \nabla_R U(jQ, kR)$

   $+ \sum_l n_l \int \nabla_R U(lS, kR) g_{iP,jQ,kR}^{lS}\, dV_S - e_k\mathbf{X} \Big) g_{iP,jQ}^{kR}\, dV_R \Big\} g_{jQ}^{iP} + \begin{Bmatrix} i \leftrightarrow j \\ P \leftrightarrow Q \end{Bmatrix}$ (52)

Let $\mathbf{v}_i^{(N)}$ be the mean-velocity vector of solute particle i when all other solute particles are assigned specified positions. Then

$$\mathbf{v}_i^{(N)} = \frac{\int v_i P_M\, d\mathbf{r}_{N+1} \cdots d\mathbf{r}_M\, d\mathbf{v}_1 \cdots d\mathbf{v}_M}{D_N}$$ (54)

which allows an evaluation of the quantity $D_N \mathbf{v}_i^{(N)}$. The physical meaning of $\mathbf{v}_i^{(N)}$ is identical to that of the mean velocity $\mathbf{v}_{jQ,kR,...}^{iP}$ defined in Section 2.1, with all ions but ion i explicitly mentioned in the subscript index. One may define a less restrictive mean-velocity vector for the ion i by the equation

$$\mathbf{v}_i^{(K)} = \frac{\int \mathbf{v}_i P_M \, d\mathbf{r}_{K+1} \cdots d\mathbf{r}_N \cdots d\mathbf{r}_M \cdots d\mathbf{v}_1 \cdots d\mathbf{v}_M}{D_K} \tag{55}$$

where $K < N$.

The following relation is readily derived:

$$D_K \mathbf{v}_i^{(K)} = V^K \int D_N \mathbf{v}_i^{(N)} \, d\mathbf{r}_{K+1} \cdots d\mathbf{r}_N \tag{56}$$

where V is the volume over which the spatial integrations are carried out.

Let us note the correspondence between the D_K K-particle function defined here and the unconditional K-particle distribution function $g^{(K)} = g^{1P,2Q,...,KR}$ defined in Section 2.1:

$$g^{(K)} = V^K D_K \tag{57}$$

By integrating the classical Liouville equation for the $N + N_s$ particle velocities and for the N_s solvent molecule positions, Ebeling derived the following exact relation:

$$\frac{\partial}{\partial t} D_N(\mathbf{r}_1, \ldots, \mathbf{r}_N, t) = - \sum_{i=1}^{N} \nabla_i (D_N \mathbf{v}_i^{(N)}) \tag{58}$$

Then, after introducing as an approximation the assumption that the external perturbation mainly affects the position of the ions and not so much their velocity, and affects neither the position nor the velocity of the solvent molecules that therefore have velocities which remain near their equilibrium values, Ebeling obtained the following result:

$$D_N \mathbf{v}_i^{(N)} = \sum_{j=1}^{N} [\omega_j \delta_{ij} \boldsymbol{\delta} + (1 - \delta_{ij}) \boldsymbol{\chi}_{ij}] \left(e_j \mathbf{X} D_N - D_N \sum_{s \neq j} U_{js} - kT \nabla_j D_N \right) \tag{59}$$

It is now remarkable to observe that Eqs. (58) and (59) are strictly identical to the whole set of the various hierarchy equations given in Table 1 and that were obtained by simple phenomenological considerations. Indeed it is easy to check that

$$\mathbf{v}_i^{(1)} \equiv \mathbf{v}^{iP}/V \tag{60}$$

just by integrating Eq. (56) over $d\mathbf{r}_2 \cdots d\mathbf{r}_N$. It is also a matter of elementary algebra, though more tedious, to show that

$$D_2 \mathbf{v}_i^{(2)} \equiv g^{iP,jQ} \mathbf{v}_{jQ}^{iP}/V^2 \tag{61}$$

by integrating Eq. (56) over $d\mathbf{r}_3 \cdots d\mathbf{r}_N$. This last result introduced in Eq. (58) leads directly to

$$\frac{\partial}{\partial t} D_2(\mathbf{r}_1, \mathbf{r}_2, t) = -\sum_{i=1}^{2} \nabla_i (D_2 \mathbf{v}_i^{(2)})$$

$$= -(\nabla_P g_{jQ}^{iP} \mathbf{v}_{jQ}^{iP} + \nabla_Q g_{iP}^{jQ} \mathbf{v}_{iP}^{jQ})/V^2$$

$$= \frac{\partial}{\partial t} g_{jQ}^{iP}/V^2 \tag{62}$$

which is the second equation of the mass evolution hierarchy given in Eq. (52) of Table 1.

These results of Ebeling are very important because they give us the link to the fundamental basis of statistical mechanics (the Liouville equation). Now the set of equations given in Table 1, which are a generalization of the Onsager diffusion-like approach to conductance, are established within the framework of a well-defined approximation and no longer belong to the domain of uncontrolled empiricism.

The Ebeling derivation that leads from the Liouville equation for the N ions plus N_s solvent molecules mixture to Eqs. (59) and (62) constitutes, in fact, a display of the implicit approximations that are made when dealing with a continuum model for the solvent. This remark may eventually help in the future to find some way to bring corrections to the present continuum treatment.

This was an essential condition that had to be met before more progress could be expected from further studies by specialists in the field of statistical mechanics. However, mathematical difficulties are such that further approximations will be necessary before solving the integral-differential problem at the present level of Table 1.

2.4. Practical Reformulation of the Continuity Equation

In the original form given by Eq. (52) of Table 1, the Ebeling equation is not directly integrable for several reasons: First, it is not a mathematically closed system, since it contains explicitly higher-order distribution functions up to the fourth order, which in turn implies that all higher-order equations are needed. We shall see in a later section how to circumvent this difficulty by introducing *ad hoc* approximations. Secondly, Eq. (52) is essentially an equation that relates two quantities, g_{jQ}^{iP} and g_{iP}^{jQ}. Given the properties derived in Section 2.1 concerning these two distribution functions, the equation can be reduced formally by setting

$$g_{ji}(\mathbf{r}) = g_{jQ}^{iP} = g_{iP}^{jQ} = g_{ij}(-\mathbf{r}) \tag{63}$$

with

$$\mathbf{r} = \mathbf{r}_P - \mathbf{r}_Q \tag{64}$$

This allows the initial equation to be transformed into a continuity equation for the quantity of $g(\mathbf{r})$ after noting, by use of the change of variable given above, that

$$\nabla_P = -\nabla_Q = \nabla$$

After substituting these properties in Eq. (41),

$$-\frac{\partial}{\partial t} g_{ji}(\mathbf{r}) = \nabla g_{ji}(\mathbf{r})[\mathbf{v}_{ji}(\mathbf{r}) - \mathbf{v}_{ij}(-\mathbf{r})] = -\frac{\partial}{\partial t} g_{ij}(-\mathbf{r}) \qquad (65)$$

with

$$\mathbf{v}_{ji}(\mathbf{r}) = \mathbf{v}_{jQ}^{iP} \quad \text{and} \quad \mathbf{v}_{ij}(-\mathbf{r}) = \mathbf{v}_{iP}^{jQ}$$

Before proceeding further in the mathematical transformation for $\mathbf{v}_{ji}(\mathbf{r})$ and $\mathbf{v}_{ij}(-\mathbf{r})$ in Eq. (65), we shall distinguish in each quantity special terms that explicitly contain one-particle direct forces, that is, the external force $e_k\mathbf{X}$, the relaxation force $\mathbf{F}^{kR}$, and the electrophoretic force $\mathbf{v}_e^{kR}/\omega_k$. These terms will be called "monitoring" terms. In order to achieve this goal it is necessary to recall an obvious property of conditional mean quantities that can be expressed as follows: Any quantity such as ζ_{jQ}^{iP} can be expressed as

$$\zeta_{jQ}^{iP} = \zeta^{iP} + \varepsilon_{jQ}^{iP}(\mathbf{r}) \qquad (66)$$

where ζ^{iP} is a constant in the case of homogeneous solutions and $\varepsilon_{jQ}^{iP}(\mathbf{r})$ is an excess quantity that vanishes as $|\mathbf{r}|$ goes to infinity. The application of this property is well known for pair distribution functions that can be written

$$g_{jQ}^{iP} = g^{iP} + h_{jQ}^{iP} \qquad (67)$$

where g^{iP} is a constant normalized to unity and h_{jQ}^{iP} is the correlation function, which obviously vanishes when the two points P and Q are infinitely far apart.

More interesting for the purpose of our goal is the application of this idea to the two-particle mean force. In fact, Eq. (34) can be rewritten as

$$\mathbf{F}_{jQ}^{iP} = \mathbf{F}^{iP} + \mathbf{f}_{jQ}^{iP} - kT\nabla_P \ln g_{jQ}^{iP} \qquad (68)$$

with

$$\mathbf{f}_{jQ}^{iP} = -\nabla_P U(iP, jQ) - \sum_k n_k \int \nabla_P U(iP, kR)(g_{iP,jQ}^{kR} - g_{iP}^{kR}) \, dV_R \qquad (69)$$

where both $\mathbf{f}_{jQ}^{iP}$ and $-kT\nabla_P \ln g_{jQ}^{iP}$ vanish as $r \to \infty$.

A similar expression may be obtained for the three-particle conditional forces, which may be rewritten as

$$\mathbf{F}_{iP,jQ}^{kR} = \mathbf{F}^{kR} + \mathbf{f}_{iP,jQ}^{kR} - kT\nabla_R \ln g_{iP,jQ}^{kR} \qquad (70)$$

with

$$\mathbf{f}_{iP,jQ}^{kR} = -\nabla_R U(kR, iP) - \nabla_R U(kR, jQ) - \sum_l n_l \int \nabla_R U(kR, lS)$$
$$\times (g_{iP,jQ,kR}^{ls} - g_{kR}^{ls}) \, dV_S \tag{71}$$

The same transformation can be applied to the two-particle velocity, followed by an extraction of all constant force contributions in the excess velocity terms.

For the electrophoretic velocity the result is

$$\mathbf{v}_{e_{jQ}}^{iP} = \mathbf{v}_e^{iP} + \boldsymbol{\chi}(P, Q, \omega_i, \omega_j)(e_j \mathbf{X} + \mathbf{F}^{jQ})$$

$$+ \sum_k n_k \int \boldsymbol{\chi}(P, R, \omega_i, \omega_k)(e_k \mathbf{X} + \mathbf{F}^{kR})(g_{iP,jQ}^{kR} - g_{iP}^{kR}) \, dV_R$$

$$+ \boldsymbol{\chi}(P, Q, \omega_i, \omega_j)(\mathbf{F}_{iP}^{jQ} - \mathbf{F}^{jQ})$$

$$+ \sum_k n_k \int \boldsymbol{\chi}(P, R, \omega_i, \omega_k)$$

$$\times [(\mathbf{F}_{iP}^{kr} - \mathbf{F}^{kR})(g_{iP,jQ}^{kR} - g_{iP}^{kR}) + (\mathbf{F}_{iP,jQ}^{kR} - \mathbf{F}_{iP}^{kR})g_{iP,jQ}^{kR}] \, dV_R \tag{72}$$

Consequently the two-particle mean velocity $\mathbf{v}_{jQ}^{iP}$ can similarly be expressed as

$$\mathbf{v}_{jQ}^{iP} = \mathbf{v}^{iP} + (\mathbf{v}_{e_{jQ}}^{iP} - \mathbf{v}_e^{iP}) + \omega_i(\mathbf{f}_{jQ}^{iP} - kT\nabla_P \ln g_{jQ}^{iP}) \tag{73}$$

where $\mathbf{v}_{e_{jQ}}^{iP} - \mathbf{v}_e^{iP}$ is the excess quantity of Eq. (72) in which the last two excess terms can be rearranged following Eqs. (68) and (70) as

$$\mathbf{v}_{e_{jQ}}^{iP} - \mathbf{v}_e^{iP} = \boldsymbol{\chi}(P, Q, \omega_i, \omega_j)(e_j \mathbf{X} + \mathbf{F}^{jQ})$$

$$+ \sum_k n_k \int \boldsymbol{\chi}(P, R, \omega_i, \omega_k)(e_k \mathbf{X} + \mathbf{F}^{kR})(g_{iP,jQ}^{kR} - g_{iP}^{kR}) \, dV_R$$

$$+ \boldsymbol{\chi}(P, Q, \omega_i, \omega_j)[\mathbf{f}_{iP}^{jQ} + kT\nabla \ln g_{ji}(\mathbf{r})]$$

$$+ \sum_k n_k \int \boldsymbol{\chi}(P, R, \omega_i, \omega_k)(\mathbf{f}_{iP,jQ}^{kR}g_{iP,jQ}^{kR} - kT\nabla_R g_{iP,jQ}^{kR}$$

$$- \mathbf{f}_{iP}^{kR}g_{iP}^{kR} + kT\nabla_R g_{iP}^{kR}) \, dV_R \tag{74}$$

The continuity equation can thus be rewritten as

$$\frac{1}{kT(\omega_i + \omega_j)} \frac{\partial}{\partial t} g_{ji}(\mathbf{r}) = \nabla g_{ji}(\mathbf{r}) \left(\nabla \ln g_{ji}(\mathbf{r}) - \frac{\mathbf{M}_1 + \mathbf{f}_{ji}(\mathbf{r}) + \mathbf{M}_h(\mathbf{r}) + \mathbf{H}_{ji}(\mathbf{r})}{kT} \right) \tag{75}$$

where the terms $\mathbf{M}_1$, $\mathbf{f}_{ji}(\mathbf{r})$, $\mathbf{M}_h(\mathbf{r})$, and $\mathbf{H}_{ji}(\mathbf{r})$ are given by

$$\mathbf{M}_1 = (\mathbf{v}^{iP} - \mathbf{v}^{jQ})/(\omega_i + \omega_j) \tag{76}$$

$$\mathbf{f}_{ji}(\mathbf{r}) = [\omega_i \mathbf{f}_{jQ}^{iP}(\mathbf{r}) - \omega_j \mathbf{f}_{iP}^{jQ}(-\mathbf{r})]/(\omega_i + \omega_j) \tag{77}$$

$$M_h(\mathbf{r}) = M_{h1}(\mathbf{r}) + M_{h2}(\mathbf{r}) \tag{78}$$

$$M_{h1}(\mathbf{r}) = \chi(P, Q, \omega_i, \omega_j)(e_j\mathbf{X} + \mathbf{F}^{jQ} - e_i\mathbf{X} - \mathbf{F}^{iP})/(\omega_i + \omega_j) \tag{79}$$

$$M_{h2}(\mathbf{r}) = \sum_k n_k(e_k\mathbf{X} + \mathbf{F}^{kR})\frac{\mathbf{X}}{X}\frac{B_{ji}(P, Q, \omega_i, \omega_k) - B_{ij}(Q, P, \omega_j, \omega_k)}{\omega_i + \omega_j} \tag{80}$$

$$B_{ji}(P, Q, \omega_i, \omega_k) = \int \chi(P, R, \omega_i, \omega_k)\frac{\mathbf{X}}{X}(g_{iP,jQ}^{kR} - g_{iP}^{kR})\,dV_R \tag{81}$$

$$B_{ij}(Q, P, \omega_j, \omega_k) = \int \chi(Q, R, \omega_j, \omega_k)\frac{\mathbf{X}}{X}(g_{iP,jQ}^{kR} - g_{jQ}^{kR})\,dV_R \tag{82}$$

$$H_{ji}(\mathbf{r}) = H_{ji}^{(1)}(\mathbf{r}) + H_{ji}^{(2)}(\mathbf{r}) \tag{83}$$

$$H_{ji}^{(1)}(\mathbf{r}) = \chi(P, Q, \omega_i, \omega_j)[\mathbf{f}_{iP}^{jQ}(-\mathbf{r}) - \mathbf{f}_{jQ}^{iP}(\mathbf{r}) + 2kT\nabla \ln g_{ji}(\mathbf{r})]/(\omega_i + \omega_j) \tag{84}$$

$$H_{ji}^{(2)}(\mathbf{r}) = \left(\sum_k n_k \int \chi(P, R, \omega_i, \omega_k)\right.$$
$$\times (\mathbf{f}_{iP,jQ}^{kR}g_{iP,jQ}^{kR} - kT\nabla_R g_{iP,jQ}^{kR} - \mathbf{f}_{iP}^{kR}g_{iP}^{kR} + kT\nabla_R g_{iP}^{kR})\,dV_R$$
$$- \sum_k n_k \int \chi(Q, R, \omega_j, \omega_k)$$
$$\left.\times (\mathbf{f}_{iP,jQ}^{kR}g_{iP,jQ}^{kR} - kT\nabla_R g_{iP,jQ}^{kR} - \mathbf{f}_{jQ}^{kR}g_{jQ}^{kR} + kT\nabla_R g_{jQ}^{kR})\,dV_R\right)\Big/(\omega_i + \omega_j) \tag{85}$$

It is to be noted, of course, that *at equilibrium* the excess hydrodynamic force $H_{ji}(\mathbf{r}) = 0$.

The continuity equation has now taken a form in which the terms M_1 and $M_h(\mathbf{r})$ are easily identified as monitoring terms, since the forces on which they depend are constants, independent of their location. The other two terms, $\mathbf{f}_{ji}(\mathbf{r})$ and $H_{ji}(\mathbf{r})$, are *excess* terms, since they depend on forces that depend on their location. The importance of this partition in the terms of the continuity equation will become clear once the linear response approximation is introduced.

2.5. Restriction to the Case of Linear Response

In conductance we are only interested with the *linear* response of the system under the external perturbation. This is what is commonly referred to as Ohm's law, which states that the net velocity of the charged particle is proportional to the external field $\mathbf{X}$. This restriction implies the use of a small electrical field, as is always the case in experimental conductance techniques where the resistance of a cell is measured. This restriction also simplifies the solution of the continuity equation. If one approximates any quantity y by

$$y = \overset{\circ}{y} + y' + O(X^2) \tag{86}$$

where $\hat{y}$ is the value at equilibrium and y' the perturbation that is proportional to the external field, one can thus neglect all the cross products of the primed terms, which would lead to higher-order contributions in X.

The continuity equation can now be restricted to

$$\frac{1}{kT(\omega_i + \omega_j)} \frac{\partial}{\partial t} g'_{ji}(\mathbf{r}) = \nabla\left(\nabla g'_{ji}(\mathbf{r}) + g'_{ji}(\mathbf{r})\nabla \ln \overset{\circ}{g}_{ji}(\mathbf{r}) - \overset{\circ}{g}_{ji}(\mathbf{r}) - \overset{\circ}{g}_{ji}(\mathbf{r})\right.$$

$$\left.\frac{\mathbf{M}_1 + \mathbf{f}'_{ji}(\mathbf{r}) + \mathbf{M}_h(\mathbf{r}) + H_{ji}(\mathbf{r})}{kT}\right) \tag{87}$$

where we have used

$$\overset{\circ}{\mathbf{f}}_{ji}(\mathbf{r}) = kT\nabla \ln \overset{\circ}{g}_{ji}(\mathbf{r}) \tag{88}$$

The monitoring terms $\mathbf{M}_1$ and $\mathbf{M}_h$ are unchanged. The perturbation quantity $\mathbf{f}'_{ji}(\mathbf{r})$ is now given by

$$\mathbf{f}'_{ji}(\mathbf{r}) = (\omega_i \mathbf{f}''^{iP}_{jQ} - \omega_j \mathbf{f}''^{jQ}_{iP})/(\omega_i + \omega_j) \tag{89}$$

with

$$\mathbf{f}''^{iP}_{jQ} = -\sum_k n_k \int \nabla_P U(iP, kR)(g'^{kR}_{iP,jQ} - g'^{kR}_{iP}) \, dV_R \tag{90}$$

At equilibrium the quantity $\mathbf{H}_{ji}(\mathbf{r})$ is zero, since the hydrodynamic force is itself a perturbation; it can consequently be expressed according to Eq. (83), in terms of perturbations only, by

$$\mathbf{H}^{(1)}_{ji}(\mathbf{r}) = \boldsymbol{\chi}(P, Q, \omega_i, \omega_j)\left[\mathbf{f}''^{jQ}_{iP}(-\mathbf{r}) - \mathbf{f}''^{iP}_{jQ}(\mathbf{r}) + 2kT\nabla \frac{g_{ji}(\mathbf{r})}{\overset{\circ}{g}_{ji}(r)}\bigg/(\omega_i + \omega_j)\right] \tag{91}$$

$$\mathbf{H}^{(2)}_{ji}(\mathbf{r}) = [W^{iP}_{jQ}(\mathbf{r}) - W^{jQ}_{iP}(-\mathbf{r})]/(\omega_i + \omega_j) \tag{92}$$

with

$$W^{iP}_{jQ}(\mathbf{r}) = \sum_k n_k \int \boldsymbol{\chi}(P, Q, \omega_i, \omega_k)(A^{kR}_{iP} - A^{kR}_{iP,jQ}) \tag{93}$$

$$A^{kR}_{iP} = kT\nabla_R g'^{kR}_{iP} - kTg'^{kR}_{iP}\nabla_R \ln \overset{\circ}{g}^{kR}_{iP} - \overset{\circ}{g}^{kR}_{iP} f'^{kR}_{iP} \tag{94}$$

$$A^{kR}_{iP,jQ} = kT\nabla_R g'^{kR}_{iP,jQ} - kTg'^{kR}_{iP,jQ}\nabla_R \ln \overset{\circ}{g}^{kR}_{iP,jQ} - \overset{\circ}{g}^{kR}_{iP,jQ} f'^{kR}_{iP,jQ} \tag{95}$$

Operating on Eq. (87) with respect to the differential operator ∇ leads finally to the following second-order differential equation:

$$\frac{1}{kT(\omega_i + \omega_j)} \frac{\partial}{\partial t} g'_{ji}(\mathbf{r}) = \nabla g'_{ji}(\mathbf{r}) + \sum_{l=1}^{2} S_l + \sum_{l=1}^{9} T_l \tag{96}$$

with

$$S_1 = -\frac{\omega_i e_i - \omega_j e_j}{\omega_i + \omega_j} \frac{\mathbf{X}}{kT} \nabla \mathring{g}_{ji}(r) \tag{97}$$

$$S_2 = -\mathring{g}_{ji}(r) \nabla \frac{\mathbf{f}'_{ji}(\mathbf{r})}{kT} \tag{98}$$

$$T_1 = -\nabla \mathring{g}_{ji}(r) \cdot \frac{\mathbf{f}'_{ji}(\mathbf{r})}{kT} \tag{99}$$

$$T_2 = -g'_{ji}(\mathbf{r}) \Delta \ln \mathring{g}_{ji}(r) \tag{100}$$

$$T_3 = -\nabla g'_{ji}(\mathbf{r}) \nabla \ln \mathring{g}_{ji}(r) \tag{101}$$

$$T_4 = -\frac{\omega_i \mathbf{F}^{iP} - \omega_j \mathbf{F}^{jQ}}{(\omega_i + \omega_j)kT} \nabla \mathring{g}_{ji}(r) \tag{102}$$

$$T_5 = -\frac{\mathbf{H}_{ji}(\mathbf{r})}{kT} \nabla \mathring{g}_{ji}(r) \tag{103}$$

$$T_6 = -\mathring{g}(r) \nabla \frac{\mathbf{H}_{ji}(\mathbf{r})}{kT} \tag{104}$$

$$T_7 = -\frac{\mathbf{M}_h(\mathbf{r})}{kT} \nabla \mathring{g}_{ji}(r) \tag{105}$$

$$T_8 = -\mathring{g}_{ji}(r) \nabla \frac{\mathbf{M}_h(\mathbf{r})}{kT} \tag{106}$$

$$T_9 = -\frac{\mathbf{v}_e^{iP} - \mathbf{v}_e^{jQ}}{(\omega_i + \omega_j)kT} \nabla \mathring{g}_{21} \tag{107}$$

where the S terms contribute to the limiting law result, and the T terms to higher-order contributions in concentration. In this presentation the original monitoring term $\mathbf{M}_1$ given by Eq. (76) and rewritten

$$\mathbf{M}_1 = \nu_{ji} \frac{\omega_i e_i - \omega_j e_j}{\omega_i + \omega_j} \mathbf{X} \tag{108}$$

with

$$\nu_{ji} = \frac{\omega_i(e_i \mathbf{X} + \mathbf{F}^{iP} + \mathbf{v}_e^{iP}/\omega_i)(\mathbf{X}/X^2) - \omega_j(e_j \mathbf{X} + \mathbf{F}^{jQ} + \mathbf{v}_e^{jQ}/\omega_j)(\mathbf{X}/X^2)}{\omega_i e_i - \omega_j e_j} \tag{109}$$

has been split into its three components S_1, T_4, and T_9, allowing an easy comparison with the continuity equations used by Pitts, Fuoss and Onsager, and Falkenhagen *et al.* We shall see, for instance, that Pitts neglects T terms for $l = 4, 5, 6, 8$, whereas Fuoss and Onsager neglect T terms for $l = 6, 8, 9$, with the wrong algebraic sign for T_4.

Falkenhagen *et al.* deliberately neglect all hydrodynamic terms (T_5-T_9), as well as T_1, T_2, and T_4, in their work on the integration of the continuity equation, for which they have, however, an exact expression. For the other terms, which are not neglected, variances are to be found owing to specific approximations made. But before proceeding further with this comparison, it is now necessary to introduce the approximations that are common to *all* derivations and necessary in order to make the integration of the above continuity equation tractable.

Equations (75), (87), and (96) are as general as the original Ebeling equation, Eq. (52), but they greatly facilitate the discussion by illustrating the meaning of the various approximations common to all the calculations to be found in the literature.

It will be useful now to point out some interesting properties of Eqs. (75), (87), and (96):

1. They are not restricted only to the case of conductance. Perturbations other than an external electrical field **X** can be envisaged. In such cases, only the monitoring terms $\mathbf{M}_1$ and $\mathbf{M}_h$ will have to be reexpressed in terms of the new irreversible process considered.
2. They can also be used for the study of transition states. They are, for instance, applicable to the study of $g_{ji}(\mathbf{r})$ during the relaxation period when the external field **X** has just been cut off, down to the time when the final equilibrium is reached.

3. The Boundary Conditions

3.1. The Boundary Condition at Infinite Distance

The continuity equation requires boundary conditions for the evaluation of the constants of integration. Two such conditions are necessary, since the continuity equation is of the second order.

One of these conditions is obvious. At infinity the perturbation of the distribution function must vanish, so that

$$g_{r \to \infty} \to \overset{\circ}{g} \to 1 \tag{110}$$

3.2. The Boundary Condition at Short Distances

The second boundary condition, though less obvious, is easily derived, however, if one analytically expresses the fact that at short interionic distances the interaction potential becomes infinite. This is done by defining G_{iP}^{jQ} by

$$g_{iP}^{jQ} \equiv \exp\left(-\frac{U^*(iP, jQ)}{kT}\right) G_{iP}^{jQ} \tag{111}$$

where

$$U^*(iP, jQ) \equiv \text{repulsion potential function}$$

Let us consider a hard-core repulsive potential

$$U^*(iP, jQ) = U^*(jQ, iP) \begin{cases} =0 & \text{for } r \geq a_{ij} \\ =\infty & \text{for } r < a_{ij} \end{cases} \tag{112}$$

When $r \geq a_{ij}$, g_{iP}^{jQ} is equal to G_{iP}^{jQ}.

Introducing the identity (111) in the original continuity equation, Eq. (65), leads to

$$-\exp\left(-\frac{U^*(iP, jQ)}{kT}\right) \frac{\partial}{\partial t} G_{ji}(\mathbf{r}) = \exp\left(-\frac{U^*(iP, jQ)}{kT}\right) \nabla\{G_{ji}(\mathbf{r})[\mathbf{v}_{ji}(\mathbf{r}) - \mathbf{v}_{ij}(-\mathbf{r})]\}$$

$$+ \{G_{ji}(\mathbf{r})[\mathbf{v}_{ji}(\mathbf{r}) - \mathbf{v}_{ij}(-\mathbf{r})]\}\nabla \exp\left(-\frac{U^*(iP, jQ)}{kT}\right) \tag{113}$$

Given the definition of U^*, follows that

$$\nabla \exp\left[-U^*(jQ, iP)/kT\right] = (\mathbf{r}/r)\delta(r - a_{ij}) \tag{114}$$

where $\delta(x)$ is the Dirac function, so that the continuity equation takes two forms, depending on the value of the distance r: (1) For $r > a_{ij}$

$$\nabla G_{ji}(\mathbf{r})[\mathbf{v}_{ji}(\mathbf{r}) - \mathbf{v}_{ij}(-\mathbf{r})] = -\frac{\partial}{\partial t} G_{ij}(\mathbf{r}) \tag{115}$$

which is identical to the initial formulation; and (2) for $r = a_{ij}$

$$\frac{\mathbf{r}}{r} G_{ji}(\mathbf{r})[\mathbf{v}_{ji}(\mathbf{r}) - \mathbf{v}_{ij}(-\mathbf{r})] = 0 \tag{116}$$

Since the distribution functions do not vanish at $r = a_{ij} + \varepsilon$ and since the two-particle velocity vectors $\mathbf{v}_{ji}(\mathbf{r})$ and $\mathbf{v}_{ij}(\mathbf{r})$ cannot be equal in magnitude for two ions i and j of different kinds, each radial component of the two vectors must go separately to zero when r decreases to a_{ij}; that is,

$$\left. \begin{array}{l} \dfrac{\mathbf{r}}{r} \cdot \mathbf{v}_{ji}(\mathbf{r}) \to 0 \\[2mm] \dfrac{\mathbf{r}}{r} \cdot \mathbf{v}_{ij}(-\mathbf{r}) \to 0 \end{array} \right\} \quad \text{when } r \searrow a_{ij} \qquad \begin{array}{l} (117) \\[4mm] (118) \end{array}$$

For practical reasons, owing to the analytical form of the continuity equation, this boundary condition is usually given the more symmetrical form

$$g_{ij}(\mathbf{r})(\mathbf{v}_{ji} - \mathbf{v}_{ij}) \cdot \mathbf{r} = 0 \quad \text{for } r = a_{ij} \tag{119}$$

However, its real physical meaning, summarized by Eqs. (117) and (118), must be kept in mind, since it will be shown that it is sometimes useful to use an approximation to this exact boundary condition, as explained below in Section 6.2.

The physical interpretation of the short-range boundary condition tells us that the radial component of the two-particle velocity vector must vanish at $r = a_{ij}$. This is, after all, an easily understood property of hard-core collisions, so long as these are ideally elastic. The collisions then have all the properties of a mere reflexion on a plane tangent to the contact spheres, so that, on the time average, the radial parts of the velocity vectors vanish, since the radial part just before each collision is opposed to that just after the collision, hence the cancellation. The same behavior must be observed even if the repulsion is of a soft-core type. Consequently, the fact that this quantity vanishes at $r = a_{ij}$ is due to the repulsion itself.

4. Approximations Leading to a Set of Differential Equations for Two-Particle Distribution Functions

The Onsager treatment, as described above, involves the combination of various hierarchy equations, which ultimately leads to a continuity equation in which distribution functions, up to the fourth order, appear explicitly. This continuity equation was first obtained by Ebeling[38] directly from the Liouville equation for the evolution of systems of ions in a continuum solvent. Whereas the derivation of the continuity equation given in Section 2 was achieved by combining equations of increasing order, starting from the lowest order of each hierarchy, in contrast, the Ebeling derivation proceeds from the highest order. As well as giving the continuity equation, the Ebeling treatment also gives the hierarchy equations themselves.

Although the Ebeling continuity equation contains explicitly no distribution function of order higher than four, its exact solution necessitates the introduction of the whole BBGKY hierarchy, since the fourth-order distribution function can be evaluated only by considering the fourth equation hierarchy which, in turn, is a function of the fifth-order distribution function. In order to circumvent this increasing degree of complexity it is necessary to introduce approximations that will decouple the continuity equation from the higher-order equations of the hierarchy. Of course, the disadvantage is a limitation of the range of validity of the final result with respect to the various experimental parameters such as electrolyte concentration.

4.1. The First Kirkwood Approximation

We shall call the simplest approximation used the first Kirkwood approximation. This approximation states that any conditional distribution function of order n can be considered as a product of $n - 1$ pair distribution functions. For instance,

$$g_{iP,jQ}^{kR} = g_{iP}^{kR} g_{jQ}^{kR} \qquad (120)$$

The physical assumption involved in such an approximation is easy to state. It implies that the occurrence of a certain multiparticle configuration is pairwise independent, so that the resulting probability of finding an ion of type k in the vicinity of two ions i and j is assumed to be equal to the product of the probabilities of finding the individual pairwise events (k in the vicinity of i and k in the vicinity of j). With this approximation the Eberling continuity equations become a closed system. However, even at this level the solution of the mathematical problem is very difficult owing to the presence in the resulting continuity equations, of higher-order convolution integrals such as

$$-\sum n_k \int \nabla_P U(iP, kR) g_{iP}^{kR} g_{jQ}^{kR} \, dV_R$$

4.2. The Second Kirkwood or Poisson–Boltzmann Approximation

Clearly other approximations are needed in order to simplify the calculation further and this is the reason for the second Kirkwood approximation. This approximation consists in neglecting all cross terms in the right-hand side of the expression

$$g_{iP}^{kR} g_{jQ}^{kR} \equiv (1 + h_{iP}^{kR})(1 + h_{jQ}^{kR}) \equiv 1 + h_{iP}^{kR} + h_{jQ}^{kR} + h_{iP}^{kR} h_{jQ}^{kR} \qquad (121)$$

However, neglecting $h_{iP}^{kR} h_{jQ}^{kR}$ in Eq. (121) is very restrictive, since it clearly leads to the omission of the effects of many configurations of orders higher than two in the final theoretical equation. To make the consequences easy to understand it is convenient to use quasichemical language and say that only ion-pair effects can be described exactly, while most of the triplets and higher clusters are now lost with the exclusion of the cross terms in Eq. (121). At equilibrium, for instance, this approximation allows the excess thermodynamic functions to be evaluated up to the second virial coefficient. Thus, even with the use of the second Kirkwood approximation, it is possible to determine the conditions for which the quasichemical model of anion–cation pair association, initially introduced by Bjerrum in 1926, is correct.

It is interesting to point out here an analytical formulation, equivalent to the second Kirkwood approximation, that is commonly referred to as the Poisson–Boltzmann approximation. In fact, once introduced into the second equation of the BBGKY hierarchy, the second Kirkwood approximation leads to

$$\mathbf{F}_{jQ}^{iP} = -kT\nabla_P \ln g_{jQ}^{iP} - \nabla_P U(iP, jQ) - \sum_k n_k \int \nabla_P U(iP, kR)(g_{jQ}^{kR} + h_{iP}^{kR}) \, dV_R$$

(122)

Considering the electrostatic part of this force gives

$$-e_i \nabla_P \psi_{jQ}^{iP} = \mathbf{F}^{iP} + \mathbf{f}_{jQ}^{iP}$$

$$= \mathbf{F}^{iP} - \nabla_P \frac{e_i e_j}{D|\mathbf{r}_P - \mathbf{r}_Q|} - \frac{e_i}{D} \sum_k n_k e_k \int \left(\nabla_P \frac{1}{|\mathbf{r}_P - \mathbf{r}_S|}\right) g_{jQ}^{lS} \, dV_S$$

(123)

where ψ_{jQ}^{iP} is the average electrical potential acting on ion i at P in the vicinity of an ion j at Q, e_i, e_j, and e_k are the electrical charges of the ions considered and D is the dielectric constant of the solvent. The quantity $\mathbf{F}^{iP}$ is a constant that represents the contribution of h_{iP}^{lS} in the integral of Eq. (122) and which can be identified with the electrostatic relaxation force acting on the ion i at P. The vectors $\mathbf{r}_P$, $\mathbf{r}_Q$, and $\mathbf{r}_S$ define the locations of points P, Q, and S, respectively, in an arbitrary reference frame. Taking the divergence of the electrical force and eliminating the charge e_i leads to

$$\Delta_P \psi_{jQ}^P = \Delta_P \psi_{jQ}^{iP}$$

$$= \frac{4\pi}{D} \sum_k n_k e_k \int \delta |\mathbf{r}_P - \mathbf{r}_S| g_{jQ}^{lS} \, dV_S$$

(124)

where the direct force and relaxation force contributions vanish in the derivation process and δ is the Dirac function. It is to be noticed first that within the framework of the second Kirkwood approximation the internal potential ψ_{jQ}^{iP} is in fact independent of the charge e_i of the ion i, so that this potential is only a function of the reference ion and not of the type of ion to which this potential applies. Finally, after integrating Eq. (124), it follows that†

$$\Delta \psi_{jQ}^P = -\frac{4\pi}{D} \sum_k n_k e_k g_{jQ}^{kP} = \frac{1}{e_i} \nabla_P \mathbf{f}_{jQ}^{iP}$$

(125)

where the summation term is nothing more than the average charge density ρ_{jQ}^P at the point P. We finally have

$$\Delta \psi_{jQ}^P = -\frac{4\pi}{D} \rho_{jQ}^P$$

(126)

which is identical to the formulation that could have been obtained directly by applying the Poisson relation to the potential ψ_{jQ}^P. However, in doing so, one must keep in mind that an approximation has been made equivalent to the second Kirkwood approximation. This is not always clearly understood.

---

† In other texts $\Delta\psi$ is commonly written as $\nabla^2\psi$.

The Poisson relation is an exact electrostatic relation, but it can be applied only to instantaneous charge distributions and electrical potentials and *not* to average quantities, as is the case in the formulation above. Applying, as Debye originally did, the Poisson relation to the average potential ψ_{jQ}^{P} eliminates, in fact, any possibility of ultimately obtaining theoretical results in which triplets and higher-order cluster effects are correctly accounted for.

The first Kirkwood approximation implies a superposition of the *mean ion-pair potentials* ψ_{jQ}^{IS} and ψ_{iP}^{IS}, whereas the second Kirkwood approximation is equivalent to the superposition of *mean forces*, which is much less exact. While this leads to mathematical simplifications, again the price to pay is a further restriction of the range of validity of the final results to rather high dilutions.

4.3. The Debye Approximation or Linearization of the Poisson–Boltzmann Equation

We shall finally mention the Debye approximation, which consists in linearizing the Poisson–Boltzmann equation, Eq. (126), giving

$$\Delta\psi_{jQ}^{P} = \kappa^2 \psi_{jQ}^{P} \tag{127}$$

where

$$\kappa^2 = \frac{4\pi}{DkT} \sum_k n_k e_k^2 \tag{128}$$

This even further restricts the range of validity of the final result, since now even the two-particle configurations are no longer adequately represented. It is well known that only the limiting laws are treated correctly within the Debye approximation; it leads to

$$\ln f_\pm = -\kappa q + O_1(c) \tag{129}$$

and

$$\Lambda = \Lambda_0 - (\alpha\Lambda_0 + \beta)c^{1/2} + O_2(c) \tag{130}$$

in the equations for the activity coefficient† and the conductance, respectively (here q is the Bjerrum length, $q = e^2/2DKt$, and α and β are the relaxation and electrophoretic limiting law coefficients). Of course, the Debye approximation can give higher-order terms in the concentration c, but none of them can be considered as exact.

In particular, the linear terms in c that have a great importance, since they contain the specificity of the Hamiltonian representing the solute ionic particles, are incorrect in the Debye formulation. These restrictions were clearly understood by Bjerrum in 1927, when he proposed a way of correcting for these deficiencies. We shall return to this problem later in Section 6.

---

† See Chapter 2 in this volume.

4.4. Introduction of the Second Kirkwood Approximation in the Ebeling Continuity Equation

The second Kirkwood approximation, once introduced in the internal perturbation force $\mathbf{f}_{jQ}^{\prime iP}$, gives

$$\mathbf{f}_{jQ}^{\prime iP} \simeq -\sum_k n_k \int \nabla_P U(iP, kR) \exp\left(-\frac{U^*(iP, kR)}{kT}\right) g_{jQ}^{\prime kR} \, dV_R \qquad (131)$$

which, in the case of binary electrolytes, reduces to

$$\mathbf{f}_{jQ}^{\prime iP} \simeq -n_i \int \nabla_P U(iP, iR) \exp\left(-\frac{U^*(iP, iR)}{kT}\right) g_{jQ}^{\prime iR} \, dV_R \qquad (132)$$

with $i \neq j$.

These approximations have to be introduced in Eq. (96) which becomes a function of the perturbation parts of the pair distribution functions.

Taking into account the identification of the second Kirkwood approximation with the Poisson–Boltzmann equation leads to the following formulation of the term S_2:

$$S_2 = -\mathring{g}_{ji}(r)\kappa^2 \sum_k n_k e_k [\omega_i e_i g_{jk}'(\mathbf{r}) + \omega_j e_j g_{ki}'(\mathbf{r})] / \sum_k n_k e_k^2 (\omega_i + \omega_j) \qquad (133)$$

In the case of binary electrolytes this gives

$$S_2 = -\mathring{g}_{ji}(r)\kappa^2 q^* g_{ji}'(\mathbf{r}) \qquad (134)$$

with

$$q^* = \frac{e_i \omega_i - e_j \omega_j}{(e_i - e_j)(\omega_i + \omega_j)} \qquad (135)$$

which, for a *binary symmetrical* electrolyte, further reduces to

$$q^* = \tfrac{1}{2} \qquad (136)$$

The second Kirkwood approximation must also be introduced in the terms $\mathbf{M}_{h_2}(\mathbf{r})$ and $\mathbf{H}_{ji}^{(2)}(\mathbf{r})$, given by Eqs. (80) and (85), which depend explicitly on three-particle distribution functions and which become greatly reduced in their complexity of formulation.

As a result, each continuity equation is a function of pair distribution functions only, so that the mathematical problem of integration is formally tractable when appropriate expressions are substituted for the equilibrium pair distribution function, a problem of equilibrium that is supposed to be solved.

5. Comparison of the Various Treatments in the Literature

5.1. The Continuity Equation

The solution of the continuity equation is the key problem for the evaluation of the relaxation forces $\mathbf{F}^{iP}$, $\mathbf{F}^{jQ}$, etc., since those quantities that are essential in the theoretical formulation of the dependence of conductance on concentration depend on the perturbations of the pair distribution functions. Although theoretical derivations of conductance are numerous, they reduce, as far as the continuity equation is concerned, to three original approximations: that of Pitts,[3] Pitts being the first to give a general derivation in 1953, that of Fuoss and Onsager,[4] which followed shortly after that of Pitts, and that of Falkenhagen et al.[5] (FEK), whose most refined result was given in 1971. All other derivations proceed either from the Fuoss–Onsager[4] approximation (Fuoss–Hsia,[39] Fuoss–Onsager–Skinner,[24] Fuoss–Chen,[34,44] Murphy–Cohen,[25,26] Carman,[42] Quint–Viallard[43] and quite recently Chen[44]) or from the FEK approximation.[5] The differences in each subgroup are the number of iterations involved in the integration itself, the equilibrium pair distribution function used, details in the one-particle electrophoretic velocity, or generalization to the case of unsymmetrical electrolytes or to the case of mixtures of any kind of ions.

5.1.1. The Hydrodynamic Terms in the Continuity Equation

In order to understand the approximations involving the hydrodynamic terms T_5–T_9 we shall briefly review the position of each set of authors in this respect.

Falkenhagen et al.[5] deliberately decided to neglect all hydrodynamic concentrations to the relaxation field, though such terms are explicitly present in the original Ebeling equation. They considered that these contributions in the final conductance equation would be small. It is to be noted here that, following a prediction by Feistel,[45] Sändig and Feistel[46] showed that the term T_8 leads in fact to a hydrodynamic $\log c$ contribution, as will be discussed in more detail below.

The Pitts approximation finds its basis in the fact that Pitts identified the two-particle electrophoretic velocities $\mathbf{v}_{e_iP}^{jQ}$ and $\mathbf{v}_{e_jQ}^{iP}$ with the corresponding one-particle quantities $\mathbf{v}_e^{jQ}$ and $\mathbf{v}_e^{iP}$, respectively. As a result, the excess terms T_5–T_8 are missing.

The Fuoss approach to the problem of the hydrodynamic contribution to the continuity equation is quite peculiar, since this author identified the two-particle electrophoretic velocity with the velocity of the supporting medium around a reference ion.[47] In other words, Fuoss used an approximation for the quantity $\mathbf{v}_{e_jQ}^P$ and not for $\mathbf{v}_{e_jQ}^{iP}$. The two problems are basically different. It can be shown how $\mathbf{v}_{e_jQ}^P$ derives from $\mathbf{v}_{e_jQ}^{iP}$ just by eliminating from Eq. (72) all contributions from the ion of type i. The result is then

$$\mathbf{v}_{e_jQ}^P = \boldsymbol{\chi}(P, Q, \omega_i = \infty, \omega_j)(e_j\mathbf{X} + \mathbf{F}^{jQ})$$

$$+ \sum_k n_k \int \boldsymbol{\chi}(P, R, \omega_i = \infty, \omega_k)(e_k\mathbf{X} + \mathbf{F}^{kR})(\mathring{g}_{jQ}^{kR} - 1)\, dV_R$$

$$+ \sum_k n_k \int \boldsymbol{\chi}(P, R, \omega_i = \infty, \omega_k)(\mathbf{F}_{jQ}^{kR} - \mathbf{F}^{kR})\mathring{g}_{jQ}^{kR}\, dV_R \qquad (137)$$

where in addition ω_i is taken as infinity in order to eliminate all contributions in ω_i in the short-range part of the coupling tensor of Eq. (49). The first term in the right-hand side of Eq. (137) is the contribution of the velocity of the reference ion itself to the velocity of the medium at P, whereas the next two terms are the contribution at P of the ionic atmosphere of the reference ion j. It is to be noted that, in addition, Fuoss and Onsager neglected the force $\mathbf{F}^{jQ}$ in Eq. (137). This oversimplified identification of $\mathbf{v}_{e_jQ}^{iP}$ with the velocity of the supporting medium at P, $\mathbf{v}_{e_jQ}^P$, led Fuoss and Onsager to neglect all terms resulting from $\nabla_P \mathbf{v}_{e_jQ}^{iP}$ and $\nabla_Q \mathbf{v}_{e_jP}^{jQ}$, since the medium was considered to be incompressible. This explains why they missed the terms T_6 and T_8, as well as the incomplete terms T_5 and T_7, as shown in Table 2.

5.1.2. The Nonhydrodynamic Terms in the Continuity Equation

In their formulation Falkenhagen et al.[5] deliberately neglected the terms T_1 and T_4, considering that their contribution to the linear concentration terms was probably very small. For $\mathring{g}_{ji}$ in T_2 and T_4 they used the direct potential instead of the mean potential as an approximation, which leads to $T_2 = 0$.

In his phenomenological approach to the evaluation of the two-particle force $\mathbf{F}_{jQ}^{iP}$ Pitts[3] considered that it was an error to take into account the one-particle force $\mathbf{F}^{iP}$ and pointed out that this was considering the same effect[48] twice. Equation (68) shows that this is not the case.

Inspection of the Fuoss–Onsager continuity equation[49] shows that the wrong algebraic sign associated with the term T_4 is used not only by these authors but also by all the other authors who started from the same equation in their calculations. It is easy to see that the T_4 term leads to a linear contribution in conductance that is nothing but $\alpha^2 c$. Indeed, in the final result of Fuoss[50] and Onsager it is possible to trace a negative $-\alpha^2 c$ contribution, and it is consequently easy to correct for this error.[51]

5.2. Comparisons of the Boundary Conditions for the Continuity Equation

At infinity all authors agree with the exact evaluation of $g_{ji}'(\mathbf{r}) \to 0$ as $r \to \infty$, but differences are, however, observed at $r = a$.

Table 2

Comparison of the Various Approximations[a] in the Formation of the Continuity Equation

$$\Delta g' + \sum_{k=1}^{2} S_k + \sum_{l=1}^{9} T_l = 0 \qquad (96)$$

Authors	$\mathring{g}_{ji}$	Short-range boundary condition	Main terms		Nonhydrodynamic terms				Hydrodynamic terms				
			S_1	S_2	T_1	T_2	T_3	T_4	T_5	T_6	T_7	T_8	T_9
Pitts[3]	DH	$g'(\mathbf{r}) \to 0$						0	0	0	0	0	(LL)
Fuoss–Hsia[82]	DH	$\mathbf{v}_{rji}(\mathbf{r}) \to 0$											
Fuoss–Chen[40]	DH	$\mathbf{v}_{rji}(\mathbf{r}) \to 0$											
Fuoss–Onsager–Skinner[24]	M	$\mathbf{v}_{rji}(\mathbf{r}) \to 0$											
Quint–Viallard[27]	DH	$\mathbf{v}_{rji}(\mathbf{r}) \to 0$											
Murphy–Cohen[25]	M	?											
Chen[28]	M	$\mathbf{v}_{rji}(\mathbf{r}) \to 0$						$-T_4$ (LL) (1)	~ (2)	0	~ (2)	0	0
Falkenhagen et al.[5]	M	$\mathbf{v}_{rji}(\mathbf{r}) \to 0$	$\mathring{g}_{ji}=1$		0	0	$\mathring{\psi}_{ji}=U_{ji}$	0	X	X	X	X	X
Lee–Wheaton[31]	DH	$\dfrac{dg'}{dr}(\mathbf{r}) \to 0$	$\mathring{g}_{ji}=1$		0	0	$\mathring{\psi}_{ji}=U_{ji}$	0	0	0	0	0	(LL) (1)
Carman[29]	DH	$g'_{ji}(\mathbf{r}) \to 0$			0	0	0	0	0	0	~ (3)	0	0
Lowest-order contributions to $\dfrac{\Delta X}{X}(c)$			$-\alpha c^{1/2}$ $+ E_1'c\log c$ $+ O(c)$		$O(c)$			$\alpha^2 c$ First echo	$O(\log c)$	$O(c)$	$-\dfrac{E_2'}{\Lambda_0}c\log c\log c$	$\dfrac{1}{2}\dfrac{E_2'}{\Lambda_0}c\log c$	$\dfrac{\alpha\beta c}{\Lambda_0}$ First echo

[a] The second column gives the approximation used for the equilibrium distribution function of Eq. (96) (DH for Debye and Hückel equation and M for Meeron equation). The third column recalls which is the short-range boundary condition used as r decreases to the contact distance a of the hard-sphere model. When no indication is given, the corresponding term was used as described in Eq. (48); otherwise the approximation is explicitly given. 0 means that the term was overlooked, whereas X means it was deliberately neglected; LL means that only the limiting law value was used in the term concerned.

The following symbols are also used. (1)—Chen corrected for the error on sign on T_4 instead of $-T_4$; (2)—the approximations used for T_5 and T_7 are detailed in Section 5-1-1; (3)—Carman neglects all hydrodynamic terms except terms T_7 and T_8 which are originally contained in his definition of $\mathbf{W}_{ji}$ by Eqs. 22 and 24 of Ref. a. However, T_8 is dropped when the $\Delta\mathbf{W}_{ji}$ is neglected later without any explanation. Finally, in T_7, $\mathbf{M}_n$ reduces to $\mathbf{M}_{n1}$ (Eq. 79) where $\mathbf{F}^{i0}$ and $\mathbf{F}^{iP}$ are neglected and where the hydrodynamic tensor $\mathbf{X}$ reduces to the classical Stokes tensor as in all derivations. a) P. C. Carman, J. S. Afr. Chem. Inst. 28, 80 (1975)

In all of Fuoss' derivations, with the exception of his latest derivation Fuoss–Onsager,[4] Fuoss–Hsia,[39] Fuoss–Chen,[40,41] Fuoss–Onsager–Skinner,[24] the exact condition given by Eq. (120) was chosen. This same condition was used by Quint and Viallard,[43] Chen,[44] and Falkenhagen et al.[5] In more recent publications Fuoss[32] chose another condition,

$$\lim_{c \to 0} g'_{ji} \neq \infty \tag{138}$$

which he finally recognized as incorrect and quite recently[51] changed for

$$g'_{ji}(\mathbf{r}) \to 0 \qquad \text{as } r \to R \tag{139}$$

where R is the radius of the Gurney solvation cosphere of the ions.

Pitts[3] and Carman[42] chose

$$g'_{ji}(\mathbf{r}) \to 0 \qquad \text{as } r \to a \tag{140}$$

In a treatment of the Gurney model following Fuoss's latest derivations Lee and Wheaton[30] chose

$$\frac{dg'_{ji}(\mathbf{r})}{dr} \to 0 \qquad \text{as } r \to R \tag{141}$$

however, they prefer to use the Pitts condition for the hard-sphere model.

In Section 6 we shall discuss the short-range condition for the Gurney model and present a much better approximation for the hard-sphere model; we will show that only the hydrodynamic condition given by Eq. (119) is exact and all the others are but approximations. The consequences of the approximation used by Pitts and Carman are not, however, very important, since the Debye–Hückel equilibrium distribution function introduced in these two cases in the continuity equation already constitutes a very crude approximation at short distances.

In Murphy and Cohen's derivation[25] it was not possible to find the boundary conditions chosen, the latter not being explicitly stated.

5.3. General Formulation of the Perturbation of the Ion-Pair Distribution Functions $g'_{ji}(\mathbf{r})$

The solution of the continuity equation, restricted to the case of linear response, can be carried out through a succession of iterations. First the equation is solved by neglecting all monitoring terms but one, and then one proceeds until all monitoring terms have been considered; the various solutions are then simply added. Inspection of the general formulation given by Eq. (87) shows that there are three such monitoring terms: the leading one, $\mathbf{M}_1$, and the two less important ones, $\mathbf{M}_{h1}$ and $\mathbf{M}_{h2}$, which are hydrodynamic in nature.

The first one can be written as

$$\frac{\mathbf{M}_1}{kT} = \nu_{ji} \frac{\omega_i e_i - \omega_j e_j}{\omega_i + \omega_j} \frac{\mathbf{X}}{kT} \qquad (142)$$

with

$$\nu_{ji} = \frac{\omega_i e_i (1 + \Delta X^i/X + \lambda_e^i/\lambda_0^i) - \omega_j e_j (1 + \Delta X^j/X + \lambda_e^j/\lambda_0^j)}{\omega_i e_i - \omega_j e_j} \qquad (143)$$

where $\Delta X^i/X$ is the relaxation contribution and λ_e^i/λ_0^i the electrophoretic contribution for the ion i. This leads to a first contribution $g'^{(1)} \equiv g_{ji}'^{(1)}(\mathbf{r})$ which can be written

$$g'^{(1)} = g_I'^{(1)} \nu_{ji} \qquad (144)$$

with

$$g_I'^{(1)} = Y_{ji}^{(1)}(r) \cos \theta \frac{\omega_i e_i - \omega_j e_j}{\omega_i + \omega_j} \frac{X}{kT} \qquad (145)$$

The hydrodynamic term $\mathbf{M}_{h1}(\mathbf{r})/kT$ can be written as

$$\frac{\mathbf{M}_{h1}}{kT} = \mu_{ji} \frac{e_j - e_i}{\omega_i + \omega_j} \frac{X}{kT} \mathbf{x}(P, Q, \omega_i, \omega_j) \frac{\mathbf{X}}{X} \qquad (146)$$

with

$$\mu_{ji} = \frac{e_j(1 + \Delta X^j/X) - e_i(1 + \Delta X^i/X)}{e_j - e_i} \qquad (147)$$

so that the corresponding contribution $g'^{(h1)}$ will be

$$g'^{(h1)} = g_I'^{(h1)} \mu_{ji} \qquad (148)$$

with

$$g_I'^{(h1)} = Y_{ji}^{(h1)}(r) \, \omega \mathrm{s} \, \theta \frac{e_j - e_i}{\omega_i + \omega_j} \frac{X}{kT} \qquad (149)$$

Similar considerations show that the contribution of the third monitoring term

$$\frac{\mathbf{M}_{h2}}{kT} = \sum_k n_k \frac{e_k X}{kT} \left(1 + \frac{\Delta X^k}{X}\right) \left(\int \mathbf{x}(P, R, \omega_i, \omega_k) \frac{\mathbf{X}}{X} (\overset{\circ}{g}_{jQ}^{kR} - 1) \, dV_R \right.$$

$$\left. - \int \mathbf{x}(Q, R, \omega_j, \omega_k) \frac{\mathbf{X}}{X} (\overset{\circ}{g}_{iP}^{kR} - 1) \, dV_R \right) \qquad (150)$$

leads to a sum of contributions of the following type:

$$g'^{(h2)} = \sum_k g_I'^{(h2)}(k) \left(1 + \frac{\Delta X^k}{X}\right) \qquad (151)$$

with

$$g_I'^{(h2)}(k) = Y_{ji}^{(h2)}(k,r)\cos\theta\,\frac{n_k e_k X}{kT} \qquad (152)$$

In the case of a binary symmetrical electrolyte ($n_i = n_j = n$; $e_i = -e_j$) for which

$$\frac{\Delta X^{iP}}{X} = \frac{\Delta X^{jQ}}{X} \qquad (153)$$

the general solution takes the following form for $g' \equiv g'_{+-}(\mathbf{r})$:

$$g' = g_I'^{(1)}\left(1 + \frac{\Delta X}{X} + \frac{\Lambda_e}{\Lambda_0}\right) + g_I'^{(2)}\left(1 + \frac{\Delta X}{X}\right) \qquad (154)$$

where

$$\Lambda_e = \lambda_e^+ + \lambda_e^- \qquad (155)$$

$$\Lambda_0 = \lambda_0^+ + \lambda_0^- \qquad (156)$$

and

$$g_I'^{(2)} = g_I'^{(h1)} + \sum_{k=1}^{2} g_I'^{(h2)}(k) \qquad (157)$$

It is to be noted that the contribution $g_I'^{(2)}$ is purely hydrodynamic, since, being dependent on the hydrodynamic monitoring term, whereas the contribution $g_I'^{(1)}$ contains both a hydrodynamic contribution $g_{Ih}'^{(1)}$ through the $\mathbf{H}_{ji}(\mathbf{r})$ hydrodynamic internal force and a purely coulombic one $g_{IC}'^{(1)}$ through $\mathbf{f}_{ji}(\mathbf{r})$ in Eq. (75),

$$g_I'^{(1)} = g_{IC}'^{(1)} + g_{Ih}'^{(1)} \qquad (158)$$

The perturbations $g_{IC}'^{(1)}$ and $g_{Ih}'^{(1)}$ are the solutions obtained from the integration of the following continuity equations:

$$\Delta g_{IC}'^{(1)} + S_1 + S_2 = -T_1 - T_2 - T_3 \qquad (159)$$

$$\Delta g_{Ih}'^{(1)} + S_1 + S_2 = -T_5 - T_6 \qquad (160)$$

As a consequence, $g_{IC}'^{(1)}$ will be the sum of four contributions, the first obtained by neglecting T terms and the three others obtained by adding the contributions of the T_1, T_2, and T_3 terms. The perturbation $g_{Ih}'^{(1)}$ is obtained by the sum of the particular solutions of the two hydrodynamic contributions issued from T_5 and T_6. The contributions due to the terms T_4 and T_9 are easily accounted for by multiplying $g_I'^{(1)}$ by ν_{ji}.

The solution for the perturbation $g_I'^{(2)}$ is obtained by solving

$$\Delta g_I'^{(2)} + T_7 + T_8 + S_2 = -T_1 - T_2 - T_3 - T_5 - T_6 \qquad (161)$$

which proceeds from Eqs. (159) and (160) after substituting the primary monitoring term S_1 with the hydrodynamic one $T_7 + T_8$.

5.4. The Relaxation Contribution

5.4.1. The Relaxation Field

A relaxation field contribution $\Delta X_I^{iP(x)}/X$ can be defined from the corresponding evaluation of $g_{I_{ik}}^{\prime(x)}$ by use of the relation

$$\frac{\Delta X_I^{iP(x)}}{X} = -\frac{1}{e_i}\frac{\mathbf{X}}{X}\frac{1}{X}\sum_k n_k \int \nabla_P U(iP, kR) g_{I_{ik}}^{\prime(x)} \, dV_R \qquad (162)$$

where the index (x) represents any of the various indices used to characterize the various functions g_I' introduced in Section 5.3.

In the following, however, we shall restrict the discussion to the case of binary symmetrical electrolytes in order to keep the formulation as simple as possible. Then

$$\frac{\Delta X^{iP}}{X} = \frac{\Delta X^{iQ}}{X} = \frac{\dot{\Delta} X}{X}$$

The general formula for the relaxation term in conductance closely follows that of g' in Eq. (154):

$$\frac{\Delta X}{X} = \frac{\Delta X_I^{(1)}}{X}\left(1 + \frac{\Delta X}{X} + \frac{\Lambda_e}{\Lambda_0}\right) + \frac{\Delta X_I^{(2)}}{X}\left(1 + \frac{\Delta X}{X}\right) \qquad (163)$$

with

$$\frac{\Delta X_I^{(1)}}{X} = \frac{\Delta X_{IC}^{(1)}}{X} + \frac{\Delta X_{Ih}^{(1)}}{X} \qquad (164)$$

where $\Delta X_I^{(1)}/X$ originates from $g_I^{\prime(1)}$, and $\Delta X_I^{(2)}/X$ from $g_I^{\prime(2)}$.

From the discussion given in Section 5.1 it follows that the various derivations of $\Delta X/X$ reduce to approximations in Eqs. (163) and (164) that are summarized in Table 3.

In the Fuoss–Onsager treatment it is to be noted the algebraic minus sign for $(\Delta X/X)_{LL}$ due to the use of the wrong algebraic sign for the term T_4 in the continuity equation, as mentioned earlier.

If the theoretical evaluation of the conductance equation is restricted to the first terms beyond the limiting law of the expansion of $\Lambda(c)$ versus concentration, it is found that the various approximations displayed above have no drastic consequences, because the hydrodynamic contributions to $\Delta X/X$ are always small, as are the cross-product terms of the limiting law. For such truncated expansion formulations $\Delta X_{IC}^{(1)}/X$ is by far the dominant contribution in the relaxation effect. This will explain that commonly, in

Table 3
Comparison of the Approximations to the Relaxation Term in Conductance[a]

Authors and approximations	General formulation $$\frac{\Delta X}{X} = \left(\frac{\Delta X_{Ic}^{(1)}}{X} + \frac{\Delta X_{Ih}^{(1)}}{X}\right)\left(1 + \frac{\Delta X}{X} + \frac{\Lambda_e}{\Lambda_0}\right) + \frac{\Delta X_I^{(2)}}{X}\left(1 + \frac{\Delta X}{X}\right)$$
Pitts (P)[3]	$$\frac{\Delta X}{X} \sim \frac{\Delta X_{Ic}^{(1)}}{X}(P)\left(1 + \frac{\Lambda_{eLL}}{\Lambda_0}\right)$$
Fuoss–Hsia[82] Quint–Viallard[27](*) Fuoss–Onsager–Skinner[24]	$$\frac{\Delta X}{X} \sim \left(\frac{\Delta X_{Ic}^{(1)}}{X}(*) + \frac{\Delta X_{Ih}^{(1)}}{X}(*)\right)\left(1 - \frac{\Delta X_{LL}}{X}\right) + \frac{\Delta X_I^{(2)}}{X}(*)$$
Lee–Wheaton (L–W)[31]	$$\frac{\Delta X}{X} \sim \frac{\Delta X_{Ic}^{(1)}}{X}(L\text{–}W) + \frac{\Delta X_I^{(2)}}{X}(L\text{–}W)$$
Carman (C)[29]	$$\frac{\Delta X}{X} \sim \frac{\Delta X_{Ic}^{(1)}}{X}(P) + \frac{\Delta X_I^{(2)}}{X}(C)$$
FEK[5]	$$\frac{\Delta X}{X} \sim \frac{\Delta X_{Ic}^{(1)}}{X}(FEK)$$
Murphy–Cohen (M-C)[25]	$$\frac{\Delta X}{X} \sim \left(\frac{\Delta X_{Ic}^{(1)}}{X}(M\text{–}C) + \frac{\Delta X_{Ih}^{(1)}}{X}(M\text{–}C)\right)\left(1 - \frac{\Delta X_{LL}}{X}\right) + \frac{\Delta X_I^{(2)}}{X}(M\text{–}C)$$
Chen (Ch)[28]	$$\frac{\Delta X}{X} \sim \left(\frac{\Delta X_{Ic}^{(1)}}{X}(Ch) + \frac{\Delta X_{Ih}^{(1)}}{X}(Ch)\right)\left(1 + \frac{\Delta X_{LL}}{X}\right) + \frac{\Delta X_I^{(2)}}{X}(Ch)$$

[a] The $\Delta X_{Ic}^{(1)}/X$ term leading to the contributions $-\alpha c^{1/2} + E_1' c \log c$ originated from $g_{Ic}''^{(1)}$. The $\Delta X_{Ih}^{(1)}/X$ hydrodynamic contribution of order $O(c)$ originated from $g_{Ih}''^{(1)}$. The $\Delta X_I^{(2)}/X$ hydrodynamic contribution originating from $g_I''^{(2)}$ resulted from the monitoring terms T_7 and T_8. The part from T_7 leads to a $(-E_2'/\Lambda_0)c \log c$ contribution, and that from T_8 to a $\frac{1}{2}(E_2'/\Lambda_0)c \log c$ contribution (this last contribution is missing in all treatments). Each evaluation of $\Delta X_{Ic}^{(1)}/X$, $\Delta X_{Ih}^{(1)}/X$, and $\Delta X_I^{(2)}/X$ varies according to the approximation used in the continuity equation and summarized in Table 2.

practice, the various equations available were in fact quite similar as far as numerical analysis of conductance data goes for 1–1 salts in water, for example. We shall return to this point in Section 6.2.

In addition to the differences in the treatments of the relaxation effect reviewed above, which arise on account of neglecting various terms in the continuity equation, other differences in the final result originate from the function used in the continuity equation for the ion-pair distribution $\overset{\circ}{g}_{ji}(r)$ at equilibrium. This quantity appears explicitly in different terms of the continuity equation. Obviously the terms in which it is most important to use as refined as possible an evaluation of $\overset{\circ}{g}_{ji}$ are the monitoring term $\mathbf{M}_1$ and, more particularly, S_1, which is the external field contribution to $\mathbf{M}_1$. It is easy to see that if one uses the crudest evaluation, $\overset{\circ}{g}_{ji} = 1$, in $\mathbf{M}_1$, $\mathbf{M}_1 \nabla \overset{\circ}{g}_{ji}$ (and S_1)

cancel, and so do $g'_{I_{ji}}(\mathbf{r})$ and the relaxation field $\Delta X_I^{(1)}/X$, which controls the limiting law. The other terms of the continuity equation are, so to say, less demanding of $\mathring{g}_{ji}(r)$ as far as their contributions to a given coefficient of the expansion (versus concentration) of the conductance equation are concerned.

Pitts[3] (1953), Fuoss and Onsager[4] (1957), Fuoss and Hsia[39] (1967), Fuoss and Chen[40] (1973), Carman[42] (1975), and Quint and Viallard[43] (1978) used the Debye–Hückel approximation to $\mathring{g}_{ji}(r)$ whereas Fuoss, Onsager, and Skinner[24] (1965), Falkenhagen et al.[5] (1971), Murphy and Cohen[25] (1970), and Chen[44] (1978) used the Meeron expression, which leads to quite significant improvements in the evaluation of the linear coefficient with respect to concentration in the conductance equation.

5.4.2. The Osmotic Contribution to the Relaxation Field

Onsager introduced this contribution in 1955 in order to take the following effect into account.[10,13]

Given the asymmetry of the anion–cation pair distribution functions in the presence of an external field, a given ion will be struck more often from behind than from the front by ions of opposite charge during its migration. The result will be a net positive contribution to conductance, which is called the osmotic part of the relaxation field $\Delta P/X$. Fuoss and Onsager evaluated this quantity for an ion of type i by

$$\Delta P^i = \int \pi^i \, dS \tag{165}$$

where π^i is the pressure in excess of that at equilibrium, which is given by

$$\pi^i = kT \sum_k n_k g'_{ik} \tag{166}$$

and

$$dS = a^2 \sin \theta \, d\theta \, d\phi \tag{167}$$

which is a surface element on a sphere of radius a around the ion of type i.

Denoting the cylindrical symmetry of the distribution function by

$$g'_{ik}(\mathbf{r}) = y_{ik}(r) \cos \theta \tag{168}$$

the result is finally

$$\Delta P^i = \frac{4\pi}{3} kT \sum_k n_k a^2 y_{ik}(a) \tag{169}$$

Two criticisms made by Valleau[52] concerning the importance of this osmotic-pressure term should be mentioned here. He points out that an ion in solution is always surrounded by a dense distribution of particles that are either other ions or solvent molecules. Consequently, the dissymmetry of counter-ions around the reference ion involves a complementary symmetry of solvent molecules. As a result, the osmotic push forward is much less important than that calculated in a continuum model, since these must be added to the corresponding deficiency of solvent collisions; altogether the real dissymmetry of the osmotic pressure should be quite small. Moreover, Valleau observes that in addition to the spatial dissymmetry of ions, there is also a spatial dissymmetry of their velocities: Thus fewer counter-ions hit in front, but they do so with greater energy owing to their drift in the opposite direction. The latter effect is, however, also quite small, according to Valleau's evaluation. Although Valleau's remarks are certainly worth close attention, for the sake of self-consistency with the continuum model for the solvent used so far we shall not introduce them in the calculations below.

Falkenhagen et al.[5] have shown that the same result, Eq. (169), can be obtained in a much more straightforward manner without having to start from a physical description of the phenomena.

If the repulsion part U_{ik}^* of the ionic interactions is introduced in the definition of the one-particle force $\mathbf{F}^{iP}$ given by Eq. (33), the following relations are obtained:

$$\mathbf{F}^{iP} = -\sum_k n_k \int \nabla_P [U_{ik}^* + U_{ik}] \exp\left(-\frac{U_{ik}^*}{kT}\right) g'_{ik}(\mathbf{r}) \, dV_1 \qquad (170)$$

$$\mathbf{F}^{iP} = \mathbf{F}_c^{iP} + \mathbf{F}_{(*)}^{iP} \qquad (171)$$

with

$$\mathbf{F}_{(*)}^{iP} = -\frac{\mathbf{X}}{X} \sum_k n_k \int \exp\left(-\frac{U_{ik}^*}{kT}\right) \frac{\partial}{\partial x} U_{ik}^* y_{ik}(r) \cos\theta \, dV_R \qquad (172)$$

Taking into account the step-potential nature of the hard-core function U_{ik}^* gives

$$\exp\left(-\frac{U_{ik}^*}{kT}\right) \frac{\partial}{\partial x} U_{ik}^* = -kT \cos\theta\, \delta(r-a) \qquad (173)$$

which leads directly to

$$\mathbf{F}_{(*)}^{iP} = \frac{\mathbf{X}}{X} \frac{4\pi}{3} kT \sum_k n_k a^2 y_{ik}(a) \qquad (174)$$

which identifies with the Fuoss–Onsager result given above. Of course, the final evaluation of this contribution depends on the calculation of the perturbation function $g'_{ik}(r)$ that is of $y_{ik}(r)$.

A particularly interesting feature of this osmotic contribution is that, when evaluated in conjunction with the use of Meeron distribution functions and the exact short-range boundary condition in the continuity equation (see Table 5), it leads to the $2e^b/b^3$ term of the $B(b)$ function which is the most important part of the $Q^{\text{rel}}(b)$ relaxation term. This observation will have great importance when combined with the echo formulation to be developed in Section 6.1, since the $B(b)$ function is nothing but the Bjerrum function, which defines the Bjerrum association constant:

$$K_A^B(a, R) = \frac{4\pi N}{1000} \int_a^R r^2 \exp\frac{2q}{r}\, dr$$

$$= 2E_1'\left[B(b) - B\left(\frac{2q}{R}\right)\right] \tag{175}$$

Ultimately the relaxation contribution $\Delta X/X$ when expanded versus concentration leads to the linearized formulation:

$$\frac{\Delta X}{X} = -\alpha c^{1/2} + (E_1' - k_h E_2'/\Lambda_0)c \log c$$

$$+ 2E_1'\left(Q^{\text{rel}}(b) + \log\frac{2\kappa q}{c^{1/2}}\right)c$$

$$+ 2E_2'\left(Q_h^{\text{rel}}(b) - k_h \log\frac{2\kappa q}{c^{1/2}}\right)c + O(c^{3/2}) \tag{176}$$

where b is the Bjerrum parameter of the restricted primitive model used in all derivations. The coefficient k_h and the function $Q_h^{\text{rel}}(b)$ depend on the approximations made in the hydrodynamic terms of the continuity equation, while the function $Q^{\text{rel}}(b)$ also depends on the approximations in the nonhydrodynamic terms in the continuity equation. In addition, the functions $Q^{\text{rel}}(b)$ and $Q_h^{\text{rel}}(b)$ depend on the approximation used for the equilibrium pair distribution function $\mathring{g}_{ji}(r)$ when solving the continuity equation, as well as on the boundary conditions. The various results for k_h, Q^{rel}, and Q_h^{rel} are given in Table 5.

5.5. The Electrophoretic Contribution

Substituting the two-particle force in Eq. (46) by its components expressed by Eq. (68) and restricting to linear response, the result for the ionic electrophoretic velocity vector leads to

$$\mathbf{v}_e^{iP} = \sum_k n_k \int \mathbf{\chi}(P, R, \omega_i, \omega_k)(e_k \mathbf{X} + \mathbf{F}^{kR})\mathring{g}_{iP}^{kR}\, dV_R$$

$$+ \sum_k n_k \int \mathbf{\chi}(P, R, \omega_i, \omega_k)(\mathring{\mathbf{f}}_{iP}^{kR} g_{iP}'^{kR} + \mathbf{f}_{iP}'^{kR}\mathring{g}_{iP}^{kR} - kT\nabla_k g_{iP}'^{kR})\, dV_R \tag{177}$$

where

$$\mathring{\mathbf{f}}_{iP}^{kR} = kT\nabla_R \log \mathring{g}_{iP}^{kR} \tag{178}$$

and

$$\mathbf{f}_{iP}^{\prime kR} = -\sum_l n_l \int \nabla_R U(kR, lS)(g_{iP,kR}^{\prime lS} - g_{iP}^{\prime lS}) \, dV_S \tag{179}$$

Introducing the Kirkwood second approximation in $\mathbf{f}_{iP}^{\prime kR}$ gives

$$\mathbf{v}_e^{iP} = \sum_k n_k \int \mathbf{\chi}(P, R, \omega_i, \omega_k) e_k \mathbf{X}\left(1 + \frac{\Delta X^k}{X}\right) \mathring{g}_{iP}^{kR} \, dV_R$$

$$+ \sum_k n_k \int \mathbf{\chi}(P, R, \omega_i, \omega_k)\left(g_{iP}^{\prime kR} kT \nabla_R \log \mathring{g}_{iP}^{kR}\right.$$

$$\left. - \mathring{g}_{iP}^{kR} \sum_l n_l \int \nabla U(kR, lS) g_{kR}^{\prime lS} \, dV_S - kT \nabla_{Rg} g_{iP}^{\prime kR}\right) dV_R \tag{180}$$

Converting the velocity vector into a conductance quantity through Eq. (43) and restricting the result to the case of binary symmetrical electrolytes gives

$$\lambda_e^i = \lambda_{e1}^i\left(1 + \frac{\Delta X}{X}\right) + \lambda_{e2}^{i(1)}\left(1 + \frac{\Delta X}{X} + \frac{\Lambda_e}{\Lambda_0}\right) + \lambda_{e2}^{i(2)}\left(1 + \frac{\Delta X}{X}\right) \tag{181}$$

where

$$\lambda_{e1}^i = z_i \mathscr{F} \frac{\mathbf{X}}{X} \sum_k n_k \int \mathbf{\chi}(P, R, \omega_i, \omega_k) e_k \frac{\mathbf{X}}{X} \mathring{g}_{iP}^{kR} \, dV_R \tag{182}$$

and $\lambda_{e2}^{i(1)}$ and $\lambda_{e2}^{i(2)}$ are both obtained from the second term of Eq. (180), in which the g_{+-}' function is replaced by Eq. (145). A definition of the one-particle electrophoretic velocity $\mathbf{v}_e^{iF}$ is introduced in Figure 4.

Finally for the molar electrophoretic conductance the result

$$\Lambda_e = \lambda_e^+ + \lambda_e^- \tag{183}$$

$$\Lambda_e = (\Lambda_{e1} + \Lambda_{e2}^{(2)})\left(1 + \frac{\Delta X}{X}\right) + \Lambda_{e2}^{(1)}\left(1 + \frac{\Delta X}{X} + \frac{\Lambda_e}{\Lambda_0}\right) \tag{184}$$

is obtained.

A comparison of the various approximations to Λ_e, relative to the general Eq. (184), is summarized in Table 4. The $\Lambda_{e2}^{(1)}$ term is sometimes referred to as the Chen term,[40,41] since Chen[28] pointed out this omission in the Fuoss–Onsager[4] derivations. However, part of this term had already been considered by Pitts[3] in 1953.

The contribution of $\Lambda_{e2}^{(1)}$ has an interesting feature, since it is this factor that leads to a $c \log c$ contribution in conductance, just like $\Delta X_{Th}^{(1)}/X$ does from $g_I^{\prime h(1)}$ defined in Eq. (149). It is interesting to notice that Pitts,[3,48] as

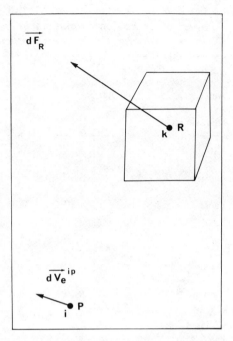

Figure 4. Definition of the one-particle mean electrophoretic velocity $\mathbf{v}_e^{iP}$. The quantity $d v_e^{iP}$ is the average contribution to the velocity of ions of type i at P originating from the force $d\mathbf{F}_R$ transmitted to the medium by the friction with the solvent of ions of all kinds in the volume element dV_R. The force is given by

$$dF_R = \sum_k n_k(e_k X + F_{iP}^{kR}) g_{iP}^{kR} \, dV_R$$

and the relation between $d\mathbf{V}_e^{iP}$ and $d\mathbf{F}_R$ is

$$d\mathbf{V}_e^{iP} = \mathbf{\chi}(P, R, \omega_i, \omega_k) \, d\mathbf{F}_R$$

where $\mathbf{\chi}$ is the hydrodynamic tensor given by Eq. (47).

well as Fuoss and Onsager, obtained a $-E_2 c \log c$ contribution, but from different origins (from $\Lambda_{e2}^{(1)}$ for Pitts' case and from $\Delta X_{Th}^{(1)}/X$ for Fuoss and Onsager's case).

This point had been clearly noticed by Carman,[55,57,58] who correctly concluded at the time that both contributions, being different, should be added, thus advocating for the first time the introduction of a $-2E_2' c \log c$ term in the conductance equation. With regard to the $c \log c$ contribution, it should be mentioned that Feistel[45] in a very original work has shown that, in fact, the net hydrodynamic contribution should be $-\frac{3}{2} E_2 c \log c$. It was shown quite recently by Sändig and Feistel[46] that the extra $+\frac{1}{2} E_2' c \log c$ is indeed a relaxation field term originating from the contribution of $H_{ji}^{(1)}(r)$ defined by Eq. (84) which appears in the T_6 term of Eq. (104) of the continuity equation, which has always been overlooked so far.

The electrophoretic contribution Λ_e, when expanded with respect to concentration, leads to the linearized formulation

$$\Lambda_e = -\beta c^{1/2} - k_e E_2' c \log c$$

$$+ 2E_2'\left(Q^{el} - k_e \log \frac{2\kappa q}{c^{1/2}}\right) c + O(c^{3/2}) \tag{185}$$

The various results for k_e and Q^{el} are summarized in Table 5.

Table 4

Comparison of the Approximations in the Electrophoretic Contribution to Molar Conductance[a]

Authors and approximations	General formulation Eq. (184) $\Lambda_e = (\Lambda_{e1} + \Lambda_{e2}^{(2)})\left(1 + \dfrac{\Delta X}{X}\right) + \Lambda_{e2}^{(1)}\left(1 + \dfrac{\Delta X}{X} + \dfrac{\Lambda_e}{\Lambda_0}\right)$
Pitts (P)[3]	$\Lambda_{e1}(P) + \Lambda_{e2}^{(1)}(P)$
Fuoss–Hsia (F)[82]	$\Lambda_{e1}(F)\left(1 + \dfrac{\Delta X_{LL}}{X}\right)$
Fuoss–Onsager–Skinner (FOS)[24]	$\Lambda_{e1}(FOS) + \Lambda_{e2}(FOS)$
Fuoss–Chen (F–C)[40]	$\Lambda_{e1}(F) + \Lambda_{e2}^{(1)}(F–C)$
Quint–Viallard (Q–V)[27]	$\Lambda_{e1}(F)\left(1 + \dfrac{\Delta X_{LL}}{X}\right) + \Lambda_{e2}^{(1)}(Q–V)$
Carman[29]	$\Lambda_{e1}(F)$
L–W[31]	$\Lambda_{e1}(L–W) + \Lambda_{e2}^{(1)}(L–W)$
Chen[28]	$\Lambda_{e1}(Ch)\left(1 + \dfrac{\Delta X_{LL}}{X}\right) + \Lambda_{e2}^{(2)}(Ch) + \Lambda_{e2}^{(1)}(Ch)$
FEK[5]	$\Lambda_{e1}(FEK) + \Lambda_{e2}^{(1)}(FEK)$

[a] Λ_{e1} is the leading term of electrophoretic conductance in $\beta c^{1/2}$. All authors who used the D–H equilibrium distribution functions for the restricted primitive model of characteristic distance a found

$$\Lambda_{e1}(F) = -\frac{\beta}{1 + \kappa a}c^{1/2}$$

For the terms $\Lambda_{e2}^{(1)}$ and $\Lambda_{e2}^{(2)}$ differences are due to the differences in the evaluations of the functions $g_I^{\prime(1)}$ and $g_I^{\prime(2)}$ summarized in Table 2 and depending on (1) the approximation used in the continuity equation, (2) the approximation used in the boundary conditions, and (3) the functions $\mathring{g}_{ij}$ used in the continuity equation. Strictly speaking, no derivation is exact, even for the contribution linear in concentration; however, a numerical analysis has shown that when $b > 2$, linear contributions practically cancel each other out.

5.6. The Fuoss–Onsager Truncated Series-Expansion Formulation of Conductance for Symmetrical Electrolytes

The final comparison of results is now easily achieved by combining the relaxation and electrophoretic contributions, leading to

$$\Lambda = \Lambda_0\left(1 + \frac{\Delta X}{X}\right) + \Lambda_e \tag{186}$$

This gives

$$\Lambda = \Lambda_0 - Sc^{1/2} + E'c \ln c + J(a)c + J_{3/2}(a)c^{3/2} + \cdots \tag{187}$$

with

$$S = \alpha \Lambda_0 + \beta \tag{188}$$

$$E' = [E'_1 \Lambda_0 - (k_e + k_h)E'_2] \tag{189}$$

$$J(a) = \sigma_1(a)\Lambda_0 - \sigma_2(a) \tag{190}$$

$$\sigma_1(a) = 2E_1\left(Q^{\text{rel}}(b) + \ln \frac{2\kappa q}{c^{1/2}}\right) \tag{191}$$

$$\sigma_2(a) = 2E'_2\left(Q_h^{\text{rel}}(b) + Q(b) - (k_h + k_e)\ln \frac{2\kappa q}{c^{1/2}}\right)c \tag{192}$$

$$J_{3/2}(a) = [\sigma_3(a)\Lambda_0 + \sigma_{4e}(a) + \sigma_{4h}(a)]c \tag{193}$$

The various coefficients can be calculated as

$$\alpha = 0.8204 \times 10^6 z^{3/2}(DT)^{3/2} \tag{194}$$

$$\beta = 82.501 \times z^{3/2}(DT)^{1/2} \tag{195}$$

$$E'_1 = 2.9422 \times 10^{12} \times z^3/(DT)^3 \tag{196}$$

$$E'_2 = 0.43329 \times 10^8 \times z^3/\eta(DT)^2 \tag{197}$$

$$b = 16.708 \times 10^{-4} \times z/aDT \tag{198}$$

$$\frac{\kappa q}{c^{1/2}} = 4.20155 \times 10^6 \times z^{3/2}/(DT)^{3/2} \tag{199}$$

where z is the absolute charge number of the ions.

Table 5 gives the necessary information to calculate k_h, k_e, Q^{rel}, Q_h^{rel}, and Q^{el} from the original publications.

The functions σ_3, σ_{4e}, and σ_{4h}, which allow the coefficient $J_{3/2}(a)$ to be evaluated, are only crudely approximated in most derivations, if not simply neglected. Table 6 gives the expressions for these coefficients for a few equations.

6. Recent Developments in Conductance

6.1. The "Echo" Formulation[22]

In the classical developments of Pitts, Fuoss, and Onsager and of Falkenhagen *et al.*, described in the previous sections, the aim was to obtain a conductance equation of the following form:

$$\Lambda = \Lambda_0 - Sc^{1/2} + Ec \log c + J(a)c + O(c^{3/2}) \tag{200}$$

Table 5

The Linear Coefficient J(a) of the Concentration Expansion of the Conductance Equation[a]

$$\Lambda = \Lambda_0 - Sc^{1/2} + [E'_1\Lambda_0 - (k_h + h_e)E'_2]\ln c + J(a)c + O(c^{3/2})$$

with

$$J(a) = \sigma_1(a)\Lambda_0 + \sigma_2(a)$$

$$\sigma_2(a) = \sigma_{2h}(a) + \sigma_{2e}(a)$$

$$\sigma_1(a) = 2E'_1\left(Q^{rel}(b) - \ln\frac{2\kappa q}{c^{1/2}}\right)$$

$$\sigma_{2h}(a) = 2E'_2\left(Q_h^{rel}(b) - k_h\ln\frac{2\kappa q}{c^{1/2}}\right)$$

$$\sigma_{2e}(a) = 2E'_2\left(Q_e(b) - k_e\ln\frac{2\kappa q}{c^{1/2}}\right)$$

$$b = 2q/a$$

g_{ii} Function Used	Equation	Q^{rel}	k_h	Q_h^{rel}	k_e	$Q^{el} \equiv Q_e$
	Pitts	$\dfrac{2}{b} + 1.7718 - \ln b$	0	0.78107	1	$\dfrac{8}{b} - 0.7674 + \ln b$
	Fuoss–Hsia		1	$-\dfrac{2}{3b} - 0.76685 + \ln b$	0	$\dfrac{8}{b} + 0.7812$
Hückel	Fuoss–Chen	$-\dfrac{1}{b^3} + \dfrac{2}{b^2} + \dfrac{2}{b} + 0.9074 - \ln b$	1		1	$\dfrac{1}{b^2} + \dfrac{6.5}{b} - 0.2676 + \ln b$
Hückel	Quint–Viallard		1	$-\dfrac{2}{3b} - 1.0168 + \ln b$	1	$\dfrac{1}{b^2} + \dfrac{6.5}{b} - 0.6596 + \ln b$

Debye and Meeron	Lee–Wheaton	Eq. (165) of Reference 31	0	Eq. (165) of Reference 31	1	Eq. (165) of Reference 31
	Carman	Eq. (34) of Reference 57	1	Eq. (34) of Reference 57	0	$\dfrac{8}{b} + 0.7812$ Eq. (11) of Reference 58
	Fuoss–Onsager–Skinner	Eq. (1.38) of Reference 24	1	Eq. (138) of Reference 24	0	$F(b)$ given by Eq. (1.38) of Reference 24
	Murphy–Cohen	Eq. (23) of Reference 25	1	Eq. (23) of Reference 25	1	Eq. (23) of Reference 25
	FEK	$B(b) - \dfrac{1}{2}\dfrac{T_1(b)}{1 - T_1(b)} - 0.036$ Eq. (127) of Reference 5	X	X	1	Eq. (130) of Reference 5
	Chen	$B(b) - 2\dfrac{T_1(b)}{1 - T_1(b)} - 1.28668$ Eqs. (34) and (44) of Reference 44	1	Eq. (33) of Reference 44	1	Eq. (29) of Reference 28

a The functions for the Pitts, Fuoss–Hsia, Fuoss–Chen, and Quint–Viallard derivation are taken from a recent compilation.[70] The functions for the Lee–Wheaton, Carman, Fuoss–Onsager–Skinner, and Murphy–Cohen equations can be evaluated from the original papers, for which precise references are given. In the FEK and Chen results functions $B(b)$ and $T_1(b)$ are given by

$$B(b) = -Ei(b) - \frac{e^b}{b}\left(1 + \frac{1}{b} + \frac{2}{b^2}\right)$$

and

$$T_1(b) = e^{-b}(1 + b + \tfrac{1}{2}b^2)$$

In the FEK derivation the contributions to Q_k^{rel}, though recognized, were not evaluated, which explains the X symbol to express the missing terms.

Table 6
Comparison of the Coefficients of the $J_{3/2}(a)$ Function[a]

	Pitts	Fuoss–Hsia	Fuoss–Chen	Quint–Viallard
σ_3	$-\dfrac{2\kappa q}{c^{1/2}}E_1'\left(\dfrac{5.1716}{b^2}+\dfrac{6.2928}{b}\right)$	$-\dfrac{2\kappa q}{c^{1/2}}E_1'\left(\dfrac{3.8284}{b^3}+\dfrac{4.4748}{b^2}+\dfrac{0.6094}{b}\right)$		$-\dfrac{2\kappa q}{c^{1/2}}E_1'\left(\dfrac{3.8048}{b^3}+\dfrac{44296}{b^2}+\dfrac{1.6094}{b}\right)$
σ_{4e}	$-\dfrac{2\kappa q}{c^{1/2}}E_2'\left(\dfrac{16}{b^2}-\dfrac{1.41424}{b}\right)$	$-\dfrac{2\kappa q}{c^{1/2}}E_2'\left(\dfrac{4}{3b^3}+\dfrac{18.6667}{b^2}+\dfrac{4.288}{b}-1.2922\right)+E_2'\dfrac{\beta}{b_{1/2}}\left(\dfrac{4}{3b}-2.219\right)$	$-\dfrac{2\kappa q}{c^{1/2}}E_2'\left(\dfrac{13.60944}{b^2}+\dfrac{0.1712}{b}\right)$	$-\dfrac{2\kappa q}{c^{1/2}}E_2'\left(\dfrac{13.60947}{b^2}-\dfrac{1.3904}{b}\right)$
σ_{4h}	$-\dfrac{2\kappa q}{e^{1/2}}E_2'\dfrac{4.2288}{b}$	$\dfrac{2\kappa q}{c^{1/2}}E_2'\left(\dfrac{2.2761}{b^2}+\dfrac{1.5405}{b}\right)$		$\dfrac{2\kappa q}{c^{1/2}}E_2'\left(\dfrac{2.2761}{b^2}+\dfrac{2.3940}{b}\right)$

[a] Pitts' and Fuoss and Hsia's functions were evaluated by Fernandez-Prini and Prue[53] and Fernandez-Prini,[56] respectively; Fuoss–Chen's σ_{4e} function was evaluated by the author[40,41]; Quint and Viallard's functions have been taken from the original publication.[27]

for which calculations were carried out in order to obtain a reliable function up to the linear term for the restricted primitive model. This purpose can be achieved only if the equilibrium distribution function used at the level of the S_1 term is of the type

$$\overset{\circ}{g}_{ii}(r) = \exp\left(-\frac{U_{ji}^*}{kT} + \frac{Z_iZ_j}{|Z_iZ_j|}\frac{2q}{r}e^{-\kappa r}\right) \quad (201)$$

where U_{ji}^* represents the infinite hard-core repulsion energy of spheres of diameter a.

However, the truncated expansion summarized by Eq. (201) proves to be quite insufficient to account for experimental data as soon as the coulombic attractions become large ($q > 0.3$ nm), as is the case for $1:1$ salts in solvents of low dielectric constant or even for $2:2$ salts in water. In these cases the observed Λ curves tend to behave very much like those for acetic acid solutions in water, where the chemical phenomenon of incomplete dissociation is well recognized. Consequently, the usual procedure is to graft empirically onto Eq. (200) the necessary complementary terms that describe the association of ions. The result is then

$$\Lambda = \gamma[\Lambda_0 - Sc^{1/2}\gamma^{1/2} + Ec\gamma \log c\gamma + J(a)c\gamma + O(c^{3/2}\gamma^{3/2})] \quad (202)$$

where γ is the fraction of free ions that follow a mass action law formulation, characterized by a so-called constant of association K_A. While this last equation is undoubtedly quite legitimate for acetic acid in water, since it is well known that in this case the anion–cation interaction does lead to a real neutral entity with a covalent bond, this is not the case for its use with ionophores where no chemical bonding occurs. While Eq. (202) is certainly valid for ionogens (electrolytes such as acetic acid in water), it was soon observed[59] that it led to systematic deviations of two kinds:

(i) The deviations $\Delta\Lambda_i = \Lambda_{i,\text{expt}} - \Lambda_{i,\text{calc}}$, after the adjustment, were not randomly distributed and were thus correlated with the concentration, the more so the lower the dielectric constant of the solution. This is due to the fact that the next term in $c^{3/2}$ of the truncated expansion is less and less negligible and has therefore to be introduced and adjusted.

(ii) The value of the distance parameter $a_{J'}$ derived from the adjusted J coefficient, increases for a given salt when the dielectric constant decreases. This trend is still more significant when the $c^{3/2}$ term is adjusted. Clearly the distance parameter a_J was not at all specific for the salt but rather for the solvent. This induced the present author[17,59] to come back to the original model of association introduced by Bjerrum[6] and to propose the theoretical equation

$$\Lambda = \gamma(\Lambda_0 - Sc^{1/2}\gamma^{1/2} + Ec\gamma \log c\gamma + J(R)c\gamma + J_{3/2}(R)c^{3/2}\gamma^{3/2} \quad (203)$$

together with the original set of equations of Bjerrum,

$$\frac{1-\gamma}{\gamma^2 c f_\pm'^2} = K_A^B = \frac{4\pi N}{1000} \int_a^R r^2 \exp\left(\frac{2q}{r}\right) dr \qquad (204)$$

$$\log f_\pm' = -\frac{\kappa q \gamma^{1/2}}{1 + \kappa R \gamma^{1/2}} \qquad (205)$$

with

$$R = \begin{cases} q & \text{if } a < q \\ a & \text{if } a \geq q \end{cases}$$

as the choice for the cutoff distance, as originally proposed by Bjerrum.

When processed with this system of equations, all systematic errors disappear and the distance parameter a_K^B, now derived from the adjusted value of K_A^B, becomes much closer to that usually expected or is at least in agreement with the same parameter derived from processing excess thermodynamic data.

Another very interesting result is that a_K^B is now quite independent of the choice for R in a range

$$q/2 < R < 2q \qquad (206)$$

In other words, changing R leads to variations in $J(R)$ [and $J_{3/2}(R)$], which are compensated by the change in K_A^B so that a_K^B remains unchanged. This is the experimental evidence that in a somewhat large range around $r = q$ the Bjerrum and Debye approximations match well owing to the fact that

$$\left(\frac{\partial K_A^B}{\partial R}\right)_{R \sim q} \simeq -\frac{1}{\Lambda_0}\left(\frac{\partial J}{\partial R}\right)_{R \sim q} \qquad (207)$$

However, these facts could not be explained until recently, when the echo effect was discovered and opened the way to the justification of the theoretical set of Eqs. (204–206).

The problem is to find which approximation in the treatment developed in the previous sections leads to the loss of the "association formulation," Eqs. (203–205).

The answer is now known and it is very simple: The loss arises when $\mathbf{M}_1$ in Eq. (76) is split into its three components S_1, T_4, and T_9 [cf. Eq. (108)] and *when the latter two terms T_4 and T_9 are either neglected or considered as small perturbation terms for which only one iteration is sufficient in the iterative integration of the continuity equation.* As described in Sections 5.3 and 5.4, the general solution for the perturbation on the anion/cation distribution function for binary symmetrical electrolytes is

$$g_{+-}' = (g_{I+-c}'^{(1)} + g_{I+-h}'^{(1)})\left(1 + \frac{\Delta X}{X} + \frac{\Lambda_e}{\Lambda_0}\right) + g_{I+-}'^{(2)}\left(1 + \frac{\Delta X}{X}\right) \qquad (208)$$

which leads to

$$\frac{\Delta X}{X} = \left(\frac{\Delta X_{Ic}^{(1)}}{X} + \frac{\Delta X_{Ih}^{(1)}}{X}\right)\left(1 + \frac{\Delta X}{X} + \frac{\Lambda_e}{\Lambda_0}\right) + \frac{\Delta X_I^{(2)}}{X}\left(1 + \frac{\Delta X}{X}\right) \qquad (209)$$

for the relaxation field contribution through use of the relation

$$\frac{\Delta X_I}{X} = -\frac{\mathbf{X}}{X}\frac{1}{e_i}\frac{1}{X}\sum_k n_k \int \nabla_P U(iP, kR) g'_{I_{ik}} dV_R \qquad (210)$$

where the homologous indices that relate the various $\Delta X_I/X$ and g'_I functions have been omitted for the sake of simplicity. Let us recall that for binary symmetrical electrolytes the summation over k reduces, in fact, to the only term $k \neq i$, so that $g'_{I_{ik}}$ reduces to $g'_{I_{+-}} = g'_{I_{-+}}$; As a consequence, $\Delta X^i/X = \Delta X^k/X = \Delta X/X$, which justifies the simplified formulation used in Eqs. (208) and (209).

We have seen in Section 5.5 that the electrophoretic contribution is the sum of two terms:

$$\lambda_e^i = (\lambda_{e1}^i + \lambda_{e2}^{i(2)})\left(1 + \frac{\Delta X}{X}\right) + \lambda_{e2}^{i(1)}\left(1 + \frac{\Delta X}{X} + \frac{\Lambda_e}{\Lambda_0}\right) \qquad (211)$$

where λ_{e1}^i corresponds to the original Fuoss–Onsager approximation, $\lambda_{e2}^{(1)}$ to the Pitts–Chen contribution. The latter originates from $\lambda g'^{(1)}_{I_{+-}}$, and $\lambda_{e2}^{i(2)}$ in a new term, hitherto neglected, that originates from $g'^{(2)}_{I_{+-}}$. Finally, for the total electrophoretic conductance there results

$$\Lambda_e = \lambda_e^+ + \lambda_e^- \qquad (212)$$

$$\frac{\Lambda_e}{\Lambda_0} = \left(\frac{\Lambda_{e1}}{\Lambda_0} + \frac{\Lambda_{e2}^{(2)}}{\Lambda_0}\right)\left(1 + \frac{\Delta X}{X}\right) + \frac{\Lambda_{e2}^{(1)}}{\Lambda_0}\left(1 + \frac{\Delta X}{X} + \frac{\Lambda_e}{\Lambda_0}\right) \qquad (213)$$

When the coulombic attraction is important, the leading contribution in these formulas is the short-range part of the relaxation effect, which reads

$$\frac{\Delta X^S}{X} = -\frac{\mathbf{X}}{X}\frac{1}{e_i X} n \int_\Omega \nabla_P U(iP, jQ) g'_{ij} dV_Q \qquad (i \neq j) \qquad (214)$$

where the integration is carried over all positions of a point Q inside a sphere Ω of radius R around a point P. It is even easy to predict that among the various contributions to $\Delta X^S/X$ the nonhydrodynamic one is by far the most important. This contribution $\Delta X_{Ic}^S/X$ comes from the term g'^1_{Ic} in Eq. (159) and is given by

$$\frac{\Delta X_{Ic}^S}{X} = \frac{2\pi}{e_i X} n \int_0^R \int_0^\pi \left(\frac{d}{dr} U_{12}\right) \cos\theta \, g'^{(1)}_{Ic12} r^2 \sin\theta \, dr \, d\theta \qquad (215)$$

Separating out in the conductance equation, Eq. (187), the hydrodynamic and nonhydrodynamic contributions of the relaxation effect, and separating out in the nonhydrodynamic contribution the short- and long-range parts, gives

$$\Lambda = \Lambda_0\left(1 + \frac{\Delta X_c^S}{X} + \frac{\Delta X_c^L}{X} + \frac{\Delta X_h}{X} + \frac{\Lambda_e}{\Lambda_0}\right) \tag{216}$$

Distinguishing the echo factor in $\Delta X_c^S/X$ gives

$$\frac{\Delta X_c^S}{X} = \frac{\Delta X_{Ic}^S}{X}\left(1 + \frac{\Delta X}{X} + \frac{\Lambda_e}{\Lambda_0}\right) \tag{217}$$

since X_{Ic}^S/X can be derived from $g_{I+-}^{\prime(1)}$, as shown in Eq. (208). Introducing Eq. (217) in Eq. (216) leads to

$$\Lambda = \frac{1}{1 - \Delta X_{Ic}^S/X} \Lambda_0\left(1 + \frac{\Delta X_c^L}{X} + \frac{\Delta X_h}{X} + \frac{\Lambda_e}{\Lambda_0}\right). \tag{218}$$

The factor

$$\frac{1}{1 - \Delta X_{Ic}^S/X} = \gamma \tag{219}$$

is a dimensionless number such that

$$0 \le \gamma \le 1 \tag{220}$$

since $\Delta X_{Ic}^S/X$ is always negative if $R > a$. The system of equations (218)–(219) can be written as

$$\begin{cases} \Lambda = \gamma\Lambda_0\left(1 + \frac{\Delta X_c^L}{X} + \frac{\Delta X_h}{X} + \frac{\Lambda_e}{\Lambda_0}\right) & (221) \\[2mm] \dfrac{1 - \gamma}{\gamma} = -\dfrac{\Delta X_{Ic}^S}{X} & (222) \end{cases}$$

This system has the advantage of making explicit use of the *unechoed* short-range part of the nonhydrodynamic relaxation coefficient only but nevertheless includes *implicitly* through γ, the total *echoed* term $\Delta X_c^S/X$. In other words, Éqs. (221) and (222) make it unnecessary for the theoretician to proceed to an infinite number of iterations in the integration of the Onsager–Ebeling continuity equations for the evaluation of the nonhydrodynamic part of the perturbation of the pair distribution function.

In fact, the echo coefficient could have been explicitly taken into account for all contributions by substituting Eqs. (209) and (213) in the conductance equation, Eq. (216). The result then reads

$$\frac{\Lambda}{\Lambda_0} = \frac{1 + \dfrac{\Lambda_{e1}}{\Lambda_0} + \dfrac{\Lambda_{e2}}{\Lambda_0}}{1 - \dfrac{\Delta X_I^{(1)}}{X}\left(1 + \dfrac{\Lambda_{e1}}{\Lambda_0} + \dfrac{\Lambda_{e2}^{(2)}}{\Lambda_0}\right) - \dfrac{\Delta X_I^{(2)}}{X}\left(1 - \dfrac{\Lambda_{e2}^{(1)}}{\Lambda_0}\right) - \dfrac{\Lambda_{e1}}{\Lambda_0} - \dfrac{\Lambda_{e2}^{(2)}}{\Lambda_0}} \tag{223}$$

Neglecting $\Lambda_{e2}^{(2)}/\Lambda_0$ and $\Delta X_I^{(2)}/X$ in Eq. (223) gives a result that corresponds to the "nonlinear" equation obtained by Ebeling, Feistel, Kelbg, and Sändig[60] (EFKS). The terms missing in the EFKS formulation are absent because in their derivation EFKS neglected the hydrodynamic monitoring $\mathbf{M}_h(\mathbf{r})$ term in the Onsager–Ebeling continuity equation, Eq. (87).

There are some advantages in choosing the formulation given by Eqs. (221) and (222) among many other possibilities offered by the echo effect. On the one hand, it is similar to a familiar formulation that has been quite successful so far, though rather empirical; this similarity will be stressed even more when the strong coupling approximation will be used to evaluate $\Delta X_{Ic}^S/X$. On the other hand, it makes a clear distinction between a numerically important short-range contribution $(\Delta X_{Ic}^S/X)$ and all the others, which decrease in numerical importance when that of the former increases. Consequently, it becomes both useful and possible to attempt to improve the accuracy of the specific term $\Delta X_{Ic}^S/X$ that is, in fact, the $g_{Ic}^{\prime(1)}$ contribution to the anion–cation pair distribution function in the short-range region $r < R$.

6.2. The Strong Coupling Approximation[21] (SCA)

When two ions of type i and j are in relative positions where strong radial interactions are important, several consequences are to be expected. First, the deviation from ideality is large so that the pair distribution functions for the two ions are far from unity in this region. Secondly, a still more important factor in the problem concerned here is to be expected: A large external perturbation will be required to modify the radial parts of the two-particle mean-velocity vectors $\mathbf{v}_{iP}^{jQ}$ and $\mathbf{v}_{jQ}^{iP}$. As a corollary, in the case where the external field X is small (Ohm's law), these quantities will remain practically unperturbed, that is, very small, since at equilibrium ($\mathbf{X} = 0$) they are strictly null.

While the first factor is obvious and well recognized, it is not properly taken into account in the Debye–Hückel treatment owing to the expansion and linearization of the Boltzmann equation. In this respect the Meeron distribution functions are a definite improvement. We shall see in Section 6.3 how this factor can be more fully taken into account. Since, as will be shown below and as a consequence of the second factor, the short-range part of $\mathring{g}_{+-}(r)$ controls the quantity $\Delta X_I^S/X$ defined in the previous section, it can then be understood why the intermediary variable γ arising from the echo effect allows a very efficient analytical formulation to be made that avoids the drastic truncation implied by the usual linearization of *all terms* versus concentration.

The second factor is less trivial but nevertheless easy to understand. In the case of strong radial interactions, as in centered Coulomb fields (or in Newtonian gravitational fields), tangential movements are "free," whereas the opposite is true for radial movements. A small external perturbation will cause a significant response in the former, whereas the latter will remain practically unchanged. A well-known example of such a situation is to be found in strongly bonded electrons in atoms: These electrons participate mainly in polarization effects ["free" only in tangential degrees of movement (θ, ϕ)], whereas weakly bonded electrons, being more "free" to move [three degrees of freedom, r, θ, and ϕ], participate almost exclusively in the passage of direct current. In another example, two spheres rolling on each other (at contact) have as the radial component of their relative velocity a vector that is obviously zero for hard-sphere reasons. However, the same quantity is also zero for a planet and its satellite; although not in contact, they are, by definition, in a situation of strong interaction. The analogy to our problem can be carried still further: The tangential response in velocity of a satellite to an external tangential perturbation will be much greater than the radial response to a radial perturbation of the same intensity. It is well known that it takes a lot more energy to increase the radial velocity than it does to increase the tangential velocity of a satellite by the same amount.

Consequently, the SCA can be summarized by

$$\mathbf{v}_{ij}(\mathbf{r}) \cdot \frac{\mathbf{r}}{r} \simeq 0 \qquad \text{for } r < R \tag{224}$$

where R is now a distance within which the ionic interactions between the two ions become strong. For practical reasons of symmetry in the analytical presentation, it is more convenient to rewrite the condition as

$$g_{ji}(\mathbf{r})[\mathbf{v}_{ji}(\mathbf{r}) - \mathbf{v}_{ij}(\mathbf{r})] \cdot \frac{\mathbf{r}}{r} \simeq 0 \simeq \text{as } r < R \tag{225}$$

The bracketed quantity in Eq. (225) has been expressed in Section 4 and leads to

$$g_{ji}(\mathbf{r})[\mathbf{v}_{ji}(\mathbf{r}) - \mathbf{v}_{ij}(\mathbf{r})] = \nabla \ln g_{ji}(\mathbf{r}) - [\mathbf{M}_1 + \mathbf{f}_{ji}(\mathbf{r}) + \mathbf{M}_h(\mathbf{r}) + \mathbf{H}_{ji}(\mathbf{r})]/kT \tag{226}$$

which must be substituted in Eq. (225). Since the strong interaction approximation is used to evaluate the nonhydrodynamic part of $g'_{ji}(\mathbf{r})$, both $\mathbf{M}_h(\mathbf{r})$ and $\mathbf{H}_{ji}(\mathbf{r})$ can be omitted. Substituting Eq. (42) for $\mathbf{M}_1$ and Eqs. (77) and (89) for $\mathbf{f}_{ji}(\mathbf{r})$, the relation

$$\frac{d}{dr} \ln g_{c_{ji}}^{(1)}(\mathbf{r}) - \left(\frac{d}{dr} v_{ji} \frac{\omega_i e_i - \omega_j e_j}{\omega_i + \omega_j} \frac{Xr \cos \theta}{kT} + \frac{d}{dr} \ln \mathring{g}_{ji}(r) - \frac{d}{dr} \frac{\phi'_{ji}(\mathbf{r})}{kT} \right) \simeq 0 \tag{227}$$

is obtained, where $\mathbf{f}'_{ji}(\mathbf{r})$ has been expressed in terms of a potential-energy function $\phi'_{ji}(\mathbf{r})$:

$$\mathbf{f}'_{ji}(\mathbf{r}) \equiv -\nabla \phi'_{ji}(\mathbf{r}) \tag{228}$$

The integration of the continuity equation, Eq. (227), is straightforward and gives

$$g^{(1)}_{C_{ji}}(\mathbf{r}) \simeq \mathring{g}_{ji}(r) \exp\left(\nu_{ji} \frac{\omega_i e_i - \omega_j e_j X r \cos\theta}{(\omega_i + \omega_j)kT} - \frac{\phi'_{ji}(\mathbf{r})}{kT}\right) \tag{229}$$

Particularizing to the case of Ohm's law by expanding and truncating to the term linear in X, omitting the echo factor ν_{ji} to obtain an evaluation of $g'_{IC_{ji}}$ instead of $g'_{C_{ji}}$, and neglecting the higher-order perturbation term $\phi'_{ji}(\mathbf{r})$ leads to

$$g'^{(1)}_{IC_{ji}}(\mathbf{r}) \simeq \mathring{g}(r) r \cos\theta \frac{(\omega_i e_i - \omega_j e_j)X}{(\omega_i + \omega_j)kT} \qquad \text{for } r < R \tag{230}$$

In the case of a binary symmetrical electrolyte Eq. (230) becomes

$$g'^{(1)}_{IC_{+-}}(\mathbf{r}) \simeq \mathring{g}_{+-}(r) r \cos\theta \frac{e_2 X}{kT} \tag{231}$$

which is, so far, the best analytical approximation to be used for the evaluation of $\Delta X^S_{IC}/X$ in Eq. (215). This result will prove to be quite efficient for the evaluation of $\Delta X^S_{IC}/X$ from Eq. (215). The result is

$$\frac{\Delta X^S_{IC}}{X} = -\frac{4\pi}{3} n \int_0^R \left(\frac{d}{dr}\frac{U_{12}}{kT}\right) r^3 \mathring{g}_{12}(r)\, dr \tag{232}$$

Integration by parts leads to

$$\frac{\Delta X^S_{IC}}{X} = -\frac{4\pi N}{1000} c \int_0^R r^2 \mathring{g}_{12}(r)\, dr$$

$$-\frac{4\pi N}{3000} c \int_0^R r^3 \exp\left(-\frac{U_{12}}{kT}\right) \frac{d}{dr}\left[\exp\left(\frac{U_{12} - \mathring{\psi}_{12}}{kT}\right)\right] dr$$

$$+\frac{4\pi N}{3000} c R^3 \mathring{g}_{12}(R) \tag{233}$$

where the cation (or anion) density n has been replaced by the molar concentration c of the binary salt, $n = (N/1000)c$, and N is Avogadro's number. Denoting the first term on the right-hand side of Eq. (233) as

$$-cG^S_{12} = -\frac{4\pi N}{1000} c \int_0^R r^2 \mathring{g}_{12}(r)\, dr \tag{234}$$

neglecting the second term of order $O(c^2)$, and transferring the last term in the long-range factor† (since it depends on the cutoff distance only) Eq. (222) then becomes:

$$\frac{1-\gamma}{\gamma} = cG_{+-}^S + O(C^2) \tag{235}$$

At this stage the problem of evaluating $\Delta X_{Ic}^S/X$ no longer belongs to the regime of linear irreversibility but to that of equilibrium behavior, since γ is now simply related to the equilibrium anion–cation distribution function $\overset{\circ}{g}_{+-}(r)$.

It is then possible to make use of recent progress made in this domain. It is possible, for example, to use the results of the hypernetted chain (HNC) calculation; however, since we are mainly seeking analytical solutions, we shall briefly summarize in the next section the present status of this question.

It should be noted, however, that Eq. (235) provides evidence that conductance is, in fact, among the most precise methods for the measurement of excess thermodynamic quantities such as activity coefficients, a result that had been ascertained and utilized on experimental grounds sometime ago. In fact, G_{+-}^S, as defined by Eq. (234), is the leading contribution to excess thermodynamic quantities in dilute solutions, and, in addition, it represents most of the factors that cause ionic specificity in dilute electrolyte solution behavior.

6.3. Use of the Activity Expansion Theorem to Reach an Approximation for the Evaluation of G_{+-}^S in Dilute Solutions[61]

In an analysis[62] concerning the activity coefficient of a binary symmetrical electrolyte the following interesting result was obtained. The activity coefficient $f_\pm$ of such an electrolyte was shown to tend to the following relations

---

† It is useful to note that the last term in Eq. (233) can be rewritten as

$$\frac{4\pi N}{3000} cR^3 \overset{\circ}{g}_{12}(R) = 2E_1' \frac{2\overset{\circ}{g}_{12}(R)}{\zeta^3} c$$

where $\zeta = 2q/R$.

We are interested in the linear contribution in c of this term; this is evaluated by substituting $\overset{\circ}{g}_{12}(R) = e^\zeta - O(c^{1/2})$. Then

$$\frac{4\pi N}{3000} cR^3 \overset{\circ}{g}_{12}(R) = 2E_1' \frac{2e^\zeta}{\zeta^3} c$$

which happens to be identical to an osmotic-pressure contribution on a sphere of radius R. It is seen that such a contribution exists as a result of the contribution given by Eq. (173) to the $\sigma_1(a)$ function, which leads to a term $2E_1'(2e^b/b^3)$, but applying to a sphere of radius a. As a consequence, substituting the function $\sigma_1(R)$ for $\sigma_1(a)$ when applying the strong coupling formulation will automatically take care of the transfer of this new term into the long-range part of the nonhydrodynamic relaxation contribution.

with decreasing concentration:

$$\begin{cases} f_\pm = \gamma f_\pm^L(R, c) + O(c^{3/2}) & (236) \\ \dfrac{1 - \gamma}{\gamma} = cG_{+-}^S & (237) \end{cases}$$

where $f_\pm^L(R, c)$ is a function that depends only on the long-range parts of the zeroth moment of the three distribution functions $(++, --, \text{ and } +-)$. The result for Hamiltonian models for anion/cation interactions that are involved here are illustrated in Figure 9.

For the sake of simplicity some terms have been omitted in the system of Eqs. (236) and (237), which concern G_{++}^S and G_{--}^S as well as excluded volume contributions (see reference (62) for the exact formulation).

The analogy with the result obtained by combining the echo effect with the SCA is obvious. In fact, at high dilutions the γ function is the same in both treatments. In a recent publication Wood, Lilley, and Thomson[63] have stimulated new interest in the Mayer activity expansion in the description of the excess thermodynamic properties of gases and solutions. Related work is described in references 64–66. This was the basis of the work of reference 64 that allowed a further advance to be made in the generalization of Eqs. (236) and (237). A space partition carried out on the cluster integrals of the Mayer expansion leads, after some elementary algebra, to the following set of expressions for the evaluation of the mean activity coefficient of a binary symmetrical electrolyte easing the restrictions of the primitive model for ions:

$$\begin{cases} f_\pm = \gamma f_\pm^L(R, c\gamma) & (238) \\ \dfrac{1 - \gamma}{\gamma} = \gamma c (f_\pm^L)^2 K_A^B + O(cK_{++}, cK_{--}, c^2 K_{+-+}, \cdots) & (239) \end{cases}$$

where

$$K_A^B = 2K_{+-} = \frac{4\pi N}{1000} \int_0^R r^2 \exp\left(-\frac{U_{+-}}{kT}\right) dr \tag{240}$$

Incidentally, this result indicates that the empirical (1926) Bjerrum formulation becomes the limiting case of the Mayer expansion at high dilutions.

The neglected terms $O(cK_{++}, \cdots)$ contain all the other integrals of the Mayer cluster expansion, each of them containing at least one short-range "bond." The function $f_\pm(R, c\gamma)$ represents the activity coefficient of a fictive binary symmetrical electrolyte at a concentration $c\gamma$ that has the characteristics of a primitive Hamiltonian model of hard spheres of diameter R. The important result in Eq. (239) is that it leads to an evaluation for G_{+-}^S that reads

$$G_{+-}^S = \gamma (f_\pm^L)^2 K_A^B + \cdots \tag{241}$$

and which can consequently be used in Eq. (235) to complete the conductance coefficient equation.

It is not without interest to look for an analytical expression for $\mathring{g}_{+-}(r)$ that would lead to the result given by Eq. (241). Obviously the Debye approximation

$$\mathring{g}_{+-}(r) = 1 + \frac{2q}{r} e^{-\kappa r} \frac{e^{\kappa a}}{1 + \kappa a} + \frac{1}{2}\left(\frac{2q}{r} e^{-\kappa r} \frac{e^{\kappa a}}{1 + \kappa a}\right)^2 \tag{242}$$

is insufficient. In this respect the Meeron distribution function

$$\mathring{g}_{+-}(r) = \exp\left(\frac{2q}{r} e^{-\kappa r}\right) \tag{243}$$

is much better, since at high dilutions the expansion of the screening factor can be truncated after its first term:

$$\mathring{g}_{+-}(r) \simeq \exp\left(-2\kappa q\right) \exp\frac{2q}{r} + \cdots \tag{244}$$

which leads to

$$G_{+-}^S \simeq (f_{\pm,\mathrm{LL}})^2 K_A^B \tag{245}$$

where $f_{\pm,\mathrm{LL}}$ is the Debye–Hückel limiting law for the activity coefficient, a result that proceeds toward Eq. (241). Clearly the formulation

$$\mathring{g}_{+-}(r) \simeq \gamma \exp\left(\frac{2q}{r} e^{-\kappa r\gamma^{1/2}}\right) + 1 - \gamma \tag{246}$$

is still better, since the same development leads to

$$G_{+-}^S \simeq \gamma \exp\left(-2\kappa q\gamma^{1/2}\right) K_A^B \tag{247}$$

Further discussion concerning these questions will be found elsewhere in more detail.[64] The final following remark is, however, appropriate: If use is made of Eq. (246) for $\mathring{g}_{+-}(r)$ in the Onsager–Ebeling equation, the functions $\Delta X_C^L/X$, $\Delta X_h/X$, and Λ_e/Λ_0 become functions of $c\gamma$ instead of c, as is obtained in the usual way (without the echo effect) with the Debye or Meeron original equation for $\mathring{g}_{+-}(r)$. As a consequence, the following set of equations is finally obtained:

$$\Lambda = \gamma(\Lambda_0 - Sc^{1/2}\gamma^{1/2} + Ec\gamma \log c\gamma$$
$$+ [\sigma_1(R)\Lambda_0 + \sigma_2(a)]c\gamma + J_{3/2}c^{3/2}\gamma^{3/2} + \cdots) \tag{248}$$

$$\frac{1 - \gamma}{(\gamma^2 c f_\pm^L)^2} = K_A^B \tag{249}$$

which, except for the function $\sigma_2(a)$, is identical with the original but empirical equation proposed by the present author in 1967. The question of the $\sigma_2(a)$ function will be discussed in Section 6.5.

6.4. Theoretical Test of the SCA on the Nonhydrodynamic Part of the Relaxation Effect

The SCA [Eq. (224)] allowed the derivation of the set of equations (248) and (249) from the original exact set of equations (221) and (222). It is important to see if it is possible to justify this result in a way that is more exact than the qualitative explanation given in Section 6.2. Of course, it will not be possible to obtain a complete quantitative justification, since this would mean that an exact derivation of Eqs. (248) and (249) is available; however, it is possible to test how the leading terms of these equations deviate from exactness. Since the leading term of the mass action law [Eq. (249)] is linear in concentration, we shall examine how the sum of the first three terms of the expansion of Eq. (248) plus the linear term of Eq. (249),

$$\frac{\Delta X_{Ic}}{X}[\text{SCA}] = -\alpha c^{1/2} + E_1 c \log c + [\sigma_1(R) - K_A^B]c \qquad (250)$$

compare with the result

$$\frac{\Delta X_{Ic}}{X} = -\alpha c^{1/2} + E_1 c \log c + \sigma_1(a)c \qquad (251)$$

obtained by the use of the exact boundary condition, but with the same continuity equation and equilibrium distribution function given by Eq. (243). Such a calculation has been already attempted indirectly by Fuoss, Onsager, and Skinner,[24] who avoided the use of the linearized Poisson–Boltzmann equation throughout their derivation. An analysis of the approximation they used for the Poisson–Boltzmann equation indicates that it is certainly consistent with the Meeron distribution function. However, it is easier to make use of a derivation that corresponds exactly to our requirement, that of Falkenhagen, Ebeling, and Kraeft, and of Kremp (EKK),[5,65] whose result is

$$\sigma_{1[\text{EKK}]}c = 2E_1'\left[-E_i(b) + \frac{e^b}{b}\left(1 + \frac{1}{b} + \frac{2}{b^2}\right) + c_1 + f_1(b) + \ln\frac{2\kappa q}{c^{1/2}}\right]c \qquad (252)$$

where

$$E_i(x) = \int_{-\infty}^{x} \frac{e^t}{t} dt \qquad (253)$$

and

$$f_1(x) = \frac{1 + x + (1/2)x^2}{2(1 + x + (1/2)x^2 - e^x)} \qquad (254)$$

It is a matter of elementary algebra to rewrite the term $-K_{AC}^B$ of Eq. (250):

$$-K_{AC}^B = 2E_1'\left[-E_i(b) + \frac{e^b}{b}\left(1 + \frac{1}{b} + \frac{2}{b^2}\right) + E_i(\zeta) - \frac{e^\zeta}{\zeta}\left(1 + \frac{1}{\zeta} + \frac{2}{\zeta^2}\right)\right]c \tag{255}$$

where

$$\zeta = 2q/R \tag{256}$$

Consequently, if use is made of the EKK function for $\sigma_1(R)$ in Eq. (250), then

$$\sigma_{1[SCA]}c = 2E_1'\left[-E_i(b) + \frac{e^b}{b}\left(1 + \frac{1}{b} + \frac{2}{b^2}\right) + c_1 + f_1(\zeta) + \ln\frac{2\kappa q}{c^{1/2}}\right]c \tag{257}$$

so that the SCA reduces to substituting $f_1(\zeta)$ for $f_1(b)$, a change that turns out to be less and less significant, compared to the other terms, as soon as b gets larger than 1.5 [$f_1(1) = -5.72$; $f_1(2) = -1.05$; $f_1(\infty) = 0$]. It must be recalled that the first two terms in Eq. (257) become numerically very important when $b > 2$. We thus understand the advantage of the echo formulation: In a simple analytical formulation (which happens to be a mass action law formula), it takes care of an infinite series expansion of large terms of alternating algebraic signs, whereas the classical treatment leads only to the first term of the expansion. Moreover, if the Debye–Hückel distribution function (or its equivalent, the linearized Poisson–Boltzmann equation) is used in the classical treatment, the result for σ_1 is the same as that obtained from Eq. (252) if one expands and truncates all the functions of b at the linear contribution in b. This result then reads

$$\sigma_{1[EKK]}(a) = 2E_1'\left(-\frac{1}{b^3} + \frac{3}{4b^2} + \frac{177}{80b} + 1.3576\right.$$

$$\left. + \ln\frac{2\kappa q}{c^{1/2}} - \ln b + O(b)\right) \tag{258}$$

Now since the $\sigma_1(a)$ function is used only as $\sigma_1(R)$ in Eq. (250), it is seen that the error introduced by using Eq. (258) instead of Eq. (252) is small so long as $R \simeq q$ and that for the long-range linear coefficient $\sigma_1(R)$ the Debye–Hückel approximation is sufficient.

These results justify the empirical generalization initially proposed in 1967 by the present author, which consisted of introducing the "fraction of free ions γ" in the initial equation where the $J(a)$ function becomes $J(R \simeq q)$, and adding a mass action law formula with a constant K_A^B that took care of all the contributions of the strong, short-range interactions when $b > 2$. The method was in fact valid in a concentration range where all the approximations that are implied hold. Experiment shows that this range is quite satisfactorily defined by

$$\kappa q \gamma^{1/2} < 0.5 \tag{259}$$

Another advantage in using the SCA in the echo formulation lies in the fact that the constant of the mass action law, Eq. (240), calls only for U_{+-}, without specifying any model Hamiltonian. Moreover, the result states that within the range of application of the SCA all models U_{+-} leading to numerically equal integrals

$$K_A^B = \frac{4\pi N}{1000} \int_0^R r^2 \exp\left(-\frac{U_{+-}}{kT}\right) dr \qquad (260)$$

are equivalent in that which concerns the behavior of the nonhydrodynamic part of the relaxation effect, which practically control the conductance coefficient completely. We shall see in the next section that the hydrodynamic contributions to conductance are quite insensitive to the type of model used and that these contributions are numerically much smaller than those studied in this section.

Finally, it is not without interest to note here that Eq. (240), which constitutes not only a justification but also a generalization of the Bjerrum concept to any kind of anion–cation interaction potential, also allows a very simple interpretation of the so-called inner- and outer-sphere complex concept.

Dividing the range $[0, R]$ into two ranges $[0, R_1]$ and $[R_1, R]$, with $0 < R_1 < R$, leads to a reformulation of Eq. (240) that reads[62]

$$K_A^B = \left[\frac{4\pi N}{1000} \int_{R_1}^R r^2 \exp\left(\frac{2q}{r}\right) dr\right]$$

$$\times \left[1 + \left[\int_0^{R_1} r^2 \exp\left(-\frac{U_{+-}^*}{kT} + \frac{2q}{r}\right) dr\right] / \int_{R_1}^R r^2 \exp\left(\frac{2q}{r}\right) dr\right] \qquad (261)$$

where all the specificity of the U_{+-} potential is supposed to arise in the range $[0, R_1]$. The above result can be identified with

$$K_A = K_{ss}(1 + K_c) \qquad (262)$$

which expresses the total "association constant" K_A as a function of K_{ss}, the association constant for the "solvent separated ion pair," and of K_c, the association constant for the "contact ion pair." For the identification to be quantitative it is sufficient to use

$$R_1 = a - d_s \qquad (263)$$

where a is the distance of closest approach and d_s the diameter of the solvent molecule. This result, which was obtained from considerations of excess properties at equilibrium,[62] is of course also valid in the case of conductance (as in any linear irreversible process proceeding from a perturbation different from an external field $\mathbf{X}$).

6.5. Discussion of the Hydrodynamic Contributions in Conductance

6.5.1. The Electrophoretic Contribution

In the following the electrophoretic contributions $\lambda_{e2}^{i(2)}$ will be neglected, since no evaluation of these quantities has ever been achieved. Consequently, in the text below λ_{e2}^{i} must be identified with $\lambda_{e2}^{i(1)}$.

In view of the discussion in Section 6.4, it must be expected that since the relative velocity of ions of opposite signs is negligible when the two ions are within a distance $R \simeq q$ of one another, the counter-ion of the ionic atmosphere inside the sphere of radius R will not contribute to the electrophoretic velocity of the reference ion. Thus only ions of the same sign in this sphere may contribute to the electrophoretic velocity. However, due to the large repulsion for such short-distance pairs, again the contribution must be negligible. Consequently, one may reasonably expect that the Λ_e/Λ_0 contribution to the conductance coefficient will be practically insensitive to the distance of closest approach a if the latter is smaller than $R \simeq q$; in fact, such is the case.

The theoretical derivation of the electrophoretic contribution to ionic conductance leads to the following results:

$$\lambda_{e1}^{i} = -\frac{\beta}{2}c^{1/2} + 2E_2'Q_{e1}^{i}(b) \tag{264}$$

$$\lambda_{e2}^{i} = -\frac{E_2'}{2}c \ln c + 2E_2'\left(Q_{e2}^{i}(b) - \tfrac{1}{2}\ln\frac{2\kappa q}{e^{1/2}}\right) \tag{265}$$

where again the functions $Q_{e1}(b)$ and $Q_{e2}(b)$ depend mainly on the equilibrium distribution functions used in the calculation. Using the Meeron distribution function [Eq. (203)] leads to

$$Q_{e1} = Q_{e1}^{i} + Q_{e1}^{j} = -\sum_{j}\Delta(b_{ij}) - \sum_{i}\Delta(b_{ji}) \tag{266}$$

where

$$b_{ij} = -\frac{Z_iZ_j}{|Z_iZ_j|}\frac{2q}{a_{ij}} \tag{267}$$

$$\Delta(x) = \int_{x}^{0}\frac{e^t - t}{t^3}dt = E_i(x) - \frac{e^x}{x^2} - \frac{e^x}{x} \tag{268}$$

and $\sum$ means summation over each possible couple of indexed species (++, --, +-, and -+).

In the case of the restricted primitive model

$$b_{+-} = -b_{++} = -b_{--} = \frac{2q}{a} = b \tag{269}$$

which leads to the result of Falkenhagen et al.[5]:

$$Q_{e1}(b) = 2\left(E_i(-b) - \frac{e^{-b}}{b^2} - \frac{e^{-b}}{b}\right) - 2\left(E_i(b) - \frac{e^b}{b^2} - \frac{e^b}{b}\right) \tag{270}$$

Obviously the function $Q_{e1}(b)$ diverges negatively with the last term when $b \to \infty$. If one turns now to the function

$$Q_{e2}(b) = Q_{e2}^i(b) - Q_{e2}^i(b) \tag{271}$$

of Falkenhagen et al., in the case of the restricted primitive model their result can be rewritten as

$$Q_{e2}(b) = A - \ln b + F_1(b) + 2\left(E_i(b) - \frac{e^b}{b^2} - \frac{e^b}{b}\right) \tag{272}$$

with

$$A = \tfrac{5}{2} - 3C - \tfrac{1}{2}\ln 2 \tag{273}$$

$$F_1(b) = \frac{1}{1 - T_1}\left(\frac{2}{b^2} + \frac{3}{b}\right) + \frac{T_1}{1 - T_1}\left(E_i(b) - \ln b - B - \frac{e^b}{b} - \frac{2e^b}{b^2}\right) \tag{274}$$

$$B = -C + 2 \tag{275}$$

$$T_1 = e^{-b}\left(1 + b + \frac{b^2}{2} + \cdots\right) \tag{276}$$

where C is Euler's constant, $C = 0.57722$. Again the function $Q_{e2}(b)$ diverges when b increases, but positively, and the sum

$$Q_e = Q_{e1} + Q_{e2} \tag{277}$$

remains a weak function of b as b increases, since the rapidly diverging terms cancel. This behavior is illustrated in Figure 5, where the Falkenhagen, Ebeling, and Kraeft functions are graphically represented. The molar electrophoretic conductance contribution

$$\Lambda_e = (\lambda_{e1}^+ + \lambda_{e1}^-)\left(1 + \frac{\Delta X}{X}\right) + (\lambda_{e2}^+ + \lambda_{e2}^-)\left(1 + \frac{\Delta X}{X} + \frac{\Lambda_e}{\Lambda_0}\right) \tag{278}$$

thus leads to the following expansion:

$$\Lambda_e(b) = -\beta c^{1/2} - E_2' c \ln c - \sigma_{2e} c \tag{279}$$

with

$$\sigma_{2e} = 2E_2'\left[Q_e(b) - \ln\frac{2\kappa q}{c^{1/2}} + \frac{\alpha\beta}{2E_2'}\right]c + O(c^{3/2}) \tag{280}$$

The contribution $\alpha\beta/2E_2' = 0.78107$ comes from the linear contribution of the cross product of the limiting laws of the first term of the r.h.s. of Eq. (278).

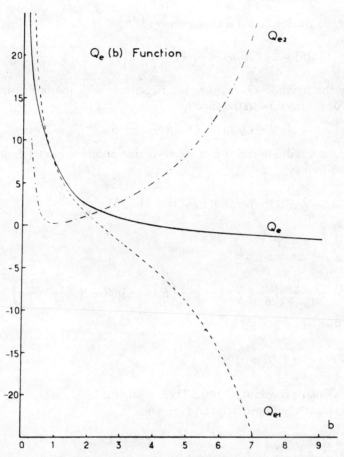

Figure 5. The electrophoretic contribution to the linear term of conductance. The electrophoretic function $Q_e(b)$ [Eq. (281)] of Falkenhagen, Ebeling, and Kraeft[5] is given by the full line. Its decomposition into the two divergent $Q_{e1}(b)$ and $Q_{e2}(b)$ functions corresponding to the Fuoss–Onsager and Chen (or Pitts) contributions respectively, is also represented. This shows how the total contribution becomes small when $b > 2$. The FEK functions were obtained by use of the Meeron distribution functions so that the three functions are exact. If use of the Debye–Hückel d.f. (DH) is made, the diverging pairs of $Q_{e1}(b)$ and $Q_{e2}(b)$ disappear and the net result for $Q_e(b)$ is not much altered. This cancellation, which in effect justifies the DH drastic approximations at short distances, is, in fact, an indirect justification of the strong-coupling approximation (SCA), since it shows that within the strong coupling sphere the relative velocity of ions is negligible, though their distribution is greatly altered by the external field.

Expanding $Q_e(b)$ with respect to b and truncating after the linear b term gives

$$Q_e(b) = \frac{1}{b^2} - \frac{31}{4b} - 1.1780 - \ln b - O(b) \qquad (281)$$

which again is equivalent to the result obtained when using the Debye–Hückel distribution function. In fact, this last result of Falkenhagen *et al.* is very similar to those of Fuoss and Chen and Quint and Viallard.

It is interesting now to avoid the use of the restricted primitive model in which there is only one specific Bjerrum parameter $b = b_{+-} = -b_{++} = -b_{--}$ and to make explicit use whenever necessary of the three different parameters characterizing the three short-range potentials defining the Hamiltonian.

The partition of the $Q_{e2}(b)$ function into long-range and short-range contributions is straightforward, since it depends only on the parameter $b \equiv b_{+-}$. Then

$$Q_{e2}(b) = Q_{e2}(\zeta) + [Q_{e2}(b) - Q_{e2}(\zeta)] \tag{282}$$

$$= Q_{e2}(\zeta) + [F_1(b) - F_1(\zeta)] + 2D(b) \tag{283}$$

where $F_1(x)$ is given by Eq. (274) and using Eq. (268)

$$D(b) = \Delta(b_{+-}) - \Delta(\zeta_{+-}) \tag{284}$$

$$\zeta_{+-} = \frac{2q}{R} = \zeta \tag{285}$$

This formulation is useful for a simple demonstration of the consequence of the SCA. This approximation does not alter the Q_{e1} function, which does not depend on the perturbation of the distribution function. It is interesting to observe that in the ultimate analysis the SCA reduces to neglecting the bracketed term in Eq. (283), that is, in approximating $F_1(b)$ by $F_1(\zeta)$. In fact, this is not a drastic approximation, since it is exactly the $F_1(x)$ function that is almost insensitive to x for $x > 1.5$, so that for $R = q$ ($\zeta = 2$) and $b > 2$ the bracketed term in Eq. (283) is practically negligible compared with $Q_{e2}(\zeta)$. The $Q_{e1}(b)$ function can also be separated into long-range and short-range contributions; however, it must be borne in mind that it is, in fact, a function of the three parameters b_{++}, b_{--}, and b_{+-}, so that the partition gives

$$Q_{e1}(b) = Q_{e1}(\zeta) + D(b_{++}) + D(b_{--}) + 2D(b_{+-}) \tag{286}$$

now with

$$D(b_{ij}) = \Delta(b_{ij}) - \Delta\left(-\frac{z_i z_j}{|z_i z_j|}\zeta\right) \tag{287}$$

In this formulation the "degeneracy" of the restricted primitive model has been lifted ($b_{++} \neq -b_{+-} \neq b_{--}$). Returning to the complete Q_e function, it is to be noted that it is, in fact, a function of b_{++} and b_{--} only;

$$Q_e = Q_e(\zeta) - D(b_{++}) - D(b_{--}) \tag{288}$$

and not of b_{+-}. This result will prove to be very important in Section 7, which deals with transference numbers. As a consequence, the Q_e now depends only on the short-range part of the U_{++} and U_{--} potential energies.

In fact, the SCA treats only the short-range part of the relative velocity of the anion and cation, and this explains the above observation. However, the dependence on b_{++} and b_{--}, which we may define as the apparent Bjerrum parameters for $++$ and $--$ interactions, is again very small when $|b_{ii}| > 1.5$, so that in ultimate analysis the SCA can be considered as justified and leading to

$$\sigma_{2e}(a) \simeq \sigma_{2e}(R) \qquad \text{when } a_{++}, a_{--}, a_{+-} < R \qquad (289)$$

Consequently, the following result is reached concerning the electrophoretic contribution to the conductance, λ_e:

$$\Lambda_e = \lambda_{e1}^+ + \lambda_{e1}^- + \lambda_{e2}^+ + \lambda_{e2}^- + \alpha\beta c + O(c^{3/2})$$

$$= -\beta c^{1/2} - E_2' c \ln c + 2E_2' \left(Q_e(\zeta) - \ln \frac{2\kappa q}{c^{1/2}} + \frac{\alpha\beta}{2E_2'} + F_1(b_{+-}) - F_1(\zeta) \right.$$

$$\left. + D(b_{++}) + D(b_{--}) \right) c + O(c^{3/2}) \qquad (290)$$

where the last three terms in the large parentheses are always very small and where $F_1(b_{+-})$ is also negligible as soon as b_{+-} becomes larger than 2. In the restricted primitive model it is *a priori* defined that $b_{++} = b_{--} = -b_{+-} = -b$; one thus understands that any attempt to identify the evaluations of this electrophoretic parameter b with that obtained from relaxation effects, which depend exclusively on b_{+-}, is doomed to failure.[66]

In principle, electrophoresis should provide complementary information concerning the system, but, unfortunately, the $D(b_{ii})$ functions are always numerically very small, so that the information they contain is effectively lost in the random noise of experimental error.

6.5.2. The Hydrodynamic Contribution to the Relaxation Effect

The evaluation of $\Delta X_h/X$ necessarily involves the evaluation of $g_I'^{(h1)}$ and $g_I'^{(h2)}$. It will be recalled that $g_I'^{(h1)}$ is the solution of the continuity equation that results from the application of the monitoring term $\mathbf{M}_1$ to the hydrodynamic term $\mathbf{H}_{ji}(\mathbf{r})$; $g_I'^{(h2)}$ results from the application of the hydrodynamic monitoring term $\mathbf{M}_h$. Up until now no evaluation of $g_I'^{(h2)}$ has been carried out. For $g_I'^{(h1)}$ there exist two evaluations from Fuoss *et al.*[10,24]: one using the Debye approximation and the other using the nonlinearized Poisson–Boltzmann equation. Since Fuoss did not use the full monitoring term $\mathbf{M}_1$ but only its unechoed approximation, the results must be identified with $g_I'^{(h1)}$ and $\Delta X_{Ih}/X$ of Eqs. (208) and (209). The result reads

$$\frac{\Delta X_{Ih}}{X} = -\frac{E_2'}{\Lambda_0} c \ln c + \sigma_{2h}(a)c + O(c^{3/2}) \qquad (291)$$

where

$$\sigma_{2h}(a) = -2E_2'\left(\frac{1}{3b^2} - 2F(b) + 0.9528 - \ln b + \ln \frac{2\kappa q}{c^{1/2}}\right) \qquad (292)$$

where $F(b)$ is a complicated function that is not very sensitive to b (cf. Figure 1 of Reference 24).

In the earlier version, where the Debye approximation was used, the Fuoss function becomes

$$\sigma_{2h}(a) \simeq -2E_2'\left(\frac{2}{3b} + 0.76685 - \ln b + \ln \frac{2\kappa q}{c^{1/2}} + O(b)\right) \qquad (293)$$

which can be considered as the truncated expansion of the previous Eq. (292) with respect to b.

Thus a peculiarity of all the hydrodynamic functions $\sigma_{2e}(a)$ and $\sigma_{2h}(a)$ is to be noted: For large values of the Bjerrum parameter b they lead to very small contributions, so that it makes little difference whether the full expression or the truncated expansion is used. This must be considered as indirect proof of the adequacy of the SCA. If the flow of counter-ions is negligible within the sphere of strong interaction, as implied by the SCA, the average contributions of these counter-ions to the hydrodynamic perturbations (Λ_e and $\Delta X_h/X$) must also be negligible. This is in fact found, since these effects practically do not depend on the distance of closest approach a as soon as this distance is smaller than the distance $R = q$ within which interactions, by definition, start being strong. Consequently, as far as these hydrodynamic effects are concerned, the linearization of the Poisson–Boltzmann equation, as made in the Debye treatment, does not lead to drastic numerical changes.

The opposite is observed for the nonhydrodynamic contribution $\sigma_1(a)$, which rapidly diverges as b increases. But factorizing out the short-range contributions, as allowed by the echo effect, takes care not only of the divergences of the linear contribution due to $\Delta X_{Ic}/X$ but also of all the terms of the series expansion through the dimensionless quantity γ, which may be considered as the "fraction of free ions." However, it must be recalled that this is only an analogy, since γ is somewhat arbitrary: Thus if the cutoff distance R is varied around its recommended value q, the quantity γ changes accordingly, but so does $\sigma_1(R)$, so that the net change in terms of Λ is completely negligible. The changes in $\sigma_{2e}(R)$ and $\sigma_{2h}(R)$ are also completely negligible, as has been shown in this section.

A state has thus been reached in the treatment of the problem where the justification of the empirical grafting of the Bjerrum concept onto conductance is realized. It is known that the formulation is only asymptotically correct when the concentration is small enough ($\kappa q \gamma^{1/2} < 0.5$) for higher-order short-range ion cluster effects to be negligible. For higher concentrations the set of equations (238) and (248) is no longer valid, and so far there is no basically sound theory available for these conditions. One possible direction for progress might be to improve the strong coupling approximation, which, of course, is valid if there is no third ion in the sphere R where the anion–cation pair considered is located. Recent contributions to extend the range of validity to higher concentrations are briefly reviewed in Appendix B.

7. Transference Numbers

Since the transference numbers of ions are directly connected with their conductance, it is important to underline here the derivation of the equation that expresses the dependence of transference numbers on concentration.

The transference number T^i of an ion i is the ratio of the ionic conductance λ^i to the molar conductance of the electrolyte:

$$T^i = \frac{\lambda^i}{\Lambda} \tag{294}$$

Consequently, all the relations necessary for the evaluation of this quantity are available in the preceding chapters. We shall deal again here with the special case of symmetrical electrolytes, since these have been widely studied in the literature.

For a long time it was felt that the transference numbers were controlled by the electrophoresis effect exclusively. In fact, all movements in the solution are induced by the external force and the ionic interaction force (the relaxation field), and they are linear in their response to these perturbations. Since the relaxation field is the same for the anion and the cation, the latter should cancel out in the ratio that defines the transference numbers, so that the only effect remaining is that of electrophoresis. These somewhat qualitative views are indeed sound, and, in its original version, such was the prediction of the theory. However, in the original version of the Fuoss–Onsager (FO) theory drastic approximations were introduced in the theory of the electrophoretic effect, which reduced to

$$\lambda_e^i = \lambda_{e1}^i \left(1 + \frac{\Delta X}{X}\right) \tag{295}$$

This equation can be obtained from the general FO equation, Eq. (181), by neglecting $\lambda_{e2}^{i(1)}$ and $\lambda_{e2}^{i(2)}$. Starting from Eq. (295), Kay and Dye[66] obtained

$$T^i = \frac{\lambda_0^i + \lambda_{e1}^i}{\Lambda_0 + \lambda_{e1}^i + \lambda_{e1}^i} \tag{296}$$

which leads to

$$T^i - \tfrac{1}{2} = \left(T_0^i - \tfrac{1}{2}\right)\Big/\left(1 - \frac{1}{\Lambda_0}\frac{\beta c^{1/2}}{1 + \kappa a}\right) \tag{297}$$

after using the FO evaluation,

$$\lambda_{e1}^{+} = \lambda_{e1}^{-} = -\frac{1}{2}\frac{\beta c^{1/2}}{1 + \kappa a} \tag{298}$$

in the special case of the restricted primitive model. However, Eq. (298) does not prove to be satisfactory, since the characteristic distances of closest approach a that could be evaluated by comparison with experimental data are found to be quite unrealistic and, further, do not agree with the values obtained from conductance, where the relaxation contribution predominates.

The conclusion of Kay and Dye was that improvements were clearly needed in the derivation of the electrophoretic factor in conductance. In fact, such an improvement was at hand when Chen[67] noted that a contribution was missing in the FO derivation. Chen proposed the introduction of a new term in Eq. (295), which then reads

$$\lambda_{e}^{i} = \lambda_{e1}^{i}\left(1 + \frac{\Delta X}{X}\right) + \lambda_{e2}^{i} \tag{299}$$

where λ_{e2}^{i} is known as the Chen contribution. Again one observes that Eq. (299) is an approximation for the complete derivation [Eq. (181)] in which $\lambda_{e2}^{i(2)}$ and the echo factor of $\lambda_{e2}^{i(1)}$ is neglected. However, using Eq. (299) leads to the unsatisfying result that the relaxation field no longer cancels in the final formulation of the transference number equation. It is to be noted that the Pitts derivation, which contains most of the λ_{e2}^{i} contribution, leads to the same observation. The new equation, Eq. (299), was used by Sidebottom and Spiro[68] in their study of transference numbers, but again the results display the same unsatisfying properties as before.

It is to be noted that the cancellation of the relaxation effect in the transference numbers of a symmetrical electrolyte is an important property (if it turns out to be correct), because the combination of a measurement of the transference number with the conductance then leads to independent evaluations of the relaxation and electrophoretic effects. This consequence was the chief motivation of Kay and Dye's study.[66] This is why it is an important reward of the echo effect to bring back into evidence this interesting property. In fact, introducing the full expression given by Eq. (181) leads, after some simple algebra, to

$$T^{i} - \frac{1}{2} = (T_{0}^{i} - \frac{1}{2})\frac{1 - (\lambda_{e2}^{i(1)} + \lambda_{e2}^{j(1)})/\Lambda_{0}}{1 + (\lambda_{e1}^{i} + \lambda_{e1}^{j} + \lambda_{e2}^{i(2)} + \lambda_{e2}^{j(2)})/\Lambda_{0}} + \delta^{i} \tag{300}$$

with

$$\delta^{i} = \frac{(1/2)[(\lambda_{e1}^{i} - \lambda_{e1}^{j}) + (\lambda_{e2}^{i(2)} - \lambda_{e2}^{j(2)}) + (\lambda_{e2}^{i(2)} - \lambda_{e2}^{j(2)})]/\Lambda_{0} + [\lambda_{e2}^{i(1)}(\lambda_{e1}^{i} + \lambda_{e2}^{i(2)}) - \lambda_{e2}^{j(1)}(\lambda_{e1}^{j} + \lambda_{e2}^{j(2)})]/\Lambda_{0}^{2}}{1 + (\lambda_{e1}^{i} + \lambda_{e1}^{j} + \lambda_{e2}^{i(2)} + \lambda_{e2}^{j(2)})/\Lambda_{0}} \tag{301}$$

where

$$i \neq j \quad \text{and} \quad \delta^i = -\delta^j \qquad (302)$$

Simplifications for the δ^i term occur when it is recalled that $\lambda_{e2}^{+(1)} = \lambda_{e2}^{-(1)} = \lambda_{e2}^{(1)}$ and $\lambda_{e2}^{+(2)} = \lambda_{e2}^{-(2)} = \lambda_{e2}^{(2)}$, whereas λ_{e1}^{+} differs only slightly from λ_{e1}^{-}, as will be seen in detail below. Thus it follows that

$$\delta^i = \frac{1}{2} \frac{\lambda_{e1}^i - \lambda_{e1}^j}{\Lambda_0} \frac{1 + 2\lambda_{e2}^{(1)}/\Lambda_0}{1 + (\lambda_{e1}^i + \lambda_{e2}^j + 2\lambda_{e2}^{(2)})/\Lambda_0} \qquad (303)$$

Recalling, as shown in Section 6.5, that

$$\lambda_{e1}^i - \lambda_{e1}^j = E_2'[D(b_{++}) - D(b_{--})] \qquad (304)$$

shows that δ^i is indeed completely negligible. As a consequence, the transference number equation reads, after expanding Eq. (300) and neglecting the $\lambda_{e2}^{(2)}$ contributions,

$$T^i - \tfrac{1}{2} = (T_0^i - \tfrac{1}{2})\left(1 - (\lambda_{e1}^i + \lambda_{e1}^j + \lambda_{e2}^{i(1)} + \lambda_{e2}^{j(1)})/\Lambda_0 + \frac{\beta^2}{\Lambda_0^2}c + O(c^{3/2})\right) \qquad (305)$$

For the restricted primitive model the result

$$T^i - \tfrac{1}{2} = (T_0^i - \tfrac{1}{2})\left\{1 + \frac{\beta}{\Lambda_0}c^{1/2}\gamma^{1/2} + \frac{E_2'}{\Lambda_0}c\gamma \ln c\gamma \right.$$
$$\left. - \left[\frac{2E_2'}{\Lambda_0}\left(Q_e(b) - \ln \frac{2\kappa q}{c^{1/2}}\right) - \frac{\beta^2}{\Lambda_0^2}\right]c\gamma + O(c^{3/2}\gamma^{3/2})\right\} \qquad (306)$$

is obtained. This result shows that the variation of transference numbers with concentration is screened by the factor $T_0^i - \tfrac{1}{2}$, so that for an electrolyte (such as KCl in water), with a limiting transference number near $\tfrac{1}{2}$, no variation is observed over a wide range of concentrations. This is one reason why it would be difficult to extract much reliable information concerning ionic interactions from transference number measurements, since the deviations for T_0^i from $\tfrac{1}{2}$ are never very large except in a few cases.

Lifting now the degeneracy of the restricted primitive model, as was done in Section 6.4, finally gives

$$T^i - \tfrac{1}{2} = (T_0^i - \tfrac{1}{2})\left\{1 + \frac{\beta}{\Lambda_0}c^{1/2}\gamma^{1/2} + \frac{E_2'}{\Lambda_0}c\gamma \ln c\gamma \right.$$
$$- \left[\frac{2E_2'}{\Lambda_0}\left(Q_e(\zeta) - \ln \frac{2\kappa q}{c^{1/2}} + F_1(b) - F_1(\zeta)\right.\right.$$
$$\left.\left. + D(b_{++}) + D(b_{--})\right) - \frac{\beta^2}{\Lambda_0^2}\right]c\gamma + O(c^{3/2}\gamma^{3/2})\right\} \qquad (307)$$

When strong interactions occur ($|b_{ij}| > 1.5$), the last four terms of the term in large parentheses are negligible and an excellent approximation is given by

$$T^i - \tfrac{1}{2} = (T^i_0 - \tfrac{1}{2})\left\{1 + \frac{\beta}{\Lambda_0}c^{1/2}\gamma^{1/2} + \frac{E'_2}{\Lambda_0}c\gamma \ln c\gamma \right.$$

$$\left. - \left[\frac{2E'_2}{\Lambda_0}\left(Q_e(\zeta) - \ln\frac{2\kappa q}{c^{1/2}}\right) - \frac{\beta^2}{\Lambda_0^2}\right]c\gamma + O(c^{3/2}\gamma^{3/2})\right\} \quad (308)$$

which predicts the linear coefficient of the transference number equation and is then practically independent of any specific short-range parameter characterizing the excess property of the electrolyte in any solvent. When no strong interactions prevail ($|b_{ij}| < 1.5$), the echo formulation is of no practical use ($\gamma = 1$), but the partition into short-range and long-range contributions is still useful in order to make the distinction clear between the various $++$, $--$, and $+-$ specific interaction contributions. The equation is then conveniently written as

$$T^i - \tfrac{1}{2} = (T^i_0 - \tfrac{1}{2})\left\{1 + \frac{\beta}{\Lambda_0}c^{1/2} + \frac{E'_2}{\Lambda_0}c \ln c \right.$$

$$- \left[\frac{2E'_2}{\Lambda_0}\left(Q_e(\zeta) - \ln\frac{2\kappa q}{c^{1/2}} + F_1(b) - F_1(\zeta)\right)\right.$$

$$\left.\left. + D(b_{++}) + D(b_{--})\right) - \frac{\beta^2}{\Lambda_0^2}\right]c + O(c^{3/2})\right\} \quad (309)$$

In this case, the functions $D(b_{ii})$ are not negligible, neither is the residual function $F_1(b)$, which depends on anion–cation specific short-range interactions. Unfortunately, few data are available for systems belonging to this category and the experimental precision here is not sufficient for a satisfactory evaluation of the linear coefficient.

Figure 6 summarizes the results of the various data processed according to these theoretical equations. The data processing[69] was done as follows. In a first step it was observed that the $c^{3/2}$ contribution was completely negligible, since adjusting the coefficient of a $c^{3/2}$ term as a third parameter proved to be without effect. A least-squares fitting was then carried out by adjusting the parameters T^i_0 and Q_e in Eq. (306) in which Λ_0 and the values for γ at each concentration were calculated from the results obtained in an adjustment carried out on conductance data for the same system. The experimental values obtained for Q_e are then plotted versus b, the Bjerrum parameter, calculated from the same conductance data. In Figure 6 the following theoretical curves are also represented:

1. The curve FEK/CHEN corresponds to the function $Q_e(b)$ of Ebeling, Krempt, and Kraeft, and that of Chen who used the Meeron pair distribution function (p.d.f.).

2. The curve FC corresponds to the derivations of Pitts, Fuoss–Chen, and Quint and Viallard, who made use of the Debye–Hückel p.d.f.
3. The curve denoted SCA cqrresponds to the SCA, which predicts a horizontal straight line for $b > 2$.

The display of the experimental points in Figure 6 helps to demonstrate why no correspondence was ever found between the parameters b obtained from transference number data and those from conductance: Firstly, the parameters do not refer to the same pair of ions and, secondly, transference numbers are practically insensitive to the apparent anion–anion and cation–cation contact distances. It is observed, in fact, that for all systems studied no specific behavior is exhibited and that the spread of the points is probably due to experimental error.

It seems clear that theoretical accuracy has now reached the highest level of precision attainable at this point. Obviously all short-range hydrodynamic contributions are practically insignificant in the present state of experimental precision. It is one of the advantages of conductance studies that very precise measurements can be made which are also very sensitive to anion–cation short-range interactions through the nonhydrodynamic part of the relaxation coefficient, namely, $\Delta X_{Ic}/X$.

We should mention here a method developed by Perié et al.[70] to check the eventual presence of systematic errors in theoretical functions or experimental data of both conductance and transference numbers. It consists in combining the transference number data with the conductance data for a given system in order to obtain an experimental evaluation of *ionic* conductances and to compare the latter with the expectations according to the theoretical equation for λ^i. This check of self-consistency must ultimately lead to the same values of reduced excess parameters such as K_A^B, for which now three evaluations are possible: one from the molar conductance Λ and two from λ^+ and λ^-, respectively. Then any systematic deviation at any level must lead to divergences between the three evaluations. A critical analysis[70] of the various data available showed that whenever such divergences occur, they must mostly be attributed to systematic errors in the experimental evaluation of transference numbers. This therefore constitutes an interesting selection method.

8. Ionic Limiting Mobilities

While the theory of the conductance coefficient, $f_\Lambda = \Lambda/\Lambda_0$, as a function of the concentration of an electrolyte in dilute solutions now seems well established, such is not the case for the theoretical interpretation of ionic limiting conductance λ_0^+ or λ_0^-. This is due to the fact that in the above theory of f_Λ the solvent can quite reasonably be considered as a continuum and that

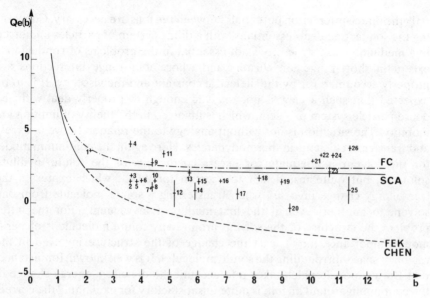

Figure 6. Plot of the electrophoretic function $Q_e(b)$ as a function of the Bjerrum parameter $b = e^2/aDkT$ for various theoretical derivations in the case of the restricted primitive model (see Table 5 for the corresponding equations). The bars correspond to the most reliable experimental evaluation of Q_e resulting from a selection of transference number data taken from Reference 69. Each evaluation is plotted as a function of the value of b obtained in the same system from conductance data. It is observed that this electrophoretic contribution is practically always negligible and centered around the prediction of the SCA, with an average scattering of the same order of magnitude as the theoretical uncertainty. Relations according to FC, FEK, and CHE are also shown.

1. $HClO_4/H_2O$		J. R. Gwyther, S. Kumarasinghe, and M. Spiro, *J. Solution Chem.* **3**, 659 (1974).
2. Bu_4NCl/CH_3NO_2		S. Blum, Ph.D. thesis, McGill University, Montreal, Canada, 1961.
3. Bu_4NBr/CH_3NO_2		See Reference 2.
4. Bu_4NBr/CF_3CH_2OH		H. M. Kusnandar, H. Bischoff, and G. Marx, private communication (1980).
5. Pr_4NCl/CH_3NO_2		See Reference 2.
6. Pr_4NBr/CH_3NO_2		See Reference 2.
7. Et_4NBr/CH_3NO_2		See Reference 2.
8. $NaCl/CH_3OH$		G. A. Vidulich, G. P. Cunningham, and R. L. Kay, *J. Solution, Chem.* **2**, 23 (1973).
9. $LiCl/CH_3OH$		See Reference 8.
10. Et_4NCl/CH_3NO_2		See Reference 2.
11. Bu_4MBr/CH_3OH (10°C)		See Reference 8.
12. Me_4NC_{IO4}/CH_3CN		C. H. Springer, J. F. Coetzee, and R. L. Kay, *J. Am. Chem. Soc.* **73**, 471 (1969).
13. $CsBr/H_2O$–dioxane (23.5%)		M. Perie, J. Perie, and M. Chemla, *Electrochim. Acta.* **21**, 739 (1976).
14. KCl/H_2O–dioxane (24.8%)		A. Fratiello and R. L. Kay, *J. Solution Chem.* **3**, 857 (1974).
15. NaI/n-propanol		M. Mirza, Ph.D. thesis, Berlin University, Berlin, 1978.

only the direct-interaction potentials between the ions are necessary. Concerning the solute, one deals essentially with a dilute system of particles immersed in a medium whose structure is not essential in the problems of conductance except in short-range ion clusters and whose long-range interactions are properly accounted for by the dielectric constant and the viscosity. It is to be expected that such a view is not realistic enough to properly deal with the dense particle system problem, which is involved in the theory of limiting ion mobility. The situation is not without analogy to the relation between excess and transfer coefficients in thermodynamics. Here again the continuum model for the solvent is adequate for excess coefficients of the solute in dilute solutions, but quite insufficient to account for $\Delta G^{\ominus}$, which expresses the free-energy change involved when transferring a solute molecule from one solvent to another (even if the first medium is a vacuum). For the latter problem the structure of the solvent around any solute molecule is of paramount importance (or at least the change of the structure involved in the region of space surrounding the solute molecule). It is well known, for instance, that Born's model, which takes into account the dielectric-energy change in the continuum around an ion, is quite unsatisfactory for explaining the correct magnitude of the observed quantity $\Delta G^{\ominus}$. It is thus easy to foresee that the same must be observed for λ_0^i if the same phenomenon is considered in the theory of limiting conductance.

So far only the continuum model has been considered quantitatively in great detail. Historically this was achieved in two steps. In the first step the hydrodynamic nature of the solvent continuum was taken into consideration. The solvent was characterized by its viscosity coefficient by applying Stokes' law to the ion considered as a hard-sphere of radius R. This leads directly to the following expression for the mobility ω_i of ions of type i:

$$(1/\omega_i)_0 = 6\pi\eta R_i \qquad \text{(perfect sticking)} \tag{310}$$

or

$$(1/\omega_i)_0 = 4\pi\eta R_i \qquad \text{(perfect slipping)} \tag{311}$$

Figure 6 (*cont.*)

16. $\frac{1}{2}MgSO_4/H_2O$	R. M. Fuoss and F. Accascina, *Electrolyte Conductance*, Wiley-Interscience, New York (1959).
17. Me_4NBr/CH_3NO_2	See Reference 2.
18. Me_4NCl/CH_3NO_2	See Reference 2.
19. Et_4NF/n-butanol	H. U. Fusban, Ph.D. thesis, D188, Berlin University, Berlin, 1974.
20. N_2Br/n-propanol	See Reference 15.
21. $LiCl/n$-butanol	See Reference 19.
22. KI/n-butanol	See Reference 19.
23. Et_4NI/n-butanol	See Reference 19.
24. Et_4NCl/n-butanol	See Reference 19.
25. $CsBr/H_2O$–dioxane (49.6%)	See Reference 13.
26. Me_4NMe_4B/n-butanol	See Reference 19.

depending on the choice of the hydrodynamic boundary condition used at the surface of the sphere R_i in the integration of the hydrodynamic continuity equation. The factor of 6 is obtained if one assumes perfect "sticking" of the continuum at $r = R_i$, and the factor 4 is obtained if, on the contrary, perfect "slipping" is assumed. Whatever choice is made, the result leads to the Walden rule, which states that

$$\lambda_0^i \eta = \text{const.} \tag{312}$$

since the limiting conductance λ_0^i is related to the mobility ω_i through

$$\lambda_0^i = \omega_i |Z_i| e\mathscr{F} \tag{313}$$

It is a matter of common experience to observe that this law is not followed; moreover, it is also well known that Stokes' radii R_i, which can be evaluated from experimental values of λ_0^i, are always far from what one may expect from intramolecular considerations or from a relation to the crystallographic radii.

In fact, it was recognized quite early by Born that, in addition to hydrodynamic friction, an electrostatic drag must also exist that is not negligible. Owing to the dielectric response of the solvent to the high electrical field of the ion, a braking force arises that still further diminishes the mobility of the ions. This is due to the fact that the relaxation time τ_s of the solvent molecule is finite, so that behind the moving ion the solvent remains polarized, whereas in front it is not yet completely polarized. As a result, the ion moves in a relaxation field because of the dielectric solvent continuum, which, so to speak, pulls the ions backward (in the direction opposite to the external field). Qualitatively, the situation is not unlike that occurring within the relaxation of the ionic atmosphere, with the difference that it is not dependent on the ionic concentration but, rather, on the structure of the solvent and on its intrinsic molecular dynamic properties summarized by the relaxation time τ_s of solvent molecule reorientation. The difficulty is, however, much increased by the fact that one has to deal now with a dense system of particles (the ion considered surrounded by the solvating molecules and other solvent molecules). However, if the continuum model is retained for the solvent and some restrictions such as incompressibility, homogeneity, and constancy of all properties in space (such as polarizability and viscosity) are assumed, a set of fundamental equations may then be obtained that can be solved for the quantity ω_i, which is of interest here. The problem will then be one of being certain that the above restrictions are consistent with the solution found. For instance, if we are beyond a situation of a linear response by the solvent continuum, then some correlated phenomena such as dielectric saturation and electrostriction must occur, which certainly alters the response of the solvent and makes the analytical solution found quite insufficient. However, taking into account these effects would make it necessary to give up the continuum model and introduce more unknown parameters. Consequently, the above restrictions are maintained and the approximation of this dielectric effect must be evaluated accordingly. However, even in this state of approximation the problem cannot be easily solved.[71,72] Thus, only recently have Hubbard and Onsager[73,74] (HO) given a complete solution after the pioneering work of Fuoss,[75] Boyd,[76] and Zwänzig.[77]

An excellent review of this work has been recently published by Evans, Tominaga, Hubbard, and Wolynes,[78] which will now be briefly summarized. The model treated by Hubbard is the incompressible continuum model for the solvent, with homogeneous viscosity and dielectric constants and no electrostriction or dielectric saturation effects, even in the immediate surroundings of the ion. Each volume element of the solvent continuum is considered as undergoing a static electrical polarization $\mathbf{P}_0$,

$$\mathbf{P}_0 = \frac{\varepsilon_0 - \varepsilon_\infty}{4\pi} \mathbf{E}_0 \tag{314}$$

with

$$\mathbf{E}_0 = \frac{|Z|e\hat{r}}{\varepsilon_0 r^2} \tag{315}$$

due to the field $\mathbf{E}_0$ of the ion. The fluid is supposed to be characterized by a single Debye dielectric relaxation time τ_s. The classical hydrodynamic equations are used as continuity equations:

$$\eta \nabla^2 \mathbf{v} = \nabla P + \mathbf{F}_{DF} \tag{316}$$

$$\nabla \mathbf{v} = 0 \tag{317}$$

where $\mathbf{v}$ is the velocity of the fluid, P is the pressure and $\mathbf{F}_{DF}$, a function of $\mathbf{P}_0$, τ_s, and $\mathbf{v}$, is the contribution of dielectric friction to the total force acting on the fluid. One important space-variable introduced by Hubbard is

$$z(r) = -r^4 \Big/ \left[\frac{e^2}{4\pi\eta} \left(\frac{\varepsilon_0 - \varepsilon_\infty}{\varepsilon_0^2} \right) \tau_s \right] \tag{318}$$

It allows much simplification in the analytical formulation of the various equations involved. The dielectric friction coefficient $1/\omega_i$ of an ion becomes a function of $z_i = z(R_i)$ through the relation

$$1/\omega_i = y(z_i)(1/\omega_i)_0 \tag{319}$$

where y is a somewhat complicated function of z_i that turns out to be a correction factor to the limiting case of the pure Stokes' law behavior summarized by Eq. (10) or (11). The various results for $y(z_i)$ will be found in the original publications. In the case of large ions ($|z_i| \gg 1$) the result can be given a simple limiting formulation by expanding the y functions versus z_i^{-1} and truncating after the first term, yielding

$$\frac{1}{\omega_i} = 6\pi\eta R_i + \frac{17}{280} \frac{\tau_s e^2}{R_i^3} \frac{\varepsilon_0 - \varepsilon_\infty}{\varepsilon_0^2} + O(z_i^{-2}) \tag{320}$$

$$\frac{1}{\omega_i} = 4\pi\eta R_i + \frac{1}{15} \frac{\tau_s e^2}{R_i^3} \frac{\varepsilon_0 - \varepsilon_\infty}{\varepsilon_0^2} + O(z_i^{-2}) \tag{321}$$

for the perfect sticking and perfect slipping cases, respectively.

This asymptotic behavior confirms the former results of Zwanzig referred to the special case of high dielectric constant solvents, with some differences in the numerical factors $\frac{17}{280}$ and $\frac{1}{15}$, which were $\frac{3}{16}$ and $\frac{3}{8}$, respectively. These differences arise on account of further approximations introduced by Zwanzig for $\mathbf{F}_{DF}$ in Eq. (316).

The complete HO result shows that $1/\omega_i$ is an infinite series of inverse powers of $|z_i|$ for large values $|z_i|$ (>1), whereas it is an infinite series of positive powers of $|z_i|$ for small values $|z_i|$ (<1).

A very interesting feature of the HO theory is that the friction coefficient $1/\omega_i$ does not decrease to zero (as in the Zwanzig or the limiting HO theories) when the size R_i of the ion tends to zero ($|z_i| \rightarrow 0$), but, rather, tends to a *finite limit*, independent of the boundary condition used (perfect slipping or perfect sticking). This result agrees *qualitatively* with experimental results, which show that the Walden product of the smaller alkali ions tends to a constant value, specific to each solvent (these ions are of course the smallest that can be studied). However, this "punctual" friction coefficient

$$\frac{1}{\omega_i}(\text{point}) = 15.624 \frac{e^2}{16\pi} \frac{\varepsilon_0 - \varepsilon_\infty}{\varepsilon_0^2} \tau_s \tag{322}$$

is not in quantitative agreement with the experimental evaluations. Comparison with experiment is conveniently achieved by plotting the ionic Walden product $\lambda_i^0 \eta$ as a function of $1/R_i$, as shown in Figure 7.

As can be seen in this example, and although theories and experiment exhibit some common qualitative features (presence of a maximum and a nonzero limit for small ions), the agreement is far from perfect from the quantitative view point. One important point is that the continuum theory cannot explain the difference observed between the behaviors of anions and cations. Obviously, structural aspects are involved here that cannot be accounted for. Since this difference is relatively large, it must be concluded that the introduction of more specific molecular parameters is now a necessary requirement for progress towards a substantial improvement in this field. Whereas the continuum theory leads to satisfying results in dilute solutions for the excess thermodynamic and transport properties, such is not the case for transfer thermodynamic properties or for the limiting mobilities.

Taking the molecular nature of the solvent into account leads to great mathematical difficulties. We deal in such a case with a very dense system of

Figure 7. Comparison of various relations for ionic limiting mobilities with some experimental data for acetone as solvent. Curves 1 and 2 refer to the perfect slipping and sticking limiting cases, respectively. Curves S refer to the Stokes law, curves Z to the Zwanzig theories, and curves HO to the Hubbard–Onsager theories, and curves HO_{LL} to their limit cases for large z_i values [Eqs. (320), (321)]. Open circles represent data for cations: from left to right, Bu_4N^+, PrN^+, Et_4N^+, Me_4N^+, Cs^+, K^+, and Na^+; solid circles represent data for anions: from left to right, Ph_4B^-, ClO_4^-, I^-, Br^-, and Cl^-.

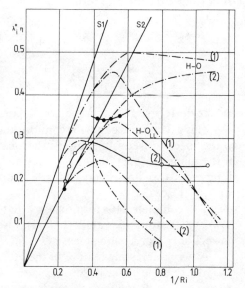

The experimental values are taken from D. F. Evans, J. A. Nadas, and M. A. Matesich, *J. Phys. Chem.* **75**, 1714 (1971). The curves correspond to the following data for acetone: $\varepsilon_0 = 20.56$, $\varepsilon_\infty = 1.9$, $\tau_s = 3.1 \times 10^{-12}$ sec, and $\eta = 0.303 \times 10^{-2}$ P.

For other examples in various solvents see Reference 78.

particles. Even in the case of pure liquids little is known at the level of pure equilibrium thermodynamics, and still less at the level of hydrodynamics. It can be foreseen that the situation will be still more complex when one ion is present in the liquid system.

Whatever is the progress observed in these approaches, the bases involved in these theories are so different in nature from what has been developed in this article that no attempt will be made to enter into any detail in this respect. We shall simply point out that present progress in this field is concerned more with opening new roads for exploration rather than with obtaining exact theoretical results of immediate quantitative utility for the experimenter. The theoretical study of the liquid state, in its full generality, is still a great challenge in our present times.

9. Conductance of Mixtures of Electrolytes of Any Kind

For a long time only the special case of simple binary electrolytes was theoretically studied. The reasons for this are easy to understand. The difficulties encountered were such that it was hard enough to complete the study in this special case! However, the problem is mainly a complexity in symbolism as far as the continuity equations are concerned. More difficult is the problem of the boundary conditions at short range. Clearly, if the treatment is limited to the case of the general restricted primitive model (all distances of closest approach being set equal, whatever the particles concerned), one cannot expect a result for which the reward will be as great as the complexity encountered, since again, but still more emphatically, all the specificity of each short-range pair configuration will be "averaged" in one single parameter b. On the other hand, introducing *ab initio* as many b_{ij} parameters as there are various possibilities is a formidable task that has so far never been attempted. This explains why only recently two derivations were carried out with the restricted primitive model: that of Quint and Viallard,[27] who used Debye and Hückel distribution functions at equilibrium, and that of Chen,[28] who used the more refined Meeron distribution function, which enables an exact form of the conductance equation to be derived up to the linear term in concentration. Both derivations use the matrix symbolism devised by Onsager and Kim,[71] whose former calculation included the limiting law only. We shall not go into more detail here, since the calculations are too voluminous, even if one simply wants to give the final results! The calculations will be found in detail in the original publications. Rather, we shall simply summarize some points that are relevant to the special developments given in this chapter.

It must first be mentioned that one has to deal with a set of continuity equations containing as many relations as there are different couples of ions i and j encountered. However, these equations will differ from one another only in their pair of indices i and j. As a consequence, there are no formal differences at the level of the formulations given by Eq. (52). However, changes arise when summations over all species k are encountered, since k no longer necessarily refers to the species i or j only. As a result, all the various continuity equations are interconnected, a difficulty that led Onsager and Kim to develop their matrix symbolism. An important point is that Eq. (75) is still a valid formulation for a given continuity equation; consequently, the echo formulation can easily be generalized to the case of mixtures. In particular, the definition of the monitoring term given by Eq. (76) is still valid. Among other differences the echo factor ν_{ji} defined by Eq. (109) will no longer be equal to Λ/Λ_0, as in the case of a single binary symmetrical electrolyte, but will turn out to be

$$\nu_{ji} = \frac{1}{Z_i} \frac{\lambda_i/Z_i - \lambda_j/Z_j}{\lambda_i^0/Z_i^2 - \lambda_i^0/Z_j^2} \tag{323}$$

Reassembling various terms in the definition of the specific conductance χ of the solution and factorizing out the short-range part of the non-hydrodynamic relaxation terms $\Delta X_{Ic}^S/X$ finally gives[72]

$$\begin{cases} 10^3\chi = \sum_i c_i\gamma_i\lambda_i^0\left(1 + \frac{\Delta X_{Ic}^{Li}}{X} + \frac{\Delta X_{Ih}^i}{X} + \frac{\lambda_i^e}{\lambda_i^0}\right) \\ \qquad + \sum_i \sum_{j\neq i} c_i\gamma_i c_j A_{ij}^S \frac{(z_i + z_j)^2}{(z_i^2/\lambda_i^0 + z_j^2/\lambda_j^0)B_{ji}} \tag{324} \\ \frac{1-\gamma_l}{\gamma_l} = \sum_{k\neq l} c_k A_{lk}^S \tag{325} \end{cases}$$

with

$$B_{ij} = \left[z_i \left(1 + \frac{\Delta X_i}{X} + \frac{\lambda_i^e}{\lambda_i^0} \right) + z_j \left(1 + \frac{\Delta X_j}{X} + \frac{\lambda_j^e}{\lambda_j^0} \right) \right] \bigg/ (z_i + z_j) \qquad (326)$$

and

$$A_{lk}^S = \frac{4\pi N}{1000} \int_0^R r^2 \mathring{g}_{lk}\, dr + O(c^{3/2}) \qquad (327)$$

This result obviously generalizes in a simple manner the set of equations (221) and (222) to the case of a mixture of ions of any kind. This set of equations presents some interesting features:

1. The second term of Eq. (324) is new if one compares it with Eq. (221). It turns out to be easily identifiable as the contribution of ion pairs ij to conductance. These pairs, when the charges are symmetrical, $(Z_i + Z_j) = 0$, do not drive the current, which explains why the equivalent contribution does not exist in Eq. (221). These pairs behave as entities that are characterized by a "limiting conductance,"

$$\Lambda_{ij}^0 = (z_i + z_j)^2/(z_i^2/\lambda_i^0 + z_j^2/\lambda_j^0) \qquad (328)$$

a result which is in close agreement with what a "chemical model" should predict, that is, if the two ions were bonded.

2. The quantities γ_l are involved in Eq. (325), which again is an obvious generalization of Eq. (222).

3. Finally, the short-range specificity of each pair (l, k) dealt with is concentrated in the term A_{ik}^S, the definition of which, deducible from the SCA, is given by Eq. (327). It can easily be anticipated from the brief summary given here that there is a trend toward a complete generalization of the special case developed in detail in the previous sections. Further work concerning this problem is still in progress. It can reasonably be hoped that one will ultimately be able to graft the echo formulation easily onto the equation of Quint and Viallard or that of Chen in a similar, though more complex, way, as was done in the case of binary symmetrical electrolytes. Again, then, the SCA will lift the degeneracy of the restricted primitive model as shown by Eqs. (325) and (327), a significant step in this field.

This matter of the theoretical study of conductance of electrolyte mixtures has recently been extended by substantial developments.[27,28,72] Such treatments will undoubtedly lead to a renewal of experimental measurements on such systems in the near future.

10. Optimization in Measurements and Data Processing

Once the "electrolyte–solvent" system to be studied has been chosen, the question arises as to which is the best way of optimizing conditions for obtaining the most reliable information on the system. This is easily achieved by a judicious choice of the concentration range studied and by an adequate extraction of the adjustable parameters of the theoretical equation.

10.1. High-Precision Measurements

10.1.1. Basic Recommendations

The reader needs hardly be reminded of the necessity of forming chemically pure products for the salts as well as for the solvent in order to measure with sufficient precision ($\sim$1% or better) the conductivity χ_0 of the solvent, which should not represent more than 2% of the lowest conductivity measured for the most dilute solution studied. The resistances of the decade-measuring bridge must also be adequately calibrated, allowing for any dependence on temperature, as well as must those of the connecting wires and of the leads of the cell, the constant for which must be known with the same precision as that desired in the required measurements; also, the temperature of the thermostat bath must be controlled to within a few thousandths of a degree C. The concentration of the solution should be evaluated by weighing, and the conversion from molality to molarity performed using the density of the solution. A microbalance is necessary for weighing the salt, the amounts of which may in some cases be as low as about 10 mg; also, a kilogram balance is needed for the solvent, which is often weighed by difference before and after introduction in the conductance cell. It is necessary to calibrate the weights, and also to correct for buoyancy.

The next problem is now to evaluate the best concentration range for measurements. It is recommended that the concentrations be linearly spaced in the range chosen in order to minimize errors in the evaluation of J or K_A. The highest value of the concentration may be derived by simple calculation, using

$$\kappa q \gamma^{1/2} = 0.5 \tag{329}$$

as a maximum value. The corresponding concentrations require a preliminary evaluation of the association constant, assuming the hard-sphere model for the ion–solvent system under study. From an evaluation of the Pauling ionic radii one obtains $a = r_+ + r_-$ and K_A from

$$K_A = \frac{4\pi N}{1000} \int_a^q r^2 \exp\left(\frac{2q}{r}\right) dr \tag{330}$$

directly or, better, by an extrapolation on an abaque where $\log K_A$ is plotted versus q for different values of a. The maximal concentration is then obtained by using

$$(c\gamma)_{\max} = 0.25 \bigg/ \frac{\kappa^2 q^2}{c} \tag{331}$$

with

$$\frac{\kappa^2}{c} = 0.25295 \times 10^{20} Z^2/DT \tag{332}$$

and

$$q = 8.354 \times 10^{-4} Z^2/DT \tag{333}$$

The corresponding value of γ is obtained from the mass action law, which may be rewritten as

$$\frac{1-\gamma}{\gamma} = K_A(c\gamma)_{max} \exp\left(-\frac{\kappa q \gamma^{1/2}}{1 + \kappa q \gamma^{1/2}}\right) \tag{334}$$

leading to

$$\gamma_{max} = \frac{1}{1 - K_A c \exp(-1/3)} \tag{335}$$

Hence the highest concentration is

$$c_{max} = \frac{(c\gamma)_{max}}{\gamma_{max}} \tag{336}$$

Of course, if the case where $q \leq a$ or when the association constant is small, the calculation is much easier, since the condition reduces to $\kappa a < 0.5$, which leads directly to

$$c_{max} = 0.25 \bigg/ \frac{\kappa^2 a^2}{c} \tag{337}$$

The concentration of the ith point (where $i = 1$ corresponds to the point for the most dilute solution) is then

$$c_i = \frac{i}{N} c_{max} \tag{338}$$

where N is the total number of points in the conductance run. There is no upper limit for N, but it is recommended not to choose $N < 5$, since there are three adjustable parameters.

10.1.2. Cells, Bridges, and Baths

Depending on the technique used to prepare the solutions (by dilution or concentration), the shape and volume of the cell may differ. In all cases it is recommended that the electrode compartment be geometrically separated from the actual bulk compartment of the cell and built in such a way that the electric field therein be as uniform as possible. This can be achieved by using a geometry that prevents the electric lines of force to expand too far away from the cylinder defined by the two electrodes. If this condition is respected, the cell constant may be expected to be independent of the concentration of electrolyte, with good precision. The usual way of avoiding polarization effects at the electrodes is to have them platinized, but this is not a necessity, since an extrapolation to zero frequency of the resistance measured versus the reciprocal of frequency (10, 5, 2, and 1 kHz) is usually sufficient to obtain a reliable value of the ohmic resistance. The internal resistance of the cell leads is obtained by measuring the resistance of the cell fitted with mercury. Details on calibration of the cell will be given in the next section.

The bridge is, of course, a very important feature of the whole conductance apparatus. It must be provided with a Wagner ground in order to eliminate all spurious capacitance effects of its various elements. Nowadays it is not a big problem to build a precision bridge, since high-quality resistors with a very low leakage capacitance are available that can constitute the ratio arms and the elements of a decade. Progress in electronics allows a variety of generators and detectors to be used. One good way to achieve very sensitive null detection is to use a phase-sensitive amplifier that is synchronized with the signal of the generator. This method allows a spectacular increase of the sensitivity of detection of the balance of the bridge.

A thermostated oil bath must be used to control the temperature of the solution in the cell to at least a few thousandths of a degree. Two possibilities are offered for achieving this goal:

1. One can use an on–off 200-W heating device controlled by a simple mercury contact thermometer of the standard type that operates on electronic relay.

It is recommended in this method that a large volume (50 liter) of well-stirred oil be used, placed in an insulated container, and cooled by circulation of a liquid that is itself thermostated to within 0.05 K. This circulating liquid should be at temperature lower than that of the main bath, with a refrigerating power smaller than the calorific power of the heating source by a factor of $\frac{1}{2}$. In this way the oil of the bath is alternatively heated and cooled with the same power ($\sim$100 W), and the result is a surprisingly constant regulation. The cost of such a thermostat is relatively low.

The prethermostat can be a relatively cheap device of the same type that, this time, is cooled by tap water if the conductance bath has to be maintained at 298 K. For lower temperatures it is necessary to use a commercial cooler as the prethermostat.

2. Another possibility is to control the temperature by a proportional, integral, differential (PID) amplifier. This gives excellent results and allows rapid changes in temperature to be made when necessary and this may be a determinative advantage. The idea is to detect any deviation of temperature by the signal of a platinum resistance bridge, to amplify this signal, and to then generate a current that is the sum of three components whose intensities

will respectively be proportional to the intensity of the signal, to the derivative of, and to the integral of the time-dependent intensity of the signal. This current is then used to feed a heater. Once the three constants of proportionality are adequately adjusted, the result is a bath temperature that is remarkably constant and insensitive to thermal perturbations. This bath must be cooled as in the case of the former on–off device. The PID control is particularly efficient for oil baths having small volumes.

10.1.3. Cell Calibrations

Until 1959 the calibration of cells was conducted by measuring the conductivity of a cell that was filled with a secondary standard solution of exact composition. Since 1933, the secondary standard solutions were those of Jones and Bradshaw,[79] who made absolute determinations with reference to mercury. The difficulty in applying this method was to prepare a solution of exact concentration of KCl in water, with only three possibilities, corresponding approximately to 1, 0.1, and 0.01 M. Another disadvantage was that a cell could be calibrated by only one solution, that is, at one value of resistance, since mostly only one value of resistance led to a value in the optimal range of the bridge ($200 < R < 10,000 \, \Omega$).

The situation changed in 1959 when Lind, Zwolenik, and Fuoss[80] proposed the calibration equation relative to the conductance of KCl in water at 298 K for concentrations $c \leq 0.01 \, M$:

$$\Lambda = 149.93 - 94.65c^{1/2} + 58.74c \log c + 198.4c \qquad (339)$$

Their proposal made possible the calibration of a cell having a constant around unity ($1 \, \Omega^{-1}$ cm) over a wide range of conductivity using a technique for measurements identical to that in an ordinary conductance run. In 1967, Fuoss and Hsia[81] extended the range of concentrations for KCl in water at 298 K up to 0.1 M with the equation

$$\Lambda = 149.936 - 94.88c^{1/2} + 25.48c \log c + 221.0c - 229c^{3/2} \qquad (340)$$

In 1968 the following equation, valid up to 0.05 M, was proposed by the present author[17,18]:

$$\Lambda = 149.89 - 94.87c^{1/2} + 58.63c \log c + 229.0c - 264.3c^{3/2} \qquad (341)$$

Also, an equation proposed in 1977 by Sändig et al., valid up to 1 M should be mentioned:

$$\Lambda = 150.000 - 99.282c^{1/2} + 135.798c - 120.788c^{3/2}$$
$$+ 57.891c^2 - 11.725c^{5/2} \qquad (342)$$

These equations differ very slightly from one another. It can be said simply that from a practical point of view the equation whose maximum concentration

value of the concentration, each time using the current values of the parameters being adjusted.

The derivatives need not be numerically exact with respect to the theoretical function for the iterative process to converge; the degree of exactness will only influence the number of iterations, which is, however, an important aspect of the problem in terms of computer time. The derivatives may be approximated by a finite increment formula:

$$\left(\frac{\partial \Lambda}{\partial P}\right)_i \sim \frac{\Delta \Lambda_i}{\Delta P} = \frac{\Lambda(P_2, c_i) - \Lambda(P_1, c_i)}{\Delta P} \tag{346}$$

where $\Delta P = P_2 - P_1$ is sufficiently small. This method is useful for $(\partial \Lambda / \partial K_A)_i$. For the other differential quantities it is more rapid to use the analytical derivatives

$$\left(\frac{\partial \Lambda}{\partial \Lambda_0}\right)_i = \gamma_i(1 - \alpha c_i^{1/2}\gamma^{1/2} + E_1 c_i \gamma_i \log c_i \gamma_i + \sigma_1(q)c_i\gamma_i) \tag{347}$$

and

$$\left(\frac{\partial \Lambda}{\partial J_{3/2}}\right)_i = \gamma_i c_i^{3/2} \gamma_i^{3/2} \tag{348}$$

All these calculations require an evaluation of γ_i, which, in turn, implies the solution of the mass action law equation. This can be achieved easily by a subroutine that searchs for the solution in the range $0 < \gamma_i \leq 1$. At each iteration k, the standard deviation is calculated,

$$(\sigma_\Lambda)_k = \left(\frac{\Sigma \Delta \Lambda_i^2}{N - 3}\right)^{1/2} \tag{349}$$

and the process is stopped when the relative improvement on σ_Λ between two successive iterations is less than a chosen value (~ 0.0001, for instance).

After each iteration new values of the parameters

$$(\Lambda_0)_k = (\Lambda_0)_{k-1} + (\Delta \Lambda_0)_k$$

$$(K_A)_k = (K_A)_{k-1} + (\Delta K_A)_k \tag{350}$$

$$(J_{3/2})_k = (J_{3/2})_{k-1} + (\Delta J_{3/2})_k$$

are calculated, so that when the process stops, the parameters have reached their optimal values. The rapidity of the process is also optimized by the use of three judiciously chosen initial values. The linear system can be solved by a matrix-inversion subroutine.

The standard deviation on each of the parameters is given by

$$\sigma_{\Lambda_0} = \sigma_\Lambda \sqrt{A_{11}}$$

$$\sigma_{K_A} = \sigma_\Lambda \sqrt{A_{22}} \tag{351}$$

$$\sigma_{J_{3/2}} = \sigma_\Lambda \sqrt{A_{33}}$$

where A_{ii} are the diagonal coefficients of the inverse of the original matrix. In the cases where only Λ_0, J, and $J_{3/2}$ are adjusted, only one iteration is sufficient for convergence, whatever the choice of the initial values, since the theoretical function is linear in the three parameters. It is convenient, in this case, to choose zero for the initial values so that the solution of the first iteration gives $(\Delta\Lambda_0)_1 = \Lambda_0$, $(\Delta J)_1 = J$, and $(\Delta J_{3/2})_1 = J_{3/2}$ directly.

Table 7 summarizes the most significant data of the least-squares adjustment of a given run. For CsBr in water the two data procedures (adjustment A of Λ_0, K_A, and $J_{3/2}$, or B of Λ_0, J, and $J_{3/2}$) can be used almost indifferently, since for this system the apparent distance of closest approach a' happens to be close to the Bjerrum cutoff distance q. As in all the adjustments displayed here, the Fuoss–Hsia[81] equation was used in its expansion version, given by Fernandez-Prini[56,89] (FHFP equation). However, given the fact that a' is smaller than q, the results obtained from procedure A must be considered as more reliable. It is interesting to note that the two series of measurements obtained by Justice *et al.* and Fuoss *et al.* lead to practically the same results. For this system $d_{+-}/kT = -0.18 \pm 0.01$ appears as the best evaluation.

In Table 8, procedure B will be chosen whenever $a' > q$, and procedure A will be chosen when $a' \leq q$.

Table 8 gives the limiting conductance Λ_0 and the Gurney parameter d_{+-}/kT for most measurements concerning $1:1$ electrolytes available in water. An interesting observation concerning alkali halides in water is that the d_{+-}/kT evaluations from conductance are very similar to the corresponding evaluations obtained from excess thermodynamic data by Rasaiah.[96] This is evidence of a self-consistency in the two kinds of theoretical treatments and hopefully of the reliability of the obtained parameters. This also shows that conductance data can lead to an evaluation of excess thermodynamic quantities. In the dilute range of concentration $(kq\gamma^{1/2} < 0.5)d_{+-}/kT$ is the only parameter necessary for the calculation of activity coefficients.

Figure 10 displays the great range of variation of the association constant of one electrolyte (tetrabutylammonium tetraphenylboride) in different solvents. For this salt an apparent distance of closest approach $a' = 6.9\,\text{Å}$ is sufficient to represent its behavior in all solvents studied.

Figures 11–19 illustrate the results for a large variety of $1:1$ electrolyte–solvent systems and help to display some interesting correlations.

The various observations made from Figures 11–19 show that even if the phenomena involved are complex, owing to the great variety of structures encountered in the ions and solvents, it is possible to sort out some simple correlations associated with these structures. This is certainly the most interesting and positive aspect of the situation due to the present state of elaboration of the conductance theory.

Table 7
Illustration of the Two Data Adjustment Procedures on a Limit Case Where Both Procedures Are Possible[a]

CsBr/H$_2$O [J. -C. Justice, R. Bury, and C. Treiner, *J. Chim. Phys.* **65**, 1708 (1968)]

$10^4 c$(mol/liter)	Λ	$\Delta\Lambda$ (A)	$\Delta\Lambda$ (B)
316.091	140.289	−0.018	−0.019
239.784	141.973	0.026	0.028
184.296	143.398	0.017	0.018
132.164	145.006	−0.012	−0.012
91.390	146.602	−0.014	−0.015
72.966	147.470	−0.015	−0.016
53.099	148.573	0.010	−0.010
29.440	150.289	0.025	0.027

$$(\kappa q \gamma^{1/2})_{max} = 0.21$$

A

$\sigma_\Lambda = 0.023$
$\begin{cases} \Lambda_0 = 155.437 \pm 0.032 \\ K_A = 0.40 \pm 0.05 \\ J_{3/2} = -246.1 \pm 24 [R_{J_{3/2}} = (3.6 \pm 0.2 \text{ Å}] \end{cases}$
$\begin{cases} a' \equiv a_K^B = (3.00 \pm 0.07) \text{ Å} \\ d_{+-}/kT = -0.19 \pm 0.02 \end{cases}$

B

$\sigma_\Lambda = 0.025$
$\begin{cases} \Lambda_0 = 155.422 \pm 0.033 \\ J = 182.8 \pm 7 \\ J_{3/2} = -156.9 \pm 39 [a'_{J_{3/2}} = (2.9 \pm 0.3 \text{ Å}] \end{cases}$
$\begin{cases} a' \equiv a_J = (2.77 \pm 0.11) \text{ Å} \\ d_{+-}/kT = -0.25 \pm 0.03 \end{cases}$

CsBr/H$_2$O [L. K. Hsia and R. M. Fuoss, *J. Am. Chem. Soc.* **90**, 3055 (1968)]
Number of points = 63 in the range $0.0027 \leq c \leq 0.1036$ mol/liter

$$(\kappa q \gamma^{1/2})_{max} = 0.38$$

A

$\begin{cases} \sigma_\Lambda = 0.018 \\ \Lambda_0 = 155.426 \pm 0.007 \\ K_A = 0.38 \pm 0.002 \\ J_{3/2} = -234.7 \pm 1 \, [R_{J_{3/2}} = (3.47 \pm 0.01) \text{ Å}] \end{cases}$
$\begin{cases} a' \equiv a_K^B = (3.05 \pm 0.01) \text{ Å} \\ \Downarrow \\ d_{+-}/kT = -0.175 \pm 0.003 \end{cases}$

B

$\begin{cases} \sigma_\Lambda = 0.015 \\ \Lambda_0 = 155.378 \pm 0.005 \\ J = 194.3 \pm 0.5 \\ J_{3/2} = -179.2 \pm 1.4 \, [a_{J_{3/2}} = (3.05 \pm 0.01) \text{ Å}] \end{cases}$
$\begin{cases} a' \equiv a_J = (2.92 \pm 0.01) \text{ Å} \\ \Downarrow \\ d_{+-}/kT = -0.21 \pm 002 \end{cases}$

---

[a] For CsBr in water at 25°C the equivalent primitive model has a characteristic distance (apparent distance of closest) a' smaller than the Bjerrum distance $q = 3.57$ Å.

In procedure A, where R is fixed equal to q, the adjusted parameters are Λ_0, K_A, and $J_{3/2}$, which leads to $a' \equiv a_K^B = (3.00 \pm 0.07)$ Å and ultimately to $d_{+-}/kT = -0.19 \pm 0.02$.

In procedure B, R is chosen equal to $a' < q$, which implies that $K_A = 0$ and the adjustment of Λ_0, J, and $J_{3/2}$ is evaluated from the Fuoss–Hsia J function, leading to $a' \equiv a_j = (2.77 \pm 0.11)$ Å, which leads to $d_{+-}/kT = -0.25 \pm 0.03$.

One observes that practically the same evaluation of the excess energy parameter d_{+-}/kT is made, with a preference to be given to the first evaluation for the sake of self-consistency.

One observes, however, that for systems such that $a' \sim q$ (that is, in the classical language, for systems slightly or not at all "associated") the final result in terms of excess parameters is more sensitive to the choice of the cutoff distance than for systems where a' is much smaller than q (strongly "associated" electrolytes) (see Figure 8).

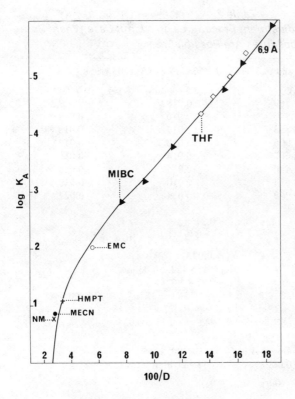

Figure 10. Association constant K_A of tetrabutylammonium-tetraphenyl boride in various solvents as a function of the dielectric constant at 25°C: ×—in nitromethane [M. A. Coplan, M. -C. Justice, and M. Quintin, *J. Chim. Phys.* **65**, 1152 (1968)], ●—in acetonitrile [J. F. Coetzee and G. P. Cunningham, *J. Am. Chem. Soc.* **87**, 2534 (1965)], +—in hexamethylphosphotriamide [C. Atlani and J. -C. Justice, *J. Solution Chem.*, **4**, 955 (1975)], ○—in ethylmethylketone [S. R. C. Hughes and D. H. Price, *J. Chem. Soc.*, 1093 (1967)], ▲—in methyl-isobutylketone–dioxane mixtures [M. Micheletti and J. -C. Justice, to be published, ◆—in tetrahydrofurane–dioxane mixtures [M. Micheletti and J. -C. Justice, to be published]. The solid curve represents the theoretical value, with an apparent distance of closest approach $a' = 6.9$ Å and a cutoff distance equal to the Bjerrum distance q, with the direct potential of the primitive (continuum) model.

10.3. Comparison with Other Procedures

In the following we shall examine some properties of the method proposed above and compare them with another method found in the literature. This discussion will be commenced by examining Figure 8, which graphically displays the influence of changing the cutoff distance R on the adjusted parameter a' and on the standard deviation σ_Λ. The run studied is for CsI in a water–dioxane mixture of dielectric constant $D = 12.89$. For this system the Bjerrum distance is $q = 21.9$ Å. The equation used was that of Fuoss and Hsia[31] in the expanded version given by Fernandez-Prini.[90] It is found that the standard deviation remains quite small over a rather wide range of R and that the a' parameter is also independent of R from $R = 8$ Å to $R > 28$ Å. Solving the adjusted value of $J_{3/2}$ for $R_{J_{3/2}}$ by use of the Fuoss–Hsia $J_{3/2}$ function shows that the inexactness of the $J_{3/2}$ function is such that $R_{J_{3/2}}$ is different from R_J as can be expected. In this case, however, there is a coincidence between R_J and $R_{J_{3/2}}$ for two values. As a consequence, adjusting Λ_0 and K_A only for various values of R imposed in the *whole* equation (that is, in J and $J_{3/2}$) would lead to an identity with our own results only for the

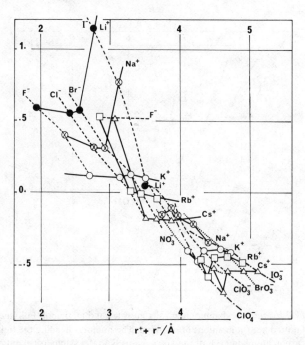

Figure 11. d_{+-}/kT for alkali metal salts in water at 298 K. Full lines join cations of one kind and dotted lines join anions of one kind. A general observation is that the larger the size, the smaller the excess parameter d_{+-}/kT. However, in a more detailed view the effect of an anion associated with a given cation is not so simple. For lithium and sodium halides very similar patterns are observed, as well as for rubidium and cesium; the case of potassium is intermediate, with no significant change with anions. For oxygenated anions a definite pattern holds for potassium, rubidium, and cesium that is somewhat distorted for sodium. Although most of the features of this diagram are still unexplained, the correlations observed must undoubtedly become a source for further insight.

For references of original data see Table 8.

two points of coincidence; everywhere else the σ_Λ observed by the second method must lead to values higher than ours. This is what is actually observed, as seen by the dotted line in Figure 8, which represents the standard deviations given by this second method. This gives a simple explanation of the two minima often quoted by authors[93,94] who systematically use the second least-squares method.

We then see the danger of proceeding to a physical interpretation of the two minima in terms of "ionic interactions." The two minima are only an artifact due to the approximate nature of the theoretical $J_{3/2}(R)$ function. If the $R_{J_{3/2}}$ curve is below the R_J straight line, then the dotted curve exhibits only *one* minimum, the abscissa of which is randomly located, depending on the experimental errors of the conductance run studied. Because the K_A values are well-defined functions of R, comparison of associated constants for two different runs is meaningless; moreover, when the $R_{J_{3/2}}$ curve is below

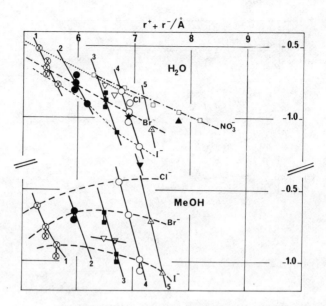

Figure 12. d_{+-}/kT for tetraalkylammonium salts in water and methanol. Full lines join cations of one kind and dotted lines join anions of one kind. The numbers heading the full lines indicate the number of carbon atoms in each of the alkyl chains R in the symmetrical ammonium cation R_4N^+. $\otimes$, $\bullet$, $\blacksquare$, $\bigcirc$, and $\triangle$ refer to Me_4N^+, Et_4N^+, Pr_4N^+, Bu_4N^+, and Am_4N^+, respectively. When several identical symbols gather on a small vertical segment, they refer to evaluations at different temperatures, the lowest at 283 K, the highest (when three are present) at 318 K. $\square$ represents the five salts from Me_4NO_3 to Am_4NO_3 from left to right. ∇ represents the $(EtOH)_4N^+$ cation associated with bromide or iodide (from left to right). ∇ represents $(EtOH)_4NI$ in methanol at 283 and 298 K, which lead to identical evaluations of d_{+-}/kT. $\star$, $\blacktriangledown$ and $\blacktriangle$ refer to triisoamylammonium bromide, iodide, and nitrate respectively. The pattern observed in the diagram seems well enough established so that one might extrapolate with confidence all missing data in a given series. Except for the $(EtOH)_4N^+$ cation in methanol, the effect of temperature is clear: Decreasing temperature lowers d_{+-}/kT. This is evidence of an effect of solvent structure that increases around the hydrophobic cation when temperature decreases. It is worth noting that all the points in water in this diagram are a continuation of the general trend observed in Figure 11.

the R_J line, the minimum of the dotted curve is always above our σ_Λ curve. This is the graphical evidence that our method eliminates the main part (if not all) of the systematic deviations.

It should be mentioned here that some equations never lead to two minima, contrary to the behavior of the Fuoss–Hsia equation, when processed with the second fitting procedure. This is often interpreted as a superiority of the former equations. One sees that in fact it only means just the opposite, that the theoretical $J_{3/2}$ function is worse, so that the coincidence between R_J and $R_{J_{3/2}}$ is never observed. One must conclude, on the contrary, that the Fuoss–Hsia $J_{3/2}$ function happens to be quite consistent with the $J(R)$ function. This is interesting to note but in fact has no importance with regard to the

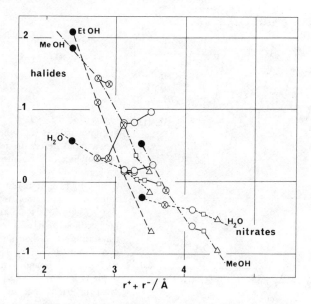

Figure 13. d_{+-}/kT of some alkali halides and nitrates in water, methanol, and ethanol at 25°C. Alkali cations are represented by the same symbols as in Figure 11. Each interrupted line joins the chlorides or nitrates in a given solvent. The solid lines starting from a chloride in a given solvent lead to the bromide and iodide in the same solvent. Very similar patterns are observed in the three hydrogen-bonded solvents with increasing trends from water to ethanol. For references of original data used, see Table 8 for water as solvent. For methanol: LiCl, NaBr, KBr, and KI, R. E. Jervis, D. R. Muir, J. P. Butler, and A. R. Gordon, *J. Amer. Chem. Soc.* **75**, 2855 (1953); NaCl and KCl, J. P. Butler, H. I. Schiff, and A. R. Gordon, *J. Phys. Chem.* **19**, 752 (1951); nitrates and RbCl, J. F. Frazer and H. Hartley, *Proc. Royal Soc.* **A109**, 351 (1925); CaCl, R. L. Kay and J. L. Hawes, *J. Phys. Chem.* **69**, 2787 (1965). For ethanol: LiCl, NaCl, and KCl, J. Graham, G. Kell, and A. R. Gordon, *J. Amer. Chem. Soc.* **79**, 2352 (1957); KBr, R. L. Kay and T. L. Broadwater, *J. Solution Chem.* **5**, 57 (1976); CsCl, J. L. Hawes and R. L. Kay, *J. Phys. Chem.* **69**, 2420 (1965).

evaluation of a', which would be evaluated just as well even if the Fuoss–Hsia $J_{3/2}$ function did not exist or was omitted. What is important is to adjust this coefficient, and not so much to have a more or less exact theory for it, at least in the present state of the theoretical treatments.

10.4. Evaluation of Short-Range Interaction Parameters

Once the association constant K_A has been evaluated with a known value of the cutoff distance R in the $J(R)$ function, it is possible to evaluate parameters that are characteristic of the short-range part of the direct anion–cation interaction potential in the solvent.

The simplest parameter is the apparent distance of closest approach a' of the two ions. It is the distance parameter of the *equivalent* primitive model.

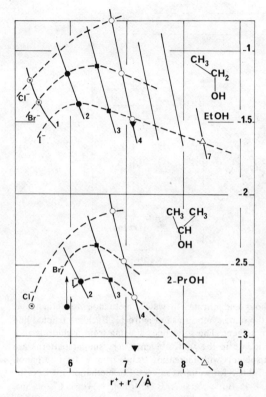

Figure 14. d_{+-}/kT for tetraalkylammonium halides in ethanol and 2-propanol. Δ refers to tetraheptylammonium iodide. For other symbols refer to Figure 12. The patterns are very similar to that observed in methanol, with only more negative values of d_{+-}/kT. Obviously the alkyl chains of the alcohols lead to a more negative translation when their numbers of carbon atoms increase. The point for Et_4NBr in 2-propanol that is clearly out of place, is to be noted; the deviation is most probably due to a small but significant error of measurement.

References of original conductance data used are the following: in ethanol—D. F. Evans and P. Gardam, *J. Phys. Chem.* **72**, 3281 (1968); in 2-propanol—M. A. Matesich, J. A. Nadas, and D. F. Evans, *J. Phys. Chem.* **74**, 4568 (1970).

The evaluation of a' is obtained by solving the equation

$$K_A = \frac{4\pi N}{1000} \int_{a'}^{R} r^2 \exp\left(\frac{2q}{r}\right) dr \tag{352}$$

The solution of this equation can be obtained by calculating K_A for various a' values until the calculated value approaches the experimental one. Integration subroutines can be utilized that use the trapezoid or Simpson's method.

A more interesting parameter is the Rasaiah–Friedman d_{+-}/kT parameter, which is obtained from a' by the relation

$$\frac{d_{+-}}{kT} = -\ln\left[\int_{a'}^{a+d} r^2 \exp\left(\frac{2q}{r}\right) dr \Big/ \int_{a}^{a+d} r^2 \exp\left(\frac{2q}{r}\right) dr\right] \tag{353}$$

where d is the width chosen for the square-mound perturbation. This parameter has the advantage of being free from the primary effect of the dimensions of the ions. This refines the comparison of both the behavior of one electrolyte in various solvents and that of various electrolytes in the same solvent. The parameter d_{+-}/kT really contains most of the specificity of the unknown interactions between anion and cation when their solvation cospheres overlap.

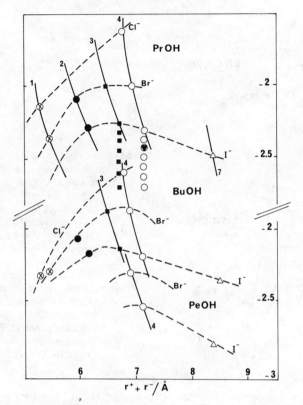

Figure 15. d_{+-}/kT for tetraalkylammonium halides in propanol, butanol, and pentanol. The effect of the alkyl chain composition of the alcohol is confirmed. The isomeric structure of 2-PrOH is more pronounced than that of BuOH (the more bulky, the more negative are the d_{+-}/kT).

The references of the original conductance data used are the following: in propanol—D. F. Evans and P. Gardam, *J. Phys. Chem.* **72**, 3281 (1968); in butanol and pentanol—D. F. Evans and P. Gardam, *J. Phys. Chem.* **73**, 158 (1969).

The vertical series of ■ and ○ in propanol refer to tetrapropyl- and tetrabutylammonium iodides data by R. Wachter, Ph.D. thesis, University of Regensburg, Federal Republic of Germany, 1973, at 25, 10, 0, −10, −20, −30, and −40°C from top to bottom. (At 25°C complete identity with Evans' results is observed.)

The importance of these parameters is thus great, since they will, in the more or less near future, be primary data on which theoreticians will have to practice their sagacity.

Since there is an infinite number of equivalent models, other parameters may be used that are at some variance with d_{+-}/kT. The conversion of one to another is straightforward so that it is not necessary to give more details here.

Figure 9 summarizes the connection between the Rasaiah–Friedman model and some more realistic models which can be envisaged. On the figure, the energy potential between an anion and a cation is plotted as a function

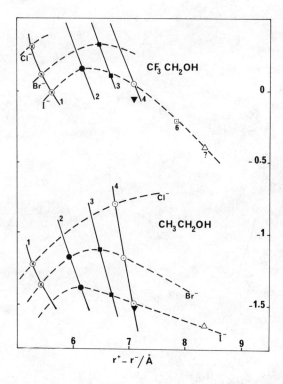

Figure 16. d_{+-}/kT of tetraalkyl-ammonium halides in ethanol and in trifluoroethanol. The substitution of CF_3 to CH_3 in ethanol does not perturb the general features of the pattern, but merely translates positively all values of d_{+-}/kT. This indicates that the pattern is specific for the hydroxyl radical (or for its structural implications as to the structure of the solvent), whereas its ordinate depends on the nature of the other end of the molecule. For references for ethanol, see Figure 14; for trifluoroethanol, D. F. Evans, J. A. Nadas, and M. A. Matesich, *J. Phys. Chem.* **75**, 1708 (1971).

of their center-to-center distance. It is assumed that the two ions are hard spheres so that this potential becomes infinitely positive when $r_{+-} < a$. Curve 1 represents the primitive model:

$$U_{+-} = \frac{Z_1 Z_2 e^2}{rD} \text{ for } r > a \tag{354}$$

This potential is certainly realistic enough for large distances but breaks down at short distances where the situation is more complex, since more specifics of each anion–cation–solvent system are dealt with. On one hand, the dielectric constant of the solvent is certainly no longer a reliable enough screening parameter for the coulombic interactions between the two ions. This screening should be less at short distance than is predicted by the bulk dielectric constant. Consequently, one may qualitatively expect to have a potential given by Curve 2 which corrects for the primitive (homogeneous continuum) model. A second perturbation is expected at short distances: when the two ions come near to contact, the discrete nature of the solvent must be considered. A simple view is that extra energy is required during the approach to expel the solvent molecules which are more or less strongly attracted by each ion by ion–dipole interaction in the most simple case. This extra energy may be represented by Curve 3 which represents in a rough way the energy to enter the second

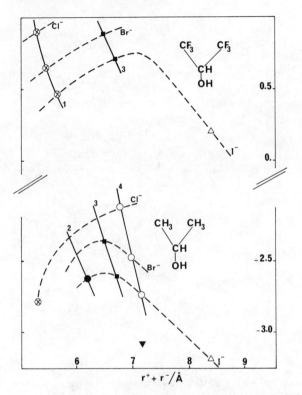

Figure 17. d_{+-}/kT of tetraalkylammonium halides in 2-propanol and hexafluoro-2-propanol. The same observation is made as in Figure 16. The positive translation due to the fluor substitution is still increased. For references for 2-propanol, see Figure 14; for hexafluoro-2-propanol, M. A. Matesich, J. Knoefel, H. Felman, and D. F. Evans, *J. Phys. Chem.* **77**, 366 (1973).

solvent cosphere, followed by the still greater energy to enter the first cosphere. As a result, the net potential is obtained by adding Curve 3 to Curve 2, leading to Curve 4. Such a result is in qualitative agreement with the results of Patey and Vallean[87] who carried Monte Carlo calculations on two ions immersed in the bulk of dipole molecules. Realistic as this model may be, the number of parameters involved in its microscopic description prevents any useful utilization in the case of the conductance technique. But given the concept of equivalence stated above, one may quite conveniently use the crude model given on the bottom right of Figure 9. There, the perturbation to the primitive model becomes a square mound (or well) with an adjustable height (or depth) and a fixed width which may, for instance, be equal to one diameter of the solvent molecule. Another equivalent model is that where the distance of closest approach a' is the adjustable parameter and where the primitive model holds for $r > a'$. The equivalence relation between these last two models is given by Eq. (353). Quite generally, the equivalence relation

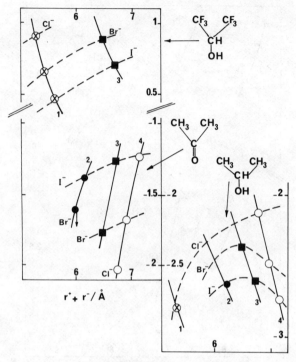

Figure 18. Comparison of d_{+-}/kT for tetraalkylammonium halides in 2-propanol, acetone, and hexafluoro-2-propanol. This comparison with acetone shows that in the latter the pattern is now changed since effect of the halide anions are reversed. However, all values of d_{+-}/kT remain negative in acetone. For references for the alcohols, cf. Figs. 14, 17; for acetone, D. F. Evans, J. A. Nadas, J. Thomas, and M. A. Matesich, *J. Phys. Chem.* **75**, 1714 (1971).

is obtained by equating the two short-range integrals:

$$\int_0^R r^2 \exp\left(-\frac{U_{+-}^{(1)}}{kT}\right) dr = \int_0^R r^2 \exp\left(-\frac{U_{+-}^{(2)}}{kT}\right) dr \qquad (355)$$

10.5. Some Results from the Literature

There are already some extensive reviews of conductance data from the literature[89–92] in which are reported the various parameters which result from the original data adjustments.

In the following, we shall simply display some results concerning readjustments of data of the literature in view of conclusions reached in the above chapters. As a consequence, only the limiting conductances Λ_0 and the parameters d_{+-}/kT will be given together with limiting transference numbers when available. There is no ambition to be exhaustive in this literature scanning. Whenever a given system has been studied by various authors only

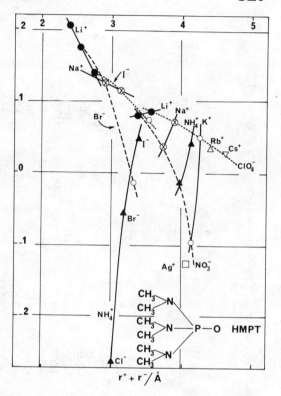

Figure 19. d_{+-}/kT in hexamethyl-phosphotriamide. The salts can easily be identified from information given on the graph. The anion associated to the lithium ion on top of the figure is chloride. References: E. M. Hanna, A. D. Pethybridge, J. E. Prue, and D. J. Spiers, *J. Solution Chem.* **3**, 563 (1974). The value for KClO$_4$ was confirmed by C. Atlani and J. -C. Justice, *J. Solution Chem.* **4**, 955 (1975).

the results which correspond to studies which satisfy, at best, the optimal criterion $\kappa q \gamma_{max}^{1/2}$ nearest to 0.5, were retained.

Table 7 summarizes the most significant data of the least-squares adjustment of a given run. For CsBr in water the two data procedures (adjustment A of Λ_0, K_A and $J_{3/2}$, or B of Λ_0, J, and $J_{3/2}$) can be used almost independently, since for this system the apparent distance of closest approach a' happens to be close to the Bjerrum cutoff distance q. As in all the adjustments displayed here, the Fuoss–Hsia[82] equation was used in its expansion version, given by Fernandez-Prini[56,90] (FHFP equation). However, given the fact that a' is smaller than q, the results obtained from procedure (A) must be considered as more reliable. It is interesting to note that the two series of measurements obtained by Justice *et al.* and Fuoss *et al.* lead to practically the same results. For this system $d_+/kT = -0.18 \pm 0.01$ appears as the best evaluation.

In Table 8, whenever $a' > q$, procedure B will be chosen, and procedure A will be chosen when $a' \leq q$.

Table 8 gives the limiting conductance Λ_0 and the Gurney parameter d_{+-}/kT for most measurements concerning 1:1 electrolytes available in water. An interesting observation concerning alkali halides in water is that the d_{+-}/kT evaluations from conductance are very similar to the corresponding evaluations obtained from excess thermodynamic data by Rasaiah.[96] This

Table 8

Values of the Limiting Conductance and of the Excess Parameter d_{+-}/kT for Various Electrolytes in Water at 298 K[a]

Ion radius (Å)	F⁻ 1.36	Cl⁻ 1.81	Br⁻ 1.95	I⁻ 2.16	NO₃⁻ 2.77	ClO₃⁻ 2.90	ClO₄⁻ 2.91	BrO₃⁻ 3.20	IO₃⁻ 3.39	IO₄⁻ 3.40
Li⁺ 0.60	92.67 / 0.59 / (1)	115.04 / 0.55 / (2)	116.97 / 0.57 / (3)	115.78 / 1.15 / (3)	110.11 / 0.22 / (4, 5)		105.97 / −0.25 / (6)			
Na⁺ 0.95	105.42 / 0.39 / (7)	126.60 / 0.31 / (8)	128.53 / 0.32 / (9)	127.07 / 0.76 / (3)	122.33 / −0.36 / (10)	114.02 / −0.11 / (11)	117.05 / −0.10 / (11)			
K⁺ 1.33	128.88 / 0.10 / (12)	149.95 / 0.11 / (8)	151.72 / 0.13 / (13)	150.59 / 0.03 / (14)	145.18 / −0.41 / (10)	137.52 / −0.38 / (11)	140.76 / −0.47 / (11)	129.33 / −0.42 / (15)	113.97 / −0.43 / (10)	127.80 / −0.56 / (16)
Rb⁺ 1.48	132.64 / 0.52 / (3)	153.64 / 0.01 / (17)	155.43 / −0.01 / (18)	154.01 / −0.04 / (19)	148.79 / −0.47 / (10)	141.11 / −0.44 / (11)		132.56 / −0.45 / (10)	117.76 / −0.50 / (10)	
Cs⁺ 1.69	132.75 / 0.51 / (3)	153.30 / −0.23 / (20)	155.43 / −0.18 / (21, 22)	154.22 / −0.19 / (22)	148.90 / −0.53 / (10)	141.02 / −0.51 / (11)		133.01 / −0.61 / (10)	117.23 / −0.55 / (10)	
Me₄N⁺ 3.47		120.54 / −0.56 / (23)	122.64 / −0.65 / (24)	121.40 / −0.72 / (23, 25)	116.16 / −0.70 / (31)					
Et₄N⁺ 4.00		109.01 / −0.67 / (26)	110.47 / −0.76 / (23)	109.08 / −0.96 / (26)	103.61 / −0.77 / (31)		99.48 / −0.95 / (26)			
Pr₄N⁺ 4.52			101.45 / −0.86 / (23)	100.28 / −1.06 / (23)	94.91 / −0.90 / (31)					
(EtOH)₄N⁺ 4.52		103.86 / −0.83 / (27)	105.32 / −0.77 / (27)							

Ion	d				
Bu_4N^+	4.94	95.61 −0.81 (23)	97.58 −0.95 (23)	96.30 −1.18 (28)	90.96 −0.96 (31)
$i\text{-}Am_3BuN^+$	4.94		96.46 −0.99 (29)	95.62 −1.33 (30)	
$i\text{-}Am_4N^+$	4.94		96.28 −0.98 (29)		89.39 −1.02 (31)
Am_4N^+	5.36		96.00 −1.08 (29)		88.93 −1.02 (31)

[a] These results were obtained by reprocessing the original data from the Fuoss–Hsia–Fernandez–Prini equation given in Tables 5 and 6 by the three-parameter procedure described in Section 10.2. The references (in parentheses) of the original data selected are given below. This choice is somewhat arbitrary, since in many cases several sets of measurements were available that gave similar results. For the sake of simplicity, in the presentation only one result was quoted for each salt, except in some rare cases, where two runs led to identical results within the precision of the figures given. The main criteria were the number of points (c, Λ) in the run, the range of concentration used (maximal in the range $\kappa q \gamma^{1/2} < 0.5$), and the standard deviation of the run: This allows the best evaluation of d_{+-}/kT, but not necessarily that of the limiting conductance, which depends on other more important external considerations (salt purity).

The parameters d_{+-}/kT are obtained through Eq. (353) from a', which in turn are evaluated from the adjusted J or K_A coefficients. The diameter d of the water molecule is taken as 2.76 Å and the distance of closest approach a is evaluated from the ionic radii given in this table.

References for Table 8
1. M. Kahlveit, Z. Phys. Chem. (Leipzig) 21, 436 (1959).
2. C. F. Nattina and R. M. Fuoss, J. Phys. Chem. 79, 1604 (1975).
3. A. D. Pethybridge and D. J. Spiers, J. Chem. Soc. Faraday Trans. 1 73, 768 (1977).
4. G. D. Parfitt and A. L. Smith, Trans. Faraday Soc. 59, 257 (1963).
5. M. V. Ramanamurty and R. C. Yadav, Indian J. Chem. 9, 1003 (1971).
6. H. Hognas, Ann. Acad. Sci. Fenn. A2 7, 145 (1969).
7. T. Mallanoo and R. H. Stokes, Aust. J. Chem. 30, 1375 (1977).
8. Y. C. Chiu and R. M. Fuoss, J. Phys. Chem. 72, 4123 (1968).
9. A. Reynolds, Ph.D. thesis, Yale, New Haven, Connecticut, 1966.
10. M. -C. Justice, R. Bury, and J. -C. Justice, Electrochim. Acta 16, 687 (1971).
11. R. Bury, M. -C. Justice, and J. -C. Justice, J. Phys. Chem. 67, 2045 (1970).
12. C. G. Swain and D. F. Evans, J. Am. Chem. Soc. 88, 383 (1966).
13. G. Jones and C. F. Bickford, J. Am. Chem. Soc. 56, 602 (1934).
14. A. d'Aprano, J. Komiyama, and R. M. Fuoss, J. Solution Chem. 1, 279 (1976).
15. I. H. Jones, J. Am. Chem. Soc. 66, 1115 (1944).
16. I. H. Jones, J. Am. Chem. Soc. 68, 240 (1946).

Table 8 (cont.)

17. R. W. Kunze and R. M. Fuoss, J. Phys. Chem. 67, 911 (1963).
18. J. E. Lind and R. M. Fuoss, J. Phys. Chem. 66, A27 (1962).
19. T. L. Fabry and R. M. Fuoss, J. Phys. Chem. 63, 971 (1964).
20. C. Treiner, J. -C. Justice, and R. M. Fuoss, J. Phys. Chem. 68, 3886 (1964).
21. J. -C. Justice, R. Bury, and C. Treiner, J. Phys. Chem. 65, 1708 (1968).
22. K. L. Hsia and R. M. Fuoss, J. Am. Chem. Soc. 90, 3055 (1968).
23. D. H. Evans and R. L. Kay, J. Phys. Chem. 70, 366 (1966).
24. B. L. Levien, Aust. J. Chem. 18, 1161 (1965).
25. M. -C. Justice and J. -C. Justice, C. R. Acad. Sci., 262, 608 (1966).
26. M. Tissier and G. Douheret, J. Solution Chem. 7, 87 (1978).
27. D. H. Evans, G. P. Cunningham, and R. L. Kay, J. Phys. Chem. 70, 2974 (1966).
28. R. W. Nartel and C. A. Krauss, Proc. Nat. Acad. Sci. USA 40, 382 (1954).
29. R. L. Kay, D. H. Evans, and G. P. Cunningham, J. Phys. Chem. 73, 3322 (1969).
30. M. Quintin and M. -C. Justice, C. R. Acad. Sci. 261, 1287 (1965).
31. Y. Prigent, M. -C. Justice, and J. -C. Justice, J. Solution Chem., to be published.

is evidence of a self-consistency in the two kinds of theoretical treatments and hopefully of the reliability of the obtained parameters. This also shows that conductance data can lead to an evaluation of excess thermodynamic quantities. In the dilute range of concentration $(\kappa q \gamma^{1/2} < 0.5) d_{+-}/kT$ is the only parameter necessary to calculate activity coefficients.

Figure 10 displays the great range of variation for the association constant of one electrolyte (tetrabutylammonium tetraphenylboride) in different solvents. For this salt an apparent distance of closest approach $a' = 6.9$ Å is sufficient to represent its behavior in all solvents studied. Figures 11–19 illustrate the results for a large variety of 1–1 electrolyte–solvent systems and help to display some interesting correlations. The various observations made from Figures 11–19 show that even if the phenomena involved are complex due to the great variety of structures encountered in the ions and in the solvents, it is possible to sort out some simple correlations associated with these structures. This is certainly the most interesting and positive aspect of the situation due to the present state of elaboration of the conductance theory.

Appendix A

It is recognized that the symbolism used in denoting the various contributions to the perturbation part of the P.d.f.'s may seem somewhat complex. However, it is the author's opinion that it is quite practical, since it allows one to perceive at a glance the origin of each of the various contributions involved, allowing a simple and immediate recollection of the implicit physical content of each term concerned.

The following summary may be found useful for a fast relocation of each term in its original physical context. The ion-pair distributions function g_{ji} referring to two ions i and j is separated into an equilibrium contribution $\overset{\circ}{g}$ and a perturbation contribution g'_{ji}:

$$g_{ji} = \overset{\circ}{g}_{ji} + g'_{ji} \tag{A1}$$

For binary symmetrical electrolytes the peturbation part g'_{ji} is itself the sum of two contributions. Each term of this sum is the product of two quantities: an *unechoed* contribution denoted by $g'_{I_{ji}}$ and a dimensionless echo factor ν_{ji} or μ_{ji} given by Eqs. (144) and (147) according to the term concerned:

$$g'_{ji} = g'^{(1)}_{I_{ji}} \nu_{ji} + g'^{(2)}_{I_{ji}} \mu_{ji} \tag{A2}$$

The various unechoed contributions $g'_{I_{ji}}$ are evaluated by solving the continuity equation in which all echo factors, ν_{ji} and μ_{ji} are set equal to unity. Inspection of the continuity equation shows that $g'^{(1)}_{I_{ji}}$ is itself the sum of a hydrodynamic contribution (which depends on the viscosity) $g'^{(1)}_{Ih_{ji}}$ and a non-hydrodynamic one (which does not depend on viscosity) $g'^{(1)}_{IC_{ji}}$, whereas $g'^{(2)}_{I_{ji}}$

is a hydrodynamic contribution denoted by $g'^{(2)}_{Ih_{ji}}$. The final result thus reads

$$g'_{ji} = (g'^{(1)}_{IC_{ji}} + g'^{(1)}_{Ih_{ji}})\nu_{ji} + g'^{(2)}_{Ih_{ji}}\mu_{ji} \tag{A3}$$

as used in Eq. (208). This system of indexing, together with its interpretation, is fully transposable to the relaxation field term

$$\frac{\Delta X}{X} = \left(\frac{\Delta X^{(1)}_{IC}}{X} + \frac{\Delta X^{(1)}_{Ih}}{X}\right)\nu_{ji} + \frac{\Delta X^{(2)}_{Ih}}{X}\mu_{ji} \tag{A4}$$

Similarly, for the ionic electrophoretic conductance λ^i_e of an ion of type i

$$\lambda^i_e = (\lambda^i_{e1} + \lambda^{(2)}_{e2})\mu_{ji} + \lambda^{(1)}_{e2}\nu_{ji} \tag{A5}$$

where λ^i_{e1} is calculated from $\mathring{g}_{ii}$ and $\mathring{g}_{ij}$ [see Eq. (182)], $\lambda^{(1)}_{e2}$ from $g'^{(1)}_{IC_{+-}} + g'^{(1)}_{Ih_{+-}}$, and $\lambda^{(2)}_{e2}$ from $g'^{(2)}_{Ih_{+-}}$.

Appendix B

Very recently a new derivation of conductance has been proposed by Ebeling *et al.*[97,98] It is based on a generalization to transport phenomena of the Mean Spherical Approximation (MSA). Badiali *et al.*[99] have shown that the MSA is variant for an approximation to the three-particle distribution functions in the second equation (31) of the BBGKY hierarchy (30–32). Ebeling's idea is to introduce the same approximation into the continuity equation (52), keeping otherwise unchanged all other approximations as summarized in Table 2 except for $\mathring{g}_{ij}$ at equilibrium for which the MSA result is used. In a second step,[98] a mass action law (MAL) is introduced in the former result in order to generalize the MSA linear conductance equation to the case of strong ionic interactions. Let us note that the MAL constant used by Ebeling differs from that (K^B_A) used by the author in the present chapter. Our constant is obtained from the application of the Echo effect to the *short range* part of the *Space-partition* of the integral of the Boltzmann function. The Ebeling constant identifies with the *higher terms* of a *series expansion partition* of the same function. In both cases, the important point is that the echo effect is applied to the part which rapidly diverges when the Bjerrum parameter becomes large.

The main point is that the MSA approximation is much more efficient than that of Debye and Hückel, so that the final result should be valid in a much larger range of concentrations as seems to be shown by the first tests presented by Ebeling and Grigo.[98]

Acknowledgments

I am thankful to Dr. M.-C. Justice for her determinant help in the elaboration of this manuscript. Her comments and suggestions contributed to many significant improvements.

I wish to pay a large tribute to Professor M. A. Coplan, who kindly agreed, on several occasions, to read large parts of this manuscript and to try, as much as this was possible, to make clear the many obscure phrases of the original English draft.

Thanks are also due to Professor R. H. Wood, whose last visit to Paris was mostly spent discussing, as well as reading and correcting, newly prepared pages.

My colleagues could not, unfortunately, correct everything and I am entirely responsible for all remaining errors.

Last, but not least, it is my pleasant duty to express my gratitude to Professor W. Ebeling for so many stimulating discussions in steady correspondence.

Auxiliary Notation

a	distance of closest approach of bare ions
a'	apparent distance of closest approach of anion and cation in the solvent
b	Bjerrum's parameter $z^2 e^2 / a' D k T$ [Eq. (198)]
c	molar concentration
d	width of the square-mound perturbation
D	dielectric constant
e	proton charge
e_i	charge of an ion of type i
$\mathbf{E}_0$	electrical field on an ion
E'_1	$= \kappa^2 q^2 / 6c$, relaxation coefficient of the Naperian log term in conductance [see Eq. (196)]
E_1	$= E'_1 / 0.43429$; same as E'_1, but with decimal log
E'_2	$= \kappa q \beta / 8 c^{1/2}$, hydrodynamic coefficient of the Naperian log term in conductance [see Eq. (197)]
E_2	$= E'_2 / 0.43429$; same as E'_2, but with decimal log
E	$= E_1 \Lambda_0 - (k_e + k_h) E_2$
E'	$= E'_1 \Lambda_0 - (k_e + k_h) E'_2$
$\mathbf{f}_{jQ}^{iP}$	mean interionic force [Eq. (69)]
$f_{\pm}$	mean activity coefficient
$f'_{\pm}$	mean activity coefficient of the free ions
$f\Lambda$	conductance coefficient $= \Lambda / \Lambda_0$
$\mathscr{F}$	Faraday constant
$\mathbf{F}^{iP}$	mean interionic force on an ion of type i
$\mathbf{F}_{jQ}^{iP}$	same as $\mathbf{F}_{jQ}^{iP}$ when an ion of type j is present at Q
g_{jQ}^{iP}	$= g_{ji}(\mathbf{r})$ distribution function relative to an ion of type i at P when an ion of type j is present at Q with $\mathbf{r} = \mathbf{r}_P - \mathbf{r}_Q$
$\mathring{g}_{ji}, g'_{ji}$	cf. Appendix A
h_{jQ}^{iP}	correlation function $h_{jQ}^{iP} = g_{jQ}^{iP} - 1$

k	Boltzmann's constant
k_e	electrophoresis numerical constant, Table 5
k_h	hydrodynamic numerical constant, Table 5
K_A	association constant (Eq. 10)
K_A^B	Bjerrum association constant (Eq. 11)
K_A^F	Fuoss association constant (Eq. 20)
K_d	dissociation constant
n_i	stoichiometric density of ions of type i in the solution
N	Avogadro's number
q	Bjerrum's distance $q = Z^2 e^2/2DkT$
q^*	constant defined in Eq. (135)
$\mathbf{r}$	unit vector in the radial direction
R	cutoff distance in the integrals along the radial variable r
S	Debye–Onsager limiting law coefficient
$s(z)$	Shedlovsky function (Eq. 15)
T	Kelvin temperature
$U(iP, jQ)$	direct potential energy of interaction of two ions of type i and j located at P and Q, respectively, in the pure solvent
$U^*(iP, jQ)$	repulsion part of $U(iP, jQ)$
$\mathbf{v}^{iP}$ or $\mathbf{v}^i$	mean velocity of an ion of type i (at P)
$\mathbf{v}_{jQ}^{iP} = \mathbf{v}_{ji}(\mathbf{r})$	same as $\mathbf{v}^{iP}$ when an ion of type j is present at Q, with $\mathbf{r} = \mathbf{r}_P - \mathbf{r}_Q$
$\mathbf{v}_e^{iP}, \mathbf{v}_{ejQ}^{iP}$	electrophoretic part of $\mathbf{v}^{iP}$ and $\mathbf{v}_{jQ}^{iP}$
$\mathbf{v}_{ejQ}^{P}$	mean velocity of the medium at P in the surroundings of an ion of type j at Q [Eq. (137)]
$\mathbf{X}$	external field
z	absolute charge number of ions of a symmetrical electrolyte; also Shedlovsky variable [in Eq. (16)]
Z_i	algebraic charge number of an ion of type i
α	relaxation coefficient of the Onsager limiting law [Eq. (194)]
β	electrophoretic coefficient of the Onsager limiting law [Eq. (195)]
γ	fraction of free ions
$\boldsymbol{\delta}$	unit tensor
δ	Dirac function
δ_{ij}	Kronecker symbol
ε_0	static dielectric constant
ε_∞	high-frequency dielectric constant
η	viscosity of the solvent
κ	reciprocal Debye's radius
$\boldsymbol{\lambda}, \boldsymbol{\lambda}'$	long- and short-range hydrodynamic tensors [see Eqs. (48, 49)]
λ^i	ionic conductance of an ion of type i [Eq. (43)]
Λ	molar conductance

Λ_0, λ_0^i	molar and ionic limiting conductance
Λ_e, λ_e^i	molar and ionic electrophoretic contributions to conductance (see Appendix A)
ξ	long range Bjerrum parameter, $\xi = \dfrac{2q}{R}$
τ_s	relaxation time of the solvent molecules
χ	conductivity of the solution [Eq. (44)]
$\mathbf{\chi}(P, Q, \omega_i, \omega_j)$	hydrodynamic coupling tensor [Eq. (47)]
ψ_{ji}	mean potential of interactions between two ions at finite concentration
ψ', ψ_{ji}'	perturbation part of the mean potential
ω_i	mobility of ion i in pure solvent [Eq. (45)]
∇	Nabla vectorial operator $= \mathbf{i}(\partial/\partial x) + \mathbf{j}(\partial/\partial y) + \mathbf{k}(\partial/\partial z)$.
$\nabla_P U_{ji}$	$(\mathbf{r}_P - \mathbf{r}_Q)$ Gradient at point P of the scalar function U ($\equiv \text{grad}_P U_{ji}$)
$\nabla_P \mathbf{v}_{ji}(\mathbf{r}_P - \mathbf{r}_Q)$	divergence at point P of the vectorial function $\mathbf{v}$ ($\equiv \text{div}_P \mathbf{v}_{ji}$)
Δf	Laplacian of function f ($\Delta f = \nabla \times \nabla f$)

References

1. L. Onsager, *Phys. Z.* **27**, 388 (1926).
2. L. Onsager, *Phys. Z.* **28**, 277 (1927).
3. E. Pitts, *Proc. R. Soc. London A* **217**, 43 (1953).
4. R. M. Fuoss and L. Onsager, *J. Phys. Chem.* **61**, 668 (1957).
5. H. Falkenhagen, W. Ebeling, and W. D. Kraeft, in *Ionic Interactions*, S. Petrucci, ed., Academic Press, New York (1971).
6. N. Bjerrum, in *Selected Papers*, Einar Munksgaard Publisher, Copenhagen (1949).
7. P. Debye and E. Hückel, *Phys. Z.* **24**, 185 (1923); or in *Collected Papers of Peter J. W. Debye*, Wiley-Interscience, New York (1954).
8. N. Bjerrum, K. Dan. Vidensk. Selskab. Math.-Fys. Medd. **7**, 7 (1926); or for a translation of pp. 1–17 into English.
9. T. Shedlovsky, *J. Franklin Inst.* **225**, 739 (1938).
10. R. M. Fuoss and F. Accascina, in *Electrolytic Conductance*, Wiley-Interscience, New York (1959).
11. R. H. Stokes, *J. Phys. Chem.* **65**, 1242 (1961).
12. R. M. Fuoss, *J. Am. Chem. Soc.* **80**, 3163 (1958).
13. R. M. Fuoss and L. Onsager, *Proc. Nat. Acad. Sci. USA* **49**, 274 (1955).
14. R. M. Fuoss, *J. Am. Chem. Soc.* **79**, 3301 (1959).
15. R. M. Fuoss and C. A. Kraus, *J. Am. Chem. Soc.* **79**, 3304 (1957).
16. R. M. Fuoss, *J. Am. Chem. Soc.* **81**, 2659 (1959).
17. J. -C. Justice, Ph.D. dissertation, University of Paris VI, CNRS N° A. O. 1558 (1967).
18. J. -C. Justice, *J. Chim. Phys.* **65**, 353 (1968).
19. See R. M. Fuoss and F. Accascina, in Electrolytic Conductance, Wiley-Interscience, New York (1959), p. 217.
20. R. M. Fuoss, *J. Phys. Chem.* **78**, 1383 (1974).
21. J. -C. Justice and W. Ebeling, *J. Solution Chem.* **8**, 809 (1979).
22. J. -C. Justice, *J. Solution Chem.* **7**, 859 (1978).

23. H. Falkenhagen, W. Ebeling, and H. G. Hertz, in *Theorie der Elektrolyte*, S. Hirzel Verlag, Leipzig (1971).
24. R. M. Fuoss, L. Onsager, and J. F. Skinner, *J. Phys. Chem.* **69**, 2581 (1965).
25. T. J. Murphy and E. G. D. Cohen, *J. Chem. Phys.* **53**, 2173 (1970).
26. T. J. Murphy, *J. Chem. Phys.* **56**, 3487 (1972).
27. J. Quint and A. Viallard, *J. Solution Chem.* **7**, 533 (1978).
28. M. -S. Chen, *J. Solution Chem.* **8**, 509 (1979).
29. P. G. Carman, *J. S. Afr. Chem. Inst.* **28**, 341 (1975).
30. W. H. Lee and R. J. Wheaton, *J. Chem. Soc. Faraday Trans. 2* **74**, 743 (1978).
31. W. H. Lee and R. J. Wheaton, *J. Chem. Soc. Trans. 2* **74**, 1456 (1978).
32. R. M. Fuoss, *J. Phys. Chem.* **79**, 525 (1975).
33. R. M. Fuoss, *J. Phys. Chem.* **81**, 1529 (1977).
34. R. M. Fuoss, *J. Phys. Chem.* **82**, 2427 (1978).
35. W. Ebeling, R. Feistel, and R. Sändig, *J. Solution Chem.* **8**, 53 (1979).
36. P. Debye and H. Falkenhagen, *Phys. Z.* **29**, 401 (1928); in *Collected Papers of Peter J. W. Debye*, Wiley-Interscience, New York (1954).
37. H. Yamakava, *J. Chem. Phys.* **53**, 436 (1970); U. Felderhof, *Physika* **89A**, 373 (1977).
38. W. Ebeling, *Wiss. Z. Univ. Rostock Math. Naturwiss. Reihe* **14**, 271 (1965); see also H. Falkenhagen, W. Ebeling, and W. D. Kraeft, in *Ionic Interactions*, S. Petrucci, ed., Academic Press, New York (1971), p. 73.
39. R. M. Fuoss and K. L. Hsia, *Proc. Nat. Acad. Sci. USA* **6**, 1550 (1967).
40. J. Barthel, J. -C. Justice, and R. Wachter, *Z. Phys. Chem.* **84**, 100 (1973).
41. E. Renard and J. -C. Justice, *J. Solution Chem.* **3**, 633 (1974).
42. P. C. Carman, *J. S. Afr. Chem. Inst.* **28**, 80 (1975).
43. J. Quint and A. Viallard, *J. Solution Chem.* **7**, 137 (1978).
44. M. -S. Chen, *J. Solution Chem.* **7**, 675 (1978).
45. R. Feistel, *Z. Phys. Chem. (Leipzig)* **259**, 2017 (1977).
46. R. Sändig and R. Feistel, *J. Solution Chem.* **8**, 411 (1979).
47. R. M. Fuoss, *J. Phys. Chem.* **63**, 633 (1959).
48. E. Pitts, B. E. Tabor, and J. Daly, *Trans. Faraday Soc.* **65**, 849 (1969).
49. See R. M. Fuoss and L. Onsager, *J. Phys. Chem.* **61**, 688 (1957), Eq. (24).
50. R. M. Fuoss, *J. Phys. Chem.* **79**, 525 (1975) (see Figure 3 or $\Delta X_3/X$ in Appendix D).
51. See R. M. Fuoss, *J. Phys. Chem.* **82**, 2427 (1978), Eq. (26).
52. J. P. Valleau, *J. Phys. Chem.* **69**, 1745 (1965); *see also* G. N. Patey and J. P. Valleau, *J. Chem. Phys.* **63**, 2334 (1975).
53. R. Fernandez-Prini and J. E. Prue, *Z. Phys. Chem.* **228**, 373 (1965).
54. R. Fernandez-Prini, in *Physical Chemistry of Organic Solvent Systems*, A. K. Covington and T. Dickinson, Eds., Plenum Press, New York (1973).
55. P. C. Carman, *J. Phys. Chem.* **74**, 1653 (1970).
56. R. Fernandez-Prini, *Trans. Faraday Soc.* **65**, 3311 (1969).
57. P. C. Carman, *J. S. Afr. Chem. Inst.* **28**, 80 (1975).
58. P. C. Carman, *J. S. Afr. Chem. Inst.* **28**, 264 (1975).
59. J. -C. Justice, *Electrochim. Acta* **16**, 701 (1971).
60. W. Ebeling, R. Feistel, G. Kelbg, and R. Sändig, *J. Non-Equilibr. Thermodynam.* **3**, 11 (1978).
61. J. -C. Justice, *J. Solution Chem.*, in press.
62. M. -C. Justice and J. -C. Justice, *J. Solution Chem.* **6**, 819 (1977).
63. R. H. Wood, T. H. Lilley, and P. T. Thomson, *J. Chem. Soc. Faraday Trans. 1* **74**, 1301 (1978).
64. J. -C. Justice, M. -C. Justice, and C. Micheletti, *Pure Appl. Chem.*, in press.
65. W. Ebeling, W. D. Kraeft, and D. Kremp, *J. Phys. Chem.* **70**, 3338 (1966); *Ann. Phys. (Leipzig)* **18**, 246 (1966).
66. R. L. Kay and J. L. Dye, *Proc. Nat. Acad. Sci. USA* **49**, 5 (1963).
67. M. S. Chen, Ph.D. dissertation, Yale University, New Haven, Connecticut, 1969.

68. D. P. Sidebottom and M. Spiro, *J. Chem. Soc. Faraday Trans. 1* **69**, 1287 (1973).
69. J. -C. Justice, J. Périé, and M. Périé, *J. Solution Chem.* **9**, 583 (1980).
70. J. Périé, M. Périé, and J. -C. Justice, *J. Solution Chem.* **9**, 395 (1980).
71. L. Onsager and S. K. Kim, *J. Phys. Chem.* **61**, 215 (1957).
72. C. Micheletti and J. -C. Justice, *J. Solution Chem.*, to be published.
73. J. B. Hubbard and L. Onsager, *J. Chem. Phys.* **67**, 4850 (1977).
74. J. B. Hubbard, *J. Chem. Phys.* **68**, 1649 (1978).
75. R. M. Fuoss, *Proc. Nat. Acad. Sci. USA* **45**, 807 (1959).
76. R. H. Boyd, *J. Chem. Phys.* **35**, 1281 (1961).
77. R. Zwanzig, *J. Chem. Phys.* **38**, 1603 (1963); **52**, 3625 (1970).
78. D. F. Evans, T. Tominaga, J. B. Hubbard, and P. G. Wolynes, *J. Phys. Chem.* **83**, 2669 (1979).
79. G. Jones and C. Bradshaw, *J. Am. Chem. Soc.* **55**, 1780 (1933).
80. J. E. Lind, Jr., J. J. Zwolenik, and R. M. Fuoss, *J. Am. Chem. Soc.* **81**, 1557 (1959).
81. R. M. Fuoss and H. L. Hsia, *Proc. Nat. Acad. Sci. USA* **57**, 1550 (1967).
82. R. Sändig, R. Feistel, J. Einfeldt, and A. Grosch, *Z. Phys. Chem.* (*Leipzig*), *in press*.
83. P. Saulnier and J. Barthel, *J. Solution Chem.* **8**, 847 (1979).
84. G. J. Janz and R. P. T. Tomkins, *J. Electrochem. Soc.* **55c**, 214 (1977).
85. G. J. Janz and R. P. T. Tomkins, in *Non-Aqueous Electrolytes Handbook*, Vol. 1, Academic Press, New York (1972).
86. R. Wachter and J. Barthel, *Electrochim. Acta* **16**, 713 (1971).
87. G. N. Patey and J. P. Valleau, *J. Chem. Phys.* **63**, 2334 (1975).
88. M. A. Coplan, Ph.D. dissertation, Yale University, New Haven, Connecticut, 1963.
89. R. Fernandez-Prini, in *Physical Chemistry of Organic Solvent Systems*, A. Covington and T. Dickinson, eds., Plenum Press, New York (1973), Chap. 5.
90. J. Barthel, *Angew. Chem. Int. Ed. Engl.* **7**, 260 (1968).
91. J. Barthel, *Fortschritte der physicalischen Chemie*, Vol. 10, Ionen in nichwasserigen Lösungen Dietrich Steinkopff, Verlag, ed., Darmstadt (1976).
92. J. Barthel, R. Wachter, and H. J. Gores, in *Modern Aspect of Electrochemistry*, Vol. 13, Temperature Dependence of Conductance of Electrolytes in Non-Aqueous Solutions, B. E. Conway and J. O'M. Bockris, eds., Plenum, New York (1979).
93. A. D. Pethybridge and S. S. Taba, *Faraday Discuss. Chem. Soc.* **64** (1978).
94. P. Beronius, *Acta Chem. Scand.* **A30**, 115 (1976).
95. R. L. Kay, *J. Am. Chem. Soc.* **82**, 2099 (1960).
96. J. C. Rasaiah, *J. Solution Chem.* **2**, 301 (1973).
97. W. Ebeling and J. Rose, *J. Solution Chem.*, **10**, 599 (1981).
98. W. Ebeling and M. Grigo, *J. Solution Chem.*, **11**, 151 (1982).
99. J.-P. Badiali and J.-C. Lestrade, *J. Phys.*, Colloq. 1978 (1), 191–195.

4

Proton Solvation and Proton Transfer in Chemical and Electrochemical Processes

S. LENGYEL and B. E. CONWAY

1. Introduction

The proton in chemistry plays a fundamental role in many processes, analogous to that of the electron in the physics and chemistry of metals, semiconductors, and plasmas. In a sense it is the "elementary particle" of chemistry and, with the electron, of electrochemistry.

In early work, interest in the proton in solution was stimulated by the observation of its unusual ionic mobility in water and its striking effects in catalyzing the hydrolysis of esters. While the importance of the hydrogen ion in chemistry was thus already recognized in some early developments of the subject, it was not until the 1920s and later that the mechanisms of its role in chemical reactions and its state in solution became at all well understood, due principally to the work of Brønsted[1] and of Bell,[2] following ideas on solvation by Goldschmidt,[3] Drude and Nernst,[4] and Born.[5] Related and later work laid the foundation of the role of the proton in organic chemistry through the understanding of acid catalysis (already treated in some detail by Goldschmidt and Udby[6]) and the perception of the role of proton transfer

S. LENGYEL • Central Research Institute for Chemistry of the Hungarian Academy of Sciences, Budapest, Hungary.
B. E. CONWAY • Chemistry Department, University of Ottawa, Ottawa, Canada.

in carbonium ion chemistry. Modern developments in "proton chemistry" have been largely concerned with processes in the *gas* phase, where mass-spectrometric studies have led to the quantitative evaluation of proton affinities and of acid–base reactions in the absence of a solvent. These observations have led to important revisions of the apparent magnitudes of inductive effects in acid–base strengths of species in aqueous solution and recognition of the critical role of steric effects in solvation of organic molecules in solution.

Of all the ions of elements involved in chemistry, the proton is in many ways the most important,[2,7] as its formation or removal is a common process in the thermodynamics, kinetics, and mechanisms of many reactions. The driving force for reactions involving protons and the high stability of the proton in many (especially hydroxylic) solvent media is closely connected with the solvation of the proton and its existence in the form of "onium"-type ions, for instance, H_3O^+, ROH_2^+, etc., to be discussed below.

The formation of the free proton from the elemental atom is distinguished by an ionization energy I_H much larger than that for a corresponding single electron removal from an atom of any other element that forms a monatomic univalent cation stable in solution. Its ionization energy of 13.595 eV, or 313.22 kcal g atom^{-1}, thus indicates that significant concentrations of free protons cannot normally be in equilibrium with H_2 and other hydrogen-containing molecules. Appreciable concentrations of protons can only arise if a corresponding large free energy of stabilization resulting from proton acceptance, or solvation in the most general sense, is involved.

The chemistry of the proton is therefore mainly concerned with (1) its existence in a solvated form, (2) its transfer in acid–base reactions, (3) its formation or consumption in redox reactions of oxyanions or oxycations in aqueous media, (4) its transfer in low-field conductance, and (5) its heterogeneous transfer in electrode processes involving molecule evolution, hydrogen-alloy formation, and other heterogeneous electrochemical reactions involving proton acceptance or proton removal, such as the reduction or oxidation of organic substances at electrodes. It is clear that in all these types of processes, including possible proton tunneling steps, the kinetics of the proton-transfer step will be determined in an important way by the thermodynamics and molecular mechanics of the solvation of the proton.

2. Proton Solvation and Characterization of the H_3O^+ Ion

2.1. The Condition of the Proton in Solution

2.1.1. Early Information

The proton in ionized aqueous solutions of acids, including autoprotolyzing pure solvents such as water, is generally regarded as existing in the form of a solvated ammoniumlike ion H_3O^+, variously called the oxonium, hydronium, or hydroxonium ion; the latter term appears to be favored in the European English-language literature, while the term *hydronium ion* is favored in the North American literature. By analogy with the term *ammonium*, it would seem preferable to adopt the name *oxonium ion* for H_3O^+, as recommended by the International Union of Pure and Applied Chemistry commission on nomenclature. Alternatively, the term *aquonium ion* might be proposed. The term *hydronium* is used for the overall hydrated proton $H_3O^+ \cdot n\,H_2O$. The question of the existence of H_3O^+ is closely bound up with the problem of the lifetime of the species, since in a number of pure solvents the proton has an anomalously high mobility and the "existence" of H_3O^+ must be defined in terms of some time average measure of the number of vibrations that the OH bonds undergo before the proton is transferred to another molecule of solvent acting as a base.

Following observations of Ostwald, the earliest evidence of the existence of a specific oxonium species arose in Goldschmidt's work[3,6] on the acid catalysis of esterifications, in which the retarding effects of small quantities of water in alcohol solutions were examined. The acid-catalyzed esterifications were regarded as proceeding by protonation of the alcohol ROH to give a corresponding alkoxonium type of ion (now presumed to be analogous to the oxonium ion, that is, ROH_2^+), which was considered to be the reactive species in the esterification reaction. The effect of water was presumed to arise on account of the decomposition of the catalytic species through the equilibrium reaction

$$ROH_2^+ + H_2O \rightleftharpoons ROH + H_3O^+ \tag{1}$$

which is in favor of H_3O^+ at the expense of ROH_2^+† owing to the smaller proton affinity (see below) of ROH compared with that of H_2O. On the basis of the equilibrium in the reaction of Eq. (1), assuming that one ROH is associated with H^+, Goldschmidt obtained quantitative agreement between theoretical predictions and experimentally observed kinetic behavior in these reactions. The assumption of any other formal stoichiometry would have destroyed this agreement. Independent evidence for a stoichiometric inner primary hydration of the proton as H_3O^+ follows from Bagster and Cooling's observation[8] that a solution of HBr in SO_2 (in which solvent water is not normally soluble) will dissolve a $1:1$ stoichiometric amount of water and is then electrolytically conducting.

---

† In the original discussion (1907)[6] of these effects the equilibrium was written

$$H^+(ROH)_n + H_2O \rightleftharpoons n\,ROH + H_2O \cdot H^+ \tag{2}$$

since the identity of the solvated proton could not be deduced from the experiments at high, effectively constant, concentrations of the alcohol.[2] However, in the later paper (1909)[3] the hydrogen ion was written explicitly as H_3O^+ in mass action expressions. There are indications that the relative acidities of ROH_2^+ and H_3O^+ depend on the composition of ROH–H_2O mixtures, as might be expected (see Section 2.9).

Although the work discussed above led to the concept of a specific oxonium species involving the hydration of H^+ by one water or alcohol molecule, it must be emphasized that a further, less specific, solvation of the oxonium entity itself is generally involved in polar solvents. This behavior will be discussed in Section 2.6.

2.1.2. Information from Studies in the Solid State

The most direct evidence for the existence of H_3O^+ as a distinct stoichiometric chemical species (albeit in the solid phase) has been obtained from the study of the so-called (mono) hydrates of the strong acids such as sulfuric acid, nitric acid, the hydrogen halides, and particularly perchloric acid. Thus it was shown by Volmer in 1924[9] that the monohydrate $HClO_4 \cdot H_2O$ was isomorphous with ammonium perchlorate and gives a similar X-ray diffraction pattern. It was therefore strongly indicated[9] that the hydrate was in fact an ionic crystal like $NH_4^+ClO_4^-$, with H^+ as the oxonium ion H_3O^+ replacing the NH_4^+ ion. Similar structures are indicated for the other acid "hydrates." More definitive information has been obtained by proton NMR studies on the hydrates by Richards and Smith[10] and by Kakiuchi and co-workers.[11] These results eliminate doubts about the constitution of the "hydrates" that have arisen on account of their apparently limited ionization in the liquid phase indicated by Raman spectra, refractivity, and cryoscopic measurements.

The NMR method is based on quantitative examination of the absorption line-broadening effects that arise on account of nuclear spin–spin interactions. The line shapes depend on the geometry of the interacting nuclei;[12] the result for two protons in H_2O and one in HNO_3 in the hydrate $HNO_3 \cdot H_2O$ would be very different from that expected for three equivalent protons in a trigonal H_3O^+ ion in the hydrate represented as $H_3O^+NO_3^-$. The observed[10,11] three-band spectrum is consistent only[12] with a trigonal pyramidal structure, H_3O^+, as in NH_3. The absorption characteristic of a triangular, three-proton configuration is, however, only observed below 143 K on account of rotational broadening, which occurs above this temperature (cf. the rotational transitions[13] in the dielectric behavior and heat capacity in ammonium salts and solid H_2S, which occur at comparable temperatures). $CF_3SO_3H \cdot H_2O$ is a similar H_3O^+ salt: $H_3O^+CF_3SO_3^-$.

2.1.3. Information from Studies in the Gas Phase

Normally H_3O^+ is not a stable, long-lived species in the gas phase. However, mass-spectrometric methods allow its detection and characterization and also give information on the extent of its further hydration by other water molecules.

The first results of this kind were given by Beckey[14] using a high applied field in a modified field-emission technique. The hydrated protons were formed from an adsorbed layer of water, and their masses characterized directly by means of a mass spectrometer. Species corresponding in mass to H_3O^+ ions were identified, together with higher complexes of two, three, and four ($H_9O_4^+$) water molecules. Field-variation experiments also indicated that the $H_9O_4^+$ species is the most stable, except, of course, for H_3O^+ itself. However, more experiments by Knewstubb and Tickner,[15] using a mass spectrometer to identify ions formed in the glow discharge of water vapor, and by Kebarle and Godbole,[15] indicate the presence of hydrated protons up to $H^+(H_2O)_5$. The most abundant ion is H_3O^+, but $H_9O_4^+$ is not noticeably more stable than other species.

Successive hydration steps from the formation of H_3O^+ to the formation of higher hydrates up to $H_9O_4^+$ or $H_{11}O_5^+$ can be quantitatively examined in a relative manner, and the consecutive reaction kinetics evaluated in terms of the time dependence of concentrations of H_3O^+, $H_5O_2^+$, $H_7O_3^+$, and $H_9O_4^+$ ions.

From studies on the temperature dependence of equilibrium concentrations of the various hydrated proton species in the gas phase, the enthalpies of successive stages of proton hydration can be evaluated.

2.1.4. Information from the Infrared Spectrum of Acid Solutions

On account of the strong infrared (IR) absorption of water itself over a wide wave band, vibration–rotation bands associated specifically with H_3O^+ in aqueous solution were not characterized until 1957 by Falk and Giguère.[16] Previously, Bethell and Sheppard,[17] in a report on the study of the IR spectrum of the nitric acid hydrate discussed above, mentioned, without giving details, that they had studied the system $HClO_4 \cdot H_2O$ in the liquid phase and had observed a spectrum that would correspond to H_3O^+, as found in the crystalline solid. However, in their published paper (reference 17) they presented substantial evidence for a characteristic H_3O^+ spectrum of both liquid and crystalline nitric acid monohydrate based on the observation of bands that could be ascribed to NO_3^- and bands, e.g., at 1134 and 1670 cm^{-1} that were comparable with those associated with a pyramidal molecule such as NH_3.

At about the same time Ferriso and Hornig[18] reported the IR spectra of H_3OCl and H_3OBr at 78 K in which the observed absorption can be due only to OH modes. In the case of the HNO_3 hydrate[17] the H_3O^+ spectrum is complicated by overlapping NO_3^- bands. As in the NMR work, the observed features of the spectrum can be interpreted in terms of a pyramidal oxonium ion. Four IR active fundamentals are expected for this structure: two bending modes and two stretching modes, with one of each being doubly degenerate. The assignments are summarized together with other data in Table 1, while Table 2 shows the frequencies for H_3O^+, OH^-, and H_2O.

Various proposals were put forward in earlier work to account for the supposed lack of an observable absorption spectrum for H_3O^+ ions, and they

Table 1

Vibrational Frequency Assignments for H_3O^+ and D_3O^+ in Crystals and Liquid Solutions

Substance	Spectrum	ν_2	ν_4	ν_3 and ν_1	Reference
			Frequencies (cm^{-1}) and assignments[a]		
$H_3O^+NO_3^-$	IR	1134	1670	2650–3380	17
$H_3O^+I^-$	IR	1060	1705	2635–3350	21
$H_3O^+Br^-$	IR	950	1705	2610–3250	21
		1060			
		1150			
$H_3O^+NO_3^-$ (liq)	IR	1308 (?)	1685	2520–3330	17
$H_3O^+Cl^-$	IR	1060	1700	2590–3235	21
		1150			
$H_3O^+F^-$	IR	1048	1705	2468–3150	21
		1150			
$H_3O^+ClO_4^-$	Raman[b]	—	1590	2400–3600	22
$H_3O^+ClO_4^-$	Raman	1175	1577	3285	23
$D_3O^+ClO_4^-$	IR	785	1255	2000–2445	21
HCl(aq, liq)	IR	1205	1750	2900^c	16
DCl(D_2O, liq)	IR	960	1400	2170^c	16
H_2O^d	IR	1645 (bend)		3440 (ν_1, ν_3)	24
D_2O^d	IR	1208 (bend)		2500 (ν_1, ν_3)	24

[a] The modes 3 and 4 are the doubly degenerate stretch and bend, respectively.
[b] Raman frequencies for ether HCl oxonium compounds have been given by Gantmacher.[24]
[c] Isotope shift factors are significantly below $2^{1/2}$. This is probably due to strong H- and D-bonding effects.
[d] Assignments for H_2O and D_2O are included for comparison.

Table 2

Frequencies[a] for H_2O, H_3O^+, and OH^- in Liquid Water

H_2O		H_3O^+		OH^-	
Frequency (cm^{-1})	Degeneracy	Frequency (cm^{-1})	Degeneracy	Frequency (cm^{-1})	Degeneracy
3440	1	3235	1	3615	1
1645	1	1150	1	477	2 (2)
3440	1	2590	2		
667	3 (L)[b]	1700	2		
		643	3 (L)		

[a] Based on the data of G. Swain and R. F. W. Bader, *Tetrahedron* **10**, 182 (1960), and B. Thornton, *J. Am. Chem. Soc.* **84**, 2474 (1962).
[b] Librational frequencies L.

were based mostly on the supposition that the lifetime of H_3O^+ was too short (on account of the rapid anomalous transfer manifested in the high mobility of the proton[19,20]—see later) for the spectrum[21,22] to be observed. Thus it was stated by Wicke *et al.*[23a] and by Ackermann[23b] that a definite characteristic absorption band for H_3O^+ could not be expected on account of the "unlimited mobility" of the proton charge in the $H_9O_4^+$ complex $(H_3O^+\cdot3H_2O)$. This would, of course, imply a lifetime comparable with the reciprocals of the vibrational frequencies.[24] However, the kinetic calculations of Conway, Bockris, and Linton,[20] which theoretically predicted a significant lifetime for the H_3O^+ ion, led Falk and Giguère[16] to a renewed effort to characterize the species by the IR spectrum of hydrogen halide and other acids in concentrated aqueous solutions. Strong broad bands were observed by these authors for several acids at about 1205, 1750, and 2900 cm^{-1}, in good agreement with previously observed bands assigned to H_3O^+ in the crystalline acid hydrates. The intensities were correctly related to hydrogen ion concentration and the correct H–D isotopic frequency shifts were observed. Comparisons made between the spectra of solutions of acids HX and corresponding metallic salts MX confirmed a characteristic spectrum for H_3O^+.

The fact that a H_3O^+ spectrum can be observed in strong aqueous acid solutions and that the high proton mobility (to be discussed in Section 4 of this chapter) seems to indicate a short lifetime for H_3O^+ is rationalized by noting that structure-breaking effects in the water solvent at high ionic concentrations diminish the anomalous mobility and will give larger lifetimes to H_3O^+ ions, as in MeOH–H_2O mixtures.[20]

2.1.5. Evidence from Molar Refraction

The molar refraction R of the proton in aqueous solution was considered by Fajans and Joos,[25] who showed that its value was similar to that of the NH_4^+ ion. Compared with values for the O_2^- and OH^- ions, and for water itself, the value for the proton is at the end of a series of consistently decreasing values on account of the increasing positive charge from O^{2-}, through water, to H^+. If the proton were free in solution, a negative value of about -0.67 cm^3 mol^{-1} would be expected. The relationship between R and the number of hydrogen atoms in the series of related isoelectronic species indicates that only by representation of the proton as H_3O^+ in aqueous solution can its value of R be put in a logical and consistent position in relation to values for chemically analogous and electronically similar compounds.

2.2. Conclusions on the Molecular Structure of the H_3O^+ Ion

The results of Richards and Smith[10] on the NMR spectrum of acid hydrates were discussed above. This work indicates a trigonal, rather flat pyramidal structure involving the protons, with a proton–proton separation of (1.72 ± 0.02) Å. The positions of the signals for the free (gaseous) oxonium ion have been calculated by Grahn[26] for the trigonal pyramid model and are in agreement with the observed absorption pattern obtained at 343 K by McLean and Mackor[27] in a solution of $HF \cdot BF_3$ in water, i.e., probably the strong acid hydrate $H_3O^+ \cdot BF_4^-$.

The exact form of the H_3O^+ unit is somewhat uncertain, owing to inadequate knowledge of the OH bond distance. On the basis of comparison between NH_3 and NH_4^+, the OH bond length in free H_3O^+ would be expected to be greater than that in water by about 0.01–0.02 Å[7] whereas in a hydrogen-bonded crystal or solvation complex an extension of 0.06–0.1 Å is indicated.[28] If the OH bond length in H_3O^+ were 0.06 Å greater than that in water, the ion would be symmetrically pyramidal, with a pyramidal angle of 115°, which is consistent with the X-ray diffraction data[29–31] for the nitric acid hydrate. Similar conclusions concerning the presence of the oxonium ions and their structure were reached for the sulfuric and perchloric acid hydrates.

The conclusions of Richards and Smith,[10] that the bond length is (at least) 1.02 Å, which leads to a pyramidal angle of 115°, are probably the most acceptable. A fuller discussion of these matters has been given in a review by Conway.[7] The IR spectroscopic results of Falk and Giguère[16] confirm the pyramidal configuration with C_{3v} symmetry, analogous to that in ammonia. These authors suggest a bond length of 1.01 Å, close to that in ice, since the OH stretching frequencies are almost the same in ice and H_3O^+; this value is in good agreement with that suggested by Richards and Smith.[10] A summary of estimates of the interatomic distances and bond angles is given in Table 3.

2.3. Theoretical Conclusions on the Molecular Structure of the H_3O^+ Ion

One of the most detailed theoretical studies on the molecular structure and total energy of the H_3O^+ ion has been made by Grahn,[30] using a molecular orbital and valence bond treatment. This work takes the subject well beyond the more empirical type of calculation carried out by Hund in 1925, where known experimental data for the spectra of certain molecules were used as a basis for calculations of proton affinities of water and bond lengths in the H_3O^+ ion.

Calculations of the total energy (i.e., including normal electronic levels) of H_3O^+ by Grahn as a function of $H-\hat{O}-H$ pyramidal bond angle ϕ led to an optimum total energy at $\phi = 120°$, i.e., for a *flat* configuration. This result is not inconsistent with the requirement of a low dipole moment indicated by the librational band in the acid hydrates discussed by Ferriso and Hornig.[32] While the total energy of -76.1820 a.u. (atomic units) at the 120° configuration is in relatively satisfactory agreement with the experimental value of -76.725 a.u., the calculated proton affinities (to be discussed below) for $\phi = 100°$, 110°, and 120° are, however, too high by 20–80 kcal mol^{-1}

Table 3
Bond Distances and Bond Angles in the H_3O^+ Ion

Authors	Molecules studied	H-H Distance (Å)	H-O Distance (Å)	H—O—H Angles	Angle between hydrogen bonds	Method
Volmer (1924)	$H_3O^+ClO_4^-$					X-ray
Hund (1925)	H_3O^+	1.11	1.05			Theory
Richards and Smith (1951)	$H_3O^+NO_3^-$	1.72 ± 0.02	1.02	115°		NMR
	$H_3O^+ClO_4^-$					
	$H_3O^+HSO_4^-$					
Kakiuchi et al. (1951)	$H_3O^+ClO_4^-$	1.58	0.98	110°		NMR
Bethell and Sheppard (1953)	$H_3O^+NO_3^-$					IR
	$H_3O^+ClO_4^-$					
Luzzati (1953)	$HNO_3 \cdot 3H_2O$		1.00, 0.85, 1.10		113°, 118°, 103°	X-ray
Ferriso and Hornig (1955)	Hydrogen halides		0.98	Tetrahedral		IR
Taylor and Vidale (1956)	$H_3O^+ClO_4^-$		0.98	107°		IR
Yoon and Carpenter (1959)	"$HCl \cdot H_2O$"	1.65	0.96	117°	110.4°	X-ray
Bishop (1965)	H_3O^+		0.95	114°26'		Theory

compared with the value that can be calculated from a thermodynamic cycle[33] (see below). Since the agreement between theory and experiment for the proton affinities cannot be considered good, the assignment of optimum bond angle from Grahn's calculation[30] remains uncertain.

A more recent theoretical calculation on H_3O^+ was given by Bishop,[34] using a single-center molecular wave function consisting of nine terms. These were symmetry-adapted combinations of Slater determinants, each being composed of Slater orbitals having "noninteger" quantum numbers. A rather flat pyramidal structure was predicted, having a HOH bond angle of 114°26′, with an OH bond distance of 0.95 Å (compare other results in Table 3).

2.4. Thermodynamics of Proton Solvation: Proton Affinity of H_2O

Two stages in proton hydration can be usefully distinguished: (1) protonation of a H_2O molecule in the gas phase to form a new stable molecule ion, H_3O^+ and (2) the further hydration of this species in the bulk water solvent. In this section we shall consider process (1).

The first reliable estimate of the proton affinity† P of an individual gaseous water molecule, based on experimental thermodynamic data, was given by Sherman,[33] who considered a thermodynamic cycle originally suggested by Grimm for the estimation of the proton affinity P of ammonia, that is, the energy change for the reaction

$$NH_{3\,(g)} + H^+_{(g)} \longrightarrow NH^+_{4\,(g)} + P_{NH_3} \text{ kcal mol}^{-1} \tag{3}$$

The calculation of this energy P_{NH_3} may be carried out by means of the cycle

$$
\begin{array}{ccc}
[NH_4X] & \xrightarrow{\ U\ } & [NH_4^+ + X^-] \\
\Big\uparrow {\scriptstyle Q_{NH_4X}} & & \Big\uparrow {\scriptstyle P_{NH_3}} \\
[\tfrac{1}{2}N_2 + 2H_2 + \tfrac{1}{2}X_2] & & [NH_3 + H^+ + X^-] \\
{\scriptstyle -\tfrac{1}{2}D_{H_2} - \tfrac{1}{2}D_{X_2}} \diagdown \quad \diagup {\scriptstyle -I_H - E_X} \\
{\scriptstyle -Q_{NH_3}} & [NH_3 + H + X] &
\end{array}
\tag{4}
$$

using data for the crystal lattice energies U of ammonium halides NH_4X, electron affinities E_X of halogen atoms, and the ionization energy I_H of hydrogen atoms, with the dissociation energies D_{H_2} and D_{X_2} for $\tfrac{1}{2}H_2$ and $\tfrac{1}{2}X_2$ molecules, respectively, and the Q terms, the heats of formation of the indicated species. The proton affinity of ammonia is then

$$Q_{NH_4} + U - Q_{NH_3} - D_{H_2} - D_{X_2} - I_H - E_X = P_{NH_3} \tag{5}$$

† The term *affinity* is usually used for this quantity, although it is not a free-energy change.

Depending on the salt NH_4X, derived P_{NH_3} values vary from -221 to -202.7 kcal mol^{-1}.

At the suggestion of G. H. Cady, Sherman[33] then calculated the proton affinity of water by reference to the existence of $H_3O^+ClO_4^-$, which is isomorphous with $NH_4^+ClO_4^-$. Since these substances involve ions of the same charge, it can be assumed that they have almost the same crystal lattice energy. Two cycles involving $H_3O^+ClO_4^-$ and $NH_4^+ClO_4^-$ can then be set up that are analogous to the cycle in (4) and involve Q_{NH_3}, Q_{H_2O}, $Q_{NH^+ClO_4^-}$, and $Q_{H_3O^+ClO_4^-}$, as well as P_{NH_3} and P_{H_2O}, and quantities such as the dissociation energy of ClO_4^- and H_2 and the electron affinity of ClO_4^-. All the quantities referring to the ClO_4^- species cancel when those in one cycle are subtracted from corresponding terms in the other cycle. Noting that it has been assumed that $U_{H_3O^+ClO_4^-} = U_{NH_4^-ClO_4^-}$, then

$$P_{H_2O} = P_{NH_3} - Q_{H_2O} - Q_{NH_3} + Q_{H_3O^+ClO_4^-} - Q_{NH_4^+ClO_4^-} \qquad (6)$$

Equation (6) leads to $P_{H_2O} = 182$ kcal mol^{-1} (at 0 K), using the value of P_{NH_3} deduced above [Eq. (5)].

In the calculations of Sherman the data used for the estimation of P_{NH_3} must be revised in the light of more modern figures. A corrected value for P_{H_2O} of 188 kcal mol^{-1} has been given by Sokolov.[35] A lower value is obtained mass spectrometrically.[36]

The standard free energy of formation of $H_3O^+_{(g)}$ from $H^+_{(g)}$ and $H_2O_{(g)}$ at 298 K and 1 atm may be estimated from the above value of the heat content change $-P_{H_2O}$ of -170 kcal mol^{-1} and an estimate† of the standard entropy of formation of $H_3O^+_{(g)}$ from $H^+_{(g)}$ and $H_2O_{(g)}$ of -27 cal K^{-1} mol^{-1}. The standard free energy for the protonation reaction $H_2O_{(g)} + H^+_{(g)} \rightarrow H_3O^+_{(g)}$ is thus -162 kcal mol^{-1}.

2.5. Overall Hydration Energy of the Proton: Process (1) + Process (2)

2.5.1. General

Until now only the properties of the simple molecule ion H_3O^+ and its heat and standard free energy of formation from $H^+_{(g)}$ and $H_2O_{(g)}$ have been considered in this chapter. In nearly all electrochemical and chemical reactions involving the proton, however, a H_3O^+ ion existing in a bulk dielectric medium, usually water, is involved. Since H_3O^+ has a radius similar to that of the potassium ion, it will be expected that the overall heat of solvation will be substantially more negative than the heat content change (which can be identified approximately with $-P_{H_2O}$) in the protonation of a single gaseous H_2O molecule.

All estimates of individual solvation energies must, of course, be based on some extra thermodynamic principle for division of the observed thermodynamic function for a pair of ions. Most schemes of conventional individual ionic free energies or enthalpies are based on an arbitrary value of zero for the function for aqueous H^+; However, it is obviously of intrinsic interest to evaluate the absolute value of the total free energy or enthalpy of hydration of H^+ and hence obtain absolute values for other ions, using the conventional scale. Various ways of estimating the individual absolute values of overall free energies or enthalpies of hydration of individual ions, and hence H^+_{aq}, have been proposed. However, a special method was applied by Halliwell and Nyburg[37] to obtain the total energy of hydration of H^+ and justifies a separate description here.

2.5.2. Method of Halliwell and Nyburg for the Evaluation of the Absolute Enthalpy of Hydration of the Proton

The value for the absolute enthalpy of the proton in water is of general interest in the thermodynamics and chemistry of aqueous ionic solutions, especially acid–base equilibria. Additionally, since the proton affinity of H_2O to form H_3O^+ can be obtained, as shown above, the hydration energy of "H_3O^+" itself can be derived. Also, knowledge of the total energy of hydration of the proton provides the basis for the conversion of data on the conventional scale of ionic enthalpies of hydration to absolute values.

The absolute value of the hydration enthalpy of H^+ can, in principle, be obtained by a method given by Latimer, but indications have been given about unsatisfactory aspects of this method, based on limitations of the Born equation for small ions.

Halliwell and Nyburg[37] reviewed the conventional enthalpy data for hydration of salt ions and presented a useful method for evaluating the absolute enthalpy of hydration of H^+, based on difference equations, i.e., for ΔH^0_s:

$$\Delta H^0_{s,+} = \Delta H^0_{s,c,+} + z_+ \Delta H^0_{s,H^+} \qquad \text{for cations} \qquad (7a)$$

and

$$\Delta H^0_{s,-} = \Delta H^0_{s,c,-} - |z_-| \Delta H^0_{s,H^+} \qquad \text{for anions} \qquad (7b)$$

where $\Delta H^0_{s,c,\pm}$ are the conventional values for cations or anions and $\Delta H^0_{s,H^+}$ is the required absolute enthalpy of hydration of H^+.

The method is related to that of Noyes.[38] However, Halliwell and Nyburg made use of Buckingham's[39] calculation, which takes into account the ion H_2O–quadrupole interaction in relation to the ion H_2O–dipole interactions. The combination of these interactions is not the same for cations and anions of the "same" radius, so that *differences* of ionic hydration energy should arise between cations and anions of the same size and charge, even though the dipole orientations with respect to the ionic charge are (supposedly) in a mirror-image relationship to one another; this was the assumption of Halliwell and Nyburg, which will be discussed further below. Other differences also arise, as indicated by NMR and IR spectroscopic studies.

In terms of the ion-quadrupole interaction,[39] the differences of enthalpy of hydration should be a reciprocal cubic function of $r_i + r_{H_2O}$. For the anions and cations that are available for such an analysis, the conventional enthalpy of hydration data plot out as shown in Figure 1. Unfortunately, owing to the generally greater radii of anions in relation to those of corresponding isoelectronic cations (e.g., F⁻ versus Na⁺ and Cl⁻ versus K⁺), overlap between the cation and anion curves of Figure 1 does not extend beyond the case of K⁺.

By extrapolation of the differences of the data in Figure 1, a difference plot can be made with respect to $(r_i + 1.38)^{-3}$, as shown in Figure 2, which can be reasonably extrapolated to zero on the reciprocal radius axis. The

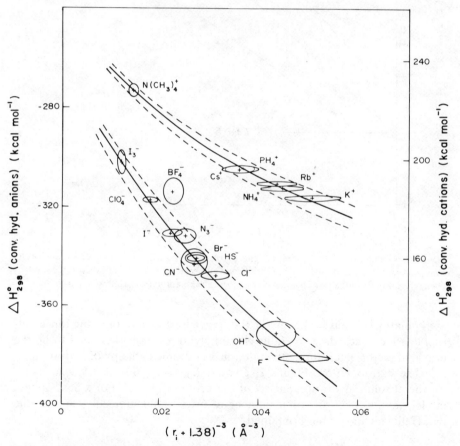

Figure 1. Plot of conventional enthalpies of hydration of ions versus reciprocal cube of ion–water intermolecular separation $r_i + 1.38$ Å (from Halliwell and Nyburg[37]).

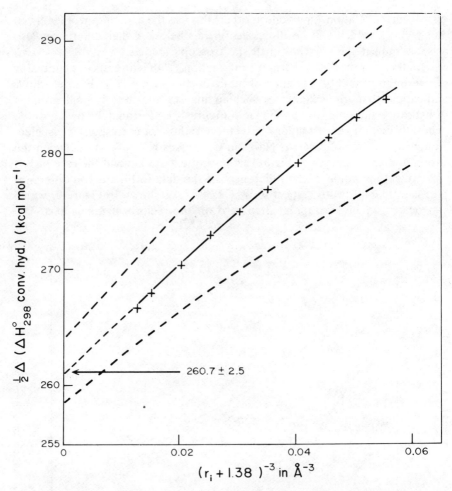

Figure 2. Evaluation of the hydration energy of the proton from a plot of differences of conventional enthalpies of hydration of cations and anions of the "same" radii versus $(r_i + 1.38)^{-3}$, giving the absolute enthalpy of hydration of H^+ by extrapolation (from Halliwell and Nyburg[37]).

extrapolation should be linear for the larger ions, where only the ion-quadrupole effect, according to Halliwell and Nyburg, can be assumed to be the principal factor determining the differences of ionic enthalpy of hydration.

The extrapolation, Figure 2, gives $\Delta H_{s,H^+} = (-260.7 \pm 2.5)$ kcal mol^{-1} for the absolute hydration energy of the proton in water. For a hard-sphere model, taking a coordination number of 6 for water molecules around the ion, Halliwell and Nyburg obtained

$$\Delta(\Delta H^0_{s,c,H^+}) = 261.1 + \frac{589}{(r_i + r_{H_2O})^3} - \frac{401.2}{(r_i + r_{H_2O})^4} \tag{8}$$

based on the equations of Buckingham.[39] The extrapolations were based only on the lower-order term in $(r_i + r_{H_2O})^{-3}$. The figures are derived from the semiempirical plot, from which the quadrupole moment θ_{22} for H_2O may be estimated as 1.4×10^{-26} esu (Buckingham quoted a value of 2×10^{-26} esu). Other coordinations were examined, as well as a soft-sphere model with a repulsive potential function for the nonelectrostatic part of the ion–water interaction.

Conway and Salomon[40] pointed out that the difference plot of Halliwell and Nyburg[37] will involve some significant error if the orientation of water at the anions and cations, discussed by Verwey,[41] is considered. In this treatment the water-dipole axis at cations is regarded as being at 52° to the radial direction, while at anions the OH bond direction is *along* this direction. Bernal and Fowler[19] also took this as the angle of orientation at anions. The justification for this model is that it allows energetically better hydrogen bonding between water molecules in the primary hydration shell and those in the neighboring bulk solvent, and specifically between the negative anion center and the H atoms of coordinated water molecules.

On the basis of such a model, the difference plot in $(r_i + r_{H_2O})^{-3}$ would not be correct, since the leading ion-dipole term in $(r_i + r_{H_2O})^{-2}$ would no longer cancel when differences of hydration energy of cations and anions of the same size and charge are considered. In fact, a factor of $1 - \cos 52°$, possibly modified by the difference of hydrogen-bonding energy with solvent molecules beyond the primary shell at cations and anions, would remain. The difficulty is compounded by the fact that a similar situation arises with the proton itself, where, as H_3O^+ in the complex $H_9O_4^+$, the configuration is regarded as being analogous to that at simple cations but involving three hydrogen-bonded water molecules.

If the ion-dipole contribution at anions and cations of the same size and charge do not cancel,[40] then the basis of the plot of Figure 2 is also called into question. This is because the ion-quadrupole term varies as $(r_i + r_{H_2O})^{-3}$, while the ion-dipole terms vary as $(r_i + r_{H_2O})^{-2}$. The Born term in $(r_i + 2r_{H_2O})^{-1}$ should still cancel.

Conway and Salomon[40] showed how these difficulties could be allowed for in a modified extrapolation plot and obtained a value for $\Delta H_{s,H^+}$ of (-267 ± 7) kcal mol^{-1}. The main uncertainty arises from a lack of any real knowledge of the mean orientation of water dipoles at cations and anions; almost certainly, the relative orientations will also change from one ion to another in a given series of anions or cations, so the basis of the simple difference plot of Halliwell and Nyburg must be questioned. However, the method itself, recognizing the role of ion-quadrupole interaction as a main factor leading to differences of hydration energy of monovalent anions and cations of comparable size, was an ingenious and useful one.

Using the method of Halliwell and Nyburg,[37] Lister, Nyburg, and Poyntz[42] later obtained a further value for the absolute heat of hydration of H^+ as -262 kcal mol^{-1}, based on the extrapolation of data for doubly charged complex ions. However, it is still open to the previous criticisms[40] made regarding the question of the mirror-image orientation of water dipoles at cations and anions.

2.5.3. Free Energy of Hydration of the Proton

Combination of the value for the absolute entropy of hydration of H^+, derived from its individual absolute partial molar entropy in aqueous solution, with the best absolute enthalpy value[40] discussed above, gives a value for $\Delta G^0_{s,H^+}$ of (-258 ± 7) kcal mol^{-1}. Some direct estimation of this quantity has also been made by Volta potential measurements of the absolute standard free energy of hydration of K^+, which gives a figure of -260.5 kcal mol^{-1} for the proton free energy of hydration.

2.6. Hydration of H_3O^+

2.6.1. Evaluation of Hydration from Experimental Data

With the proton "affinity" values mentioned earlier (182–188 kcal mol^{-1}), the above enthalpy figure gives a hydration energy of H_3O^+ in excess water in the range -85 to -79 kcal mol^{-1}.

Various studies have been made concerning the nature of the residual hydration of H_3O^+ in water, especially with regard to hydrogen bonding and formation of the hydrogen-bonded complex ion $H_9O_4^+$,[21,23] also indicated as one of several stable species by gas-phase mass spectrometry. Spectroscopic methods (see references 18 and 32) are unable to give any reliable information on this further hydration of H_3O^+, but evidence from hydration number experiments and calculations indicates a primary coordination number from 2 (compressibility method[43]) to 5 (entropy method[44]) and, by comparison with the K^+ ion, a coordination number of 4 from Bernal and Fowler's theoretical calculations.[19] The latter figure seems chemically reasonable and consistent with the hydrogen bonding (see below) that can occur at the H_3O^+ ion. Transference methods for the estimation of hydration numbers cannot, of course, be applied to hydrogen ions, owing to the anomalous mobility mechanism[19,20] involved in their transport.

A thermodynamic method involving the interpretation of partial molal heat capacities of ions was developed by Eigen and Wicke[45] on the basis of models discussed by Eucken[46] for the association structure of water; the method was applied by Wicke, Eigen, and Ackermann[23] to the study of H^+ and OH^- ions in aqueous solution. The essential features of the theory will briefly be discussed, since the results involved support the $H_9O_4^+$ complex suggested by Wicke, Eigen, and Ackermann[23] and developed by Eigen and de Maeyer[47] in discussions of the behavior of aqueous acid solutions and proton transfer.

The hydrated ions are assumed to have two principal effects on the water structure: (1) a partial "depolymerization" of the associated water, which produces a considerable negative contribution both to the apparent molal heat capacity of the water, ϕ_{C_p}, and to the apparent molal volume ϕ_V of the ion; and (2) an effect arising from the temperature dependence of the hydration of the ions. In Eucken's model the hydration is regarded as an "adsorption" of water in two layers: one, an inner layer of coordination number n, where the molecules are subjected to an electrostrictive tension and give a further negative contribution to ϕ_V, and the other, a secondary hydration shell, of variable coordination number m, where ϕ_V is increased. Positive contributions are also given to ϕ_{C_p} by the energy required for the breakdown or modification of this outer region with increases of temperature. Thus a combination of heat capacity and density measurements can lead to evaluation of the hydration number of H^+.

The apparent molal heat capacity and volume of H^+ and OH^- ions were calculated by Ackermann[23b] as a function of temperature. Comparison with the experimentally observed molar heat capacity behavior showed clearly (Figure 3) that a reasonable representation of the data in terms of n and m

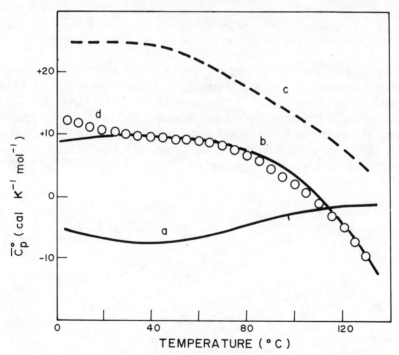

Figure 3. Calculated and experimental partial molal heat capacities of the proton in aqueous medium as a function of temperature: (a) calculated with $n = 1$ (i.e., H_3O^+), (b) calculated with $n = 4$ (i.e., $H_9O_4^+$) with temperature-dependent secondary hydration number; (c) comparative behavior of Li^+ ion; (d) $\circ$, experimental values. (From Ackermann.[23])

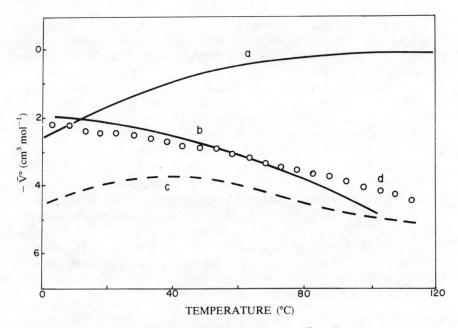

Figure 4. Calculated and experimental partial molar volumes $\bar{V}^0$ of the proton in an aqueous medium as a function of temperature. The key is the same as in Figure 3. (From Ackermann.[23b])

cannot be made theoretically by assuming that the hydration of the proton is limited to interaction with only one water molecule ($n = 1$), giving H_3O^+. A value $n = 4$ for the hydration number of H^+ (i.e., 3 for H_3O^+) is indicated, as also supported by the molar volume behavior (Figure 4). The configuration around the H_3O^+ ion is envisaged as an almost tetrahedral $H_9O_4^+$ primary shell, as originally suggested by Wicke, Eigen, and Ackermann.[23] A planar projection is shown in Figure 5. A similar situation represents the hydration of OH^- (Figure 5).

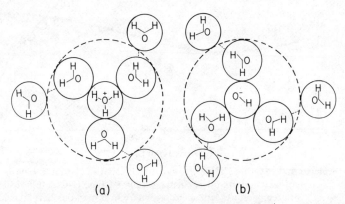

Figure 5. Projection of overall configurations of hydrated (a) H_3O^+ ions and (b) OH^- ions.

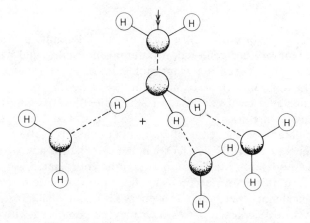

Figure 6. Configuration of $H_9O_4^+$ ion with extra electrostatically bound H_2O molecule.

An important feature of the $H_9O_4^+$ ion is that rapid interchange of the proton can occur between H_3O^+ and the surrounding three hydrogen-bonded H_2O molecules, giving a kind of delocalization of H^+. A fourth water molecule is probably electrostatically coordinated to the central H_3O^+ but cannot undergo proton exchange with it except by reorientation (Figure 6) (see later discussion on H^+ conductance).

2.6.2. The Calculations of Grahn[30] for $H_9O_4^+$

Calculations have been made by Grahn[30] for the $H_9O_4^+$ complex, following his molecular orbital and valence bond treatment of H_3O^+, discussed above. The model used is shown in Figure 6. Electron densities were obtained from wave functions calculated first for noninteracting H_3O^+ (assumed incorrectly to be a flat trigonal structure) and three further H_2O molecules.

From these electron densities the electrostatic interaction between the molecules was calculated. It is given by ion–ion terms, one-electron, and two-, three-, and four-center integrals. In this way the Coulomb interaction or "electrostatic energy" between the two systems, each supposed to remain in a state unaffected by the other, was obtained. The systems will, however, polarize each other and the corresponding changes in their configurations cause a further lowering of the total energy of the system. Grahn divided these induced changes into three parts: (1) changes in interatomic distances in the component molecules of the complex, (2) changes in electron density functions (electronic polarization), and (3) effects of exchange forces involving covalent binding in the H_3O^+—H_2O hydrogen bonds. It was concluded that factor (1) cannot be realistically considered, since the effects involved would be similar to the uncertainties in the interatomic distances in the unperturbed H_3O^+—H_2O configuration. Various simplifications were made to deal with the polarization involved in (2): neglect of the differential overlap (CNDO) between atomic orbitals centered at different molecules, and neglect of the direct interaction between the water molecules of the complex, e.g., dipole–dipole interaction (unlikely to be negligible).

The calculated total energy for $H_9O_4^+$ is -303.8401 a.u., and that for H_3O^+ discussed previously[26] was -76.1820 a.u. The energy for $H_3O^+ + 3H_2O$ calculated for the case of zero interaction is -303.6319 a.u. The hydration energy of H_3O^+ as $H_9O_4^+$ is hence -0.2082 a.u. or -130.5 kcal mol^{-1}. Since no lateral interaction between the coordinated water molecules was considered, this energy accounts for the energy of three hydrogen bonds, i.e., ~ 43.5 kcal mol^{-1} per bond. This energy is mainly "electrostatic energy" and amounts to 38.8 kcal mol^{-1}, which may be compared with the normal $O-H \cdots O$ hydrogen-bond energy of 6 kcal mol^{-1} for neutral molecule interactions. Obviously the higher $H_3O^+-H_2O$ hydrogen-bond energy is mainly an "ion-dipole" interaction. It accounts for the stability of $H_9O_4^+$.

The total hydration energy of H^+, theoretically estimated, is hence on the order of $-(170 + 3 \times 45) = 305$ kcal mol^{-1} (see reference 7 for further discussion of the theoretical calculations.) This is larger than that indicated from treatment of the experimental data (cf. reference 37).

The special stability and chemical identity of $H_9O_4^+$ is also indicated by evaluations of the hydration number;[21,23] thus various methods give an overall value of 4; 1 for H^+-H_2O as H_3O^+ plus three other coordinated H_2O molecules. A summary of these results is given in reference 7.

2.7. Deuteron Solvation

A deuterium species D_3O^+, like H_3O^+, will be expected in D_2O solution and was observed by Falk and Giguère.[16] Because the zero point energies of $O-D \cdots O$ are lower than those of $O-H \cdots O$ "hydrogen" bonds, and the moment of inertia of D_2O is larger than that of H_2O, the solvation energy of D^+ in D_2O is numerically larger than that of H^+ in H_2O.

From spectroscopic information the difference of the deuteron affinity of D_2O (to give D_3O^+) and the proton affinity of water is -1.26 kcal mol^{-1}, the D entity being the more stable. The corresponding difference for D in H_2O as H_2DO^+ and H in H_2O as H_3O^+ is -1.8 kcal mol^{-1}.

The difference of the overall solvation energy of D^+ in D_2O and H^+ in H_2O is obtained from emf measurements on cells of the type

$$|H_2/Pt|HCl \text{ in } H_2O \vdots DCl \text{ in } D_2O|Pt/D_2|$$

After correction for the substantial liquid junction potential (since the mobility of H^+ in H_2O is ~ 1.4 times that of D^+ in D_2O) an overall difference of solvation energy of -1.2 kcal mol^{-1} is obtained for $\Delta \cdot \Delta G_{s,D_3O^+-H_3O^+}$.

Noonan and LaMer[48] measured the emfs and their temperature coefficients of the cells

$$Pt/H_2|HCl(H_2O)|AgCl/Ag \text{ and } Pt/D_2|DCl(D_2O)|AgCl/Ag$$

which lead to the standard free-energy, enthalpy, and entropy changes for the overall reaction

$$\tfrac{1}{2}H_{2\,(g)} + DCl(D_2O) \rightleftharpoons \tfrac{1}{2}D_{2\,(g)} + HCl(H_2O) \tag{9}$$

This isotopic exchange reaction arising in the cell can be expressed as

$$\tfrac{1}{2}H_{2\,(g)} + D_3O^+[4D_2O]_{(sol)} + Cl^-[4D_2O]_{(sol)} + H_2O_{(liq)} + 2[4H_2O]_{(liq)}$$
$$= \tfrac{1}{2}D_{2\,(g)} + H_3O^+[4H_2O]_{(sol)} + Cl^-[4H_2O]_{(sol)} + D_2O_{(liq)} + 2[4D_2O]_{(liq)} \tag{10}$$

where the bracketed quantities $[4D_2O]_{(sol)}$ and $[4H_2O]_{(sol)}$ represent the heavy and light water molecules associated with the two isotopic oxonium ions and the Cl^- ions in the primary solvation shells of these ions, and $[4D_2O]_{(liq)}$ represents the corresponding quantities of the molecules indicated in the bulk of the free solvent. They are distinguished, since their librational behavior in bulk is different from that in the vicinity of the ions.

The following differences of thermodynamic functions for the solvation of H^+ and D^+ in H_2O and D_2O, respectively, at 298 K are obtained[48]:

$$\Delta \cdot \Delta G_{s,D^+-H^+} = -9.4 \text{ kcal mol}^{-1}$$

$$\Delta \cdot \Delta H^0_{s,D^+-H^+} = -7.6 \text{ kcal mol}^{-1}$$

$$\Delta \cdot \Delta S^0_{s,D^+-H^+} = 6.2 \text{ cal K}^{-1} \text{ mol}^{-1}$$

2.8. Partial Molal Entropy and Volume and the Entropy of Proton Hydration

Normally the standard for relative g ionic entropies of ions in solution is taken as $\bar{S}^0_{H^+,aq} = 0$. This is, of course, arbitrary, and attempts have been made to assign absolute values for the standard partial molal entropies of H^+_{aq} and Cl^-_{aq}. Various methods have been proposed. From studies on thermocells, values of $\bar{S}^0_{H^+,aq}$ from -4.7 to -6.3 ($\pm\sim4$) cal K^{-1} mol^{-1} have been recommended. With the Sackur–Tetrode standard entropy for $H^+_g = 26$ cal K^{-1} mol^{-1}, the standard entropy of solvation of the proton in aqueous solution is hence about (-31 ± 4) cal K^{-1} mol^{-1} at 298 K.

Similarly, the individual partial molal volume of the proton at infinite dilution cannot be experimentally determined, and an arbitrary scale of $\bar{V}^0_{H^+,aq} = 0$ is used. Attempts have been made to derive an absolute value for $\bar{V}^0_{H^+,aq}$ and two procedures seem satisfactory. Methods based on (1) the extrapolation of volumes of $R_4N^+X^-$ salts to zero cation size and (2) ultrasonic

Table 4
Various Estimates for the Ionic Partial Molal Volume of the Proton in Water at
298 K^a

$\bar{V}_{H^+,aq}$ ($cm^3\,mol^{-1}$)	Authors[b]	$\bar{V}_{H^+,aq}$ ($cm^3\,mol^{-1}$)	Authors
−0.2	Fajans and Johnson[1]	−4.5	Mukerjee[11]
−0.9	Padova[2]	−5.0	Millero[12]
−2.6	Glueckauf[3]	−5.1	Wirth[13]
−2.7	Eucken[4]	−5.3	Kobayazi[14]
−2.8	Noyes,[5] Panckhurst[6]	−5.4	Zana and Yeager[15]
−3.8	Bernal and Fowler,[7] Darmois,[8] Zen[9]	−5.7	Conway, Verrall, and Desnoyers[16]
−4.5	King[10]	−6.0	Couture and Laidler[17]
		−7.6	Stokes and Robinson[18]

[a] An optimum mean value of $(-5.6 \pm 0.3)\ cm^3\,mol^{-1}$ is recommended, based on selected techniques.
[b] References in Table 4:
1. K. Fajans and O. Johnson, *J. Am. Chem. Soc.* **64**, 668 (1942).
2. J. Padova, *J. Chem. Phys.* **39**, 1552 (1963); *ibid.* **40**, 691 (1964).
3. E. Glueckauf, *Trans. Faraday Soc.* **61**, 914 (1965).
4. E. Eucken, *Z. Elektrochem.* **51**, 6 (1948).
5. R. M. Noyes, *J. Am. Chem. Soc.* **86**, 971 (1964).
6. M. H. Panckhurst, *Rev. Pure Appl. Chem.* **19**, 45 (1969).
7. J. D. Bernal and R. H. Fowler, *J. Chem. Phys.* **1**, 515 (1933).
8. E. Darmois, *J. Phys. Radium* **2**, 2 (1941).
9. E. A. Zen, *Geochim. Cosmochim. Acta* **12**, 103 (1957).
10. E. J. King, *J. Phys. Chem.* **74**, 4590 (1970).
11. P. Mukerjee, *J. Phys. Chem.* **65**, 740, 744 (1961).
12. F. J. Millero, *J. Phys. Chem.* **75**, 280 (1971).
13. H. E. Wirth, *J. Mar. Res.* **3**, 230 (1940).
14. Y. Kobayazi, *J. Sci. Hiroshima Univ. Ser. A* **9**, 241 (1939).
15. R. Zana and E. Yeager, *J. Phys. Chem.* **70**, 954 (1966); *ibid.* **71**, 521 (1967); *ibid.* **71**, 4241 (1967).
16. B. E. Conway, R. E. Verrall, and J. E. Desnoyers, *Z. Phys. Chem.* (*Leipzig*) **230**, 157 (1965).
17. A. M. Couture and K. J. Laidler, *Can. J. Chem.* **35**, 207 (1957).
18. R. H. Stokes and R. A. Robinson, *Trans. Faraday Soc.* **53**, 301 (1957).
19. E. J. Millero and W. Drost-Hansen, *J. Phys. Chem.* **72**, 1758 (1968).

potentials give values for $\bar{V}^0_{H^+,aq} = (-5.5 \pm 0.5)\ cm^3\,mol^{-1}$, in excellent agreement with one another. Some results are given in Table 4.

2.9. Proton Solvation in Alcoholic Solutions

The behavior of the proton in alcoholic solutions was one of the earliest properties of the proton to be examined.[3,6] The proton exists as ROH_2^+ in alcoholic solutions, analogously to H_3O^+ in water, although the evidence for an alkoxonium ion structure similar to that of H_3O^+ is not available in such detail as that for H_3O^+. Bell[2] has given the base strengths of methanol and ethanol (relative to water as unity) as 0.13 and 0.33, respectively; based on data obtained in 1934. More recent estimates[2,23] put the values for the equilibrium constant for the reaction

$$H_3O^+ + ROH \rightleftharpoons ROH_2^+ + H_2O \qquad (11)$$

at 0.23, 0.059, and 0.037 (see reference 46) for R = Me, Et, and n-Pr, respectively. Goldschmidt's results give a value of 0.15 for the equilibrium constant of this reaction for R = Et, based on conductivity measurements. These estimates are therefore in agreement in placing the hydroxonium ion as an acid weaker than the alkoxonium ions, so that water will be the preferred protonated species in acid solutions in alcohol–water mixtures.

The distribution of protons in alcohol–water mixtures can be calculated from the equilibrium constants for reactions of the type in Eq. (11) for R = Me, Et, and n-Pr. Because the proton affinity of H_2O is higher than that of ROH, the "acid" protons available in the solution are mainly present as H_3O^+, even in solutions containing a small proportion of H_2O, e.g., 85% CH_3OH–H_2O, but changes occur near the 95% CH_3OH composition. A similar situation arises more markedly in HF; in this solvent (which is normally weakly conducting) H_2O dissolves to give a conducting solution of $H_3O^+F^-$. However, in ROH–H_2O mixtures over a wide concentration range, the relative acidities of ROH_2^+ and H_3O^+ change with composition owing to the changing solvation (selective hydration) of the two onium-type ions.

The Hammett acidity function H_0 has been studied for mixtures of various low molecular weight aliphatic alcohols ROH (R = Me, Et, n-propyl, and isopropyl) with water by Braude and Stern[49] and in various papers by Wells.[50,51,53] Generally[49-53] there are nonuniform changes of the basicity of water or of the alcohol component in dilute solutions of one component in the other. These effects are related to selective solvation of H^+ at the onium ions, steric effects, and hydrogen bonding.

3. Proton Transfer in Chemical Ionization Processes in Solution

3.1. Introductory Remarks

Three aspects of homogeneous proton transfer in solution are of interest: (1) the energetics of acid–base transfer of a H^+ entity in a single event, characterized thermodynamically by the acidity constant (pK_a) of the acid and the autoprotolysis constant of the solvent; (2) the relation between rate constants k for proton transfer in acid–base reactions and the equilibrium constant K for the corresponding protonation reaction (Brønsted "linear free-energy relation"[1,2]); and (3) the *continuous* acid–base proton-transfer events that take place in the electrolytic migration of the proton in hydroxylic media in an electric field (or in self-diffusion) corresponding to its ionic conductance or mobility.

3.2. Energetics

The thermodynamics of protonic ionization can be conveniently represented in terms of a Born–Haber cycle, as follows for an acid HA in a solvent S:

$$
\begin{array}{ccc}
\text{H}^{\cdot}_g + \text{A}^{\cdot}_g & \xrightarrow{\;I_{\text{H}^{\cdot}}+E_{\text{A}^{\cdot}}\;} & \text{H}^{+}_g + \text{A}^{-}_g \\
\Big\uparrow{\scriptstyle D_{\text{HA}}} & & {\scriptstyle +S}\Big\downarrow{\scriptstyle +S}\;\Delta G^{0}_{s,\text{H}^{+}}\;\Big\downarrow{\scriptstyle \Delta G^{0}_{s,\text{A}^{-}}} \\
\text{HA} + \text{S} & \xrightarrow[2.3RTpK_a]{\;\Delta G^{0}\;} & \text{SH}^{+} + \text{A}^{-}_s
\end{array}
\qquad \text{Cycle I}
$$

where ΔG^0 is the standard free energy of acid ionization of HA in the solvent S $(-\Delta G^0 = RT \ln K_a$, where K_a is the acid ionization constant, or $\Delta G^0 = 2.3RTpK_a)$, D_{HA} is the homolytic dissociation energy of HA, $I_{\text{H}^{\cdot}_g}$ is the ionization energy of $\text{H}^{\cdot}_g$, $E_{\text{A}^{\cdot}}$ is the electron affinity of $\text{A}^{\cdot}_g$, and the ΔG^0_s terms for H^+ and A^- are the standard free energies of solvation of these ions in the solvent S.

Cycle I gives an idea of the factors that, formally, can be considered to determine the acid strength of HA. Very often elementary discussions on the strengths of acids relate the pK_a to the bond energy D_{HA} of the molecular acid HA, e.g., in the hydrogen halide series.[1] This is evidently an over-simplification, since the solvation energy of the ions H^+ and A^- provide a major "driving force" (free-energy change) for the ionization of HA, while the $I_{\text{H}^{\cdot}}$ and $E_{\text{A}^{\cdot}}$ terms are also always important terms ($I_{\text{H}^{\cdot}}$ is, of course, a constant for all protonic acid ionizations).

While ΔG^0 (or pK_a) are often quite systematically related to D_{HA}, e.g., in a series of structurally similar organic acids or simple diatomic acids "HX," when standard entropies or enthalpies of ionization are related to D_{HA}, the relations found are much less systematic.[2] This is due to the well-known compensation effect[54] between the ΔH^0 and $T\,\Delta S^0$ quantities, which tends to obscure molecule-specific ionization behavior when the ΔG^0 (or pK_a) quantity is considered.

Also, steric factors in the solvation of the A^- entity are very important when A^- is not simply a spherical ion. The series CH_3NH_3^+, $(\text{CH}_3)_2\text{NH}_2^+$, and $(\text{CH}_3)_3\text{NH}^+$ [and, limitingly, $(\text{CH}_3)_4\text{N}^+$] provides an important example where steric[55] and so-called hydrophobic interaction effects play a major role in the thermodynamics of ionization of primary, secondary, and tertiary methylamine bases, leading to the "methylamine anomaly."

3.3. Relation to Kinetics

The kinetics of proton transfer in an acid–base reaction of the Brønsted type, e.g.,

$$
\text{HA} + \text{B} \xleftrightarrow{\;k(K_a)\;} \text{HB}^+ + \text{A}^- \tag{12}
$$

in relation to the equilibrium constant K_a for the same reaction, have been the subject of a large volume of work both in the 1930s (the work of Brønsted and of Bell) and more recently, with special acids and bases covering a wide range of pK_a values. Linear relations (so-called Brønsted relations) are found between the ln of the rate constant k and the ln of the equilibrium constant K. Thus for changes of HA in a series with a common base B (or vice versa), $\Delta \ln k = \alpha \Delta \ln K$, i.e., $d \Delta G_a^{\ddagger}/d \Delta G_a^0 = \alpha$, a constant approximately equal to 0.5 ± 0.15 for many HA–B series.

This well-known behavior is easily rationalized in terms of potential-energy diagrams and its further discussion here is outside the scope of the present chapter. Further details are to be found in reference 2 and other papers of Bell. α is analogous to the symmetry factor β for electrode processes.

3.4. Analogies to Electrons in Semiconductors

The state of the proton and the conjugate base anion, arising from ionization of a pure hydroxylic solvent ROH, can be thought of as analogous to ionization in an intrinsic semiconductor where the electron (n) and hole (p) pair states arise by internal release of an electron from the structure. In the hydroxylic solvent case the H^+ is analogous to the electron and the remaining RO^- to the hole, except that the charges of the respective entities are opposite to those in the intrinsic semiconductor case. "Intrinsic" ionization of a solvent is then represented by the well-known type of equation for autoprotolysis

$$2ROH \rightleftharpoons ROH_2^+ + RO^- \tag{13}$$

the simplest case being water, where $R \equiv H$. The behavior analogous to that in extrinsic semiconductors, where electropositive or electronegative dopants provide extra n or p states, with a different band gap, arises when an added acid or base is present in the solvent ROH. The closest proton–OH^- analogue to a semiconductor is ice. A protonic analogue "p–n junction" can be formed at the interface of an acidic and basic ion-exchange membrane.

3.5. Other Processes

Regarding other types of chemical reactions controlled by proton transfer, e.g., enolization, the related question of dynamics of the formation and dissociation of hydrogen bonds, the effect of hydrogen bonding on reaction rates, and H-bonding structural effects,[56] we refer to reviews and monographs on these topics,[57-60] as they are outside the scope of the present chapter.

3.6. Continuous Proton Transfer in Conductance

In the remaining major sections of this chapter we shall confine the presentation to a discussion of the nature and mechanisms of *electrolytic transport* of the hydrogen ion, especially for those cases where unusual or so-called "anomalous" or "excess" mobility is observed. A detailed review and discussion of the theories of anomalous mobility of the proton was given[7] by one of the present authors in 1964. Here, therefore, these theories will be referred to only in a brief and general way, but more recent advances made since 1964 will be described more thoroughly.

4. Continuous Proton Transfer in Conductance Processes

4.1. Definition and Phenomenology of the Unusual Mobility of the Proton in Solution

4.1.1. Comparison with the Mobility of Simple Ions

The limiting value of the equivalent conductance of monatomic or other simple inorganic ions in aqueous solutions at 298 K is about $35–80 \, \Omega^{-1} \, cm^2$ $equiv^{-1}$. Hydrogen and hydroxide ions are exceptions, with the much higher values of $350 \, \Omega^{-1} \, cm^2 \, equiv^{-1}$ and $192 \, \Omega^{-1} \, cm^2 \, equiv^{-1}$ respectively.[61]

For such ions, treated as spherical molecules moving in a homogeneous fluid of definite viscosity, application of Stokes' law gives reasonable values of effective hydrated radii[62,63] in the range 0.12–0.44 mm, depending on the crystal ionic radius of the ion. This, in some ways, surprising result has been commented on in many papers in the literature. However, despite obvious limitations in the applicability of a macroscopic hydrodynamic law to a microscopic situation, Stokes' law seems reasonably applicable to simple ions in one-component solvents, especially if dielectric friction (orientation relaxation) effects are allowed for.[64]

On the other hand, the mobility of the proton in water would give an effective radius some 7 times smaller than the above (lower) figure and some 10^6 times larger than the actual size of the proton itself. Since (see earlier sections in this chapter) the state of the proton in water corresponds to a molecular ion of at least the complexity of H_3O^+, the radius will be close to that of a water molecule considered as a sphere, or, more exactly, close to that of the isoelectronic ion NH_4^+ (mobility at 298 K, $73.4 \, \Omega^{-1} \, cm^2 \, equiv^{-1}$). Similar considerations apply to the mobility of OH^- in relation to an effective radius figure. Here the isoelectronic ion would be F^-.

An excess or anomalous mobility contribution may hence be defined as the ionic equivalent conductance of H^+ in the solvent concerned minus the expected conductance for a spherical ion of the same molecular size as H_3O^+, e.g., NH_4^+. Thus

$$\Delta\lambda_{H^+}^0 = \lambda_{H^+,aq}^0 - \lambda_{NH_4^+,aq}^0 \qquad (14)$$

or, for the OH$^-$ ion,

$$\Delta\lambda_{OH^-}^0 = \lambda_{OH^-,aq}^0 - \lambda_{F^-,aq}^0 \tag{15}$$

Other comparisons have been made with λ^0 values for K$^+$ and Na$^+$, e.g., by Hückel.[68] For aqueous H$^+$, $\Delta\lambda_{H^+,aq}^0$ is therefore $\sim$276.4 and $\Delta\lambda_{OH^-,aq}^0$ is $\sim$142.2 Ω^{-1} cm^2 equiv^{-1}, using the differences in Eqs. (14) and (15).

Abnormal mobility of H$^+$ is also indicated from the diffusion behavior, using Nernst's relation. Thus Roberts and Northey[65] calculated from their polarographic measurements of the limiting diffusion current a limiting value of $D_{H^+}^0 = 9.4 \times 10^{-5}$ cm^2 sec^{-1} for the diffusion coefficient of H$^+$ at 298 K. According to Nernst's formula, this corresponds to an equivalent conductance of $\lambda_{H^+,aq}^0 = 352$ Ω^{-1} cm^2 equiv^{-1}.

The difference between the equivalent conductances of H$^+$ and OH$^-$ is also important. The conductance of OH$^-$ is substantially less (see above) than that of H$^+$, as is also the corresponding $\Delta\lambda^0$.

For the H$^+$ and OH$^-$ ions, in contrast to the hydrodynamic migration of the others, a different mechanism of current transport was proposed as early as 1905.[66] In the case of other ions, ion–solvent interactions (hydration or solvation and dielectric relaxation friction effects) have to be considered in order to explain the differences between Stokes' and crystallographic radii, and other facts.

In early work, the anomalous conduction process was often referred to in terms of a so-called Grotthuss mechanism. (Several spellings of the name of Freiherr von Grotthuss appear in the literature.) However, examination of this mechanism, which was developed for the conductivity of electrolytes in Faraday's time, indicates that anomalous proton conductance does not really correspond to a Grotthuss type of process that involved consecutive dissociation events and "ion-pair" rotation for an otherwise undissociated electrolyte (cf. Faraday's view that electrolyte conductivity was due to a field-induced dissociation of the electrolyte).

Modern approaches to the understanding of the anomalous transfer mechanism treat the phenomenon in terms of a succession of acid–base proton transfers between H$_3$O$^+$ and H$_2$O (for H$^+$ mobility) or between OH$^-$ and H$_2$O (for OH$^-$ mobility). These are referred to as "prototropic transfer" mechanisms and can be represented formally by the following:

$$H_3O^+ + H_2O \rightarrow H_2O + H_3O^+ \rightarrow \cdots \qquad \text{(proton transfer to the right)} \tag{16}$$

or

$$OH^- + H_2O \rightarrow H_2O + OH^- \rightarrow \cdots \qquad \text{(proton transfer to the left)} \tag{17}$$

Figures 7 and 8 show the successive transfer and hydrogen-bond rearrangements required for a continuous sequence of proton-transfer events in an element of the hydrogen-bonded water liquid-lattice.

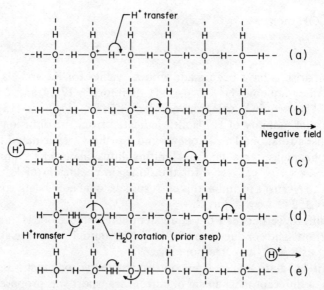

Figure 7. Schematic diagrams of successive proton-transfer events from H_3O^+ to hydrogen-bonded H_2O molecules in an element of the water lattice. Stage (c) shows the entry of a second proton from the left requiring cooperative rotations of inverted water quadrupoles for its continued passage [stages (d) and (e)] through the lattice.

The experimental facts on the existence and structure of a stable H_3O^+ entity, treated in Section 2, give some idea of its expected mobility if it were transported by the classical Stokesian mechanism. Thus the shape and size of the whole ion do not differ too much from those of the water molecule,

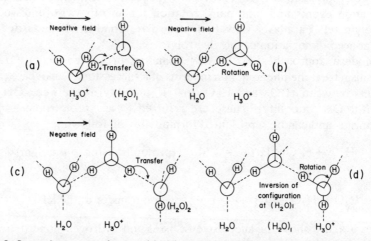

Figure 8. Successive events of stages (a), (b), and (c) in Figure 7 shown in a more accurate three-dimensional projection, with rearrangements of hydrogen-bond directions.

and, taken as a sphere, it can be assumed to be intermediate between the ions of sodium and potassium or ammonium, closer to the latter. Thus we have an idea of the crystallographic radius for H_3O^+.

On account of the anomalous mobility, it is evident that a Stokes hydrodynamic radius cannot be directly calculated from the ionic mobility $\lambda^0_{H^+,aq}$ because of the non-Stokesian mechanism of transport. However, some contribution of hydrodynamic "bodily transport" of the H_3O^+ ion is always to be expected in a proportion related to the observed ionic mobility of an isoelectronic ion such as NH_4^+, or a spherical ion such as K^+, as a fraction of the total observed mobility (see above) of H^+_{aq}. Thus two parallel mechanisms of transport of H^+_{aq} must be considered.

The hydrodynamic fraction of the total measured conductance can be estimated, starting from assumed values for the radii of ionic spheres. Lorenz[67] used radii of the H atom and the OH^- ion, calculated from atomic volumes by Reinganum's formula. A more adequate way of estimation is to take H_3O^+ as a sphere of a size intermediate between that of Na^+ and K^+, or of a size equal to that of NH_4^+, and to assume a corresponding hydrodynamic mobility (a value between 50.1 and 73.5 Ω^{-1} cm^2 equiv^{-1}).[68] This was the procedure used by Hückel.[68]

A higher value of 85 Ω^{-1} cm^2 equiv^{-1} results if we assume the hydrodynamic mobilities λ^0 of the species H_3O^+ and H_2O to be approximately equal and calculate the latter from the experimental value of the self-diffusion coefficient of the water molecule ($D = 2.25 \times 10^{-5}$ cm^2 sec^{-1})[69] by using Nernst's expression in the form

$$D = (\lambda/eF)kT \qquad (18)$$

where e denotes the electronic charge, F the Faraday constant, and k Boltzmann's constant.

A higher value for the radius of the hydrated H_3O^+ ion is obtained from the thermodynamic ionic properties (heat capacities[70,71] and activity coefficients[72]) of solutions of lithium halides and $LiClO_4$ and of salts of other alkali metals with various acid anions. This comparison shows that the behavior of the Li^+ ion comes closest to that of the H_3O^+ ion, thermodynamically speaking.

For the case of the OH^- ion a similar procedure may be adopted. Thus, if the hydroxide ion is taken as a sphere of the same radius as that of the isoelectronic fluoride ion, Stokes' law gives $\lambda^0_{OH^-} = 55\ \Omega^{-1}$ cm^2 equiv^{-1}. A fraction of the conductance of OH^- corresponding approximately to this value is due, then, to hydrodynamic migration.

The differences between the total and hydrodynamic conductances must be attributed to a special conductance mechanism for both the H_3O^+ and the OH^- ion, i.e., prototropic transfers according to the processes of Eqs. (16) and (17).

4.1.2. Temperature Effect on Anomalous Mobility

The classical mobility contribution and the special transport mechanism will have different activation energies. Therefore the relative contributions will change with temperature, depending on which mechanism has the greater energy of activation.

The temperature dependence of the total equivalent conductance (relative mobility) of H_3O^+ differs from that of other ions.[7,61,68,73-75] Since the structure of water changes with temperature, the temperature dependence of the equivalent ionic conductance cannot be fitted by a single apparent activation energy over a long range of temperature. However, if values of apparent activation energies calculated for the same temperature interval are compared, the values for H_3O^+ and OH^- are smaller than those for any other simple ion. In the interval 273–429 K the activation energy decreases with increasing temperature for all simple ions. However, this decrease is more rapid for H_3O^+ and OH^- than for the simple ions.

As was shown by Hückel,[68] the differences $\lambda_{H^+}^0 - \lambda_{Na^+}^0$ (or $\lambda_{H^+}^0 - \lambda_{K^+}^0$) can be taken approximately as the nonhydrodynamic fraction of the conductance of the hydrogen ion. The apparent activation energies calculated from these values are lower than the apparent activation energies for other ions, and their decrease with increasing temperature is rapid.

4.1.3. Pressure Effect on Anomalous Mobility

As in the case of increasing temperatures, increases of pressure cause changes in the structure of water owing to modification of the hydrogen-bond structural equilibria in this solvent. It also changes the viscosity, so that the hydrodynamic mobility contribution will be altered.

In contrast to the behavior of other ions, the equivalent conductance of the hydrogen ion increases with increasing pressure.[76-82] Correspondingly, the activation volume calculated from the pressure dependence of the abnormal contribution to proton mobility is negative.

Other aspects of the temperature and pressure effects on proton mobility will be treated in Section 5.4.3.

4.1.4. Deuterium Isotope Effects

As is well known in studies of the kinetics of proton-transfer reactions, the H–D kinetic isotope effect[2] often provides a diagnostic indication of the type of process involved. In the present problem different H–D isotope effects can be expected for the prototropic process in relation to the hydrogen-bond structural rearrangement in the water lattice, where a librational or rotational fluctuation is required, or in relation to classical hydrodynamic motion in H_2O and D_2O.

Bernal and Fowler assumed[19] that "the effective mobility of the isotope D must be less than that of the ordinary H." Since hydrodynamic and transfer mechanisms are expected to be influenced to a different degree by the isotope effect, it is important to know the ratio of the mobilities of H^+ in H_2O to D^+ in D_2O; at 298 K it is 1.42.[75] This ratio is derived from measurements of the conductances of hydrochloric acid and potassium chloride in ordinary water and in nearly pure D_2O[83] (see also the footnote on p. 836 of reference 75); alternatively, it can be derived from the diffusion coefficients calculated by the corrected form of the Ilkovič equation from polarographically measured diffusion currents.[84] The ratio derived from the latter approach[65] is 1.52 and thus differs by a small extent from the above-mentioned value. This ratio has been studied as a function of the concentration of the supporting electrolyte in solutions of some alkali halides and tetraalkylammonium bromides.[65,83]

If only a hydrodynamic mobility mechanism obtains, it would be expected that the value of $\lambda^0_{H^+,aq}/\lambda^0_{D^+,D_2O}$ would be less than the observed ratio (1.42 ~ 1.52) corresponding to the differences between the viscosities and dielectric constants of H_2O and D_2O, respectively. Supposing, on the other hand, that in the mechanism involving proton (or deuteron) transfer (usually referred to as the prototropic mechanism), it is this transfer, not a structural rearrangement, that is the rate-determining step; it would then be expected that a higher value for $\lambda_{H^+,aq}/\lambda_{D^+,D_2O}$ than that observed would arise owing to the quantal mass effect on transfer probability.[20]

4.1.5. Abnormal Proton Conductance in Other Solvents

A comparison of the equivalent conductances of other ions with those of the hydrogen ion in solvents such as methanol and ethanol[7,75] also indicates abnormal proton mobility. This abnormality is explained by consecutive processes of proton transfer and structural rearrangements, as in the water solvent. Temporary clusters of solvent molecules, bound by hydrogen bonds, appear to play an important role in this mechanism.

Figure 9 shows the equivalent conductance behavior of HCl in methanol–water mixtures at 298 K, where a conductance minimum is observed at ~80 mol % methanol, at which concentration the proton has virtually no anomalous mobility.[75]

As in the case of hydrogen and hydroxyl ions in aqueous solutions, anomalously high conductances were also found for the CH_3O^- ion in methanol, by Dempwolff[85] in the year 1904 and Tijmstra[86] in 1905, and for the pyridinium ion ($C_5H_5NH^+$) in pyridine, the formate ion in formic acid, and the acetate ion in acetic acid, by Hantzsch and Caldwell[87] in 1907. The HSO_4^- ion in pure H_2SO_4 also shows an unusually large anomalous conductance in comparison to the relatively low classical conductance that arises on account of the high viscosity of 100% H_2SO_4.

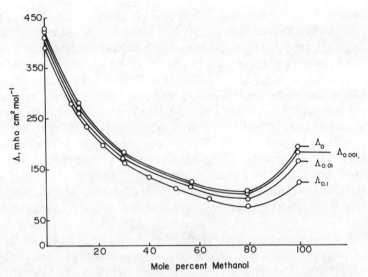

Figure 9. Conductance behavior of HCl in methanol–water mixtures at 298 K for three HCl concentrations (in mol dm^{-3}) and at infinite dilution (from Conway, Bockris, and Linton[75]).

The generalization that all ions produced by a Brønsted acid dissociation of a protic solvent molecule have anomalously high conductance was first suggested by Daneel.[66]

4.1.6. Proton Conductance in Ice

Although the conductance of ice is very much lower than that of water, K_w is much smaller so that the charge-carrier concentration is very low; combining these two factors indicates that the mobility of H^+ in ice is high, although comparison with the mobility of, for example, K^+ is difficult in the solid. Nevertheless, $NH_4^+F^-$-doped ice can be prepared, and a comparison with the NH_4^+ ion mobility made. H^+ is then found to have an appreciable anomalous mobility in ice.[7]

4.1.7. Analogous Effects in Electron Transport

Experimental evidence shows that a contribution to transport processes from a charge-transfer reaction is not limited to proton transfer. Electronic conductivity in solutions was predicted by Frumkin, as noted by Levich in his review.[88] The increase of apparent diffusion by electron transfer was measured by Ruff *et al.* in the systems ferrocene and ferricinium ion [i.e., the biscyclopentadienyl complexes of Fe(II) and Fe(III)] in alcohols,[89] ferroin and ferriin [the *tris*-1-10-phenanthroline complexes of Fe(II) and Fe(III)] ions in water,[90] and ferrous and ferric forms of cytochrome C in water.[91] These electronic contributions to ion transport arise by electron transfer between reduced and oxidized forms of a pair of redox reagent ions in parallel with their normal diffusional or electrolytic transport.

A contribution of the transfer of atoms or molecules to diffusion was also measured in systems containing sulfide and disulfide ions,[92] triiodide and iodide ions,[93] and tribromide and bromide ions[94] in aqueous solution. For systems containing antimonytrichloride and aliphatic alcohols, acids, amides of carboxylic acids, and alkyl halides, there is, in addition to proton transfer, a contribution to the electrical conductivity from the transfer of atomic chlorine.[95,96]

4.2. Solvent Structure Changes in Relation to Anomalous Proton Mobility

4.2.1. Effects of Solvent Composition

Studies on the influence of added nonelectrolytes to water on the proton mobility are aimed at testing the mechanism of anomalous proton conductance, which supposes that the rate of transfer is influenced by "structural diffusion" in the solvent lattice, i.e., rotation of hydrogen-bonded H_2O molecules or local rearrangements of the hydrogen-bonded structure (see reference 20 and the discussion in reference 75).

Thus hydrogen bonding in associated solvents seems to be an important requirement for facile transfer of a proton from H_3O^+ to H_2O, and quantum-mechanical tunneling and classical calculations for proton transfer are usually based on a model of an $O—H\cdots O$ hydrogen (or deuterium) bond. Huggins[97] was one of the first to recognize the role of a continuous hydrogen-bonded liquid structure in prototropic processes.

The addition of nonelectrolytes to aqueous electrolyte solutions affects the structure of the solvent mainly by eliminating a fraction of the hydrogen bonds and changing the size of clusters held together by these bonds, or the average degree of hydrogen bonding in the liquid.† The addition of an electrolyte, structure-modifying component to water also modifies the classical hydrodynamic component of the proton or OH^- conductivity and the stability of the proton with respect to its solvation energy.

Wulff and Hartmann[100] were the first to carry out studies on the influence of the dioxane content in hydrochloric acid solutions in water–dioxane mixtures on viscosity, dielectric constant, and equivalent conductance. Conway, Bockris, and Linton[75] investigated proton mobility in $ROH–H_2O$ mixtures (R = Me, Et, n-Pr).

Nonelectrolytes that are highly soluble in water usually contain hydroxyl groups, have a characteristic structure, and show current transport by a prototropic mechanism in their own pure state.

† Recent molecular dynamics calculations on the structure of water do not indicate distinguishable cluster states, as were considered in earlier theories, e.g., of Frank and Evans[98] and Nemethy and Scheraga[56]; rather, a continuum of variously bonded molecules is indicated, as in Pople's treatment.[99]

Several authors have studied the effect of varying the concentration of added nonelectrolytes on the electrical conductance of aqueous acid and base solutions. Some investigated the variation of conductance with the nature of the added nonelectrolytes and their concentrations. Thus the conductance of aqueous hydrochloric acid solutions containing methanol and ethanol was investigated, mainly by Goldschmidt and Dahl,[101] Walden,[102] Thomas and Marum,[103] Berman and Verhock,[104] Kortüm and Wilski,[105] Dnieprov,[106] el Aggan et al.,[107] and Tourky and Mikhail.[108] The change in conductance of aqueous hydrochloric acid solutions upon the addition of ethylene glycol has been studied by Dnieprov[106] and Kirby and Maass,[109] while the effect of glycerol has been measured by Dnieprov,[106] while the effect of glycerol has been measured by Dnieprov,[106] Conway et al.,[75] Woolf,[110] and Accascina et al.[111]; further, the effect of butanol has been investigated by Accascina et al.[112]

It has been found that the equivalent conductance of aqueous hydrochloric acid solutions decreases rapidly upon the addition of alcohols; with increasing alcohol concentration, it passes through a minimum and increases again in solutions containing very small amounts of water,[75] see Figure 9. According to Tourky et al.,[113] and Abdel Hamid et al.,[114] the conductance of hydrochloric acid varies similarly in mixtures of n-propanol, t-butanol, and water. The influence of water on proton migration in 1-pentanol has been studied by de Lisi and Goffredi.[115] The equivalent conductances of dilute ($0.01\ M$) hydrochloric acid and potassium hydroxide solutions vary with the concentration of the nonelectrolyte (methanol, ethanol, n-propanol, ethylene glycol, glycerol, and dioxane) added to aqueous solutions and show a minimum conductance around 90% nonaqueous component. These studies were made by Erdey-Gruz et al.[116,117] The results for HCl coincide qualitatively with those of former authors.[75] However, more detailed information can be derived from recent studies. The conductance minimum occurs at about 80 mol% alcohol content[75] of the solvent. On the curve for methanol, Figure 9, a sharp minimum arises, whereas it is smoothed out in the cases of n-propanol and glycerol.

In the case of KOH solutions the equivalent conductance falls rapidly with the increasing concentration of the nonelectrolyte solvent component, but a smooth minimum appears only on the curve for methanol, at about 75 mol % of the nonelectrolyte.

4.2.2. Walden's Rule and Solvent Structure Effects

The structural properties of a solvent are characterized, among other ways, by the bulk viscosity. Solvent effects in conductance are conveniently investigated by examining the applicability of Walden's rule to conductance by an electrolyte in mixed solvent media of varying viscosity.

Hydrodynamic mobilities depend on the viscosity of the solution. For the model of spherical charged particles subject to resisting forces proportional to their velocity and moving in a homogeneous fluid of definite viscosity η_0, Walden's rule,

$$\Lambda^0 \eta^0 = (0.82 \times 10^{-8})/r = \text{const.} \tag{19}$$

for a given electrolyte in a series of solvents follows from Stokes' law, where Λ^0 is the limiting conductance, η^0 is the viscosity of the pure solvent, and r is the hydrodynamic radius of the ion.

Secondly, Walden's rule applies only approximately, and then only for the ions of larger radii. This is because of ion solvation effects, which locally modify the solvent viscosity (structure-breaking effects and dielectric friction factors) and, in mixed solvents, involve selective solvation, i.e., redistribution of the components of a mixed solvent near the ions. For anomalous mobilities of H^+ and OH^- the effects are more complicated. Nevertheless, some conclusions concerning liquid structure and mechanism of transport can be drawn from the variations of Walden's product with such parameters as temperature and composition.

For HCl and KOH solutions in ROH–H_2O mixtures (ROH = methanol, ethanol, n-propanol, glycol, glycerol, and, for comparison, dioxane), the Walden product is far from constant, exhibiting maxima or minima, depending on the electrolyte or the solvent.[118,119] Glycol and glycerol solutions behave in a qualitatively different manner from the other solutions, including dioxane. The Walden products for KF and KCl were also measured[118,119] in the same solvent mixtures. These electrolytes, of course, do not exhibit any anomalous conductance; nevertheless, somewhat similar dependencies of $\Lambda^0 \eta^0$ on solvent composition are observed, with glycol and glycerol again behaving rather differently from the other solvents. For KF and KCl marked variations of the value of Walden's product are observed, due to several effects. Radii of the solvated ions may change with changing composition of the solvent, affecting the composition and structure of the first solvation shell. A dependence of the viscosity in the neighborhood of the ion on the distance from the latter may also result. It is not justified to assume the same mechanism of displacement of both the ion and the solvent particles. In an interpretation of these complications a simplified model of spherical particles moving in a continuum medium of bulk viscosity of the solution (not the solvent!) does not seem adequate.

Summarily, it is found that Walden's product increases with increasing alcohol concentrations in the lowest concentration range (glycol and glycerol seem to be exceptions if KOH is the electrolyte). Erdey-Gruz et al. consider that the decrease of mobilities is overcompensated by the increase in bulk viscosity, due to structure-strengthening effects of the alcohols. Their molecules may be assumed to fill the "holes" in the water structure. This strengthened structure is decomposed by ions to a higher degree than is the structure of pure water.

In mixtures of monoalcohols with water $\Lambda^0\eta^0$ shows a maximum between about 4 and 10% alcohol concentration; at higher concentrations it decreases below the values for solutions in pure water. The maximum is flatter for HCl and KOH than for KF and KCl. The decrease after the maximum is more pronounced than expected from viscosity changes alone. The ions strengthen the structure of the liquid and hinder displacement among its molecules. In addition, in the cases of H^+ and OH^-, the role of the prototropic mechanism decreases.

Solutions in dioxane–water mixtures exhibit similar behavior, but electrolytes in polyalcohol–water mixtures behave differently. The curve of the Walden product of HCl solutions increases at low concentrations of ethylene glycol, then passes a flat maximum, and after a minimum at about 85% shows, finally, a sudden increase. The $\Lambda^0\eta^0$ curves for HCl, KF, and KCl in glycerol solutions have a positive slope over the entire range. For HCl a particularly sudden increase in mixtures containing small amounts of water can be observed. The slopes for KF and KCl are less than that for HCl. In solutions of glycol and glycerol, where the viscosity is high, conductance decreases to a degree smaller than would correspond to the increase in bulk viscosity. For exclusively hydrodynamic conductance a strong structure-breaking effect by the ions must be supposed. The variant behavior of the HCl solution can be explained by a decrease in the role of the prototropic mechanism at moderate concentrations. However, in solutions containing small amounts of water, the proton-transfer mechanism seems to prevail even more than in a pure aqueous solution.

The $\Lambda^0\eta^0$ curves of KOH show a different behavior. In ethylene glycol–water mixtures the value of Walden's product is almost independent of the composition of the solvent. In glycerol–water mixtures it decreases at low glycerol concentrations, passes through a flat minimum, and increases moderately.

4.2.3. Transference Numbers of H^+ and OH^- in Various Solutions

Conclusions concerning the variation of the prototropic contribution to conductance with the composition of the solvent can also be drawn from transference numbers.[117,118] The dependence of transference numbers of H^+ and OH^- in HCl and KOH solutions, respectively, upon the addition of certain nonelectrolytes to water is complex, maxima or minima in the transference numbers t^+ or t^- being observed. At low concentrations of the alcohol the transference number of H^+ increases with the alcohol content, passes through a maximum, and, after decreasing and passing through a minimum, increases again in solutions containing small amounts of water. In monoalcoholic solutions the minimum is sharp, and its position is at about 90–95 mol %. The latter increase has a steep slope. The first increase may be explained by a strengthening of the structure of the liquid. The decrease beyond the maximum, on the other hand, may be due to the structure-breaking effect of the alcohol, which hinders proton transfer between H_3O^+ and H_2O. At the composition corresponding to the minimum transference number the main contribution to conductance is given by the *hydrodynamic migration* of the chemical species H_3O^+. In solvents containing small amounts of water the particular hydrogen-bonded liquid structure of the alcohol prevails, and the proton transfer between hydrogen-bonded ROH_2^+ and ROH contributes to the transport of current.

The different proton affinities of alcohol and water,[49-53] which are also dependent on the composition of $ROH-H_2O$ mixtures, is another important point. If the proton affinity of H_2O (as H_3O^+) is a maximum at $\sim$90-95% MeOH, for example, H_2O molecules act as "proton traps" in the mixed solvent and prevent the rapid transfer of H^+ from one solvent site to another neighboring one in the liquid. Obviously, continuous prototropic transfer of H^+ can take place most easily when the proton-transfer event is "symmetrical," e.g., $H_3O^+ + H_2O \rightarrow H_2O - H_3O^+$ [Eqs. (16) and (17)]. Elementary processes such as $H_3O^+ + ROH \rightarrow H_2O + ROH_2^+$ are "unsymmetrical" and the solvent sites of stronger basicity will impede continuous proton transfer.

A particular liquid structure with a hydrogen-bonded network seems to be a requirement for prototropic conductance. This view is supported by the behavior of HCl solutions in dioxane-water mixtures.[118] Apart from the region of very low concentrations, the transference number of H^+ decreases abruptly with an increasing mole percentage of dioxane, and at 60 mol % reaches a value of 0.42. In this solution the conductance of H^+ is smaller than that of Cl^-. Thus dioxane "destroys" the water structure and proton transfer is no longer possible.

In the range of small water concentrations the transference number of H^+ increases with a very high slope if the concentration of glycol or glycerol is increased. This is due to a contribution of proton transfer to the conductance that is relatively much higher than in pure water. Presumably, in a polyalcohol all the hydroxyl groups of the molecule may function in proton transfer. Such double or triple roles of a molecule may have a higher probability than rotation of the molecules in a fluid of high viscosity. Since structural rearrangement, consequent to proton transfer, is necessary[19,20] for continuous prototropic conductance, and since molecular rotation seems to be excluded by high viscosity, intramolecular rearrangement in the polyalcohol molecule or its protonated form may be assumed instead.

On addition of methanol to aqueous solutions of KOH the transference number of the OH^- ion decreases to a value of about 0.5 in pure methanol. After a slight increase and a minor maximum at 10-20 mol % alcohol, the transference number decreases also in water-ethanol and water-n-propanol mixtures with increasing concentrations of alcohol and approaches the values 0.5 and 0.56, respectively. Consequently, no successive proton transfer can be assumed between OH^- and the monoalcohol molecule. The polyalcohols behave differently. In the range of small polyol concentrations the transference number of OH⁻ decreases with increasing concentrations of alcohol, particularly in the case of glycerol. After passing a minimum, it rises again. In a solvent containing a small amount of water and a high percentage of glycerol the transference number of the OH⁻ ion is higher than in pure water. At low alcohol concentrations a considerable fraction of the hydrogen bonds is "dissolved" by alcohol, and proton transfer is therefore hindered. At higher alcohol concentrations the molecules of the alcohol themselves may participate in proton-exchange reactions.

Comparison of the conductance, $\Lambda^0 \eta^0$ and transference number curves for H^+ and OH^- leads to the conclusion that different mechanisms for proton transfer must be assumed in acid and base solutions, and that the mechanism also depends on the nature and composition of the solvent (for further details see references 20 and 119).

4.2.4. Ionic Effects on Solvent Structure and Prototropic Processes

The addition of electrolytes may have either a structure-breaking or structure-forming effect on water. However, the added ions themselves affect the electrical conductivity by their own current transport.

The observations of Woolf[120] are that the presence of "supporting" electrolytes decreases the tracer diffusion of H^+ to a higher degree than for any other ion. Correspondingly, Glietenberg et al.[121] found a particularly large effect from Li^+ ions. These effects were interpreted in terms of structural changes, which are known to be caused by ion–solvent interactions, i.e., hydration of the ions of the supporting electrolyte. These changes lead to short-range structures less favorable for proton transfer, owing to so-called structure-breaking effects.

Quantitative theories (see Section 4.4) of the anomalous conductance of the H^+ and OH^- ions usually assume infinite (or at least high) dilution and calculate limiting conductances. The presence of acids or bases at higher concentrations itself influences the structure of the solvent. Consequently, the variation of the equivalent ionic conductance with varying concentration of the electrolyte may give additional information about the mechanism of proton transfer. Transference number studies on acid and basic solutions, together with equivalent conductance data, can be used to obtain such information.

The transference numbers (relative to the solvent) of the H^+ ions in hydrochloric acid solutions in the higher concentration ranges were determined by Szabo,[122] Harned and Dreby,[123] and Lengyel et al.[124–126] by measuring the emf of concentration cells, and by Kaimakov and Fiks[127] by the moving boundary method.[128] Konstantinov and Troshin[129] determined the transference number of the H^+ ion relative to the solution in hydrochloric acid solutions over the entire concentration range above 1 mol liter⁻¹. All these studies show that the transference number of the H^+ ion (~0.84 at

$c = 1$) diminishes markedly at higher acid concentrations, and the mobility ratio $\lambda_{H^+}/\lambda_{Cl^-}$ falls from about 5.39 (at 298 K and 1 m) to about 1.8 (at 12 m).

The transference number of the OH$^-$ ion in alkali hydroxide solutions has also been studied. Lengyel et al.[130] determined the transference number of the OH$^-$ ion in aqueous KOH and NaOH solutions at 293 K in the molality ranges 1–17 and 2–17, respectively, by measuring the emf of concentration cells. In the case of KOH the value (relative to the solvent) of 0.74 found for the lowest and highest concentrations changed to a maximum of 0.78 (at 3–12 m). NaOH behaves differently, the value of 0.94 at 2 m increases to 0.98 (at 3–6 m), then diminishes to 0.86, increases again, and at $m = 14$ has a maximum of 1.00, after which it decreases to 0.59 (at $m = 17$). The extremely high (apparent) values were explained by the formation of complex ions. Troshin and Zvyagina determined the transference number of OH$^-$ relative to the solution in concentrated aqueous solutions of KOH,[131] NaOH,[132] and LiOH[133] at 298 K. Their values for KOH increased uniformly from 0.79 (at $c = 1$ molar) to 0.847 (at $c = 12$). For the transference number relative to the solution of OH$^-$ in NaOH solutions they obtained a value of 0.905 at $c = 1$, which increases to 0.911 at $c = 2$, then diminishes uniformly to 0.826 at $c = 18$ molar. No second maximum occurs. For LiOH the high value of 0.947 at $c = 1$ was found; this increases to 0.980 at $c = 4$, then drops to 0.965 at $c = 4.8$ molar. Values for higher concentrations could not be measured, owing to solubility limitations.

From the limiting equivalent conductances[40] the following limiting values can be calculated for the transference numbers at 298 K: H$^+$ in HCl, 0.823; OH$^-$ in KOH, 0.72; OH$^-$ in NaOH, 0.76. A comparison of these values with those for $m = 1$ or $c = 1$ shows a strong effect of the electrolyte, even at this quite low concentration.

4.3. Kinetics of the Proton-Transfer Event

In almost all theories of abnormal conductance (see Section 4.4) successive acid–base proton transfer, whether rate controlling or not, is assumed. The rate of proton transfer is then calculated quantum-mechanically or classically. By observation of the broadening of the corresponding NMR lines, Meiboom et al.[134–140] and other authors[141–143] determined this rate constant for proton transfer experimentally for a series of reactions. Table 5 shows a compilation of the experimental values of the second-order rate constant of proton transfer from H$_3$O$^+$ to H$_2$O and H$_2$O to OH$^-$. Table 6, on the other hand, gives rate constant values of proton transfers between an ionic and a neutral molecular species for a series of other reactions. If we rely on these rate constants, then one of the uncertain points in abnormal-conductance theories seems to be eliminated. The results of a comparison of theoretical and experimental conductances no longer depend on the uncertainties of calculations of the proton-transfer rate constant. The rates will, however,

Table 5

Values of the Second-Order Rate Constants for the
Reactions $H_3O^+ + H_2O \rightarrow H_2O + H_3O^+$ (k_1) and $H_2O + OH^- \rightarrow HO^- + H_2O$ (k_2) at 298 K (mol^{-1} liter sec^{-1})

References	$k_1 \times 10^9$	$k_2 \times 10^9$
Meiboom[134]	10.6	3.8
Löwenstein and Szöke[141]	11.0	5.5
Luz and Meiboom[135]		4.8
Luz and Meiboom[136]		3.4
Luz and Meiboom[137]	9.7	2.9
Glick and Tewari[142]a	7.9	4.5
Rabideau and Hecht[143]	8.2	4.6

[a] Calculated from $6.0 \times 10^{11} e^{-2.4/RT}$ and $1.0 \times 10^{11} e^{-2.1/RT}$, as given by the authors at 298 K.

depend on ionic or acid strength (H_3O^+ concentration), as in the conductance experiments.

4.4. Theories of Abnormal Proton Conductance

Most theories, except that of Hückel,[68] treat the problem of proton conductance in terms of a two-step mechanism that can be represented by the following:

1. Prototropic transfer in steps such as

$$H_3O^+ + H_2O \rightarrow H_2O + H_3O^+$$

Table 6

Values of the Second-Order Rate Constants at 298 K for Some Proton Reactions Involving Proton Transfer[a]

Reactants	k (mol^{-1} liter sec^{-1})	Temperature (K)	Reference
$H_2O_2 + H_3O^+ \rightarrow H_3O_2^+ + H_2O$			
$CH_3OH + H_3O^+ \rightarrow CH_3OH_2^+ + H_2O$	10^8	295	139
$CH_3OH + CH_3OH_2^+$ $\rightarrow CH_3OH_2^+ + CH_3OH$	3.6×10^9	297.8	140
$CH_3OH + OH^- \rightarrow CH_3O^- + H_2O$	2.6×10^6	295	139
$CH_3OH + CH_3O^- \rightarrow CH_3O^- + CH_3OH$	7.5×10^8	297	140
$C_2H_5OH + H_3O^+ \rightarrow C_2H_5OH_2^+ + H_2O$	2.8×10^6	295	139
$C_2H_5OH + C_2H_5OH_2^+$ $\rightarrow C_2H_5OH_2^+ + C_2H_5OH$	1.1×10^6	295	139
$C_2H_5OH + OH^- \rightarrow C_2H_5O^- + H_2O$	2.8×10^6	295	139
$C_2H_5OH + C_2H_5O^- \rightarrow C_2H_5O^- + C_2H_5OH$	1.4×10^6	295	139

[a] Reference 139 deals with solutions containing water, Reference 140 deals with solutions in pure methanol.

or

$$H_2O + OH^- \rightarrow OH^- + H_2O$$

by classical and/or quantum-mechanical tunneling.

2. A rearrangement of the hydrogen-bonded liquid lattice structure in H_2O, D_2O, and alcohols, which is required in order for successive proton-transfer events to occur through a structural element of the hydrogen-bonded solvent (this is sometimes referred to as a "structure diffusion" or "structure reorientation" process). This step was recognized by Bernal and Fowler[19] and Danneel.[66]

Processes (1) and (2) can be *coupled* if the field effect of arriving protons at a given structural element of the solvent lattice is taken into account, as in the theory of Conway, Bockris, and Linton.[75]

The relation between processes (1) and (2) is shown schematically as a two-dimensional projection in Figure 8. Three-dimensional, detailed perspective drawings were shown in Reference 75. Danneel[66] was the first to recognize the necessity for a two-step process in continuous proton transfer in water, and this view has formed the basis of most models subsequently used in quantitative treatments of mechanisms.

Theoretical treatments of the mechanism of proton transfer in conductance have aimed at making an *a priori* estimation of the specific rates of steps (1) and (2), an evaluation of the role of quantum-mechanical tunneling of H^+ (and D^+) in (1), and deducing which of steps (1) and (2) controls the observed conductance in water and hydrogen-bonded solutions, as well as in ice.

4.4.1. Early Theories, Semiempirical Treatments, and Qualitative Considerations

This section began by reminding the reader that Danneel[66] was the first to recognize the consecutive mechanism of proton transfer combined with structural reorientation (often referred to as *structural diffusion*) in the liquid. Rotation of H_2O dipoles was regarded by Danneel as the actual form of reorientation and was later treated in more detail by Bernal and Fowler[19] and Conway, Bockris, and Linton.[75]

In his attempt at a quantitative treatment, Hückel[68] assumed a combination of proton jumps from H_3O^+ to H_2O, with rotation of the resulting new H_3O^+. Neglecting the equivalence of the three protons in H_3O^+ and considering one of them fixed, he calculated the relaxation time of the rotating ionic sphere subjected to a momentum of the force acting on this proton and to the resisting force in the viscous liquid. He also calculated the proton-transfer frequency by classical vibration statistics. The ionic mobility was then related to both relaxation times by assuming a *stationary* orientation distribution resulting from the *equilibrium* orientation distribution by perturbation by the external field.

Instead of this field, the Lorentz field in a spherical cavity in the dielectric was taken into account. The inadequacy of this theory, as well as some in

Wannier's calculations[144]† relating to the same mechanism, was discussed by Conway et al.[75] The following argument should be added: because of the relatively high inertia of the rotating oxonium ion, a decrease of anomalous mobility would be expected at very high frequencies of an alternating current. However, this expectation is not fulfilled, but sufficiently high frequencies may not yet have been attained.

Assuming a two-step consecutive process involving proton transfer by quantum-mechanical tunneling, coupled with water molecule rotation, Bernal and Fowler[19] calculated the frequency of proton tunneling between two water molecules in a suitable configuration and the corresponding proton mobility, which is due to an external (Lorentz) field. The rotation frequency estimated for the water molecule as a classical isotropic free rotator has the same order of magnitude as the frequency of proton tunneling. The latter was then proposed as the rate-determining step because it seemed that the orientation of the water molecule suited for proton transfer would be presented to the H_3O^+ ion at least as frequently as this ion could utilize it.

Apart from the fact that the predicted decrease of the abnormal conductance with increasing pressure is at variance with experience, detailed analyses by Conway et al.[7,75] show that some of Bernal and Fowler's assumptions were not adequate. The tunneling frequency depends greatly on the potential-energy profile of the proton between its two alternative sites. Bernal and Fowler's rectangular well profile and its dimensions result in values for the net rate of transfer lower than the more realistic (Eckart) potential-energy curve used by Conway et al.[7,75] Also, the free-rotation model of the water molecule is incompatible with the hydrogen-bonded structure of water. For the ratio $\lambda_{H^+}^0/\lambda_{D^+}^0$ (see Section 4.14), Bernal and Fowler estimated a value of about 5 (see reference 19, p. 548); their paper[19] was received by the *Journal of Chemical Physics* on April 29, 1933. Lewis and Doody[84] published their experimental value of 1.39 at 298 K for the H^+/D^+ conductance ratio in a paper received by the *Journal of the American Chemical Society* on July 24, 1933, and published August 5, 1933. The discrepancy between the two values is experimental evidence against the assumption of proton tunneling as the *rate-determining* step, which would necessarily result[75] in much higher values for $\lambda_{H^+,aq}^0/\lambda_{D^+,D_2O}^0$.

While there were some defects in the treatment of Bernal and Fowler for anomalous proton transfer, it must be stressed that this paper was the first to put forward a detailed and impressive treatment of the problem in relation to water structure and hydrogen-bonding, and it is a historic paper in its field.

In a treatment of Huggins[97] on the state of H^+ in water (as $H_5O_2^+$; see Section 2.6) it was assumed that rapid proton-transfer events took place within

---

† Wannier's treatment, like that of Bernal and Fowler, was a very thorough one, but not without some difficulties. A detailed discussion of it was given in the earlier review (reference 7).

hydrogen-bonded chains of H_2O molecules so that a less frequent, consequently rate-limiting, proton transfer occurred between two chains at a position where the hydrogen bond was temporarily absent.[144,145,146] Thus Huggins proposed a new modification of the "structural diffusion" concept; on the other hand, his calculations and conclusions pertaining to the potential-energy profile of the hydrogen bond are difficult to accept (see pp. 118 and 119 of reference 97). A detailed treatment of H^+ transfer involving H_3O^+ rotation, rather than H_2O rotation, was given by Wannier.[144]

Using the theory of "absolute reaction rates," Stearn and Eyring[74] calculated the free energy of activation for the proton transfer between two water molecules from Hückel's estimate: $(\lambda_{H^+} - \lambda_{Na^+})$ for the abnormal conductance.[68] This activation energy was taken as the highest limit to the energy barrier over which the proton passes in the classical transfer process. Quantum-mechanical tunneling was neglected, because this tunneling, if it were rate controlling, would involve a higher value for the $\lambda_{H^+}^0/\lambda_{D^+}^0$ ratio than had been found by experiment (see Section 4.1.4). For the abnormal ionic mobility, a two-step mechanism of classical proton transfer combined with water molecule rotation in the Lorentz field was assumed. The authors seem to have intended to calculate an approximate value $(1–3.5 \text{ kcal mol}^{-1})$ of the free-energy activation for the process of proton transfer from H_3O^+ to H_2O, instead of presenting a theory of anomalous conductance. For this reason the question of which of the two successive steps is rate controlling was not considered. Reasons against the classical transfer mechanism were given by Conway[7] in his earlier review (point IV.4.V in Reference 7). Thus, Conway and Salomon[147] calculated lower (4.6) and upper (6.5) limits to the isotope ratio $\lambda_{H^+}^0/\lambda_{D^+}^0$ resulting from the classical proton-transfer step. Without presenting a quantitative calculation, Wulff and Hartmann[100] suggested a structural diffusion mechanism similar to that of Huggins.

In a semiempirical treatment based on Eucken's model[148] of stoichiometric water association, Gierer and Wirtz[149] assumed activated classical proton transfers within the hydrogen-bonded ionic aggregates, one transfer in $(H_5O_2)^+$ and three simultaneous transfers in $(H_9O_4)^+$, contributing to the abnormal conductance. The pre-exponential factor can be obtained by combining the Nernst and Einstein expressions for diffusion.

The average square displacement of the transfers was called the _structure factor_ and was linearly composed of Eucken's temperature-dependent mole fractions of $(H_2O)_2$ and $(H_2O)_4$. With these as functions of temperature, the product of the exponential and pre-exponential terms could be well fitted to the experimental temperature dependence of the abnormal conductances $\lambda_{H^+}^0 - \lambda_{Na^+}^0$ and $\lambda_{OH^-}^0 - \lambda_{Cl^-}^0$, respectively, over the long range 0–579 K by a single value of the activation energy (2.44 kcal mol^{-1} for H^+ and 3.0 kcal mol^{-1} for OH^-, with the same structure factor for both). Gierer and Wirtz[149] considered a two-step mechanism of structure reorientation (decomposition of hydrogen bonds and the formation of new ones) and proton transfer;

however, their treatment allows both to be the rate-controlling step. An attempt was made to explain the pressure dependence of the abnormal conductance and the ratio $\lambda_{H^+}^0/\lambda_{D^+}^0$ as well. For some problems with the Gierer and Wirtz theory, see the review by Conway.[7]

In order to interpret anomalies occurring in the interpretation of the IR spectra of aqueous solutions, Wicke *et al.*[150] suggested a modification of Gierer and Wirtz's basic assumptions[149] on the structure of ion hydrates. They regarded $(H_9O_4)^+$ as a complex (H_3O^+, plus the first hydration shell) stable to about 443 K, surrounded by a second hydration shell of water molecules, hydrogen bonded to those in the first shell. The number of these bonds decreases with increasing temperature to zero in the lower temperature ranges (below 433 K). Within $(H_9O_4)^+$ the protons are assumed to be mobile. To explain the abnormal H^+ mobility on the basis of this hydration model Wicke *et al.*[150] and Eigen and de Maeyer[58,151] assumed (in the lower temperature range) very fast protonic charge fluctuation inside the hydrogen-bonded $(H_9O_4)^+$ complex, as in Hermans' cage[152] (see below), and considered the formation and decomposition of hydrogen bonds at the periphery of the complex as the rate-limiting process. In the higher temperature range (>433 K) decomposition of hydrogen bonds in the first hydration shell (i.e., in the complex $H_9O_4^+$) leads to a further decrease of abnormal mobility. These conclusions were based mainly on analyses of the temperature dependence of the thermal conductivity of water and electrolyte solutions[153] and of the temperature and pressure dependence of the equivalent conductance of the hydrogen ion. Some support for this view is based on the reaction distance (0.75–1.25 nm) of the recombination reaction $H_3O^+ + OH^-$, estimated from the rate constant measured by Eigen and Schoen.[154] Thus, it has been stated that, "this distance corresponds to the dimensions of the hydration sphere and indicates that the proton is mobile *within* its H-bonded hydration structure" (reference 151, p. 518).

In order to explain the experimental value of the isotopic ratio $\lambda_{H^+}^0/\lambda_{D^+}^0$ (see Section 4.1.4) Darmois and Sutra[155] treated the protons as *freely* moving particles of an ideal gas with an average kinetic energy proportional to temperature and applied classical metallic conduction equations! For the rather serious inadequacy of this theory see the comments of Conway *et al.*[7,75] Earlier Hermans[152] had suggested the application of the cage theory of liquids to free protons oscillating in cells of the liquid and making classically activated jumps between the cells (cf. the state of solvated electrons).

4.4.2. The Theory of Conway, Bockris, and Linton[75]

Rejecting H_3O^+ rotation (Hückel's mechanism) because of the equivalence of the three protons and the flatness of the ion and neglecting the possibility of simultaneous proton transfers within hydrogen-bonded ionic aggregates, Conway *et al.*[75] gave a thoroughgoing analysis of classical and quantum-mechanical (tunneling) proton transfer from H_3O^+ to H_2O, pure thermal rotation, and the rotation of water molecules accelerated by the field of the H_3O^+ ion. Classical transfer was found to be about three times, and tunnel transfer two orders of magnitude, faster than corresponds to the observed abnormal proton mobility. Thermal reorientation of water molecules is insufficiently rapid. Rotation of the water molecule hydrogen bonded to a H_3O^+ and to three other H_2O molecules (i.e., rotation of one water molecule in the complex $H_9O_4{}^+$) was proposed as the rate-determining reaction, which is preceded and succeeded by fast proton transfers.

The calculated value of the abnormal mobility of H^+ ($\lambda_{calc}^0 = 270\ \Omega^{-1}\ cm^2\ mol^{-1}$) is in good agreement with the experimental mobility ($\lambda_{H^+}^0 - \lambda_{Na^+}^0$ or $\lambda_{K^+}^0$). For the isotopic ratio $\lambda_{H^+}^0/\lambda_{D^+}^0$ the square-root mass ratio $(m_D/m_H)^{1/2} = \sqrt{2}$ was obtained, also in good agreement with the experimental value (see Section 4.1.4). In the case of the OH^- ion the rotation of the water molecule is accelerated by the average OH^- ion–dipole force, because the hydrogen-bonded $(H_7O_4)^-$ hydrate structure contains two protons fewer than the $(H_9O_4)^+$ structure, and, consequently, proton–proton repulsion forces acting during the rotation of a water molecule in $(H_9O_4)^+$ do not occur in $(H_7O_4)^-$. Hence, the ratio of the abnormal conductances $\lambda_{OH^-}^0/\lambda_{H^+}^0 = \sqrt{\frac{1}{4}} = 0.5$ was calculated, also in good agreement with experiment.

Temperature and pressure dependence of the mobility were also qualitatively explained.

In *a priori* calculations such as those of Conway, Bockris, and Linton[75] the model used is of decisive importance. It comprises the geometry of the hydrogen-bonded systems H_3O^+—H_2O, H_2O—OH^-, $(H_9O_4)^+$, and $(H_7O_4)^-$, bond lengths, bond energies, potential-energy profiles of the proton transfer and of the rotation of the water molecule, and other parameters.

The basis of the calculation was Pople's model[99] (Figure 10) of the hydrogen bond in the water structure. The distances d, d_{OH}, and the proton transfer distance $2l$ are critical. For further details of the model and calculations the reader is referred to the original paper[75] and review.[7] It should perhaps be added here that according to the calculations of Conway, Bockris, and Linton, "the time taken in the transfer process is about 10^{-14} sec, while the time an H_3O^+ ion has to 'wait' to receive a favorably oriented water molecule is about 2.4×10^{-13}," i.e., some 24 times more. This is evidence not only for the effective existence of the species H_3O^+ but also for its significant hydrodynamic contribution to the conductance, because the time necessary to accelerate the ionic sphere (H_3O^+) in the viscous medium to final velocity was estimated by Hückel[68] to be about 10^{-14} sec, i.e., a small fraction of its lifetime.

The paper of Conway, Bockris, and Linton was soon followed by Eigen and de Maeyer's important review of self-dissociation and proton charge transport in water and ice.[151] In this review the authors critically examined some of the approximations made by Conway *et al.*[75] In calculating the

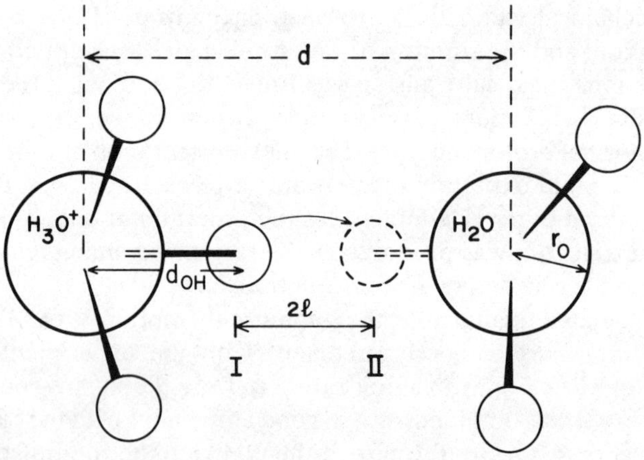

Figure 10. Model of hydrogen-bonded H_3O^+ ion for proton-transfer rate calculation (after Reference 70 and based on Pople's model for the hydrogen-bond in water[99]).

tunneling frequency of the protons, instead of a Maxwell–Boltzmann energy distribution, as used in Eqs. (12) and (17) of Reference 75, a summation of the discrete quantum states is recommended by Eigen and de Maeyer.[151] However, this does not affect the order of magnitude of the proton tunneling frequency relative to that of the structure reorientation frequency, as pointed out by Conway and Bockris[156] in their reply (see p. 118 of Reference 7). Thus the use of quantum statistics equally proves structure reorientation to be the rate-controlling step.

Despite the reasons given[7,75,156] for the potential barrier used and the proton jump distance chosen by Conway et al.,[75] there is no direct experimental or theoretical evidence for the correctness of this choice, as was pointed out by Eigen and de Maeyer.[151,157] In the light of more recent developments on the structure of water, a more rigorous treatment of the rotation kinetics is desirable, e.g., as in spin-lattice relaxation effects in NMR behavior.

4.4.3. More Recent Developments in the Experimental and Theoretical Study of Abnormal Proton Conductance

From the effect of nonelectrolytes dissolved in water on prototropic conduction, it can be concluded that the presence of isolated water molecules (Section 4.2.1) is not sufficient for the transport of electricity by this mechanism. It requires the presence of some complexes (aggregates, associated molecular or ionic species, polymers, lattice-like regions) formed from a certain number of water molecules by hydrogen bonds, through which easy transfer of protons may take place. The decrease in size of such complexes hinders prototropic conduction, and monomer water molecules are inadequate for this mechanism. This is supported by the fact that the ratio $(\lambda_{HCl}^0 - \lambda_{KCl}^0)/\lambda_{KCl}^0$, which provides some approximate empirical information on the extent of prototropic conduction, is 2.26, 1.84, and 1.07 at 273, 298, and 373 K, respectively. Thus the share of the prototropic mechanism in the transport of current is strongly diminished by the increase of temperature, which is probably due to the loosened connection between water molecules at elevated temperatures; i.e., the size† of the regions bound together by hydrogen bonds decreases, and the degree of structure in the liquid water is diminished.

With increasing pressure, however, the share of prototropic conduction increases. According to Hamann and Strauss,[78,79] the value of the above-mentioned ratio at 298 K increases from 1.85 under atmospheric pressure to 21.5 under 3000 bars. Thus pressure has a *structure-forming* effect on water, in the sense that it assists the proton exchanges by providing some of the repulsion energy needed to bring the oxygen atoms close enough together for a facile proton switch to occur. A different interpretation was given by Horne and Courant[158] and Lown and Thirsk,[159] however—namely, that "high pressure reduces the extent of H-bonded association in water, thereby facilitating the rotation of water molecules which is regarded as the rate-determining step in the proton transfer mechanism." According to Horne *et al.*,[76] who measured the specific conductances of aqueous KCl, KOH, and HCl solutions as functions of hydrostatic pressure and temperature over the ranges 278–318 K and 1–6900 bars and calculated the variation of the corresponding apparent activation energies with pressure, "at lower pressures the rotation of water molecules appears to be the rate-controlling step, but above about 1380 bars the proton jump becomes rate-determining," particularly because of the breaking of hydrogen bonds (see Reference 158).

The water structure-destroying effect of dioxane was mentioned in Section 4.2.1. Other phenomena also indicate that H_2O molecules, separated from the normal liquid structure of water, do not participate in prototropic conduction. For example, Gusev and Palei[160] and Gusev[161] investigated the effect of nonelectrolytes on the conductance of aqueous solutions of electrolytes. They have inferred that water molecules, being in the hydrate spheres of ions (i.e., to a certain extent separated from the original liquid structure of water), do not participate in the prototropic conduction mechanism.

With respect to the role of "monomeric" water molecules in a prototropic conduction mechanism, Horne and Courant[158] have drawn dissimilar conclusions from their conductance measurements in aqueous hydrochloric acid solutions in the temperature range 272–283 K. They found the apparent activation energy of proton conduction to be approximately constant, unlike the activation energies of other ions, which exhibit a maximum at the

---

† Modern molecular dynamics calculations on water properties and structure indicate[189] that two- or multistate models[162] of water structure in hydrogen-bonded aggregates are unnecessary. Discrete significant structures[19,56] are not required in the model.

temperature of maximum density. According to these authors, vacancies determining the hydrodynamic migration can be formed in both the Frank–Wen-type[162] molecular clusters and the monomeric water around them, whereas the rotation of water molecules required for prototropic conduction can occur only in the monomeric state. In general, this latter mechanism is assumed to predominate, but below 275 K the share of hydrodynamic conductance increases. However, monomeric states are unrealistic.[189,190]

The ratio of prototropic to hydrodynamic conductance depends in a complicated way on the temperature, pressure, and the concentration of the base or acid. Lown and Thirsk[159,163] measured the electric conductance of aqueous solutions of KOH, NaOH and LiOH within the ranges 298–473 K, 1–3000 bars, and 0.1–6.68 m and of orthophosphoric acid at 298 K within the ranges 1–3000 bars and 1–15.7 m. The conductance due to proton transfer is reduced progressively with increasing concentration in the 1–6.8 mol liter^{-1} range in LiOH, NaOH, and KOH solutions, and hydrodynamic migration increases considerably.

This has been explained by the fact that an increasing fraction of the water molecules becomes bound to the ions in the hydrate spheres of the ions with increasing ionic concentration, and these molecules thus become unsuitable for the conduction via the proton transfer of hydroxide ions. The prototropic conduction mechanism breaks down in KOH solutions at concentrations lower than in LiOH solutions, which can be ascribed to some ionic association in the latter case. In KOH solutions the Walden product becomes more and more independent of temperature and pressure with increasing concentration. This indicates the suppression of the prototropic mechanism. At low concentrations the conductance of phosphoric acid solutions increases with increasing pressure. At medium concentrations it decreases, and in the range 11.7–15.7 mol liter^{-1} it becomes independent of pressure. This behavior can be attributed to the fact that water molecules bound in the hydration spheres do not participate in the prototropic mechanism. Lown and Thirsk point out that both prototropic and hydrodynamic mechanisms are accompanied by the rupture of hydrogen bonds. This may explain why, somewhat surprisingly, the Debye–Hückel–Onsager theory holds to a good approximation equally well for dilute solutions of both KCl and HCl.

Roberts and Northey[65] pointed out that ions such as, for example, K^+ increase fluidity and the self-diffusion of water but reduce proton mobility. They compiled values of the orientation time of the water molecule in pure water and in diamagnetic salt solutions, as estimated by McCall and Douglass[164] and Hertz and Zeidler[165] from NMR studies. KBr and KI were found to reduce the relaxation time of orientation. The facts pointed out by Roberts and Northey are in accordance with the statement of Conway et al.,[75] that the *thermal* (as opposed to the field-induced) rotation of the water molecules cannot be the rate-determining step in abnormal conductance of the hydrogen ion.

Assuming proton transfer to be the rate-determining step in the abnormal diffusion of the H^+ ion, Ruff and Friedrich[166] calculated the diffusion coefficient from the value of Luz and Meiboom[137] for the second-order rate constant. The value obtained for the abnormal fraction of the diffusion constant turned out to be equal to the directly measured value of the total diffusion coefficient. From this, Ruff and Friedrich drew the conclusion that H_3O^+ does *not* move hydrodynamically at all. However, the basic equation of their theory[167] contains a factor due to erroneous averaging, as was shown by Lengyel.[168]

4.4.4. The Quantum–Mechanical Theory of Weidemann and Zundel[169]

In their quantum-mechanical theory Weidmann and Zundel used a model based on IR spectroscopic studies.[170–173] The kernel of the model is an ion hydrate grouping $(H_5O_2)^+$ in which the H_3O^+ ion is bound to a H_2O molecule by a hydrogen bond.† Proton transfer between them occurs and a symmetrical, double minimum potential-energy curve of the proton, similar to that of Conway *et al.*,[75] is assumed. The $(H_5O_2)^+$ grouping, on the other hand, is linked to four water molecules by hydrogen bonds. These hydrogen bonds are normally "bent" and thus hinder proton tunneling from the $(H_5O_2)^+$ to neighboring H_2O molecules. However, for short periods they will be linearized by thermal motion. During these periods two of the three protons of the H_3O^+ [in the $(H_5O_2)^+$] become equivalent, and in addition to the original transfer, a transfer to one of the H_2O molecules bound to $(H_5O_2)^+$ may occur. If, finally, the bond within the original $(H_5O_2)^+$ bends, it will be found that $(H_5O_2)^+$ has moved. This process is the structural diffusion as defined by Weidemann and Zundel. This structural diffusion is influenced by an external field that polarizes $(H_5O_2)^+$, in the sense that the tunneling (excess) proton will be found with higher probability in one of its alternative sites. The site lying in the field direction will be preferred. This increases the probability of structural diffusion in the field direction (see References 19 and 75) and results in net charge transfer.

For the symmetrical double minimum potential well in a hydrogen bond the shift of the split zero level was quantum-mechanically calculated. With the value $\nu = 10^{13}$ sec^{-1} for the tunneling frequency, and the transition dipole moment calculated using a linear dimension of 0.2 nm and a charge of $\frac{1}{2}e_0$ (Zundel[173] and Weidemann and Zundel[174]), an extremely small shift was obtained as a function of the external electrical field. Then, from this shift, the shift of the weights of the two alternative sites ("proton boundary structures") of the proton and the induced dipole moment and the polarizability

---

† Note that the $H_9O_4^+$ entity is the hydrated H^+ species favored by Ackermann *et al.*[70,71] and indicated, among other hydrates of H^+, by vapor-phase mass spectrometry (see Section III).

of the hydrogen bond were evaluated. Finally, by relating the abnormal mobility to the induced dipole moment and polarizability of the hydrogen bond, averaging over all orientations, and using the value $s = 0.25$ nm for the mean displacement by a structural diffusion step, Weidemann and Zundel obtained the value $\nu_s = 1.3 \times 10^{12}$ sec^{-1} for the number of structural diffusion steps per second if no external field was acting. This value agrees reasonably with that of Eigen, who, comparing mechanisms of proton transport in ice and water, estimated a value of about 0.3×10^{12} sec^{-1}. The conclusion is made that the frequency of the structural diffusion step is about one order of magnitude smaller than the proton tunneling frequency and that the small shift of the weights of the alternative sites of the proton in the hydrogen bond, due to the external field, is the rate-determining factor in abnormal conductance.

4.4.5. Theories Not Involving Prototropic Transfers

Perrault[175] assumed that hydrogen ion transport occurs exclusively through hydrodynamic motion. However, he considered that only a small (activated) fraction C^*/C of the ions take part in such a motion. He regarded the activation energy $E_{H^+}^{\ddagger}$ in $C^*/C = e^{-E_{H^+}^{\ddagger}/kT}$ as the desolvation energy of the proton and evaluated this, semiempirically, from the apparent activation energies of the electrical conductance and viscosity of pure water, and the ionic product and the limiting conductances of the ions H^+ and OH^-. He obtained a value of $E_{H^+}^{\ddagger} = 0.260$ eV. Consequently, the concentration C^* is $e^{-10.1} \doteq 4 \times 10^{-5}$, which is less than the stoichiometric concentration of the hydrogen ion. This very low value, of course, must be combined with a very high value of mobility in order to obtain the experimental value of the limiting specific conductance. Perrault's theory is at variance with all the critical experimental and theoretical considerations regarding anomalous proton mobility.†

In a treatment by Zatsepina[176] it was assumed that the H_3O^+ ion is *not* linked to water molecules; i.e., she assumed that *no* hydrogen bonds between the H_3O^+ ions and water molecules occur in water or in aqueous acid (or base) solutions! In Zatsepina's model[176] the supposed "hydrophobic" H_3O^+ species experience translational motion in the holes of the hydrogen-bonded structure of water, between collisions, with an average velocity v corresponding to the kinetic energy of an ideal gas molecule.

Despite the agreement between some experimental findings and the conclusions from Zatsepina's theory, there is undeniable evidence of strong interaction between the H_3O^+ ion and water (see Section 2.6). Thus it is difficult to imagine free translational motion with collisions in a condensed

---

† Proton transfer is an exchange reaction in which the proton is transferred, for example, from a H_3O^+ ion to a neighboring (colliding or encountering) water molecule [reaction equation of Eq. (16)].

phase. It is also difficult to visualize a notion in which the size of the particle and the mean-free-path length are of the same order of magnitude.

4.5. General

4.5.1. Theory of the Contribution of Transport Reactions to Overall Transport Processes

By transfer of the proton between two encountering particles, the particles change places. Consequently, a displacement of the ion and molecule (in opposite or approximately opposite directions) results. In contrast to hydrodynamic motion ("bodily transport") of the ion, displacement by proton transfer is not preconditioned by change of the translational coordinates of the surrounding particles (e.g., the solvent molecules). If the exchange reaction is sufficiently fast and frequent, the displacement by transfer may have a velocity higher than that for hydrodynamic movement.

In general, by an exchange reaction of the type

$$AX + B \rightleftarrows A + XB \tag{20}$$

the particle X (electron, proton, atom, radical, molecule, ion, or energy quantum) will frequently be transferred between two colliding (encountering) particles.

If the rate of the forward reaction

$$k_f C_{AX} C_B \tag{21}$$

differs from that of the reverse reaction

$$k_r C_A C_{XB} \tag{22}$$

then the exchange reaction may contribute to the fluxes of the transport processes in the solution. However, the additional condition of *continuous repetition* of excess transfer in one direction must be fulfilled. Thus, to make successive transfer possible, a second fast elementary process is needed. Rotation of XB, for example, may be such a process as in the water case. A consecutive mechanism of dissociation–association of ions pairs (under the influence of the external field) followed by their rotation was proposed as a general mechanism for electrolytic conduction in solutions by Grotthuss. Similar double mechanisms are usually referred to as Grotthuss *chain mechanisms*, even if they differ somewhat from the original model. (See pps. 365 and 370.)

Theories of proton-transfer contributions to transport processes in liquids were reviewed above. Electron transfer between two atoms was examined in a quantum-mechanical theory of electronic conduction by McCrea.[177] The same method of calculation was used by Bernal and Fowler[19] for proton transfer in aqueous solutions.

Ruff and Friedrich[167] generalized the theory of the contribution of exchange reactions (if rate controlling) to transport processes for transferred particles (atoms, radicals, ions, electrons, energy quanta) and for isothermal diffusion, electric conductance, and thermal diffusion. However, an erroneous factor arises in the equation connecting flux and the reaction rate.[168]

The contribution of an exchange reaction to a flux can be calculated if certain conditions are imposed:

1. The exchange reaction controls the rate of the transport process.
2. Transfer occurs only at a definite distance (close contact) of the colliding particles.
3. The collision is linear, and the displacements of the particles are parallel.
4. External and thermodynamic forces have the same or opposite directions.
5. The orientation of the direction of displacements and collision with respect to the forces is homogeneous.

In the case of liquid water containing an excess of protons as H_3O^+, the experimental and theoretical considerations indicate that condition (1) does not apply, so that the exchange reaction contribution within the flux does not determine the flux. However, in the treatment of Conway, Bockris, and Linton[75] the flux of protons in a hydrogen-bonded element of the water structure is determined in a *coupled* way by the local structure reorganization event, caused by the arrival of a proton (and the consequent presentation of a local field) at an otherwise unfavorably oriented water molecule in the hydrogen-bonded liquid lattice.

In ice, on the other hand, owing to the much lower concentration of proton defects, evidence from the experimental conductance and the charge-carrier concentration leads to the conclusion[178] that it is the actual charge-transfer process that is rate controlling and characterizes the observed conductance behavior. A full discussion of the mechanism of proton conductivity in ice is given in References 75 and 151 and the rotational process was treated by Gränicher et al.[179]

5. Proton Transfer in Electrode Processes

The previous sections of this chapter have been concerned with homogeneous proton-transfer events and the state of H^+ homogeneously in solution. An important class of proton-tranfer events arises *heterogeneously* in electrode processes involving H_2 evolution and ionization, electrochemical hydrogenation, and underpotential deposition and ionization of adsorbed H.

The discharge of solvated protons in various media in the cathodic H_2 evolution reaction, or in electrochemical hydrogenation, is one of the most elementary and fundamental processes in electrode kinetics and has constituted a model reaction for consideration of the mechanism and molecular mechanics of electron- and atom-transfer types of redox reaction at electrode interfaces.

The rate of discharge of a proton at an electrode (cathode) interface bears many similarities to proton transfer in a Brønsted acid–base reaction. The electrode metal surface, at which the discharged H^+ is chemisorbed, can be regarded as the center of variable and controllable base strength (determined by the electrode surface charge or the potential relative to the potential of zero charge), and, in fact, at liquid metals the excess charge per atom of the surface is a measurable thermodynamic quantity derivable from electrocapillary or integrated capacitance measurements. By varying the potential by ΔV V, the Fermi level is changed by ΔVF in energy, making electrons more or less available for the proton transfer and neutralization in the electrochemical discharge process

$$H_3O^+ + M + e_M \text{—} H_2O + MH_{ads} \tag{23}$$

or, in alkaline solution,

$$H_2O + M + e_M \text{—} OH^- + MH_{ads} \tag{24}$$

In some cases electrochemical proton transfer from H_3O^+ or H_2O may go directly into the metal, e.g., as at Pd, rather than onto a site at the metal surface. Then the final state of the H_3O^+ discharge event is a "H" entity that retains a partial charge. The discharge event is then effectively a phase transfer of a proton from a solvent environment to a metallic one.

The dependence of the free energy of activation $\Delta G^{\ddagger}$ of these processes on the potential change ΔV provides one of the best examples of "linear free-energy relations" of the Brønsted type in physical chemistry. Thus changes of $\Delta G^{\ddagger}$ and $\delta \Delta G^{\ddagger}$ are equal to $\beta \Delta VF$, where F is the Faraday constant and β is an electrochemical Brønsted coefficient, usually constant and close to 0.5 over a wide range of ΔV; a similar value, near 0.5, is also found for homogeneous acid–base proton-transfer kinetics and equilibria (but with a rather lower constancy of the Brønsted factor).

The relation between homogeneous acid–base Brønsted relations and the electrochemical case was noted originally by Brønsted and Ross-Kane[180], and treated further by Frumkin[181] and Conway.[182] Thus increasing cathodic potential at an electrode effectively increases the base character or electron donicity of the electrode surface. This is equivalent to raising the initial-state potential-energy curve for the proton-transfer process at the electrode upwards. A similar effect arises in a Brønsted acid–base reaction with increasing base strength of the donor component. The electrochemical charge-transfer or symmetry factor β, which has a value of ~ 0.5, is thus closely analogous to the Brønsted factor "α" for homogeneous proton-transfer processes.

Thus, information on the state of the proton as "SH^+" in a solvent S is of great importance not only in the treatment of the kinetics and molecular mechanism of homogeneous acid–base proton transfer but also in electrode reactions in which proton discharge or ionization takes place (H_2 evolution, electrochemical H penetration into metals, electrochemical hydrogenation, and fuel-cell ionization of H_2).

Referring to the energy profile diagrams of Figure 11, it should be mentioned that, in the case of electrochemical processes, change of the state of the reacting H_3O^+ particle, e.g., by change of solvation energy, does *not* change the energy of activation as it would in a regular chemical reaction according to the Brønsted representation, because a compensating effect arises on account of an exactly equivalent change of the metal/solution potential difference. Both these effects tend to change the activation energy by α times the energy change of the initial state, but to equal and opposite extents. The "electrochemical" activation energy is therefore independent of the solvational state of the proton in the initial state of the reaction in the case of an electrochemical proton-transfer event, but the "chemical" component of this activation energy changes in the usual way, according to Brønsted's principle.

Comparison of the kinetics of H^+ discharge from unhydrated H_3O^+ in pure liquid $CF_3SO_3^-H_3O^+$ (mp = 307 K) and from H_{aq}^+ as "$H_9O_4^+$" in 1.0 M aqueous $CF_3SO_3^-H^+{}_{aq}$ was recently made by Conway and Tessier:[183] Discharge from H_3O^+ itself in the acid "hydrate" is some 50 times slower than that from 1 M $H^+{}_{aq}$ and, when the comparison is corrected for concentration differences (pure liquid $CF_3SO_3^-H_3O^+$ is ~10 M), the rate of discharge for H_3O^+ is found to be ~450 times slower. Activation energies and H–D kinetic

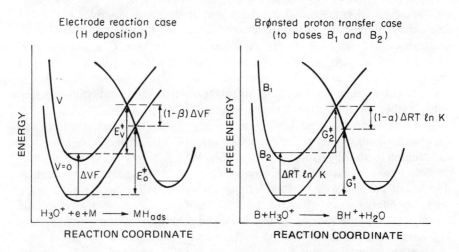

Figure 11. Comparison of the effect of an electrode potential change ΔV on the potential-energy profile and activation energy for a heterogeneous electrochemical proton-transfer process with chemisorption, with the effect of change of base strength $\Delta RT \ln K$, in a homogeneous Brønsted acid–base proton-transfer process in terms of a free-energy profile diagram. Note the relation between the electrochemical symmetry or transfer coefficient β and the Brønsted factor α.

isotope effects are also appreciably different. The observed kinetic differences were attributable to the different transition states involved in proton discharge and H transfer from H_3O^+ and $H_9O_4^+$ (with excess H_2O) and the associated solvent vibrational modes in the solvation of H^+ in the latter case, allowing for anion proximity in the case of discharge from $CF_3SO_3^-H_3O^+$.

The molecular mechanism of proton discharge, electron transfer, and H atom transfer to high overpotential metal surfaces was treated in detail by Dogonadze, Kuznetsov and Levich[184] in an important paper in 1968. They took into account the quantized vibrations of OH bonds in H_3O^+ (or H_2O). Noting that for OH frequencies of $\sim 3300\ cm^{-1}$ the associated quantum $h\nu$ is $\sim 420\ kJ\ mol^{-1}$, i.e., very much greater than kT, they concluded that OH bond activation would be negligible at ordinary temperatures.

They proposed a "double" adiabatic activation mechanism involving electron and H atom transfer in which outer-sphere activation, through fluctuations in solvent vibrations, influenced both the proton- and electron-transfer events, which, of course, must be coupled.

It is difficult to see, in this theory, why the reacting "OH" bonds are considered not to be directly activatable, as "high-energy" bonds can be in gas-phase reactions. The "high-energy" end of the Boltzmann distribution is always available for activation, with a probability $\exp(-\Delta H^\ddagger / RT)$, and it is the large $\Delta H^\ddagger$ ($\sim 840\ kJ\ mol^{-1}$ for discharge at Hg) that makes the proton discharge exchange current density so low ($\sim 10^{-13}\ A\ cm^{-2}$) at Hg at 298 K. Thus, calculations show[191] that the fraction of O—H bonds which would have to be activated to give the order of magnitude of typical electrodic currents is given by the Boltzmann factor, utilizing the parameters appropriate to give electrodes. Correpondingly, Bockris and Matthews[192] showed that a variation of the tritium–hydrogen separation factor occurs with overpotential on metals such as mercury, and that this fact may be quantitatively interpreted by taking into account the easier tunneling of the lighter isotope. However, there must then be protons available at various energies corresponding to the penetration of the proton of a barrier appropriate state.

Further, the known, unusually large heat capacity of H_2O or partial molar heat capacity of H^+_{aq} (as $H_9O_4^+$; see Ackermann et al.[70,71]) must mean that liquid lattice modes are available for these species at quantal energies much less than those corresponding to $\nu \simeq 3300\ cm^{-1}$. Indeed, a broad range of coupled vibrational/intermolecular and lower-frequency librational modes characterize the IR and Raman spectra of liquid water. With such a system of polyatomic vibrations, it is easy to conceive how energy can be channeled from closely coupled, solution lattice modes involving hydrogen-bonded H_3O^+ into the dissociating reacting OH bond from which H transfer occurs, as in the theory of unimolecular reactions[185] involving polyatomic molecules. It should be mentioned that under certain conditions at noble metals such as Pt, Rh, and Ir, and under special conditions at Ru and Pd, which adsorb as well as absorb H, the *individual stage* of proton transfer and chemisorption [reaction equations of

Eqs. (23) and (24)] can be studied separately by means of charging curves or from the pseudocapacitance measured in a linear potential-sweep experiment. At the above metals, chemisorbed H can be deposited at potentials *positive* to the reversible potential for the process $2H_3O^+ + 2e \rightleftarrows H_2 + 2HO$, in which molecular H_2 is evolved, so that the steps of Eq. (23) or (24) can be *individually* studied. Both the kinetics (from the s_0 value[186]) of these H chemisorption processes arising from proton transfer and the state, e.g., volume and entropy, of the deposited H can be evaluated as described in References 187 and 188.

Acknowledgment

The authors of this article, especially S. L., gratefully recognize the important work of the late Professor T. Erdey-Gruz in this field and his earlier contributions, on which parts of this review are based.

References

1. J. N. Brønsted, *Z. Phys. Chem.* **108**, 185 (1924); *J. Am. Chem. Soc.* **49**, 2654 (1927); **52**, 1394 (1930); *Trans. Faraday Soc.* **24**, 630 (1928); **25**, 59 (1929); *Chem. Rev.* **5**, 231 (1928).
2. R. P. Bell, e.g., see *The Proton in Chemistry*, Cornell University Press, Ithaca, New York (1959); see also W. J. Albery, in *Progress in Reaction Kinetics*, G. Porter, ed., Pergammon Press, Oxford (1967), Chap. 9, p. 353.
3. H. Goldschmidt, *Z. Elektrochem.* **15**, 4 (1909); **20**, 475 (1914); *Z. Phys. Chem.* **89**, 131 (1914).
4. P. Drude and W. Nernst, *Z. Phys. Chem.* **15**, 77 (1894).
5. M. Born, *Z. Phys.* **1**, 45 (1920).
6. H. Goldschmidt and O. Udby, *Z. Phys. Chem.* **60**, 728 (1907).
7. B. E. Conway, *Modern Aspects of Electrochemistry*, Vol. 3, J. O'M. Bockris and B. E. Conway, eds., Butterworths, London (1964), Chap. 2, p. 43.
8. M. Bagster and T. M. Cooling, *J. Chem. Soc.* 693 (1920).
9. M. Volmer, *Ann. Chem. (Leipzig)* **440**, 200 (1924).
10. R. Richards and M. Smith, *Trans. Faraday Soc.* **47**, 1261 (1951).
11. S. Kakiuchi, I. Shono, J. Matsu, and K. J. Kigoshi, *J. Phys. Soc. Jpn.* **7**, 102 (1952); *J. Chem. Phys.* **19**, 1009 (1951).
12. E. R. Andrew and R. Bersohn, *J. Chem. Phys.* **18**, 159 (1950).
13. W. F. Giauque and J. Wiebe, *J. Am. Chem. Soc.* **50**, 101 (1928); **51**, 1441 (1929).
14. R. Beckey, *Proceedings of the Fourth International Congress on Electron Microscopy*, Springer, Berlin (1958); *Z. Naturforsch* **15a**, 822 (1960).
15. P. Kebarle and E. W. Godbole, *J. Chem. Phys.* **39**, 1131 (1963); see also B. Knewstubb and A. Tickner, *ibid.* **38**, 464 (1963).
16. M. Falk and P. A. Giguère, *Can. J. Chem.* **35**, 1195 (1957); for a recent discussion on the state of H_3O^+ see P. A. Giguère, *J. Chem. Educ.* **56**, 571 (1979).
17. G. Bethell and N. Sheppard, *J. Chim. Phys.* **50**, C72, C118 (1953).
18. A. Ferriso and M. Hornig, *J. Am. Chem. Soc.* **75**, 4113 (1953).
19. J. D. Bernal and R. H. Fowler, *J. Chem. Phys.* **1**, 515 (1933).
20. B. E. Conway, J. O'M. Bockris, and H. Linton, *J. Chem. Phys.* **24**, 834 (1956).
21. T. Suhrmann and M. Breyer, *Z. Phys. Chem.* **B23**, 193 (1933).
22. T. Suhrmann and M. Wiedersich, *Z. Elektrochem.* **57**, 93 (1950); *Z. Anorg. Chem.* **273**, 166 (1953).

23. (a) E. Wicke, M. Eigen, and T. Ackermann, *Z. Phys. Chem. N.F.* **1**, 340 (1954); (b) T. Ackermann, *Discuss. Faraday Soc.* **24**, 180 (1957).
24. Y. Gantmacher, *Acta Physicochim. URSS* **12**, 786 (1940).
25. M. Fajans and I. Joos, *Z. Phys. Chem.* **23**, 1, 31 (1924).
26. M. Grahn, *Ark. Fys.* **21**, 81 (1962); **19**, 147 (1961).
27. I. McLean and K. Mackor, *J. Chem. Phys.* **34**, 2207 (1961).
28. K. Bonner and B. Hofstadter, *J. Chem. Phys.* **6**, 531 (1938); **6**, 534 (1938); **6**, 540 (1938); see also R. Davies and J. Sutherland, *J. Chem. Phys.* **6**, 755 (1938).
29. S. Luzzati, *C.R. Acad. Sci. Paris* **230**, 101 (1950); *Acta Crystallogr.* **4**, 239 (1951); **6**, 157 (1953).
30. M. Grahn, *Ark. Fys.* **21**, 1 (1962); **21**, 13 (1962).
31. A. Yoon and S. Carpenter, *Acta Crystallogr.* **12**, 17 (1959).
32. A. Ferriso and M. Hornig, *J. Chem. Phys.* **23**, 1464 (1955); see also J. V. Lundgren and J. W. Williams, *J. Chem. Phys.* **58**, 788 (1973); for more recent neutron diffraction information on H_3O^+ structure.
33. K. Sherman, *Chem. Rev.* **11**, 98 (1932); see also P. Grimm. *Handb. Phys.* **27**, 518 (1924).
34. D. M. Bishop, *J. Chem. Phys.* **43**, 4453 (1965).
35. I. Sokolov, *Hydrogen Bonding, Symposium, Ljubljana*, P. Hadzi, ed., Pergamon Press, Oxford (1959), p. 402; see also Y. Kandratyev and I. Sokolov, *Zh. fiz. Khim.*, **29**, 1265 (1955).
36. Y. Tal'rose and M. Frankevich, *Dokl. Akad. Nauk SSSR* **111**, 376 (1956).
37. F. H. Halliwell and N. C. Nyburg, *Trans. Faraday Soc.* **59**, 1126 (1963).
38. R. M. Noyes, *J. Am. Chem. Soc.* **86**, 971 (1964).
39. A. D. Buckingham, *Discuss. Faraday Soc.* **24**, 151 (1957).
40. B. E. Conway and M. Salomon, *Chemical Physics of Ionic Solutions*, B. E. Conway and R. G. Barradas, eds., Wiley, New York (1956), Chap. 24.
41. E. J. W. Verwey, *Recl. Trav. Chim. Pays. Bas.* **61**, 127 (1942); **60**, 887 (1941).
42. M. M. Lister, S. C. Nyburg, and R. B. Poyntz, *J. Chem. Soc. Faraday Trans. 1* **70**, 685 (1974).
43. M. Passynskii, *Acta Physicochim. URSS* **8**, 835 (1938).
44. T. Ulich, *Z. Elektrochem.* **36**, 497 (1930); *Z. Phys. Chem.* **108**, 141 (1934).
45. M. Eigen and E. Wicke, *Z. Elektrochem.* **55**, 534 (1951).
46. A. Eucken. *Z. Elektrochem.* **51**, 6 (1948); *Nachrichten Ges. Wiss. Göttingen Math. Phys. Kl.* 33 (1947).
47. E.g., M. Eigen and L. de Maeyer, *Proc. R. Soc. London* **A247**, 505 (1958).
48. P. Noonan and V. K. LaMer, *J. Phys. Chem.* **43**, 247 (1939).
49. E. A. Braude and E. Stern, *J. Chem. Soc.* 1976 (1948).
50. C. F. Wells, *Nature* **196**, 770 (1962); **201**, 606 (1964).
51. C. F. Wells, *Discuss. Faraday Soc.* **29**, 219 (1960).
52. L. Salomaa, *Acta Chem. Scand.* **11**, 125 (1957).
53. C. F. Wells, *Trans. Faraday Soc.* **61**, 2194 (1965); **62**, 2815 (1966).
54. K. J. Laidler, *Trans. Faraday Soc.* **55**, 1725 (1959); **55**, 1734 (1959).
55. D. H. Everett and W. F. K. Wynne-Jones, *Proc. R. Soc. London* **A177**, 499 (1940); *Trans. Faraday Soc.* **37**, 373 (1941).
56. G. Nemethy and H. A. Scheraga, *J. Phys. Chem.* **66**, 1773 (1962).
57. S. N. Vinogradov and R. H. Linnell, *Hydrogen Bonding*, Van Nostrand, New York (1971), Chap. 10.
58. M. Eigen, *Z. Angew. Chem.* **75**, 489 (1963); *Angew Chem. Int. Ed. Engl.* **3**, 1 (1964).
59. E. S. Amis, *Solvent Effects on Reaction Rates and Mechanism*, Academic Press, New York (1966).
60. G. C. Pimental and A. L. McClellan, *Annu. Rev. Phys. Chem.* **22**, 347 (1971).
61. *Handbook of Chemistry and Physics*, 54th ed., Chemical Rubber Co., Cleveland, Ohio (1973), p. D132.
62. V. M. Goldschmidt, *Skr. Nor. Vidensk. Akad. Oslo 1* **7**, (1926).

63. L. Pauling, *J. Am. Chem. Soc.* **49**, 765 (1927).
64. R. Zwanzig, *J. Chem. Phys.* **38**, 1603 (1963); **38**, 1605 (1963); cf. R. H. Boyd, *J. Chem. Phys.* **35**, 1281 (1961).
65. N. K. Roberts and H. L. Northey, *J. Chem. Soc. Faraday Trans. 1* **70**, 253 (1974).
66. H. Danneel, *Z. Elektrochem.* **11**, 125 (1905).
67. R. Lorenz, *Z. Phys. Chem.* **73**, 252 (1910).
68. E. Hückel, *Z. Elektrochem.* **34**, 546 (1928).
69. J. Tamas, S. Lengyel, and J. Giber, *Acta Chim. Acad. Sci. Hung.* **38**, 225 (1963).
70. T. Ackermann, *Discuss. Faraday Soc.* **27**, 180 (1957).
71. E. Wicke, M. Eigen, and T. Ackermann, *Z. Phys. Chem. N.F.* **1**, 340 (1954).
72. E. Högfeldt, *Acta Chem. Scand.* **14**, 1597 (1960).
73. J. D. Bernal and R. H. Fowler, *J. Chem. Phys.* **1**, 515 (1933).
74. A. E. Stearn and H. Eyring, *J. Chem. Phys.* **5**, 113 (1937).
75. B. E. Conway, J. O'M. Bockris, and H. Linton, *J. Chem. Phys.* **24**, 834 (1956).
76. R. A. Horne, B. R. Myers, and G. R. Frysinger, *J. Chem. Phys.* **39**, 2666 (1963).
77. A. J. Ellis, *J. Chem. Soc. London* **1959**, 3689 (1959).
78. S. D. Hamann, *Physico-Chemical Effects of Pressure*, Butterworths, London (1957).
79. S. D. Hamann and W. Strauss, *Trans. Faraday Soc.* **51**, 1684 (1955).
80. S. Glasstone, K. Laidler, and H. Eyring, *The Theory of Rate Processes*, McGraw-Hill, New York (1941), p. 570.
81. R. Zisman, *Phys. Rev.* **39**, 151 (1932).
82. S. Taumann and M. Tofante, *Z. Anorg. Chem.* **182**, 353 (1929).
83. G. N. Lewis and T. C. Doody, *J. Am. Chem. Soc.* **55**, 3504 (1933).
84. N. K. Roberts and H. L. Northey, *J. Chem. Soc. Faraday Trans. 1* **68**, 1528 (1972).
85. C. Dempwolff, *Phys. Z.* **5**, 637 (1904).
86. S. Tijmstra, *Z. Phys. Chem.* **49**, 345 (1904).
87. A. Hantzsch and K. S. Caldwell, *Z. Phys. Chem.* **58**, 575 (1907).
88. V. G. Levich, Present state of the theory of oxidation–reduction (bulk and electrode reactions), in *Advances in Electrochemistry and Electrochemical Engineering*, Vol. 4, P. Delahay, ed., Wiley-Interscience, New York (1961), pp. 249–371.
89. I. Ruff, V. J. Friedrich, K. Demeter, and C. Csillag, *J. Phys. Chem.* **75**, 3303 (1971).
90. I. Ruff and M. Zimonyi, *Electrochim. Acta* **18**, 515 (1973).
91. I. Ruff, M. Zimonyi, and P. Kovács, private communication.
92. I. Ruff, A. Hegedüs, and V. J. Friedrich, *Magy. Kem. Foly.* **79**, 133 (1973).
93. I. Ruff, V. J. Friedrich, and K. Csillag, *J. Phys. Chem.* **76**, 162 (1972).
94. I. Ruff and V. J. Friedrich, *J. Phys. Chem.* **76**, 2957 (1972).
95. Yu. Ya. Fialkov and Yu. A. Karapetyan, *Ukr. Khim. Zh. Russ. Ed.* **37**, 398 (1971).
96. V. P. Basov, Yu. A. Karapetyan, and A. D. Krysenko, *Elektrokhimiya* **10**, 420 (1971).
97. M. L. Huggins, Undergraduate thesis (1919).
98. H. G. Frank and M. Evans, *J. Chem. Phys.* **13**, 507 (1945).
99. J. Pople, *Proc. R. Soc. London* **A205**, 177 (1951).
100. P. Wulff and H. Hartmann, *Z. Electrochem.* **47**, 858 (1941).
101. H. Goldschmidt and P. Dahl, *Z. Phys. Chem.* **108**, 121 (1924); **114**, 1 (1925).
102. P. Walden, *Z. Phys. Chem.* **108**, 341 (1924).
103. L. Thomas and E. Marum, *Z. Phys. Chem.* **143**, 191 (1929).
104. I. I. Berman and F. H. Verhock, *J. Am. Chem. Soc.* **67**, 1330 (1945).
105. G. Kortüm and H. Wilski, *Z. Phys. Chem. N.F.* **2**, 256 (1954).
106. G. F. Dnieprov, *Chem. Zap. LGU* **40**, 19 (1939).
107. A. M. el Aggan, D. C. Bradley, and W. Wardlaw, *J. Chem. Soc.* **1958**, 2092 (1958).
108. A. R. Tourky and S. Z. Mikhail, *Egypt. J. Chem.* **1**, 1, 13, 187 (1958).
109. P. Kirby and O. Maass, *Can. J. Chem.* **36**, 456 (1958).
110. L. A. Woolf, *J. Phys. Chem.* **64**, 500 (1960).

111. F. Accascina, A. d'Aprano, and M. Goffredi, *Ric. Sci. Parte 2* **34**, 443 (1964).
112. F. Accascina, R. de Lisi, and M. Goffredi, *Electrochim. Acta* **15**, 1209 (1970).
113. H. R. Tourky, S. Z. Mikhail, and A. A. Abdel Hamid, *Z. Phys.* **252**, 289 (1973).
114. A. A. Abdel Hamid, M. F. Ragaii, and I. Z. Slim, *Z. Phys. Chem.* **254**, 1 (1973).
115. R. de Lisi and M. Goffredi, *J. Chem. Soc. Faraday Trans 1* **72**, 787 (1974).
116. T. Erdey-Grúz, E. Kugler, K. Balthazár-Vass, and I. Nagy-Czakó, *Acta Chim. Acad. Sci. Hung.* **79**, 169 (1973).
117. T. Erdey-Grúz, E. Kugler, and L. Majthényi, *Electrochim. Acta* **13**, 947 (1968).
118. T. Erdey-Grúz and I. Nagy-Czakó, *Acta Chim. Acad. Sci. Hung.* **67**, 283 (1971).
119. T. Erdey-Grúz, *Transport Phenomena in Aqueous Solutions*, A. Hilger, London (1974).
120. L. A. Woolf, *J. Phys. Chem.* **60**, 481 (1960).
121. D. Glietenberg, A. Kutschker, and M. Stackelberg, *Ber. Bunsenges. Phys. Chem.* **72**, 562 (1968).
122. Z. Szabó, *Magy. Kem. Foly.* **40**, 52 (1935).
123. H. S. Harned and E. C. Dreby, *J. Am. Chem. Soc.* **61**, 3113 (1939).
124. S. Lengyel, J. Giber, and J. Tamás, *Acta Chim. Acad. Sci. Hung.* **32**, 429 (1962).
125. J. Giber, S. Lengyel, J. Tamás, and P. Tahi, *Magy. Kem. Foly.* **66**, 170 (1960).
126. S. Lengyel, in *Electrolytes*, B. Pesce, ed., Pergamon Press, London (1962), p. 208.
127. E. A. Kaimakov and V. B. Fiks, *Zh. Fiz. Khim.* **35**, 1777 (1961).
128. B. P. Konstantinov and V. P. Troshin, *Zh. Prikl. Khim.* (Moscow) **35**, 2420 (1962).
129. B. P. Konstantinov and V. P. Troshin, *Izv. Akad. Nauk SSSR*, **22**, 2104 (1966).
130. S. Lengyel, J. Giber, Gy. Beke, and A. Vértes, *Acta Chim. Acad. Sci. Hung.* **39**, 357 (1963).
131. V. P. Troshin and E. V. Zvyagina, *Elektrokhimiya* **8**, 505 (1972).
132. V. P. Troshin and E. V. Zvyagina, *Elektrokhimiya* **8**, 1712 (1972).
133. V. P. Troshin and E. V. Zvyagina, *Elektrokhimiya* **8**, 586 (1972).
134. S. Meiboom, *J. Chem. Phys.* **34**, 375 (1961).
135. Z. Luz and S. Meiboom, *J. Chem. Phys.* **39**, 366 (1963).
136. Z. Luz and S. Meiboom, *J. Am. Chem. Soc.* **86**, 4766 (1964).
137. Z. Luz and S. Meiboom, *J. Am. Chem. Soc.* **86**, 4768 (1964).
138. M. Anbar, A. Loewenstein, and S. Meiboom, *J. Am. Chem. Soc.* **80**, 2630 (1958).
139. Z. Luz, D. Gill, and S. Meiboom, *J. Chem. Phys.* **30**, 1540 (1959).
140. E. Grunwald, C. F. Jumper, and S. Meiboom, *J. Am. Chem. Soc.* **84**, 4664 (1962).
141. A. Löwenstein and A. Szöke, *J. Am. Chem. Soc.* **84**, 1151 (1962).
142. R. E. Glick and K. C. Tewari, *J. Chem. Phys.* **44**, 546 (1966).
143. S. W. Rabideau and H. G. Hecht, *J. Chem. Phys.* **47**, 544 (1967).
144. G. Wannier, *Annalen der Physik* **24**, 545, 569 (1935).
145. M. L. Huggins, *J. Phys. Chem.* **40**, 723 (1936).
146. M. L. Huggins, *J. Am. Chem. Soc.* **53**, 3190 (1931).
147. B. E. Conway and M. Salomon, in *Chemical Physics of Ionic Solutions*, B. E. Conway and R. G. Barradas, eds., Wiley, New York (1966), p. 541.
148. A. Eucken, *Gött. Nachr. Math. Phys. Kl.* **1946**, 39 (1946); *Z. Elektrochem.* **52**, 255 (1948).
149. A. Gierer and K. Wirtz, *Annalen der Physik* **6**, 257 (1949).
150. E. Wicke, M. Eigen, and T. Ackermann, *Z. Phys. Chem. N.F.* **1**, 340 (1954).
151. M. Eigen and L. De Maeyer, *Proc. R. Soc. London A* **247**, 505 (1958).
152. J. J. Hermans, *Recl. Trav. Chim. Pays Bas* **58**, 917 (1939).
153. M. Eigen and J. Schoen, *Z. Elektrochem.* **59**, 483 (1955).
154. M. Eigen and J. Schoen, *Z. Elektrochem.* **59**, 483 (1955).
155. E. Darmois and M. Sutra, *C. R. Acad. Sci. Paris* **222**, 1286 (1946).
156. B. E. Conway and J. O'M. Bockris, *J. Chem. Phys.* **31**, 1133 (1959).
157. M. Eigen and L. De Maeyer, *J. Chem. Phys.* **31**, 1134 (1959).
158. R. A. Horne and R. A. Courant, *J. Phys. Chem.* **69**, 2224 (1965).
159. D. A. Lown and H. R. Thirsk, *Trans. Faraday Soc.* **67**, 132 (1971).

160. N. I. Gusev and P. N. Palei, *Zh. Fiz. Khim.* **45**, 1164 (1971).
161. N. I. Gusev, *Zh. Fiz. Khim.* **45**, 1164, 2238, 2243 (1971).
162. H. S. Frank and W.-Y. Wen, *Discuss. Faraday Soc.* **24**, 133 (1957).
163. D. A. Lown and H. R. Thirsk, *Trans. Faraday Soc.* **67**, 149 (1971).
164. D. W. McCall and D. C. Douglass, *J. Phys. Chem.* **69**, 2001 (1965).
165. H. G. Hertz and M. D. Zeidler, *Ber. Bunsenges. Phys. Chem.* **67**, 774 (1963); **68**, 621 (1964).
166. I. Ruff and V. Friedrich, *J. Phys. Chem.* **76**, 2954 (1972).
167. I. Ruff and V. Friedrich, *J. Phys. Chem.* **75**, 3297 (1971).
168. S. Lengyel, *Magy. Kém. Foly.* **80**, 187 (1974).
169. E. G. Weidemann and G. Zundel, *Z. Naturforsch.* **25a**, 627 (1970).
170. G. Zundel and H. Metzger, *Z. Phys. Chem. N.F.* **58**, 225 (1968).
171. G. Zundel and H. Metzger, *Z. Phys. Chem. N.F.* **59**, 225 (1968).
172. G. Zundel and H. Metzger, *Z. Naturforsch.* **22a**, 1412 (1967).
173. G. Zundel, *Hydration and Intermolecular Interaction*, Academic Press, New York (1969).
174. E. G. Weidemann and G. Zundel, *Z. Phys.* **198**, 288 (1967).
175. G. Perrault, *C. R. Acad. Sci. Paris* **252**, 3779 (1961).
176. G. N. Zatsepina, *Zh. Strukt. Khim.* **10**, 211 (1969).
177. W. H. McCrea, *Proc. Cambridge Philos. Soc.* **24**, 438 (1928).
178. B. E. Conway and J. O'M. Bockris, *J. Chem. Phys.* **28**, 354 (1958).
179. S. Gränicher, J. Paccard, P. Scherrer, and H. Steinemann, *Discuss. Faraday Soc.* **23**, 50 (1957); *Proc. R. Soc. London* **A247**, 453 (1958).
180. J. N. Brønsted and N. L. Ross-Kane, *J. Am. Chem. Soc.* **53**, 3024 (1931).
181. A. N. Frumkin, *Z. Phys. Chem.* **160**, 116 (1932).
182. B. E. Conway, in *Progress in Reaction Kinetics*, G. Porter, ed., Pergamon Press, Oxford, (1967), Chap. 10, p. 399.
183. B. E. Conway and D. Tessier, *Int. J. Chem. Kinet.* Laidler Festschrift Volume, **13**, 925 (1981).
184. R. R. Dogonadze, A. M. Kuznetzov, and V. Levich, *Electrochim. Acta.* **13**, 1025 (1968).
185. N. B. Slater, *Proc. Cambridge Philos. Soc.* **35**, 56 (1939); *Proc. R. Soc. London* **A194**, 112 (1948).
186. H. Angerstein-Kozlowska and B. E. Conway, *J. Electroanal. Chem.* **95**, 1 (1979).
187. B. E. Conway and J. C. Currie, *J. Chem. Soc. Faraday Trans. 1*, **74**, 1390 (1978).
188. B. E. Conway, H. Angerstein-Kozlowska, and W. B. A. Sharp, *J. Chem. Soc. Faraday Trans. 1*, **74**, 1373 (1978).
189. F. H. Stillinger and A. Rahman, *J. Chem.Phys.* **60**, 1545 (1974); and *J. Chem. Phys.* **61**, 4937 (1974).
190. B. E. Conway, *Ionic Hydration in Chemistry and Biophysics*, Elsevier Publishing Co., Amsterdam (1981).
191. J. O'M. Bockris and R. K. Sen, *J. Res. Inst.Catalysis, Hokkaido Univ.* **21**, 55 (1973).
192. J. O'M. Bockris and D. Matthews, *J. Chem. Phys.* **44**, 479 (1966).

5

Structure and Thermodynamics of Molten Salts

G. N. PAPATHEODOROU

1. Introduction

Studies of molten inorganic salts have been motivated for a long time by a practical interest in their thermodynamic and electrochemical properties, and, recently, in their potential uses in various systems related to new energy sources and energy conservation systems. Old and newly developed industrial processes utilize molten salts as solvents for electrolysis in metal production.[1] Molten salts have also been used as catalysts in coal-liquefaction processes[2] and as reaction media in the preparation of chemicals.[3] Other applications include uses in high-temperature molten-salt batteries[4] and fuel cells and as media for thermal-energy storage in both solar-energy and nuclear-fusion systems.[5,6]

From a fundamental point of view, molten salts provide an important testing ground for theories of liquids, solutions, and plasmas. The understanding of liquids on the microscopic level has benefited from both experimental (e.g., neutron scattering) and theoretical (e.g., computer simulation) studies on molten salts.[7]

The early research on molten salts consisted of investigations of phase diagrams and measurements of melting points and heats of fusion of single-component salts. Systematic investigations in the field started in the 1940s

G. N. PAPATHEODOROU • Chemical Engineering Division, Argonne National Laboratory, 9700 South Cass Avenue, Argonne, Illinois 60439. Present address, Department of Chemical Engineering, University of Patras, Patras, Greece.

and peaked in the mid-1960s. Several monographs appeared during that period regarding various aspects of the field.[8–13] More recent developments have been summarized in books,[14,15] a series of National Bureau of Standards publications,[16] and certain review articles.[17–20]

In the present chapter the structural and thermodynamic aspects of single-component molten salts and multicomponent salt mixtures are outlined. The purpose of the article is to serve as an introduction to the field and to present the more recent characteristic developments, results, and views in the field. Although numerous references are given, the article cannot be considered as an extensive review of the subject matter. Exceptions are Sections 3.1 and 4.2 regarding neutron and Raman scattering from melts, where most of the available information (up to autumn 1979) is included.

2. Molten Salts as Liquids

From a physicochemical point of view, molten salts are a class of liquids having many microscopic and macroscopic properties similar to the corresponding properties of other (molecular, atomic) liquids. In a simplified way, molten salts have been described as liquids composed of positive and negative ions that interact mainly via the strong (repulsive or attractive) long-range Coulomb potential. The experimental and theoretical evidence accumulated during the past 20–30 years, however, shows that, with the exception of a few cases, molten salts show individual or group peculiarities which complicate the understanding of their morphology. The difficulties in understanding the nature, properties, and structure of molten salts reflect our inability to understand fully the liquid state in general. These difficulties are made apparent in the following comparison of the liquid state with the somewhat simplified concepts of the gaseous and crystalline states.

For a gas at high temperature and low density the interactions between the molecules or atoms are considered negligible, and the internal energy of the gas is determined by the vibrational, rotational, translational, and electronic energies of each particle. Thus a partition function can easily be computed, and the thermodynamic and transport properties of the gas can be derived by the methods of statistical mechanics and the kinetic theory of gases.

For a crystalline solid the molecules vibrate about an equilibrium position and have negligible translational energy. The energy and partition function for the crystal can be formulated by knowing the translational symmetry of the solid, and the fundamental properties (e.g., Einstein's specific heat) can thus be calculated.

The stability of the molten form relative to the crystal is determined by the difference in free energy ($\Delta G = \Delta H - T \Delta S$) between the two phases. Increasing temperature increases the thermal motions of the particles and introduces disorder. The great randomness in liquids increases the entropy

($T \Delta S$) term enough to overcome the enthalpy (ΔH) term, so that the crystal melts when $T(S_{\text{liq}} - S_{\text{sol}}) = (H_{\text{liq}} - H_{\text{sol}})$. The disorder in liquids does not completely destroy the regularity of the structure; the density of the particles remains high and the cohesive forces are strong. Thus it is rather difficult to calculate the energy of the liquid system and to devise a simple acceptable partition function for the melt.

2.1. The Pair Potential

A basic assumption used in the development of theories of simple liquids[21] is that the forces between particles are central, i.e., they act through the centers of gravity, and that they are composed of forces acting between pairs of particles. The latter implies that for a liquid of N particles (molecules, atoms, ions) the configurational energy U can be represented as the sum of pair interactios Φ_{ij} between particles i and j:

$$U = \frac{1}{2} \sum_i^N \sum_j^N \Phi_{ij} \qquad (1)$$

This concept neglects three-, four-, and so on body interactions. Thus for a system of three particles i, j, and l the potential energy is not really $\Phi_{ij} + \Phi_{jl} + \Phi_{il}$ alone but also includes a "small" three-body interaction Φ_{ijl} which is neglected. The pair interaction concept is known not to be strictly true; however, the inaccuracies that result from it are likely to be negligible, relative to inaccuracies arising from the approximations used in the theoretical developments or to uncertainties accompanying the experimental results.

For a simple melt composed of positive and negative ions the pair potential is either attractive (Φ_{+-}) or repulsive (Φ_{++} and Φ_{--}). The nature of the pair ionic interactions has been thoroughly investigated in the past, and as a result of the efforts of many, including Born, Mayer, Pauling, and Tosi, the general form of the potential can be written

$$\Phi_{ij} = A_{ij} e^{-B_{ij}r} + Z_i Z_j / r - P_{ij}/r^4 - C_{ij}/r^6 - D_{ij}/r^8 \qquad (2)$$

A schematic representation of the form of Φ_{ij} as a function of the inter-particular distance r for both like and unlike ions is given in Figure 1a. The terms on the right side of Eq. (2) represent these different contributions:

$A_{ij} e^{-B_{ij}r}$. This is the short-range repulsive potential, first used for ionic crystals by Born and Mayer. It has been shown by quantum-mechanical calculations[22] that the A_{ij} and B_{ij} values are such that the function rises almost vertically at short interionic distances and that the hard-core or rigid-sphere model is adequate.

$Z_i Z_j / r$. This is the long-range "charge–charge" potential (coulombic term) between the ions with charges Z_i and Z_j.

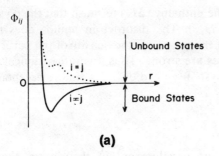

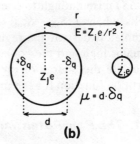

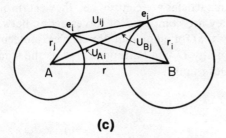

(a)

(b)

(c)

Figure 1. (a) Isolated ion-pair potential functions for unlike ($i \neq j$) and like ($i = j$) ions. (b) Induced dipole $[+\delta q - (-\delta q)]$ by ion Z_i on ion Z_j. (c) Electron–electron and electron–nucleus interactions accounting for the London forces between ions A and B at distance r.

P_{ij}/r^4. This is a charge-dipole term arising from the interaction of the electrons of the ion i (or j) with the charge of the ion j (or i). This term can be calculated quantum-mechanically[22] or from a simple classical polarization model by assuming that ion i induces a dipole moment on ion j and vice versa. With reference to Figure 1b, the induced dipole moment on ion j with polarizability a_j is $\mu = a_j E = d\delta q$ and the interaction potential (with $d \ll r^2$) is

$$[Z_i e\, \delta q/(r + 0.5d)] - [Z_i e\, \delta q/(r - 0.5d)] \simeq (Z_i e)^2 a_j/r^4$$

In a similar way, the polarization effect on ions j and i can be derived and the total polarization energy becomes

$$P_{ij}/r^4 = [(Z_i e)^2 a_j + (Z_j e)^2 a_i]/r^4$$

C_{ij}/r^6. This is the dispersion or induced dipole-induced dipole interaction. This term was approximated quantum-mechanically by London and is attributed to interactions of the ion i (or j) with the electrons of ion j (or i). With reference to Figure 1c, all interactions U_{A1}, U_{B2}, and U_{12} are taken into account and the approximation gives

$$C_{ij} = \tfrac{3}{2} a_i a_j I_i I_j/(I_i + I_j)$$

where the a's are the polarizabilities and I_i and I_j are the ionization potentials of the *ions*.

D_{ij}/r^8. This is the dipole–quadrupole interaction. This term, as well as other higher-order terms, the magnitudes of which diminish with distance as

r^{-8}, or faster terms (e.g., quadrupole–quadrupole interactions, etc.), is usually neglected. The D_{ij}/r^8 term, however, has been used as a part of the pair potential in computer simulations of molten salts, and the values of D_{ij} have been estimated by various researchers.[23]

Apart from the "ionic character" terms given in Eq. (2), the pair potential may also be affected by attractive contributions derived from partial covalent bonding and from effects influencing the relative stabilities of unfilled electronic states of the ion (e.g., splitting of the d electrons by the crystal field). The existence of contributions due to covalency has been recognized by NMR[24] and thermodynamic[25] measurements. The effect of d-electron stability on the cohesive energy of the melt has been also reflected in thermodynamic measurements of transition metal chlorides.[26,27]

Finally, it must be emphasized that in many cases an approximate correlation of the experimental data on molten salts with the theory of liquids can be achieved by devising and using an "effective" potential that is appropriate for interpreting a specific property. Two commonly used effective potentials are the following:

1. The effective potential used in the theory of corresponding states for correlating the thermodynamic properties of the melts. This potential will be discussed later in the chapter.

2. The Tosi–Fumi potential, which has been used for calculating the cohesive energies of the alkali halide crystals[28] and for most molecular dynamics calculations. It is believed that for the alkali halide melts the Tosi–Fumi potential occupies a position similar to that of the Lennard–Jones potential for rare gases.[23] Extensive discussions of this potential and its applications to computer simulations can be found elsewhere.[7,23]

2.2. The Radial Distribution Function

The knowledge of the pair potential permits, in principle, the evaluation of the configurational energy and the partition function Q_N. For a molten salt composed of μ kinds of ions with N_i molecules of the ith kind the total number of ions is $N = \sum_{i=1}^{\mu} N_i$ and the partition function is

$$Q_N = \left[1 \bigg/ \prod_{i=1}^{\mu} N_i! \left(\frac{h}{m_i kT} \right)^{(3/2)N_i} \right] \int \cdots \int e^{-U/kT} d^3 r_1 \, d^3 r_2 \cdots d^3 r_N \quad (3)$$

where U is the energy of interaction as defined in Eq. (1) and $d^3 r_i$ is the volume element associated with the center of mass of molecule i with mass m_i. Knowledge of Q_N and the use of statistical-mechanical theories permit, in principle, the calculation of all desired properties of the melt. This, however, has not been achieved in practice with reasonable accuracy.

A convenient way to describe the properties of a liquid is in terms of the radial distribution function (or pair correlation function). For a hypothetical liquid of N particles in a volume V having no cohesive forces, the distribution of the particles can be assumed to be random. In such a case the probability $\rho^2(\mathbf{r}_1, \mathbf{r}_2)$ of finding a particle in the volume element $d^3\mathbf{r}_1$ at $\mathbf{r}_1$ and, simultaneously, another particle at $d^3\mathbf{r}_2$ at $\mathbf{r}_2$ is a function of the distance between the particles ($\mathbf{r} = \mathbf{r}_2 - \mathbf{r}_1$):

$$\rho^2(\mathbf{r}) = (N/V)\left[\frac{(N-1)}{V}\right] \simeq (N/V)^2$$

In reality, however, the strong cohesive forces in the liquid correlate the positions of the particles and a correction factor $g(r)$ should be applied to the random probability $(N/V)^2$. This factor is, by definition, the pair correlation function or radial distribution function (RDF)

$$\rho^2(\mathbf{r}) = (N/V)^2 g(\mathbf{r})$$

In terms of the RDF, the expression

$$(N/V)g(\mathbf{r})\,d^3r = 4\pi r^2 (N/V)g(\mathbf{r})\,dr$$

defines the probability of finding a particle in a spherical shell of thickness dr at a distance r from the origin. Computer simulated plots of $g(r)$ versus r (Figure 2) show finite maxima of probabilities, indicating the average distances of first, second, etc., neighbors of a given particle.[29] The short-range order of the liquid is also indicated by the sampling character of $g(r)$. Additional calculations, based on the area under each maximum, give the average number of particles at that distance.

Knowledge of the RDF for a simple monatomic liquid helps determine its structural characteristics, and, in principle, an accurate analytical function of $g(r)$ can be used to calculate the configurational partition function Z and thus all the properties of the liquid:

$$\frac{\partial \ln Z}{\partial T} = \frac{2\pi k T^2 N^2}{V} \int r^2 g(r)\Phi(r)\,dr \qquad (4)$$

where $\Phi(r)$ is the pair potential.

The study of more complex and real liquids, however, is not as straightforward. Exceptions are a few halide salts (e.g., NaCl, KCl, BaCl₂), where accurate measurements of the RDF by neutron scattering techniques[30–32]

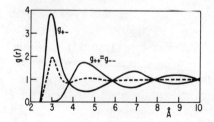

Figure 2. Radial distribution functions for a model of molten KCl. Dashed lines represent the average function (reproduced from Reference 29).

and computer simulations of the molten state permit a determination of the structural properties and a calculation of the thermodynamic quantities for the melts. For most molten salts and salt mixtures the correlation of their properties from first principles (i.e., interionic potential and distances) is too far from reality. Thus the accumulation of experimental data and the development of semiempirical[33] and operational[34] models are needed for the formulation of a more general theory and a better understanding of these liquids.

2.3. Methods of Characterization

Today's understanding of the structural properties, the nature of bonding, and the macroscopic properties of molten salts is derived from systematic investigations using various spectroscopic and thermodynamic methods.

Spectroscopic methods have been used mainly for obtaining structural information and examining the type of chemical bonding between the ions in the molten state. Table 1 summarizes most of the spectroscopic methods that have been applied and gives selective references as a starting point for a further study. The work on infrared absorption (or emission) and on magnetic behavior is limited to a few investigations. Raman and electronic absorption spectroscopy are the most widely used spectroscopic techniques, especially for investigating molten-salt mixtures and solutions.

The thermodynamic methods are aimed at understanding the macroscopic behavior of the molten state and are often correlated through theory with microscopic properties, thus enabling conclusions about the bonding and structure of the melt to be drawn. The most commonly employed methods are presented in Table 2. Extensive literature is also available on measurements of densities, compressibilities, surface tensions, and viscosities of molten salts.[16]

3. Single Salts

3.1. Structure

For the most part, structural information on single-component molten salts has been deduced directly from spectroscopic scattering experiments and, in certain cases, indirectly from a variety of investigations of the thermodynamic properties of the melts and their solutions or mixtures. Two well-accepted conclusions are (1) that anions in the melt are surrounded by cations and vice versa and (2) that the anion–cation distances in the melt are shorter than those in the corresponding crystal. Before the mid-1970s it was also believed that an anion in the melt has fewer next-nearest neighbors than in the crystal. However, recent experimental and theoretical advances in the

Table 1

Spectroscopic Methods for Investigating Molten Salts

Technique	Effect[a]	Measured quantity	Information	Remarks
A. Scattering				
Neutron diffraction References 7, 35	Neutrons scattered by nuclei; $\lambda = 0.1$ nm, $\tau = 10^{-18}$ sec (averaged over vibrational motions)	Angular distribution of the intensity of the scattered neutrons	Radial distribution function [$g(r)$] of different isotopic species; partial structure factors; internuclear distances; correlation length; number of neighbor atoms	Isotope substitution resolves distribution functions of polyatomic (polyionic) liquids; very useful method but expensive and limited to few places in the world
X-ray diffraction References 36, 37	X rays scattered by electrons; $\lambda = 0.01$–0.1 nm, $\tau = 10^{-18}$ sec (averaged over vibrational motions)	Angular distribution of the intensity of the scattered electrons	Radial distribution function and anion–cation distances	Accurate distribution functions can be obtained only from monatomic liquids (metals)
Vibrational Raman References 38, 39	Inelastic scattering of light due to polarizability changes on vibration; $\lambda =$ visible laser lines, $\tau = 10^{-14}$ sec	Intensity of scattered light as a function of energy (0–~ 4000 cm^{-1})	Energy and symmetry of Raman-active vibrational states	Symmetry considerations are used to characterize short-range order and give structural information
B. Absorption Spectra				
Vibrational infrared Reference 39	Interaction of radiation with vibrating molecular dipoles; $\lambda = 4$–0.025 mm, $\tau = 10^{-13}$ sec	Absorption (or emission) of radiation as a function of energy (0–4000 cm^{-1})	Energy of infrared-active vibrational states	Symmetry considerations are used to characterize short-range order and give structural information
Electronic absorption References 40, 41	Interaction of radiation with electronic states of the atoms; $\lambda = 2500$–200 nm, $\tau = 10^{-15}$ sec	Absorption of radiation as a function of energy (4000–50,000 cm^{-1})	Electronic states of atoms. Mainly d–d ligand field transitions	Used mainly for transition metal ions dissolved in molten media; structural information on order around the dissolved ion and the host melt is obtained

C. Magnetic Measurements

NMR, References 24, 42	Interaction of radio-frequency radiation with nuclear spin states of an atom in a magnetic field; $\tau = 10^{-9}$ sec	Resonance absorption of radio frequency in a variable magnetic field	Nuclear spin states of atom; relaxation processes	The chemical shift (Ramsey's theory) is used to characterize the distribution of electrons and the type of bonding
EPR, Reference 43	Interaction of microwave radiation with electron spin states of an atom in a magnetic field	Resonance absorption of microwaves in a variable magnetic field	Electron spin states of atom	Character of bonding and structure of environment; correlation time for spin-exchange process; fine structure of ground electronic state
Magnetic susceptibility, Reference 44	Interaction of electron spin states (permanent or induced) with the magnetic field	Force exerted on the atom in a variable magnetic field	Ground electronic spin states of atom	Dia- or paramagnetic properties that depend on the local ligand field and some structural information are obtained

[a] The wavelength of radiation used is indicated by λ; the approximate time of interaction is τ.

Table 2
Thermodynamic Methods for Investigating Molten Salts

Technique	Measured quantities	Information
Phase diagram, cryoscopy (Reference 45)	Temperature, thermal halts, chemical analysis; identification of phases of solutions in equilibrium with a solid	Determination of transition temperatures and miscibilities and construction of phase diagram; estimation of chemical potentials and partial entropies of mixing
Calorimetry (Reference 46)	Heat evolved by phase changes or by dissolving or mixing molten salts	Integral and/or partial enthalpies and heat content; heat of fusion
Galvanic cells (Reference 47)	Electromotive force in molten-salt mixtures	Chemical potential, partial enthalpies, and entropies of mixing
Equilibrium vapor pressure (References 48, 49)	Partial pressures of vapors over melt at different temperatures	Type of vapor species in the vapor; chemical potential in the melt; heats of vaporization and fusion

field of neutron scattering from melts[50,51,52] as well as applications of computer simulations[31,32,53] have yielded new results that alter some of the early beliefs.

The present section briefly describes the application of neutron and X-ray scattering techniques for determining the structure of molten metal halide salts and presents compilations of structural information on many such single-component salts. Similar measurements are not available for other kinds of salts, with the exception of some X-ray data on oxyanion melts. In certain cases, however, Raman spectroscopy has proved useful in determining the symmetry of oxyanions and the coordination symmetry of cations in strongly covalent melts (e.g., Al_2Cl_6, $ZnCl_2$). Structural information on such melts derived from X-ray and/or Raman spectroscopy is also included here.

3.1.1. Neutron and X-Ray Scattering on Metal Halides

In a typical neutron or X-ray scattering experiment the scattered intensity $I(\theta)$ is measured as a function of the scattering angle θ. For a melt composed of two types of ions the RDFs $g_{\alpha\beta}(r)$ for the various ion pairs (i.e., g_{+-}, g_{++}, g_{--}) are correlated with the scattered intensity through their Fourier transforms. The resulting quantities are known as partial structure factors $S_{\alpha\beta}$:

$$S_{\alpha\beta}(k) = \delta_{\alpha\beta} + 4\pi n \int_0^\infty [g_{\alpha\beta}(r) - 1]\frac{\sin kr}{kr}r^2\, dr \tag{5}$$

where n is the number of ions of either species per unit volume and $k = 4\pi \sin\theta/\lambda$. The scattered intensity per unit volume is given by

$$I(\theta) = n[f_+^2 S_{++}(k) + f_-^2 S_{--}(k) + 2f_+ f_- S_{+-}(k)] \tag{6}$$

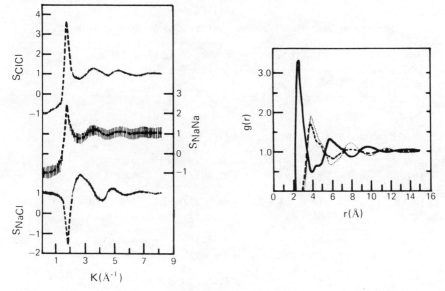

Figure 3. Partial structure factors for liquid NaCl (left); the RDF for liquid NaCl:——— $g_{+-}(r)$, $\cdots g_{--}(r)$, - - - $g_{++}(r)$ (right) (reproduced from Reference 37).

where the coefficients f_+ and f_- have different dependences on k for different types of scattering experiments.

For neutron scattering f_+ and f_- represent the mean scattering length of the individual ions; they are independent of k but depend on the isotopic state of the scattering nuclei. Thus, in experiments on different isotopes, f_+ and f_- may vary enough to give different scattering intensities for the same element. These intensities, in turn, can be used to construct three linear equations of the type of Eq. (6), whose solutions are the partial structure factors. For melts containing more than two ions additional linear equations and isotope substitution experiments are required to derive the partial structure factors.

The isotope substitution experiments have been applied to molten salts only in recent years and have contributed valuable information regarding the interionic distances and coordination numbers of the different ions in the melt.[50–52,54–56] Typical results from these experiments are presented in Figure 3 for the NaCl melt, where both the partial structure factors and the RDFs are presented. Table 3 summarizes the structural results obtained from the few available experiments.

Other neutron scattering experiments that do not use the isotope substitution method[57] have been reported since 1960. The total RDFs as determined by these measurements give reliable anion–cation interionic distances but tend to underestimate the coordination numbers. The results of these

Table 3
Structural Data from Isotopic Neutron Scattering Experiments

System and temperature	Unlike ions		Like ions				Ref.
	r_{+-}	n_{+-}	r_{++}	n_{++}	r_{--}	n_{--}	
NaCl							
liq. 1100 K	2.6	5.8 ± 0.1	3.7	13.0 ± 0.5	3.8	13.0 ± 0.5	50
sol. ~300 K	2.95	6	4.17	12	4.17	12	
KCl							
liq. 1073 K	3.06	6.1	4.84	~16	4.82	~16	51
sol. 1049 K	3.26	6	4.61	12	4.61	12	
RbCl							
liq. 1023 K	3.18	6.9 ± 0.3	4.86	13.0 ± 2.0	4.80	14.0 ± 2.0	54
sol. 300 K	3.41	6	4.82	12	4.82	12	
CsCl							
liq. 973 K	3.38	~5.9	3.85	~15	3.85	~15	
sol. 920 K	3.57	6	5.05	12	5.05	12	
AgCl							
liq. 783 K	2.6	4.3	3.40	3.1			
liq. 1083 K	2.55	2.7	3.15	4.1	3.15	2.7	55
sol. 300 K	2.97	6	4.07	12	40.7	12	
CuCl[a]							
liq. 713 K	2.3				3.7		52
sol. 300 K	2.34				3.22	8	
BaCl$_2$							
liq. 1298 K	3.1	7.7 ± 0.2	4.9	14.0 ± 2.0	3.9	7.0 ± 1.0	56
sol. (cubic) 300 K	3.18	8	5.19	6	3.67	12	

[a] Discussion on the structural peculiarities of CuCl is given in Reference 7.

experiments have been included in Table 4, where X-ray data as well as thermodynamic data are presented.

For X-ray scattering experiments[36] f_+ and f_- depend strongly on k and fall to very low values at high k. With current available knowledge the partial structure factors [Eq. (6)] cannot be separated and thus the amount of structural information obtained is definitely limited. Owing to the first strong intensity (Figures 2 and 4) of the g_{+-} pair distribution function, the total $g(r)$ measured by X rays gives reliable anion–cation distances. However, the coordination numbers derived therefrom are inaccurate, since they were deduced from total scattering patterns and are thus averages or more than one pair distribution function. Table 4 summarizes most of the presently available X-ray scattering data.

Table 4

Structural Information on Selected Metal Halides and Their Enthalpy and Volume Changes on Fusion

Salt	Mp^a K	ΔH_f (Kcal/mol)^a	$(\Delta V_f/V_s) \times 100^a$	Anion–Cation distance (Å)	Cation coordination number	References and remarks
LiF	1121	6.47	29.4	1.85 (2.013)^b	3–3.7 (6)^b	58; X ray
NaF	1268	8.03	27.4	2.30 (2.50)	4.1 (6)	59; X ray
KF	1131	6.75	17.2	2.65 (2.80)	4.9 (6)	60; X ray
LiCl	883	4.76	26.2	2.4–2.47 (2.66)	3.5–4 (6)	57, 61; X ray, neutron
LiBr	823	4.22	22.4	2.68 (2.85)	5.2 (6)	57; X ray
CsBr	909	5.64	26.8	3.5 (372)^c	4.6 (6)^c	62; X ray
LiI	742	3.5	20	2.85 (3.12)	5.6 (6)	57; X ray
NaI	933	5.64	18.6	3.15 (3.39)	4.0 (6)	57; X ray
CsI	899	5.64	28.5	3.85 (3.94)^c	4.5 (6)^c	57; X ray
BeF$_2$	816	6.53		1.59 (1.54)	4 (4)	63, 64; X ray
CaF$_2$	1684	7.1		2.35 (2.36)	5.8 (8)	59; X ray
PbCl$_2$	768	4.4		2.92 (3.05)^d	8 (9)	61; X ray
MnCl$_2$	923	9.0		(2.59)	~4 (6)	65, 66, 27; X ray, Raman, and thermodynamic considerations
FeCl$_2$	950	10.3		(2.53)	~4 (6)	26, 27; thermodynamic considerations
CoCl$_2$	989	7.9		(2.51)	~4 (6)	26, 27; thermodynamic considerations
NiCl$_2$	1303	18.1		(2.50)	~4 (6)	27, 67; thermodynamic considerations
ZnCl$_2$	591	2.4	11.6	(2.30)	~4 (4)	68–70; electronic absorption and Raman
AlCl$_3$	465	8.5	87	2.06, 2.21^e (2.325)	4 (6)	71, 72; X ray and Raman
YCl$_3$	987		2	(2.632)	~6 (6)	73; Raman

^{a} The thermodynamic data were collected from references 9, 10, and 13.
^{b} The number in parentheses indicates solid-state data.
^{c} Data for the high-temperature NaCl-type structure.
^{d} Solid PbCl$_2$ has a V_h^{16} structure, and nine equivalent Pb–Cl distances (from 2.67 to 3.23 Å) are present.
^{e} Molecular liquid with "bridged" and "terminal" anion–cation distances.

In Table 4 structural information is also presented that was derived from other kinds of spectroscopic measurements (i.e., YCl_3, $AlCl_3$, $ZnCl_2$, $MnCl_2$)[68-73] and indirectly from thermodynamic correlations of the ligand field stabilization energies with the enthalpies of fusion of transition metal halides (i.e., $MnCl_2$, $FeCl_2$, $CoCl_2$, $NiCl_2$).[27,28,67]

The decrease in the anion–cation distances in the melt relative to the solid is evident from the data in Tables 3 and 4. For the alkali chlorides and $BaCl_2$ (Table 1) the *coordination numbers in the melt appear to be remarkably close to those in the solid*. This conclusion contradicts early beliefs and is in agreement with the results of computer simulations of melts.[29,32] Furthermore, a comparison[50,56] of the experimentally determined structure factors with computer-simulated structure factors based on pair potentials of the type of Eq. (2) has shown no evidence of covalency or polarization effects in these melts. Finally, both experimental (Figure 4) and simulated (Figure 2) pair distribution functions indicate that charge cancellation occurs at ~3–4 ionic diameters from a given ion.

For the salts, other than alkali halides, listed in Tables 3 and 4 some structural peculiarities can be noticed. Thus for AgCl and CuCl there are indications that molecular-type structures are formed at high temperatures.[55,74] However, Raman spectroscopic measurements[75] and NMR measurements of relaxation times[42] are not consistent with these molecular-type models. The BeF_2 and $ZnCl_2$ melts show network-type structures involving MX_4 tetrahedral coordination.[63,68,69] The salts YCl_3 and $AlCl_3$ are of particular interest, since as crystalline solids they are isostructural (C_2^{3h}), but upon melting the molar volume of $AlCl_3$ increases by ~90% and a molecular liquid (Al_2Cl_6)[71,72] is formed, whereas the YCl_3 liquid behaves in a way similar to that of $BaCl_2$ (i.e., small molar volume changes and little change in coordination geometry[73]). Density measurements have indicated[76] that $FeCl_3$ melts exhibit structural changes similar to those occurring in ionic melts.

As in the case of $AlCl_3$, molecular liquid formation has been found to occur for many low-melting halides like BiX_3,[77] InX_3,[78,79] GaX_3,[80] SbX_3,[77] AlX_3,[81,82] HgX_2,[83] and SnX_4,[79] where X is Cl, Br, or I. The $SbCl_3$ melt has been recently investigated by X-ray and neutron diffraction[77b] and by Raman spectroscopy.[77c] A certain order appears where $SbCl_3$ (C_{3v}) molecules have a head-to-tail alignment in the melt. It should be pointed out however, that certain molten molecular salts (i.e., BiX_3 and HgX_2) tend to dissociate to molten ionic salts. Precise pressure- and temperature-dependence measurements of the conductivity of molten $BiCl_3$ and HgX_2, as well as P–V–T data, have shown extensive ionization of the molecular liquid at high pressures (Kbar range) and temperatures.[84,85]

Raman,[65] X-ray,[66] and thermodynamic considerations[27,67] have indicated that the transition metal chlorides (Table 4) exhibit drastic coordination changes upon melting. Covalent and ligand field effects appear to be important for these melts.

3.1.2. Oxyanion Melts

Structural investigations by X-ray or neutron scattering of nonhalide melts have been very limited. Low melting point oxyanion melts, however, have been thoroughly studied by Raman (and infrared) spectroscopy, and extensive review articles are available.[38,39] In these studies the behavior of the fundamental vibrational frequencies of the oxyanion is examined in terms of possible distortions by interactions with neighboring ions and/or with the overall molten-salt electrostatic field.

X-ray[86] and neutron scattering[87] measurements on ANO_3 (A = Li, Na, K, Rb, Cs, Ag, Tl) salts have shown that the N–O and O–O distances in NO_3^- are ~0.125 and 0.215 nm, respectively. Within experimental error these distances are not affected by the counter-ion A^+ in the melt. Raman and infrared (IR) measurements on a series of ANO_3 and $M(NO_3)_2$ (M = Mg, Ca, Pb, Sr, Ba, etc.) melts have shown that the NO_3^- geometry is close to D_{3h}, with structural perturbations that arise from the "cage" of the counter-ion and which lead to splitting of the vibrational modes of NO_3^- (including ν_1).[38,39] Intepretation of the data, however, in terms of distortions alone was found to be inadequate, and lattice-type modes have been proposed for these melts.[39]

3.1.2.2. Carbonates and Chlorates

The alkali metal salts with the oxyanions CO_3^{2-} and ClO_3^- show structural behavior similar to that of nitrates. X-ray scattering data are available for the A_2CO_3 (A = Li, Na, K) melts only.[88] Vibrational spectroscopy has indicated that both internal- and external- (lattice) type modes are present in these melts.[39]

3.1.2.3. Sulfates and Perchlorates

In X-ray scattering experiments[88,89] the tetrahedral (T_d) nature of the SO_4^{2-} anion in alkali metal sulfate melts has been demonstrated and the S–O and O–O distances have been measured to be 0.15 and 0.245 nm, respectively. Raman spectroscopic measurements, however, show strong C_{3v} distortions for the SO_4^{2-} ion in these melts.[39] The structure of the perchlorate ion in $LiClO_4$ appears to be similar to that of the sulfate ion.[39]

3.1.2.4. Tungstates

Molten Na_2WO_4 and $Na_2W_2O_7$ have been investigated by X-ray scattering.[90,91] Tetrahedral WO_4^{2-} ions, with a W–O distance of ~0.172 nm (relative to 0.18 nm in the solid), have been shown to be present in the melt. The melt structure of $Na_2W_2O_7$ is considered to be similar to that of the solid, with the possible existence of both tetrahedral WO_4 and octahedral WO_6 units in the melt.

3.1.3. Oxide Melts

Very little is known about the structure of the high-temperature-melting oxides of the type M_2O, MO, M_2O_3, MO_4, M_2O_5, etc. Limited information is available only for glass-forming oxides (i.e., SiO_2, B_2O_3, P_2O_5, As_2O_3), where the glassy state data[92] can infer the liquid oxide structure. It is well accepted that apart from the ankylosis (freezing out) of the translational motions, the melt structure is similar to that of the glass. In a few cases, however, this correlation may not be correct, since certain liquids (e.g., P_2O_5)[93] have been found to exhibit *allotropic* modifications where network-type and molecular liquids can form on melting. Details about the structure of oxide glasses can be found in a variety of review articles.[92,94] Two types of oxides will be briefly discussed here.

3.1.3.1. MO₂

Representative of this type of oxide is SiO_2, which has been thoroughly investigated as a glass.[64,95] Thermodynamic measurements of SiO_2 in solutions with other oxides and in slags[96] and series of spectroscopic investigations[36] have been used for probing the structure of the melt. A networklike liquid has been proposed in which $[SiO_4]$ tetrahedral units are linked to each other by oxygen bridges. A similar structure has been proposed for GeO_2.[36]

3.1.3.2. M₂O₃

The glass-forming oxides B_2O_3, P_2O_5, and As_2O_3 are also of the network type, where $[MO_3]$ units with $\sim C_{3v}$ symmetry are linked to each other by bridging $[O_2M]$ molecular units, which preserve a certain (e.g., C_{2v}) symmetry. Raman spectroscopic investigations of As_2O_3[97] and B_2O_3[98] in the glassy and liquid (melt) state have shown that the symmetries of the basic units $[MO_3]$ and $[O_2M]$, and their vibrational modes, are very similar in both phases. Figure 4 shows that the vibrational modes of glassy As_2O_3 do not change drastically as the material passes through the glass transition ($T_g \sim$ 450 K) into the liquid state. In the low-frequency range, however, ($0 < \omega <$ 50 cm^{-1}) a temperature-dependent band in the glass disappears in the liquid, yielding a broad "Rayleigh wing." This low-frequency band has been shown[97,98] to arise from acoustical-type modes in the glass, and its intensity depends on the density of vibrational states. These types of modes are probably absent in the melt.

3.2. Thermodynamic Correlations

In recent years theoretical studies of molten salts have followed a dual approach, where pure statistical-mechanical and computer-simulation methods are used to calculate the thermodynamic and transport properties and to elucidate the structure of the melts. The statistical-mechanical methods are based on fundamental equations developed in theories of simple

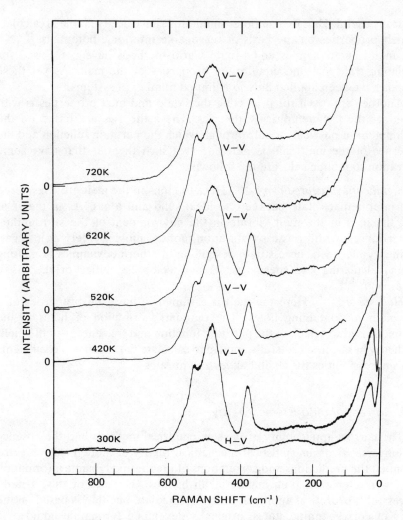

Figure 4. Raman spectra of As_2O_3 above and below the glass transition temperature at 430 K (reproduced from Reference 97).

liquids[99,100] and attempt to derive analytical expressions for calculating properties of the melt in terms of a given interionic potential, using the smallest number of parameters. Current advances include the mean spherical approximation[101] and the restricted primitive model[102] approaches, as in the treatment of aqueous ionic solutions (Chapter 2). These theories are in their early stages, and although they have some predictive ability for the properties of melts, they are restricted to very simplified salt models and are not yet practically useful. Thus far computer simulations have been reliable in calculating melt properties on the basis of reasonable interionic potentials.[29,31,32] Both molecular dynamics and Monte Carlo methods have been used for calculating transport and equilibrium properties of the melt. As yet these methods have been applied only to a limited number of systems.

Another approach for predicting the liquid and melt properties is with the aid of the "semiempirical" theories. These theories are based on the development of physical models through which the partition function and the properties of the melt can be derived. Two such theories that have some application to molten salts are the following:

Significant structure theory.[103] Here the ions in the melt are considered in an intermediate state between an ideal solid and a gas. As in the solid state, the ions in the melt vibrate in the environment of the surrounding counter-ions, but still possess translation motion, moving freely as in a gas. The theory has been successful in describing the thermodynamics of melting and in calculating the specific heats and compressibilities of the alkali halides.[103,104,105]

Hole theory.[106] Here the melt is assumed to have the properties of a solid medium containing holes. The random distribution of holes in the medium is used to calculate the partition function and free energy of the melt. The theory has been successfully applied to calculate the expansion coefficients and compressibilities for alkali halides and nitrates.[106,107]

3.2.1. Corresponding-States Theory

The narrow applicability of the semiempirical methods and the absence of a universal rigorous statistical-mechanical theory of molten salts have so far limited the predictions and correlations of the thermodynamic information to a few molten salts (i.e., mainly alkali halides). Because of this, a more "successful" theoretical approach is an intermediate one that is based on the theory of corresponding states, originally developed for simple liquids.[108] The application and development of the theory for molten salts[109] has, as an objective, the correlation of the thermodynamic properties of melts within a certain class (e.g., alkali halides, oxides, etc.). The theory was found to be important not only because of its applicability to different classes of single-component salts but also because of its use in developing "structureless" models for correlating the thermodynamics of mixing of molten salts. The following is an outline of the basic principles of the theory; its applicability to molten-salt mixtures[110] will be discussed in Section 4.1.1.

The aim of the theory of corresponding states is the derivation of a generalized equation of state for molten salts. This is achieved by a dimensional analysis of the configurational integral of Eq. (3). The model ionic melt used assumes local order in the melt, where only short-range interactions between ions are important. Short-range dispersion and polarization interactions are neglected and the many-body interactions are accounted for by an effective dielectric constant u. The pair potential Φ_{ij} of Eq. (2) is replaced by a hard-core spherical charge model:

$$\Phi_{+-} = \begin{cases} \infty & r \le d \\ z_{(+)}z_{(-)}e^2/ur, & r > d \end{cases}$$

$$\Phi_{++} = z_{(+)}z_{(+)}e^2/ur, \qquad r > 0 \tag{7}$$

$$\Phi_{--} = z_{(-)}z_{(-)}e^2/ur, \qquad r > 0$$

where d is a cutoff (rigid-sphere) parameter taken as the sum of the ionic radii of the two types of ions present ($d = r_{(+)} + r_{(-)}$). The model assumes that anions are surrounded by cations and vice versa; thus contact between like-charged ions is not possible and the Φ_{++} or Φ_{--} potentials are simplified by retaining the coulombic form alone for all distances.

The configurational integral [Eq. (3)] is transformed by use of the dimensioness distance

$$r^* = r/d \tag{8}$$

and the pair potentials are expressed by

$$\Phi_{+-} = \begin{cases} \infty \\ (z_{(+)}z_{(-)}e^2/ud)\Phi^*(r^*) \end{cases}$$

$$\Phi_{++} = [(z_{(+)}e)^2/ud]\Phi^*(r^*) \tag{9}$$

$$\Phi_{--} = [(z_{(-)}e)^2/ud]\Phi^*(r^*)$$

where $\Phi^*(r^*)$ is defined as the dimensionless potential $1/r^*$.

For melts containing equally charged ions, ($|z_{(-)}| = |z_{(+)}| = z$), the inter-molecular potential U [Eq. (3)] can be written as $U = (z^2 e/ud)U^*$ and the configurational integral (Z) becomes a function F of the reduced dimension-less quantities temperature (τ) and volume (θ), and of the number of particles N:

$$\tau = dukT/(ze)^2$$

$$\theta = v/d^3 \tag{10}$$

$$Z = d^{6N}F(\tau, \theta, N)$$

With the use of these equations and standard statistical thermodynamic relations, a variety of reduced thermodynamic properties can be derived. For example, it has been shown[109] that the reduced vapor pressure (π) and reduced surface tension (Σ) are related to the pressure (P) and surface tension (σ) of the melt by

$$\pi = \pi(\tau, \theta) = [d^4 u/(ze)^2]P \tag{11}$$

$$\Sigma = \Sigma(\tau, \theta) = [d^3 u/(ze)^2]\sigma \tag{12}$$

with the first part of Eq. (11) representing a generalized equation of state for molten salts.

One of the implications of the theory is that the reduced quantities at the melting point, τ_m, θ_m, and π_m, are universal constants. In Table 5 the reliability of the model is tested for consistency in certain groups of salts.

The dimensional analysis for unequally charged ions in the melt ($|z_{(+)}| \neq |z_{(-)}|$) has also been carried out, but with a new definition of the reduced temperature,

$$\underline{\tau} = dukT/e^2 \qquad (13)$$

and the resultant equation of state has the form

$$\underline{\pi} = \underline{\pi}(\underline{\tau}, \theta, z_{(+)}, z_{(-)}) = (ud^4/e^2)P \qquad (14)$$

The reduced pressure is now a function of the charges of the ions and can be considered a universal function, but only with respect to a particular group of salts. For example, one can compare uni-divalent salts, uni-trivalent salts, etc. This is done in Table 5 for the MF_2 and LCl_3 classes of salts.

Table 5

Reduced Melting Temperatures and Corresponding-State Vapor Pressures for Several Melts

Salt	T_m K	Reduced melting temperature τ_m or $\tau_m \times 10^2$	Reduced pressure $\pi_{1.3}$ at $T = 1.3T_m$ (mm/K$^4 \times 10^{19}$)
LiF	1121	1.35	1.5
LiCl	887	1.36	2.5
LiBr	823	1.35	1.3
LiI	718	1.30	0.3
NaCl	1074	1.81	30
KCl	1045	1.80	38
KbCl	980	1.95	40
CsCl	918	1.91	39.5
MgO	3073	0.96	10
CaO	2873	1.03	28
SrO	2733	1.04	36
BaO	2198	0.90	37
MgF$_2$	1536	1.93	5.4
CaF$_2$	1691	2.43	8.4
SrF$_2$	1673	2.54	10.4
BaF$_2$	1563	2.57	18.3
LaCl$_3$	1143	1.96	13.1
CeCl$_3$	1095	1.86	17.2
PrCl$_3$	1091	1.84	14.3
NdCl$_3$	1057	1.77	2.5

4. Salt Mixtures

A successful approach for the study of the properties of molecular liquids is the systematic investigation of the properties of their mixtures and solutions.[111,112] Thermodynamic measurements in particular have contributed to understanding the nature of forces between molecules in both the component liquids and the mixture. For the same purpose a variety of molten-salt mixtures has been investigated by various techniques. The technological advances in handling corrosive materials and the availability of sophisticated electronic and optical equipment has made possible the measurement of thermodynamic quantities (e.g., enthalpies of mixing) and spectroscopic properties (e.g., vibrational spectra) of melt mixtures and solutions.

Depending on the number of component salts and the charge of the ions in each component, various categories of molten-salt mixtures can be considered. Among the many possibilities, and in order of increasing complexity, are the following types:

i. Equally *charge-symmetrical* binary systems with a common ion, e.g., NaCl–NaBr, CaO–MgO.

ii. Unequally *charge-symmetrical* binary systems with a common ion, e.g., $MgCl_2$–$CaCl_2$, Na_2SO_4–K_2SO_4.

iii. *Charge-unsymmetrical* binary systems with a common ion, e.g., $NaNO_3$–$Pb(NO_3)_2$, KF–YF$_3$.

iv. *Reciprocal* binary mixtures in which two salts having no common ion are mixed to form a melt with four ions. Depending on the ion charges, reciprocal mixtures can be *symmetrical* or *unsymmetrical*, e.g., KCl–NaBr, $CaCO_3$–Na_2SO_4.

v. *Ternary* (or higher) systems in which three (or more) salts are mixed in any of the above combinations (i–iv).

In the present section the development of basic concepts of thermodynamics of simple molten-salt mixtures (i.e., mixtures of types i–iii) will be outlined. The trends and systematics of the thermodynamics will be presented and discussed in terms of the structural particularities (e.g., complexing) in these melts. Most of the structural information of the mixtures has been derived from vibrational (Raman) and electronic absorption spectroscopy. These methods and their results, which reveal the structure of selected systems, will also be outlined.

4.1. The Thermodynamics of Mixing

The concept of an "ideal" molten-salt mixture was introduced by Temkin,[113] using a model that assumes random distribution of the different types of cations among the cations and, similarly, random distribution of the anions among the anions. The thermodynamic functions of mixing for the Temkin model are

$$\Delta H_m^i = \Delta U_m^i = \Delta V_m^i = 0, \qquad \Delta G_m^i = -T \, \Delta S_m^i \tag{15}$$

and the entropy of mixing is calculated on the basis of random distribution of the ions. For example, for mixtures containing the monovalent salts AX and BY the total entropy is related to the total number of possible equivalent and distinguishable configurations in the mixture and can be expressed as

$$-\Delta S_m^i/R = N_A \ln x_A' + N_B \ln x_B' + N_X \ln x_X' + N_Y \ln x_Y' \tag{16}$$

where N_A, N_B, N_X, and N_Y are the numbers of moles of the indicated ions in the mixture and the x''s are the ionic fractions defined as

$$x_A' = N_A/(N_A + N_B), \qquad x_B' = N_B/(N_A + N_B) \\ x_X' = N_X/(N_X + N_Y), \qquad x_Y' = N_Y/(N_X + N_Y) \tag{17}$$

The partial molar quantities in the Temkin model can be obtained by differentiating Eqs. (15) and (17). Thus the chemical potential of one component (e.g., AX) in the mixture is

$$\Delta\mu_{AX} = RT \ln x_A' x_X' \tag{18}$$

and the activity is

$$\alpha_{AX} = x_A' x_X' \tag{19}$$

For other "Temkin"-type molten-salt mixtures involving different charges and/or common ions, equations similar to Eqs. (16)–(19) have been derived and reported elsewhere.[114,115]

Experimental evidence, however, shows that for most mixtures Eqs. (15)–(19) are not valid. When two salts are mixed, an exchange of ions of different sizes and/or charges will occur, and forces such as coulombic forces and polarization [i.e., Eq. (2)] will change. This was illustrated by Forland[114] for the energy of mixing of simple common anion mixtures (e.g., LiCl–KCl).

Assuming that the structure and coordination changes on mixing are minor, then the coulombic repulsive forces between the cations will influence the energy of mixing. Thus, with reference to Figure 5,

$$\Delta E = \frac{2e^2}{d_1 + d_2} - \frac{e^2}{2d_1} - \frac{e^2}{2d_2} \approx -\frac{2e^2}{d_1 + d_2}\left(\frac{d_1 - d_2}{d_1 + d_2}\right)^2 \tag{20}$$

and the coulombic part of the energy of mixing is negative (exothermic). For the same model the contributions of polarization and dispersion have been shown[11] to be positive (endothermic).

The Temkin and Forland models give an oversimplified picture of a molten-salt mixture. Systematic measurements of the thermodynamic functions ΔH_m, $\Delta\mu_{AX}$, and the partial enthalpies $\Delta\bar{H}$ and entropies $\Delta\bar{S}$ have shown

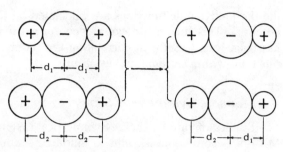

Figure 5. Changes in cation–cation repulsion energy by mixing common anion melts.

a more complex behavior. All functions depend on composition and can be either positive or negative and may even change sign with variation of the composition.

The basic experimental thermodynamic techniques for investigating salt mixtures are listed in Table 2. Chemical potentials of solute species have been measured by cryoscopy and vapor equilibrium studies, but the main data sources are galvanic cells, whereby $\Delta\mu$ can be measured accurately over the whole composition range. Formation[116] cells and concentration[117] cells with or without transport have been used. Cells with selective diaphragms for a particular ion have been employed for high melting point mixtures.[118] Temperature-dependence studies of the cell's emf allow the evaluation of partial enthalpies and entropies.

The only thermodynamic functions measured directly by calorimetry are the integral and partial enthalpies (ΔH_m, $\Delta\bar{H}$). Over 200 systems have been measured, with most of the data being produced in Kleppa's laboratory.[46]

A fully developed theory of molten-salt mixtures should, in principle, be able to justify the nonzero values of the thermodynamic quantities, predict their order of magnitude, and give a full account of the composition dependence of these quantities. Hitherto the theoretical approaches used for rationalizing and correlating the thermodynamics of mixing have not been overly successful. Apart from a few crude models, such that that of Forland, two basic procedures have been employed. In the first a structural model is introduced for each component and the mixture, and the kinds of interactions and coordination numbers of the ions involved are postulated. The thermodynamic quantities are calculated as a function of "energy parameters," ionic fractions, and coordination numbers. These procedures have been described in detail by Forland[114]; more recently structural models have been proposed by Ostvold,[119] Pelton and Thompson,[120] and Gaune-Escard et al.[121] All structural models, however, are developed and applied to a specific salt mixture and, in a few cases, to salt mixtures with a common component (e.g., $MgCl_2$–ACl; $A = Li\cdots Cs$).

The second procedure takes the theory of corresponding states into account in the formulation of the model. A structureless salt is adopted and the thermodynamic functions are calculated on the basis of an effective pair potential. This theory, known as "conformal ionic solution theory,"[110,122] has been found useful in correlating the thermodynamic (mainly enthalpy) functions within certain families of salts.

4.1.1. Charge-Symmetrical Systems

Studies of simple, charge-symmetrical systems of the type AX–BX and MX_2–NX_2 containing a common anion or cation have served as a basis for revealing the solution behavior of molten-salt mixtures. Accurate galvanic-cell and cryoscopy measurements have been available for many years.[123] However, systematic investigations are relatively new and, for the most part, have been conducted by calorimetric measurements. Comprehensive studies of complete families of mixed anion and mixed cation systems have been carried out (Table 6). The measurements have shown that the ΔH_m for these systems can in general be expressed by a parabolic type of equation of the form

$$\Delta H_m = x_1 x_2 (a + b x_2 + c x_1 x_2) \qquad (21)$$

where x_1 and x_2 are the mole fractions of the two components and a, b, and c are coefficients determined by a least-squares fit of the data.

Figure 6 shows the different shapes of the enthalpy of mixing curves observed for these systems. The right side of Figure 6 shows the *enthalpy interaction parameter*

$$\lambda = \Delta H_m / x_1 x_2 \qquad (22)$$

which is a slowly varying function of composition [i.e., the term in parentheses in Eq. (21)], revealing details that cannot be easily recognized from the ΔH_m

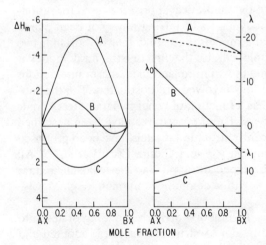

Figure 6. Typical enthalpies of mixing curves and interaction parameters for charge-symmetrical binary systems.

Table 6
Selective Symmetrical Fused Salt Systems Investigated by High-Temperature Calorimetry

System	$\lambda_{0.5}$ (kcal/mol)	System	$\lambda_{0.5}$ (kcal/mol)
Alkali Nitrates[124]		(Cu–Rb)Br	−6.79
(Li–Na)NO$_3$	−0.46	(Cu–Cs)Br	−8.42
(Li–K)NO$_3$	−1.81		
(Li–Rb)NO$_3$	−2.56	(Cu–Ag)Cl	−0.82
(Li–Cs)NO$_3$	−3.00	(Cu–Ag)Cl	−0.31
(Na–K)NO$_3$	−0.44		
(Na–Rb)NO$_3$	−0.87	*Alkali Hydroxides*[128]	
(Na–Cs)NO$_3$	−1.26	(Li–Na)OH	−1.66
(K–Rb)NO$_3$	−0.06	(Li–K)OH	−3.98
(K–Cs)NO$_3$	−0.13	(Li–Rb)OH	−3.73
(Rb–Cs)NO$_3$	−0.01	(Li–Cs)OH	−3.57
		(Na–K)OH	−0.56
		(Na–Rb)OH	−0.87
Alkali Chlorides[125]		(Na–Cs)OH	−0.76
(Li–Na)Cl	−0.78	(K–Rb)OH	+0.05
(Li–K)Cl	−4.25	(K–Cs)OH	+0.29
(Li–Rb)Cl	−5.06	(Rb–Cs)OH	+0.06
(Li–Cs)Cl	−6.08		
(Na–K)Cl	−0.52		
(Na–Rb)Cl	−0.81	*Alkali Fluorides*[129]	
(Na–Cs)Cl	−1.08	(Li–Na)F	−1.80
(K–Rb)Cl	+0.02	(Li–K)F	−3.85
(K–Cs)Cl	+0.19	(Li–Rb)F	−4.85
(Rb–Cs)Cl	+0.08	(Li–Cs)F	−3.86
		(Na–K)F	−0.08
		(Na–Rb)F	+0.09
Halides of Ag, Tl, Cu[126,127]		(K–Rb)F	+0.08
(Ag–Li)Cl	−1.82		
(Ag–Na)Cl	−1.09		
(Ag–K)Cl	−2.08	*Alkali Iodides*[130]	
(Ag–Rb)Cl	−3.59	(Li–Na)I	−0.99
(Ag–Cs)Cl	−5.08	(Li–K)I	−3.06
		(Li–Rb)I	−4.17
(Tl–Li)Cl	−0.49	(Li–Cs)I	−5.06
(Tl–Na)Cl	+1.74	(Na–K)I	−0.55
(Tl–K)Cl	+1.28	(Na–Rb)I	−0.89
(Tl–Rb)Cl	+0.72	(Na–Cs)I	−1.36
(Tl–Cs)Cl	−0.06	(K–Rb)I	−0.02
		(K–Cs)I	−0.01
(Cu–Li)Cl	+1.02	(Rb–Cs)I	+0.02
(Cu–Na)Cl	−1.52		
(Cu–K)Cl	−6.16	*Alkali Sulfates*[131]	
(Cu–Rb)Cl	−7.92	(Li–Na)$_2$SO$_4$	−2.34
(Cu–Cs)Cl	−9.95	(Li–K)$_2$SO$_4$	−6.05
		(Li–Rb)$_2$SO$_4$	−7.10
(Cu–Li)Br	+0.9	(Li–Cs)$_2$SO$_4$	−6.9
(Cu–Na)Br	−1.26	(Na–K)$_2$SO$_4$	−1.05
(Cu–K)Br	−5.30		

Table 6 (continued)

System	$\lambda_{0.5}$ (kcal/mol)	System	$\lambda_{0.5}$ (kcal/mol)
$(Na–Rb)_2SO_4$	−1.40	*Mixed Halides–Halides*[135]	
$(Na–Cs)_2SO_4$	−0.94	$(Cl–I)Li$	+0.41
$(K–Rb)_2SO_4$	−0.01	$(Cl–Br)Li$	+0.04
$(K–Cs)_2SO_4$	+0.22	$(Br–I)Li$	+0.19
$(Rb–Cs)_2SO_4$	+0.09		
		$(Cl–I)Na$	+0.46
		$(Cl–Br)K$	+0.08
		$(Br–I)K$	+0.15
Alkali Carbonates[132]			
$(Li–Na)_2CO_3$	−0.64	$(Cl–I)K$	+0.33
$(Li–K)_2CO_3$	−1.89	$(Cl–Br)K$	+0.05
$(Li–Rb)_2CO_3$	−1.80	$(Br–I)K$	+0.10
$(Li–Cs)_2CO_3$	−2.20		
$(Na–K)_2CO_3$	−0.32	$(Cl–I)Rb$	+0.26
$(Na–Rb)_2CO_3$	−0.48	$(Cl–Br)Rb$	+0.03
$(Na–Cs)_2CO_3$	−0.49	$(Br–I)Rb$	+0.09
$(K–Rb)_2CO_3$	∼0.00		
$(K–Cs)_2CO_3$	+0.02	$(Cl–I)Cs$	+0.16
$(Rb–Cs)_2CO_3$	+0.01	$(Cl–Br)Cs$	+0.01
		$(Br–I)Cs$	+0.07
		Alkaline Earth Chlorides[136]	
		$(Mg–Ca)Cl_2$	+0.88
Metaphosphates[133]		$(Mg–Sr)Cl_2$	−0.40
$(Li–Na)PO_3$	−0.86	$(Mg–Ba)Cl_2$	−2.12
$(Li–K)PO_3$	−3.24	$(Ca–Sr)Cl_2$	−0.47
$(Li–Rb)PO_3$	−4.18	$(Ca–Ba)Cl_2$	−2.09
$(Li–Cs)PO_3$	−5.01	$(Sr–Ba)Cl_2$	−0.43
$(Na–K)PO_3$	−0.62		
$(Na–Rb)PO_3$	−1.06	*Transition Metal Chlorides*[25]	
$(Na–Cs)PO_3$	−1.39	$(Mg–Mn)Cl_2$	+0.38
		$(Mg–Fe)Cl_2$	+0.30
		$(Mg–Co)Cl_2$	+0.45
		$(Ca–Mn)Cl_2$	+0.14
Mixed Halides–Nitrates[134]		$(Ca–Fe)Cl_2$	+0.41
$(NO_3–Cl)Li$	+0.21	$(Ca–Co)Cl_2$	+1.48
$(NO_3–Cl)Na$	+0.40	$(Cd–Mn)Cl_2$	+0.08
$(NO_3–Cl)K$	+0.21	$(Cd–Fe)Cl_2$	+0.09
$(NO_3–Cl)Rb$	+0.12	$(Mn–Co)Cl_2$	+0.34
$(NO_3–Cl)Cs$	+0.17	$(Mn–Fe)Cl_2$	∼0.0
$(NO_3–Cl)Tl$	+0.74	$(Fe–Co)Cl_2$	+0.160
$(NO_3–Cl)Ag$	−1.13		

curves. The limiting values of the interaction parameters at $x_2 \to 0$ or $x_2 \to 1$ are of particular interest, because they are related to the partial enthalpies of solution of the component salts:

$$\lim_{x_2=0} \lambda = \lambda_0 = \Delta\bar{H}_{BX}, \qquad \lim_{x_2=1} \lambda = \lambda_1 = \Delta\bar{H}_{AX} \qquad (23)$$

The reported values of c [Eq. (21)] have been found to be small relative to a. The *asymmetry* in the interaction parameter $b = \lambda_1 - \lambda_0$ also has small values, and, to a first approximation, the enthalpy of mixing can be reasonably specified by the value of a. Thus most charge-symmetrical systems show a "regular solution"-like behavior,[111] and $\Delta H_m \approx a x_1 x_2$.

Early attempts to rationalize Eq. (21) for the alkali nitrates were based on the coulombic repulsive model of Forland (Figure 5) and/or on the changes in the polarization forces on mixing.[137] All negative enthalpies of mixing of the alkali nitrates can be related approximately to the size parameter $\delta = (d_1 - d_2)/(d_1 + d_2)$:

$$\Delta H_m \approx x_1 x_2 \lambda_{0.5} = x_1 x_2 A \delta^2 \tag{24}$$

where $\lambda_{0.5}$ is the interaction parameter at $x_1 = x_2 = 0.5$ and A is a constant characteristic of nitrate melts. An analogy between Eqs. (24) and (20) can thus be recognized.

Further insight into the dependence of ΔH_m (or λ) on the size parameter for the nitrates, as well as for most of the systems listed in Table 6, has been given by developments of the conformal ionic solution theory.[110,138,139] The theory is based on the model ionic melt utilized in the theory of corresponding states [(Eq. (7)], but with a modified pair potential so as to include polarization and dispersion forces:

$$\Phi_{+-} = \begin{cases} \infty & r \leq d \\ -z_{(+)}z_{(-)}e^2/ur + \xi V(r), & r > d \end{cases} \tag{25}$$

$$\Phi_{ii} = z_{(i)}z_{(i)}e^2/ur + \xi V(r), \qquad r > 0, i = + \text{ or } -$$

where $\xi V(r)$ is a potential term representing one or more of the polarization–dispersion interactions in Eq. (2). Thus for dipole–dipole interactions $\xi = C_{ij}$ and $V(r) = 1/r^6$. In developing the theory, the term $V(r)$ is considered as a perturbation to the coulombic potential. For the binary systems AX–BX, with no polarization interactions present (i.e., $\xi = 0$), the conformal ionic solution theory allows the calculation of the thermodynamics of mixing in terms of the size parameter of a *reference salt*[110,138]:

$$\Delta V_m = x_1 x_2 (d_1 - d_2)^2 \left(\frac{\partial \Lambda}{\partial P} \right)_T$$

$$\Delta G_m = x_1 x_2 (d_1 - d_2)^2 \Lambda$$

$$\Delta S_m = x_1 x_2 (d_1 - d_2)^2 \left(\frac{\partial \Lambda}{\partial T} \right)_P \tag{26}$$

$$\Delta F_m = x_1 x_2 (d_1 - d_2)^2 \left[\Lambda - \left(\frac{\partial \Lambda}{\partial \ln P} \right)_T \right]$$

$$\Delta H_m = -x_1 x_2 (d_1 - d_2)^2 T^2 \left(\frac{\partial (\Lambda/T)}{\partial T} \right)_P$$

In these equations Λ is a complicated integral function of the potential energy and d_1 and d_2 are the size parameters of the two salts. In correlating the data with the theory, a reference salt is chosen (e.g., $LiNO_3$) and the dependence of ΔH_m (or of any other thermodynamic function) on $d_1 - d_2$ is examined for all systems containing the reference salt (e.g., $LiNO_3$–ANO_3; A = Na, K, Rb, Cs).

When $\xi V(r)$ is considered, then relations more complex than those of Eq. (26) are derived. For the enthalpy of mixing, relative to that of the reference salt, at least two more terms are added[139]:

$$\Delta H_m = x_1 x_2 [U_0 + U_1(d_1 - d_2) + U_2(d_1 - d_2)^2] \tag{27}$$

where the U's are complex functions related to certain coulombic and/or polarization interactions.

Experimental measurements of the enthalpies of mixing of alkali halides have shown that the U_1 term is negligible for most systems and that the measured enthalpy can be accurately represented by a semiempirical relation of the type

$$\Delta H_m = x_1 x_2 (U_0^{++} + D\delta_{12}^2), \qquad \delta_{12} = (d_1 - d_2)/d_1 d_2 \tag{28}$$

where a normalized size parameter (δ_{12}) is used. The positive term U_0^{++}, which differs from system to system, is attributed to dipole–dipole interactions between nearest neighbor cations and is in agreement with the theoretical prediction [Eq. (27)]. The positive constant D was found to be the same for families of binary mixtures containing a common salt.

Figure 7 shows the validity of the "semiempirically derived" Eq. (28). The calculation of U_0^{++} is based on a crude model in which the changes in the dipole–dipole (van der Waals–London) interactions that occur in the "mixing" process $AXA + BXB \rightarrow 2AXB$ are approximately estimated[111] by $U_0^{++} \approx (\sqrt{U_{AX}^{++}} - \sqrt{U_{BX}^{++}})^2$, with U_{AX}^{++} and U_{BX}^{++} being the dipole–dipole interactions in the pure components [i.e., $U_{ij}^{++} \sim C_{ij}/d^6$; Eq. (2)]. For the systems summarized in Figure 7 the values of the coefficient D [Eq. (28)] were found to increase in the sequence $F^- < Cl^- < Br^- < I^-$. Furthermore, for certain fluoride systems the data cannot readily be expressed by Eq. (28) and a new linear term in δ_{12} [similar to the U_1 term suggested by Eq. (27)] needs to be added.

Linear correlations similar to Eqs. (27) and (28) were also found for halides containing highly polarizable ions of Ag, Tl, and Cu,[126,127] the alkali hydroxides,[128] alkali sulfates,[131] alkali nitrates,[124] and alkaline earth chlorides.[136] Deviations, however, were observed for the chlorides of transition metals,[25] alkali metaphosphates,[133] and certain common cation mixtures.[135,140]

As a result of these studies the following general remarks can be made:

1. The magnitude and sign of the enthalpies of mixing of symmetrical binary systems arise from the competition of coulombic and dispersion interac-

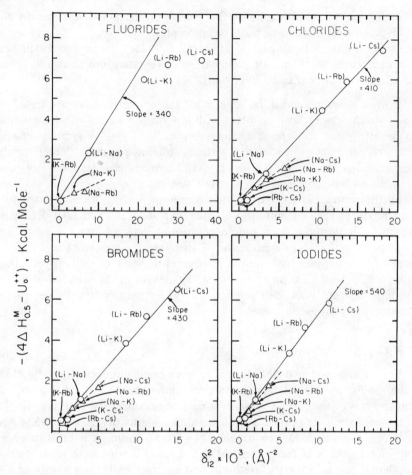

Figure 7. Dependence of the corrected interaction parameter at 50% mole fraction $(4\Delta H^M_{0.5} - U^{++}_0)$ on the size parameter δ_{12}. ⊙—systems with Li as smaller cation; △—systems with Na^+ as smaller cation; ○—systems with K^+ as smaller cation; □—systems with Rb^+ as smaller cation (reproduced from Reference 130).

tions and their changes upon mixing. Temperature variation does not appear to have an effect on the enthalpy of mixing.

2. For most melts with a common anion, the contribution of dispersion forces arises from the dipole–dipole interactions between second-nearest neighbors (i.e., the U_0 term) and, to some extent, from the cation–anion polarization energy, which affects the U_2 (or D) term of Eq. (27) (or Eq. (28)).

3. For fluoride melts and for melts containing highly polarized ions the nearest-neighbor cation–anion polarization interactions make contributions that give rise to the U_1 term.

4. For common cation melts the coulombic term is overcome by the dipole–dipole terms, yeilding positive enthalpies.

5. For melts containing transition metal chlorides, coulombic and dispersion interactions alone cannot justify their systematics; covalent and ligand field effects also appear to be important.

It must be emphasized, however, that the above remarks are based on enthalpy data alone. Discrepancies within a certain family of systems may also be attributed to volume changes (expansion) and/or entropy effects, which are presumably associated with the mixing process. Unfortunately, determinations of volume, free-energy, and entropy changes have not been systematic and are confined only to a few systems.

Furthermore, it is noteworthy that computer simulations of the structure of salt mixtures[141,142] (limited to equimolar NaCl–LiCl and LiCl–KCl mixtures) have indicated that a simple conformal ionic solution theory alone cannot account for the dependence of structure on concentration. Recent attempts at computer simulation of the enthalpies of mixing of alkali halides[143,144] tend to support this view. Extension of the simulation model to include concentration dependence may well give important clues to the understanding of these mixtures.

4.1.2. Charge-Unsymmetrical Systems

During the past two decades there have been numerous investigations of charge-unsymmetrical binary systems, and a large area of molten-salt chemistry has evolved. The present discussion is concerned with the simpler and systematically examined common anion systems of the type NX_n–AX, where N is a divalent (M) or trivalent (L) metal cation, A is most often an alkali metal, and X is the anion (halide). These binaries have been studied by both emf and calorimetric methods, and accurate enthalpies of mixing and chemical potentials are available. Calculations using these last two quantities provided very good data on the partial entropies of mixing. Table 7 lists representative unsymmetrical systems that have been investigated by galvanic and calorimetric methods.

4.1.2.1. The Interaction Parameter

The general shape of the integral enthalpy function of unsymmetrical systems is analogous to that presented in Figure 6a. Most of the enthalpies that have been measured have been negative, but certain weakly interacting systems have shown positive values. The enthalpy interaction parameter, however, depends strongly on composition. For a family of systems NX_n–AX (A = Li, Na, K, Rb, Cs) the variation of λ with x_2 (where x_2 is the mole fraction of the NX_n component) may be a smooth but fast changing function (e.g., Figure 8a) or a more complex function possessing minima at certain

Table 7

Selective Systematic Studies of Common Anion Charge Unsymmetrical Binary Systems

System	Reference number for ΔH and $\Delta \bar{H}$	Reference number for $\Delta \mu$	Reference number for ΔS	λ [a]
$Ca(NO_3)_2–ANO_3$	145			S
$Sr(NO_3)_2–ANO^3$				S
$Ba(NO_3)_2–ANO_3$				S
$PbCl_2–ACl$	146			M
$MgCl_2–ACl$	147	118	118	M
$MnCl_2–ACl$	26	148, 149	148	M
$FeCl_2–ACl$	26	150, 151, 152		M
$CoCl_2–ACl$	26	151, 153, 154	154	M
$NiCl_2–ACl$	67	155		M
$ZnCl_2–ACl$	156	157		M
$CaCl_2–ACl$	118	118	118	S
$SrCl_2–ACl$	118	118	118	S
$BaCl_2–ACl$	118	118	118	S
$MgBr_2–ACl$	118	118	118	M
$ZnBr_2–ACl$	156			M
$BeF_2–AF$	158a	158b		M
$MgF_2–AF$	159			M
$CaF_2–AF$	159			S
$SrF_2–AF$	159			S
$BaF_2–AF$	159			S
ZnF_2AF	160			M(?)
$CaF_2–AF$	161			S(?)
$AlCl_3–ACl$		162		
$LaCl_3–ACl$	163	163	163	M
$CeCl_3–ACl$	164			M
$GdCl_3–ACl$	165			M
$LCl_3–NaCl$	165			S
$YCl_3–ACl$	166	166	166	M
$AlF_3–AF$	167	168	167	M
$LaF_3–AF$	169			M(?)
$YbF_3–AF$	169			M
$YF_3–AF$	169			M

[a] Systems with A = K, Rb, and Cs, which show a minimum in λ, are denoted by "M." Systems with a smoothly varying λ are denoted with "S."

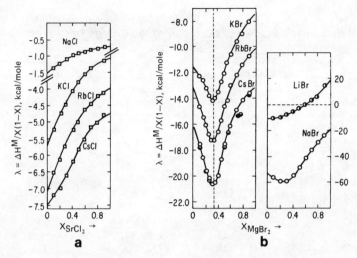

Figure 8. Enthalpy of interaction parameters for (a) the $SrCl_2$–ACl binaries and (b) the $MgBr_2$–AB_2 binaries[118] (reproduced from Reference 118).

compositions (e.g., Figure 8b). Examination of families of mixtures with different NX_n components shows variations that are depicted schematically in Figure 9 for both divalent and trivalent salts. Thus for the LiX–NX_n binaries almost linear behavior has been observed, whereas for the CsX–NX_n binaries a minimum in λ appears at $x_2 \sim 0.33$ or $x_2 \sim 0.25$ for the divalent and trivalent systems, respectively. For NX_n binaries with the remaining alkali halides (Na, K, Rb), intermediate behavior similar to that shown in Figure 8b has been found.

All families of systems in Table 7 that have interaction parameters similar to those in Figure 9 and 8b are identified by M (*minimum*) in the column under λ, whereas systems that change smoothly, as in Figure 8a, are identified by S. For certain fluoride binaries[159–161] intermediate behavior has been also observed.

Application of the conformal ionic solution theory to unsymmetrical binaries leads to complex equations.[170,171] When dispersion interactions are considered[110] as part of the ionic potential [e.g., Eq. (25)], the theoretical result appears to be impractical, since the important terms and their relation to size and polarization parameters are difficult to recognize. Oversimplification of the model by assuming that only coulombic interactions occur ($\xi = 0$) leads to the equation

$$\Delta H_m = A + \{[1 + x_2(n - 1)]B + C\}(d_1 - d_2) + E(d_1 - d_2)^2 \quad (29)$$

where again A, B, C, and E are complicated integral functions depending implicitly on the size of the ions (d_1, d_2), mole fraction (x_2), and valence (n)

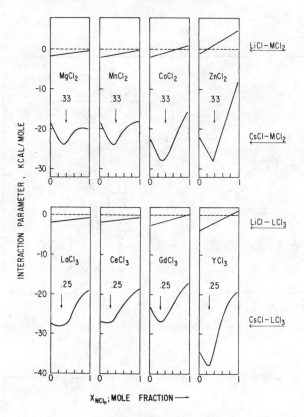

Figure 9. Systematics of the enthalpy interaction parameter of binary molten salts (data obtained from References 26, 147, 156, 163, 164, and 166).

of the cation N. It has also been shown that the limiting interaction parameter for $x_2 \to 0$ (i.e., λ_0) varies linearly with the size parameter $d_1 - d_2$ (or δ_{12}):

$$\lambda_0 = \alpha + \beta(d_1 - d_2) \qquad (30)$$

This last result has been frequently found to hold for divalent–monovalent mixtures.[26,67,118,145–147] Figure 10 demonstrates the "straight" lines found for the λ_0 of transition metal[26] and alkaline earth chlorides.[118] However, deviations from linear behavior have been observed for some of the LiX salts (Figure 10b) as well as for other families of binaries.

Recent measurements on a series of lanthanide halide–alkali halide binary melts[163–165] and of certain binary fluoride melts[167,169] have indicated that Eq. (30) cannot be used to correlate the enthalpy data. Addition of a quadratic term[163] or use of the ionic charge in the size parameter[169] have been recommended. Finally, it must be emphasized that the simple conformal ionic theory does not predict any explicit mole fraction dependence of the

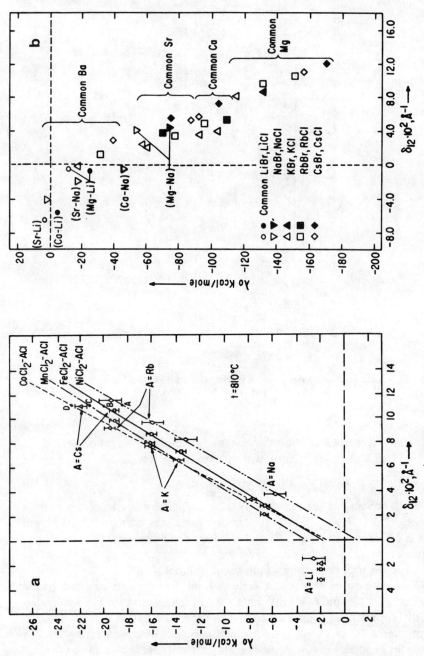

Figure 10. Dependence of the limiting interaction parameters on the size parameter $\delta_{12} = (d_1 - d_2)/(d_1 d_2)$ for the transition metal[26] and alkaline earth chlorides.[118] Note that the $d_1 d_2$ product does not drastically affect δ_{12} and can be incorporated in the b coefficient of Eq. (30) (reproduced from References 26 and 118).

interaction parameter and thus cannot account for the appearance of the minimum in λ.

4.1.2.2. Partial Molar Quantities

Measurements of the chemical potential for each of the component salts in charge-unsymmetrical binaries yield negative values, which become more exothermic with increasing concentration and by substituting the A^+ cation from Li to Cs. Figure 11a shows the variation of the chemical potential and partial molal enthalpy of the alkali chlorides for the family of binaries with $MgCl_2$.[118] The partial molal enthalpies are usually measured from integral enthalpy data by the method of intercepts. Similar dependencies of $\Delta\mu_{AX}$ and $\Delta\bar{H}_{AX}$ have been observed for other families of unsymmetrical binaries, with

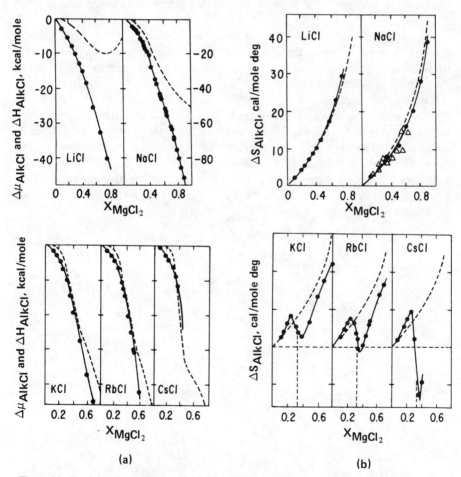

Figure 11. Partial molar quantities of the $MgCl_2$–ACl binaries (reproduced from Reference 118).

the values of $\Delta\mu_{AX}$ varying from about $-1\,\text{kcal mol}^{-1}$ for the $CaCl_2$–$LiCl$ system (at $x_2 \sim 0.4$) to high exothermic values of -5–$-10\,\text{kcal mol}^{-1}$ for systems possessing a deep minimum in λ, e.g., $CoCl_2$–$CsCl$, AlF_3–KF (at $x_2 \sim 0.4$). Calculations of partial entropies by combining the partial quantities in Figure 11a give the mole fraction dependency of $\Delta\bar{S}_{AX}$ in solution.

Figure 11b presents the results for the $MgCl_2$–ACl system[118]; the broken line represents the ideal partial entropy of the Temkin model [Eq. (16)]. Random mixing presumably occurs in mixtures containing $LiCl$ and $NaCl$. However, for systems involving K, Rb, and Cs the partial entropy is a S-Shaped curve possessing inflection points at compositions $x_2 \sim 0.30$–0.35.

Changes of the divalent salt in the MX_2–AX systems give rise to more negative entropies and sharper S curves for systems like $CoCl_2$–ACl, whereas binaries having no minimum in λ show a smoother dependence of $\Delta\bar{S}_{AX}$ on x_2. This is depicted in Figure 12 for three representative binaries with KCl.

Depending on the size of the trivalent cation, binaries of the type LX_3–AX may or may not show sharp changes in the partial entropy quantities. For those systems possessing a S-shaped ΔS_{AX} entropy curve (Figures 12 and 13) the inflection points occur at $x_2 \sim 0.20$–0.25, i.e., at the composition where the interaction parameters show pronounced minima. This correspondence is clearly indicated in Figure 13, where both the λ and $\Delta\bar{S}_{ACl}$ for the YCl_3–ACl binaries[166] are presented. In comparison with the Temkin model, both positive and negative deviations from ideality have been observed for the LX_3–AX systems. The early lanthanides (e.g., $LaCl_3$–ACl[163]) give partial entropies closer to those of the ideal mixture, while strongly interacting systems, e.g., AlF_3–AF,[167] YCl_3–ACl,[166] show very pronounced negative deviations.

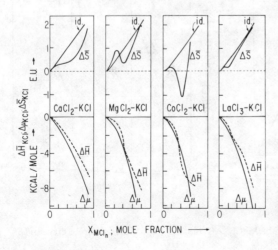

Figure 12. Schematic presentation of the systematics of the partial molar quantities in binary systems (data from References 118, 154, and 163).

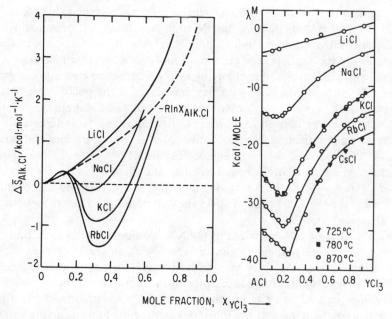

Figure 13. Partial entropies and interaction parameters in the YCl_3–ACl binary (reproduced from Reference 166).

4.1.2.3. Complexing

It is evident from the above results and discussion that the deviations from ideality and the minimum in the interaction parameter for the charge-unsymmetrical binaries cannot be described in terms of simple statistical-mechanical theories[110] alone. These effects have been hitherto semiquantitatively attributed to extreme ordering effects, especially in those melts containing K, Rb, and Cs. On the basis of simple structural models,[26,118,172] it has been proposed that the minimum in λ and the appearance of inflection points in $\Delta \bar{S}_{AX}$ are indicative of "complex" configurations in these mixtures. Thus for the MX_2–AX binaries complexes with stoichiometry MX_4^{2-} have been proposed, and calculations based on detailed structural models have been carried out[118] to reproduce the S-shaped curves for the partial entropy and interaction parameters, with minima at $x = 0.33$. For the LX_3–AX binaries the proposed stoichiometry[163] is determined by the minimum of λ, i.e., LX_6^{3-}, but detailed structural model calculations have not yet been done. This view of "complexing" has been substantiated by a series of spectroscopic investigations, which are presented in the following section.

The concept of "complexing" in molten-salt mixtures is somewhat different from that for soluble species in aqueous or nonaqueous solvents. According to the neutron scattering studies (e.g., Table 3), the two components of a NX_n–AX binary (e.g., $BaCl_2$ and KCl) possess certain structures for which an average number of first- and second-nearest neighbors can be defined. In the molten mixture the M and A cations will presumably be surrounded by the common anions and have A and/or N cations in their second coordination sphere. For mixtures dilute in NX_n there will be an excess of A cations and the probability of having two N^{n+} as second-nearest neighbors is small. This probability will be further diminished if the polarization forces (e.g., polarizing power, $1/r$; polarizability, α) in the competition for the common anions are different. According to the Lumsden[137] and Forland[114] models, strong polarization forces will favor N–X–A configurations and, when excess of A is available, the first-neighbor cations of N will be A alone. Thus, depending on charge, steric requirements, and ligand covalent effects, a number of anions (e.g., 4, 6) will surround the N^{n+} cation and the counter cations will be only A^+. In this respect the "complex" is defined by both its first and second coordination spheres.

With increasing concentration of the NX_n component, the probability of two N cations as second neighbors increases and, in terms of the second coordination sphere, "complexing" diminishes and/or new binuclear "complexes" may be formed. Thus at concentrations where there is not a sufficient quantity of A cations to surround the "complex," the microstructure in the liquid will start changing and the thermodynamic functions of mixing will be altered. The observed variations of λ and $\Delta \bar{S}_{AS}$ in Figures 8, 11, and 13 indicate that such changes occur in the composition range 0.3–0.35 and 0.2–0.25 for the MX_2–AX and LX_3–AX binaries, respectively.

It should be emphasized, however, that the absence of "complexing" in mixtures rich in NX_n and/or in mixtures involving weakly interacting systems (i.e., without a minimum in λ) does not necessarily imply restrictions on the coordinations of the N cation (e.g., Mg^{2+} may be tetrahedrally coordinated in both the $LiCl$–$MgCl_2$ and $CsCl$–$MgCl_2$ systems).

4.1.2.4. Ligand Field Effects

For melts containing ions of the transition metals (e.g., M = Mn, Fe, Co, Ni, Cu, Zn) the thermodynamics of mixing reflect a degree of covalent character in the cohesive energy[25] and are expected to show effects associated with the splitting of the d electrons of M^{2+} in the electrostatic environment of the anions. It is well established[173] that for solids and complexes in solutions the d-electron splitting leads to additional energy gain, known as ligand field stabilization energy (LFSE). To a first approximation, this splitting varies with the symmetry of the environment around the cation, and thus in a process where coordination changes occur the thermodynamic stability will also be affected (apart from the electrostatic energy changes) by the LFSE. This last effect is illustrated in Figure 14a, where the enthalpies of solid reactions for formation of $CsMCl_{3(s)}$ and $Cs_3MCl_{5(s)}$ from $MCl_{2(s)}$ and $CsCl_{(s)}$ are shown.[27] It is known from crystallographic investigations that the site symmetry of M in Cs_3MCl_5 is nearly tetrahedral, whereas in $CsMCl_3$ and MCl_2 it is near

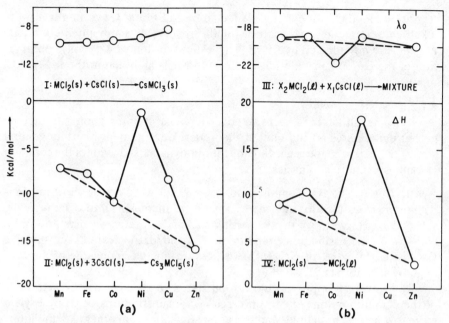

Figure 14. Parallel plots of the enthalpies of solid reactions with the limiting interaction parameters and enthalpies of fusion.

octahedral (M ≠ Cu). Thus reaction I in Figure 14a does not involve any coordination changes and the variation of the enthalpies of reaction are small. In contrast, reaction II takes place with an octahedral–tetrahedral change of site symmetry, and the LFSE is strongly reflected in the enthalpy.

Figure 14b gives parallel plots for the interaction parameter (λ_0) associated with mixing process III and for the enthalpies of fusion of the MCl_2 solids.[26,27,156] By a comparison of Figures 14a and 14b, it can be implied that, on melting, MCl_2 coordination changes occur that are similar to the ones observed for reaction II; thus melting is accompanied by a "site" symmetry change from octahedral (in the solid) to tetrahedral (in the liquid). On the other hand, the limiting interaction parameter, which reflects the properties of the highly ordered states (i.e., of "complexing") of dilute solutions of MCl_2 in CsCl (reaction III, Figure 14b), varies in a way that is similar to that observed in the formation of Cs_3MCl_5 (reaction I). This would imply that no drastic changes in the coordination of the M cation occur on mixing; i.e., the MCl_4^{2-} "complex" suggested for these mixtures[148,154] has a nearly tetrahedral coordination, which is also similar to the coordination of the M cation in pure MCl_2 liquid.

For the remaining MCl_2–ACl binary mixtures that contain alkali metal ions other than Cs the thermodynamic behavior is similar. Two conclusions can be drawn from the above observations: (1) The cohesive energy of pure transition metal chloride melts is affected by LFSE and (2) upon mixing with alkali chlorides, the gross features of the relative enthalpy interaction parameters are not affected by LFSE. The last conclusion would imply conservation of the cation coordination in the pure MCl_2 melt and in the MCl_2–ACl mixtures.

4.1.2.5. Systematics

As a result of the extensive investigations of interaction parameters, partial quantities, and the limited theoretical developments, certain general statements can be made regarding the thermodynamic behavior of the common anion unsymmetrical binaries:

1. Relative to the Temkin model, the NX_n–AX systems show *negative* deviations from ideality that *increase* with (a) increasing size of A (e.g., by substituting Na with K), (b) decreasing size of N^{n+} (e.g., by substituting La with Y), (c) increasing polarizability of N^{n+} and/or A^+ (e.g., by substituting Ca^{2+} with Ma^{2+} and/or Na^+ with Ag^+, and (d) increasing charge of the M^{n+} (e.g., by substituting Ca^{2+} with La^{3+}).

No universal empirical rule can be stated relative to the characteristics of the common anion. In certain cases[166,169] substitutions of Cl by F yield more exothermic enthalpies and increasing deviations; in others[118] the interaction parameter was found to increase in the sequence Cl < Br < I.

2. The appearance of S-type curves and inflection points on the partial entropies and of minima in the interaction parameter is indicative of "complexing" in the mixture. The reverse, however, is not valid, and in cases like the LiX–NX_n binaries, where no minima have been seen in λ, tetrahedral,[69] octahedral,[73] and/or a distribution of geometries have been observed spectroscopically.

3. Thermodynamic considerations alone cannot elucidate the structure of the complexes, and supplementary spectroscopic information is required. Thus for the strongly interacting systems with divalent ions the interaction parameters show minima at $x_2 < 0.3$, and complexes with a stoichiometry of MX_4^{2-} are anticipated. For trivalent systems the minima appear at $x_2 < 0.25$ and the complex has a MX_6^{3-} stoichiometry. Raman and electronic absorption spectroscopic studies have confirmed this and indicated nearly tetrahedral or octahedral symmetries for these stoichiometries, respectively.

4. For a family of systems the interaction parameter becomes more exothermic and its asymmetry $(\lambda_0 - \lambda_1)$ increases in the sequence Li < Na < K < Rb < Cs. For binaries having component salts with high viscosities and tendencies for networklike structures (e.g., ZnX_2[156]) high asymmetry values have been observed. Weakly interacting binaries of such melts with LiX show positive values of λ at low LiX concentrations.

5. Variations of temperature do not change the values and shape of the interaction parameter. For certain systems,[26,118,156] however, small temperature effects have been observed, but no regularities can be deduced.

Hitherto only the structural models[118,172] and the polarization concepts of Lumsden[137] can give a crude semiquantitative account of the phenomenological observations listed above. The conformal ionic solution theory appears to have no practical applicability for these systems. Computer simulations have been carried out only for the $MgCl_2$–KCl systems,[174,175] with a pair potential obscuring the importance of polarization forces and resulting in nonrealistic structures in the mixture. Similar simulations with the use of a more realistic pair potential might be more enlightening.

4.2. Spectroscopy and Structure

For single salts neutron scattering studies have provided perhaps the most direct information on the melt structures. Since very few researchers have access to the equipment needed for such work, only a limited number of salts have been investigated, and so far there have been no attempts to apply this technique to molten-salt mixtures. For a simple common anion binary six partial structure factors [Eq. (5)] have to be resolved from the scattering pattern, and complicated computational methods are required for deriving the six RDFs. A full understanding of the bonding in the mixture will additionally require a choice of pair potential, selected by successive comparisons of computer-simulated results with the neutron scattering data.

Because of these difficulties, other spectroscopic techniques (Table 1) have been used for studying the structure and bonding in melt mixtures. In many cases the choice of the technique to be applied depends on the system under observation. If, for example, systems containing transition metal ions are to be investigated, spectroscopy related to the d-electron states and ligand field splittings may be used (i.e., electronic absorption, EPR, and magnetic measurements). On the other hand, Raman spectroscopy is easily applied to transparent liquids and may give direct information on the nature and chemical species present in the melt. The following section presents and discusses the findings obtained with three principal spectroscopic techniques used for molten-salt investigations: (1) vibrational (mainly Raman) spectroscopy, (2) electronic absorption spectroscopy, and (3) magnetic measurements.

These techniques have also been used successfully in studies of solute species in different melts. Coordination geometries and oxidation states of unusual solute species have been substantiated by both spectrophotometric and Raman measurements. Solvent melts, where the "acidity" can be controlled by varying the melt composition, have established a new field of solution chemistry. Such studies are outside the main purpose of this chapter, which is concerned with the structure of the *bulk* melt, but certain review articles that address the field of molten-salt solution chemistry are available.[176-178]

4.2.1. Vibrational Spectroscopy

Infrared and Raman spectroscopies have long been effective means of investigating the vibrational frequencies and molecular characteristics of poly-atomic molecules and ions. Raman spectroscopy was applied to low melting point salts soon after the experimental discovery of the Raman effect. Limited studies of melts at high temperatures were carried out in the 1950s and 1960s by Bues,[179,180] Janz and James,[181] Young and Irish,[182] and others.[183,184] Relatively complicated experimental designs for combining the mercury arc light source and the furnace were required, and the spectra were mainly activated by photographic techniques. Descriptions of the experimental arrangements and the early data have been reported in several review articles.[185–187]

The introduction of visible-radiation lasers as excitation sources has drastically simplified the technique and permitted measurements of Raman spectra of corrosive melts at high temperatures.

The application of IR spectroscopy to the study of melts is limited to a few salts. IR absorption reflectance and emission methods have been developed and successfully used up to ~873 K.[188] These techniques, however, are more elaborate relative to that of laser Raman spectroscopy, and only a few such studies have been reported in the past decade.

4.2.1.1. Vibrations of Molecules, Lattices, and Melts

An isolated n-atom molecule has $3n$ degrees of freedom and $3n - 6$ vibrational degrees of freedom (or $3n - 5$ for a linear molecule). The collective motions of the atoms moving with the same frequency and in phase with all other atoms give rise to the normal modes of vibration. Examples of such modes are given in Figure 15 for octahedral and tetrahedral molecules. The arrows are used to indicate the instantaneous relative velocities and amplitudes of motion associated with each atom. The determination of the form of the normal modes for any molecule requires, in principle, the solution of the equation of motion appropriate to the n-atom system. Established methods for such analyses are available.[189,190] Molecular symmetry and methods of group theory[189] are important in deriving the symmetry properties of the normal modes. With the aid of the character tables for point groups and the symmetry properties of the normal modes, the "selection rules" for Raman (R) and IR activity may be derived.[191] Results for some selected octahedral and tetrahedral molecules are exemplified in Figure 15. For a molecule with a *center* of symmetry, e.g., LX_6, no Raman active modes are IR active, whereas for the MX_4 tetrahedral molecule some modes are simultaneously IR and Raman active.

The vibrational properties of a group of isolated molecules change drastically when these molecules are "condensed" to form a crystalline solid. The influence of the neighboring "molecules" and the surrounding crystalline

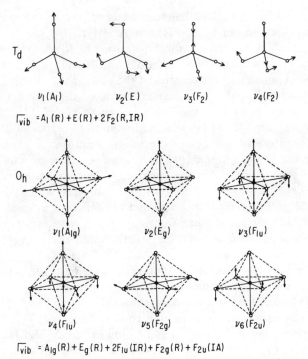

$$\Gamma_{vib} = A_1(R) + E(R) + 2F_2(R,IR)$$

$$\Gamma_{vib} = A_{1g}(R) + E_g(R) + 2F_{1u}(IR) + F_{2g}(R) + F_{2u}(IA)$$

Figure 15. Normal modes of vibration in tetrahedral (T_d) and octahedral (O_h) molecules.

lattice will alter the vibrational modes of the molecules. The long-range order correlates the atoms in the crystal, and the vibrations are described in terms of lattice waves rather than by free molecular modes.

For example, the vibrations of a simple linear diatomic lattice (Figure 16) give rise to two "transverse" types of waves, known as the optical and acoustical modes. Changes of dipole moment of the optical modes make them IR active, but no polarizability changes occur and no Raman activity is

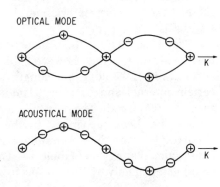

Figure 16. Optical and acoustical waves in a linear diatomic lattice.

expected for both modes (alkali halide crystals behave in a similar way; they show IR activity, but no first-order Raman activity).

For a real solid the lattice vibrations give rise to various "transverse" and "longitudinal" branches, where the IR and Raman activities are determined by the symmetry characteristics of the system and the space group of the crystal. Tables 8 and 9 give two examples of the different vibrational modes and activities for the Cs_2NaYX_6[73] and Cs_2MgCl_4[192] solids. The number of normal modes are determined from the number of atoms in the Bravais lattice. Motions of ions or molecules related to each other within the unit cell give rise to the external (or lattice) modes, which are subdivided into *translational* and *libration* (rotation) modes. *Translational* vibrations arise from *translational* motions of the atoms in the unit cell with respect to each other (i.e., the cell remains stationary) and are different from the *acoustical* modes, which are produced when the translational motions of the atoms in the unit all occur in the same direction (i.e., the whole unit moves in a specific direction).

When a molecule (or molecular ion) is introduced into the lattice, splitting of the degenerate *internal* (molecular) vibrations generally occurs; this is due to lower site symmetry and/or to interactions with internal vibrations of other molecules and ions within the same unit cell of the crystal. This is shown by the "correlation diagram" for the $MgCl_4^{2-}$ ion in the Cs_2MgCl_4 crystal. For the YX_6^{3-} ion in Cs_2NaYX_6, however, no splitting occurs, since the molecular symmetry remains undistorted in the highly symmetrical space group of the crystal.

The Raman spectra of $Cs_2NaYCl_{6(s)}$ in Figure 17 show, as predicted from Table 8, four bands, three of which are due to the internal modes of the YCl_6^{3-} octahedral $[\nu_1(A_{ig}) = 285 \text{ cm}^{-1}, \nu_2(E_g) = 223 \text{ cm}^{-1}$, and $\nu_5(F_{2g}) = 128 \text{ cm}^{-1}]$ and one of which is attributed to an external translatory mode $[\nu_{\text{lattice}} (F_{2g}) = 47 \text{ cm}^{-1}]$. In the case of $Cs_2MgCl_{4(s)}$ (Figure 18) not all the predicted Raman modes have been observed,[192] but the four characteristic frequencies of the $MgCl_4^{2-}$ tetrahedral structure are the predominant modes at room temperature $[\nu_1(A) = 271 \text{ cm}^{-1}, \nu_2(E) \sim 112 \text{ cm}^{-1}, \nu_3(F_2) \sim 370 \text{ cm}^{-1}, \nu_4(F_2) \sim 145 \text{ cm}^{-1}]$. Thus the spectra for both compounds support the presence of discrete octahedral and tetrahedral units in Cs_2NaYCl_6 and Cs_2MgCl_4, respectively.

Upon melting, the long-range order and space symmetry of the solid are destroyed. In principle, however, the vibrational modes of the liquid can be considered as the long-wavelength limit of the solid vibrations, and thus certain internal and/or external modes may be present in the vibrational spectrum of the melt. The possible existence of "external" (quasilattice) modes in melts has been recognized by the early observations of Wilmshurst and Senderoff,[183] and since then it has been the subject of several investigations, mainly aimed at oxyanion melts.[39] The "internal" modes in melts have been investigated mainly by Raman spectroscopy in a variety of melt mixtures. The

Table 8

Distribution of Normal Modes for Cs_2NaYCl_6 Solid

Space group F_m3_m (O_h^5)

Four molecules per unit cell; one per Bravais lattice

$3N = 30$ normal modes

$\Gamma_{3N} = \Gamma_{acoustic} + \Gamma_{translatory} + \Gamma_{libration} + \Gamma_{internal}$

$\Gamma_{acoustic} = F_{1u}(IR)$

$\Gamma_{translatory} = F_{2g}(R) + 2F_{1u}(IR)$

$\Gamma_{libration} = F_{1g}(IA)$

$\Gamma_{internal} = A_{1g}(R) + E_g(R) + F_{2g}(R) + 2F_{1u}(IR) + F_{2u}(IA)$

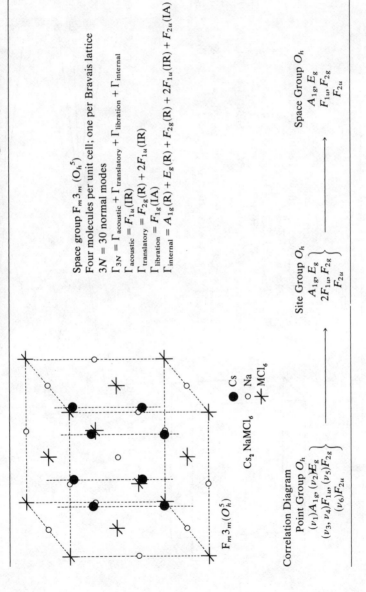

Cs₂NaMCl₆

● Cs
○ Na
✳ MCl₆

$F_m3_m(O_h^5)$

Correlation Diagram

Point Group O_h	Site Group O_h	Space Group O_h
$(\nu_1)A_{1g}, (\nu_2)E_g$	A_{1g}, E_g	A_{1g}, E_g
$(\nu_3, \nu_4)F_{1u} (\nu_5)F_{2g}$	$2F_{1u}, F_{2g}$	F_{1u}, F_{2g}
$(\nu_6)F_{2u}$	F_{2u}	F_{2u}

Table 9

Distribution of Normal Modes for Cs₂MgCl₄ Solid

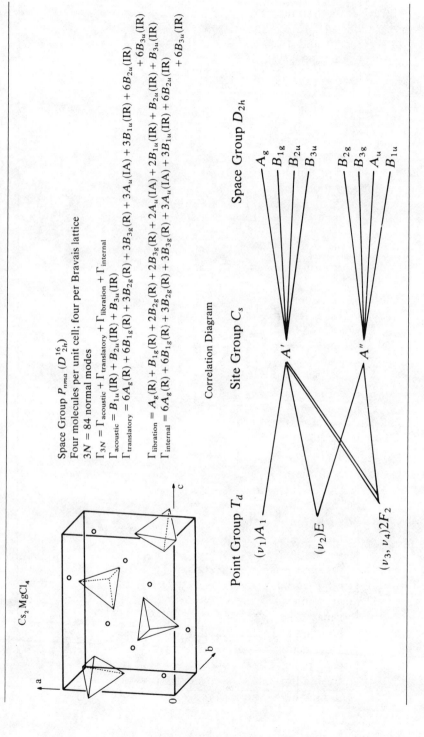

Space Group P_{nma} (D_{2h}^{16})
Four molecules per unit cell; four per Bravais lattice
$3N = 84$ normal modes

$\Gamma_{3N} = \Gamma_{acoustic} + \Gamma_{translatory} + \Gamma_{libration} + \Gamma_{internal}$

$\Gamma_{acoustic} = B_{1u}(IR) + B_{2u}(IR) + B_{3u}(IR)$

$\Gamma_{translatory} = 6A_g(R) + 6B_{1g}(R) + 3B_{2g}(R) + 3B_{3g}(R) + 3A_u(IA) + 3B_{1u}(IR) + 6B_{2u}(IR)$
$+ 6B_{3u}(IR)$

$\Gamma_{libration} = A_g(R) + B_{1g}(R) + 2B_{2g}(R) + 2B_{3g}(R) + 2A_u(IA) + 2B_{1u}(IR) + B_{2u}(IR) + B_{3u}(IR)$

$\Gamma_{internal} = 6A_g(R) + 6B_{1g}(R) + 3B_{2g}(R) + 3B_{3g}(R) + 3A_u(IA) + 3B_{1u}(IR) + 6B_{2u}(IR)$
$+ 6B_{3u}(IR)$

Correlation Diagram

Point Group T_d Site Group C_s Space Group D_{2h}

$(\nu_1)A_1$

$(\nu_2)E$

$(\nu_3, \nu_4)2F_2$

A'

A''

A_g
B_{1g}
B_{2u}
B_{3u}

B_{2g}
B_{3g}
A_u
B_{1u}

Cs₂MgCl₄

objective of these studies is the determination and characterization of possible discrete species (i.e., "complexes") in the melt.

Figures 17 and 18 show that with increasing temperature the Raman bands of the solid become broader and certain modes (e.g., the ν_1 modes for both species) shift slowly to lower energies. Upon melting, the lattice modes disappear and the predominantly observed Raman bands are those attributed to octahedral (YCl_6^{3-}) and tetrahedral ($MgCl_4^{2-}$) species. The existence of such species agrees with the thermodynamically determined stoichiometries in the YCl_3–ACl and $MgCl_2$–ACl (A = K, Rb, Cs) binary melts at which "complexes" have been found to arise (see Figures 11, 12, and 13).

It should be emphasized, however, that in a condensed phase, such as a melt, the vibrational modes of the species cannot be treated as though they were isolated, that is, without accounting for the perturbations due to the environment. Thus, in certain common anion mixtures of the type NX_n–AX (A = K, Rb, Cs) and at low concentrations of NX_n, the forces within the species (e.g., MCl_4^{2-}, YCl_6^{3-}) are stronger than the forces between the species and the neighboring ions, and it is then reasonable to consider the isolated species or complex ion model in interpreting the spectra. On the other hand, when the interactions with the neighboring cations (e.g., A = Li or in mixtures rich in NX_n) are strong, the formation of discrete complex species will be drastically perturbed, and the observed vibrational frequencies *may* then be the result of lattice-type modes.

Experimental evidence, for example, for the $AlCl_3$–ACl (A = K, Na, Cs) molten binaries[79-81] indicates that at compositions $X_{AlCl_3} < 0.5$, $AlCl_4^{-}$ tetrahedra are Raman active. At higher compositions with $X_{AlCl_3} \sim 0.66$, $AlCl_4^{-}$ tetrahedra share two common chlorines to give $Al_2Cl_7^{-}$, and additional strong Raman bands appear in the spectra, with frequencies expected from a vibrating "isolated" molecular ion $Al_2Cl_7^{-}$. On the other hand, for the YCl_3–$CsCl$ binaries the Raman spectra indicate[73] that at $X_{YCl_3} < 0.25$, the YCl_6^{3-} octahedra are present as discrete species in these melts. At higher X_{YCl_3} mole fractions the increasing number of Y ions in the melt requires that YCl_6^{3-} octahedra share chlorines, and one might expect isolated dimer species to be found in the melt (e.g., $Y_2Cl_7^{-}$). However, experimentally one observes, in this case, broadening of the YCl_6^{3-} bands with decreasing CsCl (or KCl) concentration and a continuous shift to higher frequencies of a new band whose intensity is not greatly affected by composition.[73] This new band has been attributed to lattice-type modes in these mixtures.

Finally, it is noteworthy that the vibrational modes measured by Raman spectroscopy may arise from short-lived local structures in the melt. If the lifetime of the structure is long enough (10^{-12} sec) so that there is time for vibration and interaction with the exciting light and if the structure has "bonds" (with a nonzero polarizability derivative), then Raman activity may arise. However, diffusion times in melts are of the order of 10^{-11} sec, and thus the local structure may not maintain its identity for long before exchanging ions

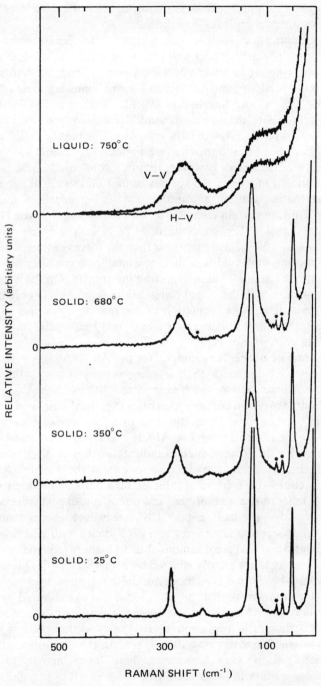

Figure 17. Raman spectra of solid and liquid Cs_2NaYCl_6 (reproduced from Reference 73).

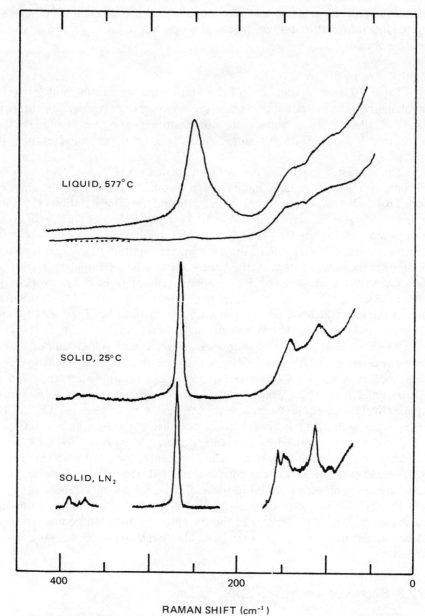

Figure 18. Raman spectra of solid Cs_2MgCl_4 from liquid–nitrogen temperatures to above the melting point of the compound (reproduced from Reference 192).

with its immediate environment. It appears from the above discussion that caution is required in interpreting Raman (and IR) spectra of melts; otherwise misleading information on the structural properties of the melts and mixtures may be obtained.

4.2.1.2. Melt Mixtures

Table 10 gives a summary of the vibrational spectra of melt mixtures involving mainly metal halide and oxyanion components. The list contains most of the available data, but other studies may be found in the literature.[39,185-187] Data for single salts have been presented earlier, in Section 3.1.1.

The suggested species for the metal halide mixtures NX_n–AX corresponds to compositions rich in alkali halide and in certain cases excludes the binaries with LiCl and/or NaCl. In reviewing the data and the interpretation of spectra given in the references listed in Table 10, certain similarities among different groups may be recognized. Thus halide mixtures having molecular melts as components (e.g., $SbCl_3$–$AlCl_3$) show vibrational bands that, to a first approximation, are a superposition of the bands due to the individual components. To some extent a similar behavior occurs for melt mixtures of a network-type liquid (e.g., $ZnCl_2$) and a molecular liquid (e.g., Al_2Cl_6). Structural modifications of at least one of the compounds (i.e., $SbCl_3$ or $ZnCl_2$) are expected and have been shown to exist, in certain cases.[77c]

When a molecular-type halide melt is mixed with alkali halides, more than one discrete species, in equilibrium with one another, have been observed (e.g., species in $AlCl_3$–KCl melts). Varying the composition alters the relative Raman band intensities arising from the species and can be used to characterize (qualitatively) the equilibrium. For nonmolecular melts (e.g., $MgCl_2$, YCl_3) in mixtures with alkali halides, usually one discrete species (e.g., $MgCl_4^{2-}$, YCl_6^{3-}) can be recognized by Raman spectroscopy. Networklike melts tend to remain polymeric at low alkali halide concentrations, but with increasing AX, breakage of the network occurs, leading to the formation of species with coordination geometries similar to those of the pure polymeric melt.

Finally, the majority of the vibrational spectra of oxyanion melts are attributed to the fundamentals of the oxyanion in distorted geometries and to certain lattice-type modes that have also been found in the pure components.

4.2.2. Electronic Absorption Spectroscopy

During the past 25 years tremendous research activity has been developed regarding the electronic $d \leftarrow d$ and $f \leftarrow f$ spectra and coordination chemistry of solute ions in aqueous and nonaqueous solvents. Among the nonaqueous solvents molten salts have proven to be a good choice for a relatively simple solvent, allowing, in many cases, studies of solute ions over a wide temperature

Table 10

Summary of Vibrational Studies of Molten-Salt Mixtures

Molten mixture	References[a,b]	Suggested species	Remarks
CuCl–ACl (A = Li, Cs)	193, 194 (IR)	Lattice modes; four coordinated $(CuCl_4)_n$ species	Pure CuCl appears to have Cu coordinated to four chlorides
AgCl–CsCl AgX–KX (X = Br, I)	193	Polymeric four coordinated species; lattice(?)	
MgX₂–KX (X = Cl, Br, I) MgCl₂–CsCl	195, 192	MgX_4^{2-} ($Mg_2Cl_7^-$ as minor species)	Force constant calculations of MX_4^{2-} in melt; in pure $MgCl_2$ melt Mg is probably four coordinated
MnCl₂–ACl (A = Li, Na, K, Rb, Cs)	196	$MnCl_4^{2-}$	
CuCl₂–ACl (A = Li, K)	195 (IR)	$CuCl_4^{2-}$ and lattice modes	
ZnCl₂–ACl (A = Li, K, Rb, Cs) ZnCl₂–CsCl ZnBr–KBr	69, 197 198	$ZnCl_4^{2-}$ $(ZnCl_2)_n$ $ZnBr_4^{2-}$	Networklike liquids with Zn in tetrahedral coordinations
CdCl₂–ACl (A = Li, K, Cs) CdBr₂–KBr Cd(NO₃)₂–KX (X = Cl, Br)	199, 200 201, 202	$CdCl_4^{2-}$, $CdBr_4^{2-}$, $CdCl_3^-$	Pure cadmium halides possess fluctuating local structures based on octahedral coordination of Cd^{2+}
HgCl₂–ACl (A = K, NH₄) HgI₂–LiI–KI HgX₂–HgX₂'(X, X' = Cl, Br, I)	83, 203 204, 205 206	$HgCl_4^{2-}$, $HgCl_3^-$, $HgCl_2$ HgI_2, HgI_3^- $HgXX'$ (molecular)	Pure mercury halides appear to be molecular
SnCl₂–ACl (A = Li, Na, K, Cs) SnBr₂–CsBr	207, 208a 208b	$(SnCl_2)_n$ polymers and $SnCl_n^{2-n}$ monomers	Polymeric aggregates in melt are similar to those in the solid structure

Table 10 (continued)

Molten mixture	References[a,b]	Suggested species	Remarks
PbX_2–KX (X = Cl, Br)	209	PbX_n^{2-n} species	
BeF_2–LiF–NaF BeF_2–AF (A = Li, Na)	210, 211	BeF_4^{2-}, $Be_2F_7^{3-}$, $Be_3F_{10}^{4-}$	Force constant calculations for BeF_4^{2-}; species in equilibrium
BF_3–AF (A = Li, Na, K, Rb, Cs)	212, 213	BF_4^-	Force constant calculations for BF_4^- and effect of counter-ion
AlF_3–AF (A = Li, Na, K, Cs)	214, 215	AlF_4^-, AlF_6^{3-}	Equilibrium between the species varies with composition and temperature
$AlCl_3$–ACl (A = Li, Na, K, Rb, Cs) $AlBr_3$–NaBr AlI_3–CsI AlX_3–InX (X = Cl, Br, I)	72, 81, 216, 217	AlX_4^-, $Al_2X_7^-$	Species in equilibrium
$AlCl_3$–$MnCl_2$	196c		Octahedral coordination for Mn
$GaCl_3$–CsCl	80	$GaCl_4^-$, $Ga_2Cl_7^-$, $Ga_nCl_{3n+1}^-$ ($n \geq 3$)	Species in equilibrium
$InCl_3$–ACl (A = Li, Na, K, Rb, Cs) InX_3–InX (X = Cl, Br)	78, 216, 217 218, 219	$InCl_4^-$, $In_2Cl_7^-$	Species in equilibrium; mixed valance states for In_2X_4
$AuCl_3$–KCl	193	$AuCl_4^-$	
$LaCl_3$–ACl (A = Na, K, Cs)	220, 221	$LaCl_6^{3-}$	
YCl_3–ACl (A = Li, Na, K, Cs)	73	YCl_6^{3-}	Lattice-type modes in melt mixtures with high YCl_3 concentrations

System	Reference	Species	Remarks
ZrF_4–MF, ThF_4–MF	221	ZrF_6^{2-}, ZrF_8^{4-} (?), ThF_8^{4-} (?)	Coordination numbers of Th appear to be higher than those of Zr
$SbCl_3$–KCl, $BiCl_3$–ACl (A = K, Cs)	77	$SbCl_n^{(n-3)}$(?), $BiCl_n^{(n-3)}$(?)	Probably high polymeric species are formed
$ZnCl_2$–$AlCl_3$	198	$(ZnCl_2)_n$, Al_2Cl_6	"Molecular" mixture; structural modifications of $ZnCl_2$
$SbCl_3$–$AlCl_3$, $BiCl_3$–$AlCl_3$, $SbCl_3$–$AlCl_3$–NaCl	77a, 77c / 77c	$SbCl_3$, Al_2Cl_6 / $BiCl_3$, Al_2Cl_6	"Molecular" mixture; structural modifications of $SbCl_3$
$SnCl_2$–A($AlCl_4$)	222	$(SnCl_2)_n$, $AlCl_4^-$, $SnCl_3^-$	"Molecular" mixture Probably monomeric $SnCl_2$ units are present in the mixture
AIO_3–ANO_3 (A = Li, Na, K, Rb, Cs, Ag, Ti)	223	IO_3^- (C_{3v}?)	The IO_3^- in melts may tend to be planar
Na_2MoO_4–KCl, Na_2WO_4–KCl	188 (R, IR)	MoO_4^{2-}, WO_4^{2-}	Solutions of oxides (MO_3, WO_3) in these melts have been also studied
Li_2CO_3–K_2CO_3, A_2CO_3–ACl (A = Li, Na, K), $LiCO_3$–$CaCO_3$	224, 225	CO_3^{2-}	Splitting of the CO_3^- modes has been observed
K_2SO_4–MSO_4 (M = Zn, Cu, Co, Ni)	226 (R, IR)	SO_4^{2-}	SO_4^{2-} symmetry is lowered from T_d to C_{3v}
B_2O_3–Li_2O	227	BO_3^{2-} and polymers	

[a] Mostly Raman studies; IR studies are noted by (IR).
[b] References within the paper cited may include studies of the same system by others.

range and in different solvent compositions. The characterization of solute species in molten salts by electronic absorption spectroscopy gave rise to a new field in chemistry where unusual entities and oxidation states can be investigated.[176,178,228] It was, for example, in molten-salt phases where, for the first time, the spectrum of $NiCl_4^{2-}$ and its tetrahedral coordination were established.[229] Spectra of uncommon oxidation states of ions, which were difficult to stabilize in ordinary solvents, have also been studied rather extensively in molten salts.[230]

Investigations of solutions of d^n cations in halide oxyanion melts have shown that the size, charge, and polarizability of the anions play an important role in determining the bonding and geometry of the species. The cations in the second coordination sphere were also found to be of equal importance for the characterization of the equilibria between the different solute coordination geometries. Extensive reviews of spectral studies of transition metals and lanthanides in a large number of solvent melts are available.[176–178,228]

Absorption spectroscopy, however, has been scarcely used for investigating the *structural* properties of the bulk melt. Fused salts containing cations with ligand field-sensitive spectra are in general dark, with high light absorbances, which make their study very difficult.[231–233] The typical careful and characteristic study in this respect is the work of Smith *et al.*,[232] regarding the coordination geometries in the $CsCl$–$NiCl_2$ binary mixtures at $NiCl_2$ concentrations up to 60 mol %. The results obtained from this study (Figure 19) have shown that at compositions up to 20 mol % the Ni^{2+} spectra are due to the tetrahedral $NiCl_4^{2-}$ chromophore, a result that was later substantiated by the thermodynamic behavior of this system.[67] At higher $NiCl_2$ contents, however, the $NiCl_4^{2-}$ tetrahedra appear to either exhibit a con-

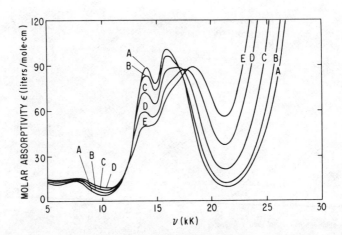

Figure 19. Electronic absorption spectra of $CsCl$–$NiCl_2$ mixtures at 1137 K. Compositions in mole percent $NiCl_2$: (a) 20.0; (b) 30.0; (c) 39.9; (d) 49.9; (e) 60.0 (reproduced from Reference 233).

tinuous distortion with increasing concentration and/or lead to an equilibrium of species, one of which is $NiCl_4^{2-}$ (T_d) and the other(s) is not identifiable from its spectrum. Spectral changes observed upon melting the existing solid compounds, Cs_3NiCl_5 (Ni^{2+} in $\sim T_d$ site symmetry) and $CsNiCl_3$ (Ni^{2+} in $\sim O_h$ site symmetry), in the binary $CsCl-NiCl_2$ mixture have further established tetrahedral $NiCl_4^{2-}$ coordinations at high CsCl concentrations.[233]

Another method for studying bulk properties of the melt by absorption spectroscopy is with the use of light-absorbing probe ions in the melt. A solute cation with a ligand field spectrum sensitive to the environment is chosen (e.g., Ni^{2+}, Co^{2+}) and its coordination geometry is examined at different melt compositions. The changes observed are then correlated with the solvent structure. For example, by comparing the spectra of Ni^{2+} (or Co^{2+}) centers in solid and molten $MgCl_2$[234] and $ZnCl_2$,[68,234,235] the coordination geometries of Mg and Zn in the melt have been characterized. Studies using probe ions are again limited to only a few systems,[68,234,235] although reports of numerous studies aimed at the coordination chemistry of the solute ion (and not the structure of the melt) occasionally contain vague statements regarding the solvent structure. It should be emphasized, however, that owing to complex interactions[68,236] between the probe ion and the solvent, the interpretation of the data may be difficult and less conclusive than that obtained in direct measurements of the solvent spectra.

4.2.3. Magnetic Measurements

Magnetic measurements have not been widely applied to molten salts, although the few available data have positively contributed to the understanding of bonding in the melts. The three techniques to be discussed here are those listed in Table 1.

NMR measurements in molten salts were initiated by Hafner and Nachtrieb[24,237] for studies of thallium halide and oxyanion melts. Measurements of the "chemical shifts" of ^{205}Tl in a series of binaries TlX–AX (A = Li, Na, K, Rb, Cs; X = Cl, Br, I) have shown that the resonance frequency varies linearly with the AX mole fraction. The chemical shift was found to depend on the size of the counter cation and the nature of the anion. On the basis of a simple electrostatic model, it was shown that the observed changes can be attributed to changes in the covalency of the Tl–X bonding. Thus, relative to the pure $TlX_{(l)}$, the binary mixture with LiX or NaX decreases the covalent interactions, whereas for melts containing NaX, KX, or CsX the Tl–X covalency is increased.

More recently NMR spectra have been reported for aluminate melts,[238] for pure $CuCl_{(l)}$,[42a] and the liquid (and glassy) B_2O_3 and B_2S_3.[42b] The AAl_2Cl_7 (A = Na, K) and CuCl melts were found to be consistent with an ionic liquid description. The data for the borate glasses were interpreted as indicating that with increasing temperature, fragmentation of the network structure occurs.

Electron paramagnetic (spin) resonance (EPR–ESR) spectroscopy has been mainly used to study transport properties and magnetic interactions in the melts. Early studies were limited to dilute chloride solutions[239,240] of low melting point salts having Mn^{2+} as a probe cation.

However, in a recent far-reaching study EPR measurements have been reported in corrosive fluoride melts at temperatures up to 1473 K.[43] The EPR linewidths of MnF_2, $KMnF_3$ $RbMnF_3$, $CsMnF_3$, and of a MnF_2–$KMnF_3$ mixture have been measured from room temperature to temperatures above the melting points of the salts. Abrupt changes in the linewidth on melting the $AMnF_3$ salts were observed, but pure MnF_2 melted without such change. The magnitude of the linewidth in the melts corresponded to a correlation time for spin exchange or motion (and/or chemical exchange) of the order of 10^{-12} sec. Melt mixtures of $KMnF_3$–MnF_2 show a linewidth intermediate in magnitude between that of $KMnF_3$ and MnF_2, indicating a rather continuous change in the magnetic and structural properties on going from the pure liquid difluoride to the mixture.

The techniques for magnetic susceptibility measurement at high temperatures are well established from methods developed for the study of liquid metals and alloys.[241] Very few investigations, however, have been reported for molten salts. Russian workers have recently been involved in measuring the changes of diamagnetic properties of melts and solid halides at elevated temperatures.[242] Diamagnetic changes upon melting are indicative of differences in bonding between the solid and the melt.[243]

Magnetic susceptibility measurements of paramagnetic (e.g., transition metal ions) compounds and aqueous solutions have been used to characterize the symmetry properties of the ground ligand field state of the ions and to thus indirectly determine the coordination symmetry of the ion.[244]

The first published studies regarding the paramagnetic properties of molten salts are those of Trzebiatowski and collaborators.[245,246] A characteristic example indicating the usefulness of the method is the study of the paramagnetic properties of the Cs_3NiCl_5 and $CsNiCl_3$ compounds in the $CsCl$–$NiCl_2$ binary, which has been carried out from liquid-helium temperatures to 1200 K (Figure 20). As the temperature is increased, the $Cs_3NiCl_{5(s)}$ is slowly decomposed to $2CsCl$ and $CsNiCl_3$[27] and its magnetic susceptibility (χ_M) changes from values that are close to the theoretical χ_M for Ni^{2+} in T_d geometry to values near the O_h geometry. At 640 K, however, $Cs_3NiCl_{5(s)}$ is re-formed and a drastic increase in χ_n occurs, with M values that correspond to tetrahedral coordination for Ni^{2+}. On melting the $Cs_3NiCl_{5(s)}$ at 820 K, no drastic changes of χ_M can be observed. Measurements of the $CsNiCl_{3(s)}$ show however, (Figure 20) that at its melting point of 1031 K abrupt changes occur, during which the χ_M^{-1} values drop from values above the theoretical curve for O_h geometry to values close to the theoretical one for T_d geometry. Thus the magnetic data for these compounds agree with the findings obtained with electronic adsorption spectroscopy, which (Section

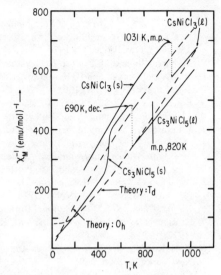

Figure 20. Temperature dependence of χ_M^{-1} for solid and liquid Cs_3NiCl_5 and $CsNiCl_3$ compounds (data from Reference 246).

4.2.2.) support a T_d geometry of Ni^{2+} in the CsCl-rich mixture. Similar magnetic susceptibility changes on melting have been observed for the MCl_2–KCl (M ≡ Mn, Co, Ni) binary systems and their compounds.[247]

Acknowledgement

This work was performed under the auspices of the Division of Basic Energy Sciences of the U.S. Department of Energy. The writing of this chapter was originated from a series of lectures given at the Institute of Inorganic Chemistry (NTH), University of Trondheim (1977) and at the Chemistry Department A, The Technical University of Denmark (DTH). Many thanks to Professors H. Oye (NTH), T. Ostvold (NTH) and N. J. Bjerrum (DTH) for their hospitality.

References

1. *Practical Techniques for Saving Energy in Chemical, Petroleum, and Metals Industry*, Sitting Noyes Data, New Jersey (1977), p. 72.
2. (a) A series of patents have been reported, see, e.g., F. W. Richardson, *Oil from Coal*, Noyes Data, New Jersey (1975). (b) S. Yoshikawa and S. Nomura, Developments in fused salt chemistry and petrochemistry, *J. Pet. Technol.* **6**, 608 (1978).
3. D. H. Kerridge, *Pure Appl. Chem.* **41**, 355 (1975).
4. (a) E. J. Cairns and R. K. Steunenberg, High-temperature batteries, in *Progress in High Temperature Physics and Chemistry*, Vol. 5, C. A. Rouse, ed., Pergamon Press, New York (1973); (b) G. Halpert in *Techniques of Electrochemistry*, E. Yeager and A. J. Salkind, eds., Wiley, New York (1978), p. 291; (c) J. L. Sudworth, High temperature batteries, in *Electrochemical Power Sources: Primary and Secondary Batteries*, M. Barak, ed., Institute of Electrical Engineering, London (1980).
5. *Molten Salt Thermal Energy Storage Systems: Salt Section*, H. C. Maru and J. F. Dullea, NTIS, Springfield, Virginia (1976).

6. (a) M. Taube, *Fast Reactors Using Molten Salt Chloride Salts as Fuels*, Eidegen Inst. fuer Reaktorforsch, Wuerenlingen, Switzerland (1978); (b) J. R. Keiser, ORNL Report No. TM-5783, Oak Ridge, TN 1977.

7. M. Parrinello and M. P. Tosi, *Rev. Nuovo Cimento* **2**, 1 (1979).

8. Yu. K. Delimarskii and B. F. Markov, *Electrochemistry of Fused Salts*, English translation, Sigma Press, New York (1961).

9. *Fused Salts*, B. R. Sundheim, ed., McGraw-Hill, New York (1964).

10. *Molten Salts*, M. Blander, ed., Wiley, New York (1964).

11. J. Lumsden, *Thermodynamics of Molten Salt Mixtures*, Academic Press, New York (1966).

12. H. Bloom, *The Chemistry of Molten Salts*, W. A. Benjamin, New York (1967).

13. G. J. Janz, *Molten Salt Handbook*, Academic Press, New York (1967).

14. *Ionic Interactions*, Vols. 1 and 2, S. Petrucci, ed., Academic Press, New York (1971).

15. *Advances in Molten Salt Chemistry*, J. Braunstein, G. Mamantov, and G. P. Smith, eds., Plenum Press, New York (Vol. 1, 1971; Vol. 2, 1973; Vol. 3, 1975); Vol. 4 (1980).

16. *J. Phys. Chem. Ref. Data*, National Bureau of Standards Publication.

17. E. Rhodes, in *Water Aqueous Solutions*, R. A. Horne, ed., Wiley-Interscience, New York (1972), p. 175.

18. D. H. Kerridge in *Chemistry of Non-Aqueous Solvents*, J. J. Lagowski, ed., Academic Press, New York (1978), p. 269.

19. *Characterizations of Solutes in Non-Aqueous Solvents*, G. Mamantov, ed., Plenum Press, New York (1978).

20. *Proceedings of the International Symposium on Molten Salts*, Washington, D.C., J. P. Pemsler, J. Braunstein, K. Nobe, D. R. Morris, and N. E. Richards, eds., Electrochemical Society (1976).

21. C. A. Croxton, *Introduction to Liquid State Physics*, Wiley, New York (1975).

22. F. H. Stillinger, Jr., in *Molten Salts*, M. Blander, ed., Wiley, New York (1964), p. 1.

23. M. J. L. Sangster and M. Dixon, *Adv. Phys.* **25**, 247 (1976).

24. N. H. Nachtrieb, in *Molten Salts: Characterization and Analysis*, G. Mamantov, ed., Marcel Dekker, New York (1969).

25. G. N. Papatheodorou and O. J. Kleppa, *J. Chem. Phys.* **51**, 4624 (1969).

26. G. N. Papatheodorou and O. J. Kleppa, *J. Inorg. Nucl. Chem.* **33**, 1249 (1971).

27. G. N. Papatheodorou, *J. Inorg. Nucl. Chem.* **35**, 465 (1973).

28. M. P. Tosi and F. G. Fumi, *J. Phy. Chem. Solids* **21**, 45 (1964).

29. L. V. Woodcock and K. Singer, *Trans. Faraday Soc.* **67**, 12 (1971).

30. J. Dupuy and A. J. Dianoux, eds., *Microscopic Structure and Dynamics of Liquids*, Plenum Press, New York (1978).

31. L. V. Woodcock, in *Advances in Molten Salt Chemistry*, Vol. 3, J. Braunstein, G. Mamantov, and G. P. Smith, eds., Plenum Press, New York (1975), p. 1.

32. S. W. de Leeuw, *Mol. Phys.* **36**, 103 (1978); **36**, 765 (1978).

33. H. Reiss, S. W. Mayer, and J. L. Katz, *J. Chem. Phys.* **35**, 820 (1961).

34. H. Bloom and I. K. Snook, in *Modern Aspects of Electrochemistry*, Vol. 9, B. E. Conway and J. O'M. Bockris, eds., Plenum Press, New York (1974), p. 159.

35. J. E. Enderby, in *Microscopic Structure and Dynamics of Liquids*, J. Dupuy and A. J. Dianoux, eds., Plenum Press, New York (1978), p. 301.

36. H. Ohno and K. Furukawa, *Kinzoku Butsuri* **3**, 129 (1978); **3**, 263 (1978).

37. H. Ohno, K. Furukawa, K. Tanemoto, Y. Takagi, and T. Nakamura, *J. Chem. Soc. Faraday Trans.* 1 **74**, 804 (1978).

38. D. E. Irish, in *Ionic Interactions*, Vol. 2, S. Petrucci, ed., Academic Press, New York (1971).

39. J. P. Devlin, in *Advances in Infrared and Raman Spectroscopy*, R. J. H. Clark and R. E. Hester, eds., Heyden, New York (1976).

40. K. E. Johnson and J. R. Dickinson, in *Advances in Molten Salt Chemistry*, Vol. 2, J. Braunstein, G. Mamantov, and G. P. Smith, eds., Plenum Press, New York (1973), p. 83.

41. S. V. Volkov and K. B. Yatsimizski, *Spectroscopy of Fused Salts*, Naukova Dumka, Kiev (1977).
42. (a) J. B. Boyce and J. C. Mikkelsen, *J. Phys. C* **10**, L41 (1977); (b) M. Rubinstein, *Phys. Rev. B* **14**, 2778 (1976).
43. E. Dormann, D. Hone, and V. Jaccarino, *Phys. Rev. B* **14**, 2715 (1976).
44. M. V. Smirnov and V. Ya. Krudyakov, *Itogi Nauki Tekh., Rastvory. Rasplavy* **2**, 172 (1975).
45. (a) *Applications of Phase Diagrams in Metallurgy and Ceramics*, Vols. 1 and 2, G. C. Carter, ed., *Nat. Bur. Stand. U.S. Spec. Publ.* 496, (1978); (b) Y. Doucet, in *Experimental Thermodynamics*, Vol. 2, B. Le Neindre and B. Vodar, eds., Butterworths, London (1975), p. 835.
46. O. J. Kleppa, in *Thermodynamics in Geology*, D. G. Fraser, ed., D. Reidel, Dordrecht (1977), p. 279.
47. J. Braunstein and H. Braunstein, in *Experimental Thermodynamics*, Vol. 2, B. LeNeindre and B. Vodar, eds., Butterworths, London (1975), pp. 901–952.
48. H. H. Emons, G. Bräutigam, and R. Tomas, *Chem. Zvesti* **30**, 773 (1976).
49. H. Linga, K. Motzfeldt, and H. Oye, *Ber. Bunsenger. Phys. Chem.* **82**, 568 (1978).
50. F. G. Edwards, J. E. Enderby, R. A. Howe, and D. I. Page, *J. Phys. C* **8**, 3483 (1975).
51. J. Y. Derrien and J. Dupuy, *J. Phys. (Paris)* **36**, 191 (1975).
52. D. I. Page and K. Mika, *J. Phys. C* **4**, 3034 (1971).
53. F. Lantelme, P. Turq, B. Quentiec, and J. W. E. Lewis, *Mol. Phys.* **28**, 1537 (1974).
54. E. W. J. Mitchell. P. F. J. Poncet, and R. J. Stewart, *Philos. Mag.* **34**, 721 (1976).
55. J. Y. Derrien and J. Dupuy, *Phys. Chem. Liq.* **9**, 71 (1976).
56. F. G. Edwards, R. A. Howe, J. E. Enderby, and D. I. Page, *J. Phys. C* **11**, 1053 (1978).
57. H. A. Levy and M. D. Danford, in *Molten Salts*, M. Blander, ed., Wiley, New York (1964), p. 109.
58. F. Vaslow and A. H. Narten, *J. Chem. Phys.* **59**, 4949 (1973).
59. J. Zarzycki, *J. Phys. Radiat. Suppl. Phys. Appl.* **18**, 65 (1957).
60. T. Sugawara, *Sci. Rep. Res. Inst. Tohoku Univ. Ser. A* **3**, 39 (1951).
61. H. Ohno, M. Yoroki, K. Furukawa, Y. Takagi, and T. Nakamura, *J. Chem. Soc. Faraday Trans. 1* **74**, 1861 (1978).
62. P. G. Mikolaj and C. J. Pings, *Phys. Chem. Liq.* **1**, 93 (1968).
63. A. Rahman, R. H. Fowler, and A. H. Narten, *J. Chem. Phys.* **57**, 3010 (1972).
64. A. H. Narten, *J. Chem. Phys.* **56**, 1905 (1972).
65. K. Tanemoto and T. Nakamura, *Chem. Lett.* **4**, 351 (1975).
66. H. Ohno, K. Tanemoto, Y. Takagi, K. Furukawa, and T. Nakamura, *J. Chem. Soc. Faraday Trans. 1* **74**, 804 (1978).
67. G. N. Papatheodorou and O. J. Kleppa, *J. Inorg. Nucl. Chem.* **32**, 889 (1970).
68. W. E. Smith, J. Brynestad, and G. P. Smith, *J. Chem. Phys.* **52**, 3890 (1970).
69. R. B. Ellis, *J. Electrochem. Soc.* **113**, 485 (1966).
70. C. A. Angell and J. Wong, *J. Chem. Phys.* **53**, 2053 (1970).
71. R. L. Harris, R. E. Wood, and H. L. Ritter, *J. Am. Chem. Soc.* **73**, 3161 (1951).
72. H. A. Oye, E. Rytter, P. Klaeboe, and S. J. Cyvin, *Acta Chem. Scand.* **25**, 559 (1971).
73. G. N. Papatheodorou, *J. Chem. Phys.* **66**, 2893 (1977).
74. J. G. Powder, *J. Phys. C* **8**, 895 (1975).
75. B. Gilbert, K. W. Fung, and G. Mamantov, *J. Inorg. Nucl. Chem.* **37**, 921 (1975).
76. H. A. Andreasen, N. J. Bjerrum, and N. H. Hansen, *J. Chem. Eng. Data* **25**, 236 (1980).
77. (a) K. W. Fung, G. M. Begun, and G. Mamantov, *Inorg. Chem.* **12**, 53 (1973); (b) E. Johnson, A. H. Narten, W. E. Thiessen, and R. Triolo, *Discuss. Faraday Soc*, **66**, 287 (1978); (c) R. Huglen, G. Mamantov, G. Begun, and G. P. Smith, *J. Raman Spectrosc.*, **9**, 188 (1980).
78. H. A. Oye, E. Rytter, P. Klaeboe, *J. Inorg. Nucl. Chem.* **36**, 1925 (1974).
79. R. E. Wood and H. Ritter, *J. Am. Chem. Soc.* **74**, 1760 (1952).
80. H. A. Oye and W. Bues, *Acta Chem. Scand. Ser. A* **29**, 489 (1975).
81. G. M. Begun, C. R. Boston, G. Torsi, and G. Mamantov, *Inorg. Chem.* **10**, 886 (1971).

82. C. Boston, in *Advances in Molten Salt Chemistry*, Vol. 1, J. Braunstein, G. Mamantov, and G. P. Smith, eds., Plenum Press, New York (1971), p. 129.

83. G. J. Janz and D. W. James, *J. Chem. Phys.* **38**, 902 (1962); **39**, 905 (1963).

84. K. Tödheide, in *Proceedings of the International Symposium on Molten Salts*, Washington, D.C., J. P. Pemsler, J. Braunstein, K. Nobe, D. R. Morris, and N. E. Richards, eds., Electrochemical Society (1976), p. 20.

85. B. Cleaver, *Annu. Rep. Prog. Chem. Sect. A* **71**, 141 (1975).

86. K. Furukawa, *Sci. Rep. Res. Inst. Tohoku Univ. Ser. A* **12** 150 (1960).

87. K. Suzuki and Y. Fukushima, *Z. Naturforsh.*, **329**, 1438 (1977).

88. J. Zarzycki, *Discuss. Faraday Soc. London* **32**, 38 (1961).

89. H. Ohno and K. Furukawa, *J. Chem. Soc. Faraday Trans.* 1 **74**, 795 (1978).

90. M. Miyake, K. Okada, S. Iwai, H. Ohno, and K. Furukawa, *J. Chem. Soc. Faraday Trans.* 2 **74**, 1141 (1978).

91. K. Okada, M. Miyake, S. Iwai, H. Ohno, and K. Furukawa, *J. Chem. Soc. Faraday Trans.* 2 **74**, 1880 (1978).

92. J. Wong and C. A. Angell, *Glass: Structure by Spectroscopy*, Marcel Dekker, New York (1976).

93. R. L. Cormia, J. D. Mackenzie, and D. Turnbull, *J. Appl. Phys.* **34**, 2245 (1963).

94. H. Bottger, *Phys. Status Solidi B* **62**, 9 (1974).

95. R. J. Bell, N. F. Bird, and P. Dean, *J. Phys. C* **1**, 299 (1968).

96. T. Forland, in *Fused Salts*, B. R. Sundheim, ed., McGraw-Hill, New York (1964), p. 63.

97. G. N. Papatheodorou and S. A. Solin, *Phy. Rev. B* **13**, 1741 (1976).

98. R. Shuker and R. W. Gammon, *J. Chem. Phys.* **55**, 4784 (1971).

99. H. C. Andersen, *Annu. Rev. Phys. Chem.* **26**, 145 (1975).

100. J. C. Hansen and I. R. McDonald, *Theory of Simple Liquids*, Academic Press, New York (1976).

101. E. Waisman and J. L. Leibowitz, *J. Chem. Phys.* **56**, 3086 (1972); **56**, 3093 (1972).

102. B. Larsen, *J. Chem. Phys.* **65** 3431 (1976); Ph. D. thesis, University of Trondheim, Trondheim, Norway, 1979.

103. H. Eyring and M. S. John, *Significant Liquid Structures*, Wiley, New York (1969).

104. M. J. Gillan, in *Thermodynamics of Nuclear Materials*, Vol. 1, IAEA, Vienna (1975), p. 269.

105. I. G. Murgulescu, Rev. Roum. Chim. **14**, 965 (1969).

106. J. O'M Bockris and N. E. Richards, *Proc. R. Soc. London* **241A**, 44 (1957).

107. H. Bloom and J. O'M Bockris, in *Fused Salts*, B. R. Sundheim, ed., McGraw-Hill, New York (1964), p. 1.

108. K. S. Pitzer, *J. Chem. Phys.* **7**, 583 (1939).

109. H. Reiss, S. W. Mayer, and J. Katz, *J. Chem. Phys.* **35**, 820 (1961).

110. K. D. Luks and H. T. Davis, *Is'EC Fundam.* **6**, 194 (1967).

111. J. H. Hildebrand and R. L. Scott, *Regular Solutions*, Prentice-Hall, Englewood Cliffs, New Jersey (1962).

112. I. Prigogine and R. DeFay, *Chemical Thermodynamics*, Longmans, Green, New York (1954).

113. M. Temkin, *Acta Physicochim. USSR* **20**, 411 (1945).

114. T. Forland, in *Fused Salts*, B. R. Sundheim, ed., McGraw-Hill, New York (1964), p. 63.

115. J. Braunstein, in *Ionic Interactions*, Vol. 1, S. Petrucci, ed., Academic Press, New York (1971), p. 180.

116. See, e.g. A. S. Kucharski and S. N. Flengas, *J. Electrochem. Soc.* **119**, 1170 (1972).

117. See, e.g., K. A. Romberger, J. Braunstein, and R. E. Thoma, *J. Phys. Chem.* **76**, 1154 (1972).

118. T. Ostvold, *High Temp. Sci.* **4**, 51 (1972).

119. T. Ostvold, *Acta Chem. Scand.* **23**, 688 (1969).

120. A. D. Pelton and W. T. Thompson, *Can. J. Chem.* **48**, 1585 (1970).

121. M. Gaune-Escard, J. C. Mathieu, P. Desré, and Y. Doucet, *J. Chim. Phys.* **69**, 1390 (1972).

122. H. Reiss, J. L. Katz, and O. J. Kleppa, *J. Chem. Phys.* **36**, 144 (1962).
123. J. H. Hildebrand and E. J. Salstrom, *J. Am. Chem. Soc.* **54**, 4257 (1932).
124. O. J. Kleppa and L. S. Hersh, *J. Chem. Phys.* **34**, 351 (1961).
125. L. S. Hersh and O. J. Kleppa, *J. Chem. Phys.* **42**, 1309 (1965).
126. L. S. Hersh, A. Navrotsky, and O. J. Kleppa, *J. Chem. Phys.* **42**, 3752 (1965).
127. (a) P. Dantzer and O. J. Kleppa, *J. Chim. Phys.* **71**, 217 (1974); *Inorg. Chem.* **12**, 2699 (1973); (b) P. Dantzer, Ph. D. thesis, Université de Nice, Nice, France, 1973.
128. B. K. Andersen and O. J. Kleppa, *Acta Chem. Scand. Ser. A* **32**, 595 (1978).
129. (a) J. L. Holm and O. J. Kleppa, *J. Chem. Phys.* **49**, 2425 (1968); (b) K. C. Hong and O. J. Kleppa, *J. Chem. Thermodyn.* **8**, 31 (1976).
130. M. E. Melnichak and O. J. Kleppa, *J. Chem. Phys.* **57**, 5231 (1972).
131. T. Ostvold and O. J. Kleppa, *Acta Chem. Scand.* **25**, 919 (1971).
132. B. K. Andersen and O. J. Kleppa, *Acta Chem. Scand. Ser. A* **30**, 751 (1976).
133. H. C. Ko and O. J. Kleppa, *Inorg. Chem.* **10**, 771 (1941).
134. O. J. Kleppa and S. V. Meschel, *J. Phys. Chem.* **67**, 668 (1963); **67**, 2750 (1963).
135. M. E. Melnichak and O. J. Kleppa, *J. Chem. Phys.* **57**, 5231 (1972).
136. G. N. Papatheodorou and O. J. Kleppa, *J. Chem. Phys.* **47**, 2014 (1967).
137. J. Lumsden, *Discuss. Faraday Soc.* **32**, 138 (1961).
138. H. T. Davis and J. McDonald, *J. Chem. Phys.* **48**, 1644 (1968).
139. H. T. Davis and S. A. Rice, *J. Chem. Phys.* **41**, 14 (1964).
140. D. I. Marchidan, *Int. Rev. Sci. Phys. Chem.* **10**, 309 (1975).
141. D. Larsen, T. Forland, and K. Singer, *Mol. Phys.* **26**, 1521 (1973).
142. M. Pirrinello and M. P. Tosi, *Rev. Nuovo Cimento* **25B**, 242 (1975).
143. M. L. Saboungi and A. Rahman, *J. Chem. Phys.* **65**, 2393 (1976); **66**, 2773 (1977).
144. D. J. Adams and I. R. McDonald, *Mol. Phys.* **26**, 1521 (1977).
145. O. J. Kleppa, *J. Phys. Chem.* **66**, 1668 (1962).
146. F. G. McCarty and O. J. Kleppa, *J. Phys. Chem.* **68**, 3846 (1964).
147. O. J. Kleppa and F. G. McCarty, *J. Phys. Chem.* **70**, 1246 (1966).
148. T. Ostvold, *Acta Chem. Scand.* **26**, 2788 (1972).
149. A. S. Kucharski and S. N. Flengas, *J. Electrochem. Soc.* **119**, 1170 (1972).
150. C. Beusman, ORNL Report No. 2323, Oak Ridge, TN 1957.
151. J. Josiak, *Rocz. Chem.* **44**, 1875 (1970); **46**, 1029 (1972).
152. H. Kühn, M. Wittner, and W. Simon, *Electrochim. Acta* **24**, 55 (1979).
153. D. C. Hamby and A. B. Scott, *J. Electrochem. Soc.* **117**, 319 (1970).
154. Y. Dutt and T. Ostvold, *Acta Chem. Scand.* **26**, 2743 (1972).
155. D. C. Hamby and A. B. Scott, *J. Electrochem. Soc.* **115**, 705 (1968).
156. G. N. Papatheodorou and O. J. Kleppa, *Z. Anorg. Allg. Chem.* **401**, 132 (1973).
157. R. J. Robertson and A. S. Kucharski, *Can. J. Chem.* **51**, 3114 (1973).
158. (a) J. L. Holm and O. J. Kleppa, *Inorg. Chem.* **8**, 207 (1969); (b) J. Braunstein, K. A. Romberger and R. Ezell, *J. Phys. Chem.* **74**, 4383 (1970).
159. (a) O. J. Kleppa and K. C. Hong, *J. Phys. Chem.* **78**, 1478 (1974); (b) K. C. Hong and O. J. Kleppa, *J. Phys. Chem.* **82**, 1596 (1978).
160. O. J. Kleppa and M. Wakihara, *J. Inorg. Nucl. Chem.* **38**, 715 (1976).
161. M. Wakihara and O. J. Kleppa, *High Temp. Sci.* **9**, 35 (1977).
162. H. Ikeuchi and C. Krohn, *Acta Chem. Scand. Ser. A* **28**, 48 (1974).
163. G. N. Papatheodorou and T. Ostvold, *J. Phys. Chem.* **78**, 181 (1974).
164. G. N. Papatheodorou and O. J. Kleppa, *J. Phys. Chem.* **78**, 176 (1974).
165. F. von Dienstbach and R. Blachnik, *Inorg. Allg. Chem.* **412**, 97 (1975).
166. G. N. Papatheodorou, O. Waernes, and T. Ostvold, *Acta Chem. Scand. Ser. A* **33**, 173 (1979).
167. K. C. Hong and O. J. Kleppa, *J. Phys. Chem.* **82**, 176 (1978).
168. E. W. Dewing, *J. Electrochem. Soc.* **123**, 1289 (1976).

169. K. C. Hong and O. J. Kleppa, *J. Phys. Chem.* **83**, 2589 (1979).
170. H. T. Davis, *J. Chem. Phys.* **41**, 2761 (1964).
171. H. T. Davis, *J. Phys. Chem.* **76**, 1629 (1972).
172. A. D. Pelton, *Can. J. Chem.* **49**, 3919 (1972).
173. H. L. Schläfer and G. Glieman, *Ligand Field Theory*, Wiley, New York (1969), p. 150.
174. B. R. Sundheim and L. V. Woodcock, *Chem. Phys. Lett.* **15**, 191 (1972).
175. B. R. Sundheim, in *Proceedings of the International Symposium on Molten Salts*, Washington, D.C., J. P. Pemsler, J. Braunstein, K. Nobe, D. R. Morris, and N. E. Richards, eds., Electrochem. Society (1976), p. 48.
176. G. P. Smith, *Molten Salts*, M. Blander, ed., Wiley, New York (1964), p. 427.
177. S. L. Holt, in *Ionic Interactions*, Vol. 1, S. Petrucci, ed., Academic Press, New York (1971), p. 126.
178. K. E. Johnson and J. R. Dickinson, in *Advances in Molten Salt Chemistry*, Vol. 2, J. Braunstein, G. Mamantov, and G. P. Smith, eds., Plenum Press, New York (1973).
179. W. Bues, *Z. Anorg. Allg. Chem.* **279**, 104 (1955).
180. W. Bues, *Z. Phys. Chem. (Frankfurt am Main)* **10**, 1 (1957).
181. G. J. Janz and D. W. James, *J. Chem. Phys.* **35**, 5089 (1961).
182. T. F. Young and D. E. Irish, *Annu. Rev. Phys. Chem.* **13**, 435 (1962).
183. J. K. Wilmshurst and S. Senderoff, *J. Chem. Phys.* **35**, 1078 (1961).
184. G. Walrafen, D. E. Irish, and T. F. Young, *J. Chem. Phys.* **37**, 662 (1962).
185. R. E. Hester, in *Advances in Molten Salt Chemistry*, Vol. 1, J. Braunstein, G. Mamantov, and G. P. Smith, eds., Plenum Press, New York (1971), p. 1.
186. D. W. James, in *Molten Salts*, M. Blander, ed., Wiley, New York (1964), p. 507.
187. R. E. Hester, in *Raman Spectroscopy*, H. A. Szymanski, ed., Plenum Press, New York (1967).
188. T. W. Cape, V. A. Maroni, P. T. Cunningham, and J. B. Bates, *Spectrochim. Acta Part A* **32**, 1219 (1976).
189. F. A. Cotton, *Chemical Application of Group Theory*, Wiley-Interscience, New York (1971).
190. E. B. Wilson, Jr., J. C. Decius, and P. C. Cross, *Molecular Vibrations*, McGraw-Hill, New York (1955).
191. G. Herzberg, *Spectra of Diatomic Molecules*, D. Van Nostrand, New York (1950).
192. M. H. Brooker, *J. Chem. Phys.* **63**, 3054 (1975).
193. B. Gilbert, K. W. Fung, and G. Mamantov, *J. Inorg. Nucl. Chem.* **37**, 921 (1975).
194. J. K. Wilmshurst, *J. Chem. Phys.* **39**, 1779 (1963).
195. (a) V. A. Maroni, E. J. Hathaway, and R. J. Cairns, *J. Phys. Chem.* **75**, 155 (1970); (b) V. A. Maroni, *J. Chem. Phys.* **55**, 4789 (1971).
196. (a) K. Tanemoto and T. Nakamura, *Chem. Lett.*, 351, (1975); (b) *Jpn. J. Appl. Phys.* **12**, 2161 (1978); (c) W. Bues, L. El-Sayed, and H. A. Øye, *Acta Chem. Scand. Ser. A.* **31**, 461 (1977).
197. (a) J. R. Moyer, J. C. Evans, and G. Y. S. Lo, *J. Electrochem. Soc.* **113**, 157 (1966); (b) W. Bues and W. Brockner, *Z. Phys. Chem.* **88**, 290 (1974).
198. G. M. Begun, J. Brynestad, K. W. Fung, and G. Mamantov, *Inorg. Nucl. Chem. Lett.* **8**, 79 (1972).
199. M. Tanaka, K. Balasubramanyam, and J. O'M. Bockris, *Electrochim. Acta* **8**, 621 (1963).
200. V. A. Maroni and E. J. Hathaway, *Electrochim. Acta* **15**, 1837 (1970).
201. J. H. R. Clarke, P. J. Hartley, and Y. Kuroda, *J. Phys. Chem.* **76**, 1931 (1972).
202. J. H. R. Clarke, P. H. Hartley, and Y. Kuroda, *Inorg. Chem.* **11**, 29 (1972).
203. G. J. Janz and J. D. E. McIntyre, *J. Electrochem. Soc.* **109**, 842 (1962).
204. G. J. Janz and T. R. Kozlowski, *J. Chem. Phys.* **39**, 843 (1963).
205. G. J. Janz, C. Baddiel, and T. R. Kozlowski, *J. Chem. Phys.* **40**, 2055 (1964).
206. V. A. Maroni and E. J. Cairns, in *Characterizations of Solutes in Non-Aqueous Solvents*, G. Mamantov, ed., Plenum Press, New York (1978), p. 231.
207. (a) J. H. R. Clarke and S. Solomons, *J. Chem. Phys.* **47**, 1823 (1967); (b) T. Kirkerup, P. Klaeboe, and H. A. Øye, *J. Inorg. Nucl. Chem.* **41**, 189 (1979).

208. (a) E. J. Hathaway and V. A. Maroni, *J. Phys. Chem.* **76**, 2796 (1972); (b) W. Bues, M. Somer, and W. Brockner, *Z. Anorg. Allg. Chem.* **435**, 119 (1977).
209. V. A. Maroni, *J. Chem. Phys.* **54**, 4126 (1971).
210. A. S. Quist, J. B. Bates, and G. E. Boyd, *J. Phys. Chem.* **76**, 78 (1972).
211. L. M. Toth, J. B. Bates, and G. E. Boyd, *J. Phys. Chem.* **77**, 216 (1973).
212. A. S. Quist, J. B. Bates, and G. E. Boyd, *J. Chem. Phys.* **54**, 4896 (1971).
213. J. B. Bates and A. S. Quist, *Spectrochim. Acta Part A* **31**, 1317 (1975).
214. B. Gilbert, G. Mamantov, and G. M. Begun, *J. Chem. Phys.* **62**, 950 (1975).
215. S. K. Ratkje and E. Rytter, *J. Phys. Chem.* **78**, 1499 (1974).
216. (a) N. R. Smyrl, G. Mamantov, and L. E. McCurry, *J. Inorg. Nucl. Chem.* **40**, 1489 (1978); (b) E. Rytter, Ph. D. thesis, University of Trondheim, Trondheim, 1974.
217. P. L. Radloff and G. N. Papatheodorou, *J. Chem. Phys.* **72**, 992 (1980).
218. J. H. R. Clarke and R. E. Hester, *J. Chem. Phys.* **50**, 3106 (1969); *Inorg. Chem.* **8**, 1113 (1969).
219. K. Ichikawa and K. Fukushi, *J. Chem. Soc. Faraday Trans.* 1 **79**, 291 (1980).
220. (a) V. A. Maroni, E. J. Hathaway, and G. N. Papatheodorou, *J. Phys. Chem.* **78**, 1134 (1974): (b) G. N. Papatheodorou, *Inorg. Nucl. Chem. Lett.* **11**, 463 (1975).
221. (a) L. M. Toth, A. S. Quist, and G. E. Boyd, *J. Phys. Chem.* **77**, 1384 (1973); (b) L. M. Toth and G. E. Boyd, *J. Phys. Chem.* **77**, 2654 (1973).
222. T. Kirkerud, P. Klaeboe, and H. A. Oye, *J. Inorg. Nucl. Chem.* **41**, 189 (1979).
223. V. Maroni and E. J. Hathaway, *J. Inorg. Nucl. Chem.* **34**, 3049 (1972).
224. V. A. Maroni and E. J. Cairns, *J. Chem. Phys.* **52**, 4915 (1970).
225. J. B. Bates, M. H. Brooker, A. S. Quist, and G. E. Boyd, *J. Phys. Chem.* **76**, 1565 (1972).
226. R. E. Hester and K. Krishnan, *J. Chem. Phys.* **49**, 4356 (1968).
227. W. Bues, G. Foster, and R. Schmitt, *Z. Anorg. Allg. Chem.* **344**, 148 (1966).
228. K. B. Yatsimirski, *Pure Appl. Chem.* **49**, 115 (1977).
229. D. M. Gruen and R. M. McBeth, *J. Phys. Chem.* **63**, 392 (1959).
230. N. J. Bjerrum, in *Characterization of Solutes in Nonaqueous Solvents*, G. Mamantov, ed., Plenum, New York (1978), p. 251.
231. A. Trutia and M. Musa, *Phys. Status Solidi* **8**, 663 (1965).
232. G. P. Smith, C. R. Boston, and J. Brynestad, *J. Chem. Phys.* **45**, 829 (1966).
233. C. R. Boston, J. Brynestad, and G. P. Smith, *J. Chem. Phys.* **47**, 3193 (1967).
234. H. A. Oye and D. M. Gruen, *Inorg. Chem.* **4**, 117 (1965).
235. C. A. Angell and D. M. Gruen, *J. Inorg. Nucl. Chem.* **29**, 2234 (1967).
236. J. Brynestad and G. P. Smith, *J. Am. Chem. Soc.* **92**, 3198 (1970).
237. S. Hafner and N. H. Nachtrieb, *Rev. Sci. Instrum.* **35**, 680 (1964); *J. Chem. Phys.* **40**, 2891 (1964); **42**, 631 (1965).
238. U. Anders and J. A. Plainbeck, *J. Inorg. Nucl. Chem.* **40**, 387 (1978).
239. (a) L. Yaunus, M. Kukk, and B. R. Sundheim, *J. Chem. Phys.* **40**, 33 (1964); (b) B. R. Sundheim, J. Flato, and L. Yaunus, *J. Chem. Phys.* **51**, 4132 (1969); (c) M. I. Pollack and B. R. Sundheim, *ibid.* **78**, 1957 (1974).
240. T. B. Swanson, *J. Chem. Phys.* **45**, 176 (1966).
241. P. W. Selwood, *Magnetochemistry*, Wiley-Interscience, New York (1956).
242. M. V. Smirnov, V. Ya. Kudyakov, *Itogi Nauki Tekh. Rastvory.`Rasplavy* **2**, 172 (1975) (this is a review article containing diamagnetic susceptibility data on many metal halides and their mixtures).
243. G. N. Papatheodorou, *Inorg. Nucl. Chem. Lett.* **14**, 249 (1978).
244. For details see, e.g., E. Koening, Magnetic properties of coordination and organo-metallic transition metal compounds, in *Landholt*, Zahlenwerte and Funktionen Gruppe II/2 (1966).
245. W. Trzebiatowski and J. Mulak, *Bull. Acad. Pol. Sci. Ser. Sci. Chim.* **13**, 759 (1965).
246. R. Wojciechowska, J. Mulak, and W. Trzebiatowski, *Bull. Acad. Pol. Sci. Ser. Sci. Chim.* **18**, 127 (1970).
247. K. Tanemoto, T. Nakamura, and T. Sata, *Chem. Lett.*, 911 (1973).

Index